Conférence Moshé Flato 1999

MATHEMATICAL PHYSICS STUDIES

VOLUME 21

Conférence Moshé Flato 1999

Quantization, Deformations, and Symmetries
Volume I

edited by

Giuseppe Dito
Université de Bourgogne,
Dijon, France

and

Daniel Sternheimer
CNRS and Université de Bourgogne,
Dijon, France

KLUWER ACADEMIC PUBLISHERS
DORDRECHT / BOSTON / LONDON

A C.I.P. Catalogue record for this book is available from the Library of Congress.

ISBN 978-90-481-5550-7

Published by Kluwer Academic Publishers,
P.O. Box 17, 3300 AA Dordrecht, The Netherlands.

Sold and distributed in North, Central and South America
by Kluwer Academic Publishers,
101 Philip Drive, Norwell, MA 02061, U.S.A.

In all other countries, sold and distributed
by Kluwer Academic Publishers,
P.O. Box 322, 3300 AH Dordrecht, The Netherlands.

Printed on acid-free paper

Table of contents

Volume II

IV Contributed papers

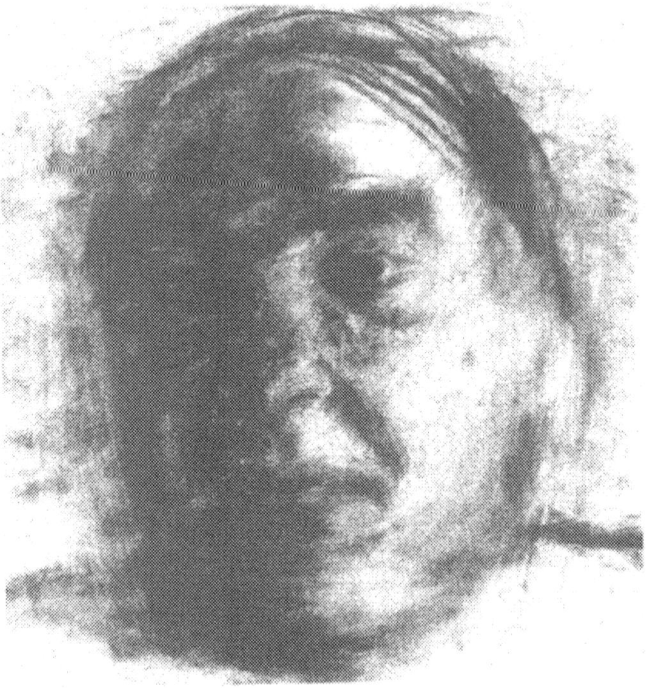

Drawing of Moshé by YAN Pei-Ming, artist and professor at Ecole Nationale des Beaux-Arts de Dijon, from memory and based on a number of photographs, including two by Margo Weinstein in 1994, one of which can be seen on: http://www.math.berkeley.edu/~alanw/moshe.html.

Preface

Daniel Sternheimer

These two volumes constitute the Proceedings of the "Conférence Moshé Flato 1999". Their spectrum is wide but the various areas covered are in fact strongly interwoven by a common denominator, the unique personality and creativity of the scientist in whose honor the Conference was held, and the far-reaching vision that underlies his scientific activity.

In 1928, there appeared the first edition of Hermann Weyl's seminal book, "Gruppentheorie und Quantenmechanik". Its English edition was a bedside book for Moshé. Symmetries, covariant field equations, and quantization are the main ingredients in that book. These notions played a fundamental role in the development of theoretical physics in the twentieth century, and their interactions with mathematics were at the origin of new avenues of research. In the seventies Moshé added to this list a fourth ingredient, deformation theory. Thereby he was able to explain, on the conceptual level, the evolution of physics from one domain of scale to another — in particular quantization. This addition turned out to be seminal also in mathematics.

With the present set of two volumes the reader will be able to take stock of the present state of the art in a number of subjects at the frontier of current research. It represents a covering of Flato's manifold interests by strongly intertwined contributions. It will be of interest to many researchers and graduate students in a variety of domains, ranging from abstract mathematics to theoretical physics and other applications. Some parts will be accessible to good undergraduate students, and even to persons with a minimum of scientific knowledge and enough curiosity.

Moshé Flato, born September 17, 1937, founder in 1968 of the "Laboratoire de Mathématique physique" in Dijon and of "Letters in Mathematical Physics" in 1975, died suddenly from cerebral hematoma in Paris, November 27, 1998. At the request of many in the mathematics and mathematical physics community, a Conference was organized in his honor. The idea was to have a Conference which Moshé would have been happy to attend, which shows the many domains of mathematics, mathematical physics and physics that were close to his heart. It took place in Dijon, September 5 through September 8, 1999. Its various components were attended by about 150 French and foreign scientists, in addition to a number of other participants. Indeed, Moshé would have enjoyed this conference immensely. (A follow-up conference takes place in September 2000, a "EuroConference"; it is dedicated to the most recent progress in the main areas treated in the 1999 Conference, which contributed to trigger those dramatic developments.)

The present Proceedings start with a more personal touch on Moshé Flato, the man and the scientist. It will be of interest to the many who have known him, to those who wish they had, and to historians of Science. A small section is dedicated to applications of sciences in insurance and finance, in relation with Moshé's activity as founder of the Scientific Council of Union des Assurances de Paris. At a time when these applications become increasingly fashionable, the indications that they give will be of interest also to mathematicians and physicists.

The bulk of the two volumes is the 'physical mathematics' part where the reader will find the above-mentioned four ingredients in various combinations, fundamental mathematical developments based on them, and challenging interpretations of physical phenomena. Most of the authors are leading scientists and all contributions were duly refereed. Even those mathematicians and physicists who do not (at present) work actively in the areas covered should find a lot of interest in these volumes. The Table of Contents is self-explanatory, so we shall not enter here into further details.

As coordinator of the Conference, I would like to thank all those who helped us make this Conference and these Proceedings possible, and responded so rapidly and efficiently to our requests. They include colleagues from all over the world, administrators at the local, national, and international level in a variety of capacities, and anonymous benefactors. A detailed list would be a "who is who" of Moshé's friends and he would have hated the idea, so we shall omit it. Last but not least, I would like to thank my co-editor for the fantastic job he did using his scientific and linguistic knowledges, combined with superb TEXpertise, to produce these camera-ready volumes.

Moshé Flato,
the man and the scientist

Program of the inaugural meeting

Under the auspices of Édith Cresson, former Prime Minister

Chaired by Hubert Curien, Academy of Science, former Minister of Research

Sunday September 5, Cellier de l'Espace Grévin, Dijon

15:30 Opening by Bernard Laurin, President of the University
Speech by Hubert Curien, Academy of Sciences

Messages from
Édith Cresson, read by Hubert Curien
Yvette Chassagne, read by Chantalle Reimeringer
Clarice Schwinger and Kay & Leon Cooper, read by Noriko Sakurai
Ed Witten, read by Ruth Boulet-Kerins
GU ChaoHao (Shanghai and Academia Sinica), read by RAO BoPeng

Speeches by:
Huzihiro Araki (Tokyo and Kyoto)
Jerôme Lichnerowicz (André's son)
Bernard Loiseau (Chef of *La Côte d'Or*, Saulieu)
Bengt Nagel (KTH, former president of the Nobel Committee for Physics)
Konrad Osterwalder (Rector of ETH, Zürich)

Music:
Chopin: *Nocturne in C Minor and Andante, Pianato, Grande Polonaise*
interpreted by Eugen Indjic, piano

Université de Bourgogne's medal awarded to I. Flato-Izakson, A. Connes, M. Kontsevich, Y. Aharonov, and B. Nagel.

17:00 *Short break*

Speeches by Irit Flato-Izakson and Joseph Sack, with slide show
Moshé d'après lui-même, in English, by Daniel Sternheimer

Montage video:
Extracts from the *Colloque Louis de Broglie,* November 6–7, 1987
Polac's TV program *Faut-il brûler le CNRS ?* November 1985
The scientific council of UAP, September 1987
Seminar in Dijon, 1995

Music:
Michio Miyagi: *Haru no Umi* and Kozaburo Hirai: *Sakura Sakura*
interpreted by Tomoko Ojima, piano

Songs by Civia Millender and Margalit Izakson, accompaniment at the guitar by Évariste:
Golden Jerusalem (Naomi Shemer, 1967); *Ma mie* (Évariste);
Song for peace (lyrics by Yair Rosenblum, sung by Y. Rabin just before his assassination)

Short break

19:00 *Apéritif, followed by a buffet by Lenôtre (and Loiseau)*
23:00 *End*

Speeches and Messages

delivered at the Conference by personalities, speakers or session chairs, and contributed to the Proceedings.

SHALOM HAVER

Moshé Flato, arrivé comme Professeur à l'Université de Bourgogne le 1er janvier 1968, y a gravi tous les échelons de sa carrière universitaire, que la mort a prématurément interrompue à l'automne de 1998.

Mais chacun sait que son parcours, administrativement linéaire, n'a pas été pour lui-même, son entourage et ses collègues, ce long fleuve tranquille, souvent décrit. Sa personnalité marquée, sa liberté et son originalité créatrice, ne pouvaient en faire un professeur «standard» et discret.

Non, il fut pour les Présidents de l'Université l'un de ceux, heureusement rares, qui pimentent l'exercice de leur fonction. Mais il n'aurait été que cela que nous n'aurions pas été amenés à témoigner aujourd'hui dans les premières pages d'un recueil scientifique qui lui est dédié.

En effet, son apport dans le domaine de la Physique Mathématique, qu'il a si brillamment authentifié, s'est construit à partir des fondations qu'il a bâties dans notre Université. Nous voulons rappeler la création du Laboratoire de Physique Mathématique, creuset de sa créativité, et la naissance en 1975 du Journal scientifique «Letters in Mathematical Physics» dont la réputation est maintenant bien établie. Aujourd'hui, notre Université bénéficie d'une place de choix dans la reconnaissance internationale de son pôle en Sciences mathématiques. C'est à une génération de fondateurs intrépides, dont il fut, qu'elle le doit. En hommage à leur œuvre, l'Université souhaite que les continuateurs s'obligent à la même ambition.

C'est peut-être une équation personnelle singulière «non triviale» que Moshé Flato a cherché à résoudre dans toute son activité et sa réflexion. Sa personnalité s'est rapidement affirmée dans le choc des cultures et des idées de l'après guerre, et dans la recherche d'une certaine stabilité dans un environnement particulièrement instable. Il a façonné sa carrière scientifique par son propre acharnement à la réussir. Il a dû à tous âges et constamment démontrer, prouver, convaincre. Mais sa curiosité sans limites et sans frontières fut à la fois intuitive et savante, logique et aléatoire, rigoureuse et passionnée et s'exprima sous des formes aussi bien populaires qu'élégantes. Chacun de nous peut retrouver dans ses souvenirs trace de cette étrange «cuisine», qui fit de lui cet attracteur libre et controversé.

Il ne nous revient pas de juger son œuvre mais le succès de ses publications, du journal LMP, et la somme de témoignages émouvants du monde entier que son décès imprévu et prématuré a suscitée, prouvent l'intensité de son facteur d'impact.

G. Dito and D. Sternheimer (eds.), Conférence Moshé Flato 1999, Vol. 1, 3 – 22.

Ces témoignages viennent, a posteriori, donner raison à ceux qui, malgré bien des écarts à l'orthodoxie administrative, budgétaire et pédagogique, et quelques éclats de voix contestataires, ont aimé retrouver sa démarche, sa science, sa culture et ses passions.

De fait, l'Université, Dijon et la Bourgogne, où il aimait poser le sac, ont vibré de sa forte présence, et ont accueilli régulièrement les discussions scientifiques prolongées de ses amis qu'il y faisait converger.

Nous sommes aujourd'hui orphelins de ce collègue hors pair et nous gardons tous depuis ce 27 novembre 1998, comme une richesse, tout ce que nous avons partagé avec lui.

Shalom Haver
Salut, camarade...

Bernard Laurin
Président de l'Université de Bourgogne

Gilles Bertrand
Vice-Président du Conseil scientifique
Ancien Président de l'Université (1988–1993)

* * *

Message de Madame Edith Cresson, ancien Premier Ministre
lu par Monsieur Hubert Curien

Madame la Présidente Yvette Chassagne, Messieurs les Professeurs Daniel Sternheimer et Jacques Simon, Mesdames et Messieurs les Professeurs, Mesdames, Messieurs,

Permettez-moi tout d'abord de vous dire combien je suis présente avec vous par la pensée, même si les obligations de mon emploi du temps ne m'ont malheureusement pas permis de me rendre à Dijon aujourd'hui.

J'ai été très honorée et émue que l'on m'ait proposé d'accorder mon patronage à cette session inaugurale.

Cela fera bientôt un an que le Professeur Moshé Flato fut enlevé à notre affection à l'âge de 61 ans. Son décès brutal et prématuré fut pour nous tous un véritable choc. Ses amis sont aujourd'hui réunis pour lui rendre hommage. Et nous pouvons mesurer à quel point ces amis sont nombreux, originaires des quatre coins de la planète et venant des horizons les plus variés, pas seulement scientifiques.

Car là était l'une des grandes qualités de Moshé Flato. Sa forte personnalité, imaginative et démonstrative, laissait rarement indifférent. Son caractère ouvert et entier fut à l'origine de nombre d'amitiés indéfectibles.

Pour ma part, j'avais eu l'occasion de rencontrer Moshé Flato à plusieurs reprises, notamment comme Ministre de l'Industrie et à l'époque où, Ministre des Affaires Européennes, j'avais créé les GEM, les Groupes Européens de Mobilisation. Je l'avais revu, plus récemment, à Bruxelles dans le cadre de mes fonctions de Commissaire européen.

A chacune de ces rencontres, il m'avait fortement impressionnée par son ouverture d'esprit, par son inventivité, par la justesse de ses analyses et par l'étendue de ses connaissances.

Mais, si la personnalité si attachante de Moshé Flato conduit à le rendre inoubliable dans nos cœurs, il n'en reste pas moins, bien entendu, que sa principale contribution est son œuvre scientifique dans le domaine de la mathématique physique. Cela fait aujourd'hui plus de trente ans depuis la fondation à Dijon du Laboratoire de Mathématique Physique, et près de 25 ans depuis le lancement de *Letters in Mathematical Physics*. Son œuvre lui survivra sous l'impulsion de ses collègues, mais elle n'en est pas moins aujourd'hui orpheline.

Nous avons tous pu noter comment ses travaux, non seulement étaient d'une grande qualité scientifique, mais aussi débouchaient sur des applications concrètes, par exemple en matière d'assurance et de finance à travers l'appréciation du risque et la modélisation. Je suis particulièrement sensible à cet aspect : la science n'existe pas pour elle-même, mais elle doit servir le progrès économique et social. C'est notamment pour cette raison que j'ai conçu et lancé le cinquième programme-cadre de recherche de façon à recentrer la recherche communautaire sur la croissance, la compétitivité, l'emploi et les besoins sociaux, en rupture avec l'approche traditionnelle organisée selon les domaines et les disciplines scientifiques.

Cette nouvelle approche encourage l'interdisciplinarité, qui a été matérialisée dans le cinquième programme-cadre par la notion d'action-clé. Moshé Flato était en quelque sorte un précurseur grâce à sa double formation de physicien et de mathématicien.

Mais, au-delà de sa contribution purement scientifique, Moshé Flato a su s'ériger en exemple pour la communauté scientifique. J'en veux pour preuve la haute tenue de *Letters in Mathematical Physics* ou bien la création et l'animation l'Association Internationale de Physique Mathématique.

Demain commencera la partie scientifique de cette conférence. Compte tenu de l'intérêt scientifique du programme, je suis convaincue que Moshé Flato aurait beaucoup aimé y participer. Même si son absence physique se fera particulièrement sentir, il restera plus que jamais présent dans notre mémoire.

* * *

Hubert Curien
Vice-Président de l'Académie des Sciences, Ancien Ministre de la Recherche

Quelle belle initiative prise par nos collègues, et tout spécialement par Daniel Sternheimer, de nous réunir pour parler de Mathématique Physique en honorant la mémoire de notre ami Moshé Flato!

Moshé incarnait la spontanéité et l'imagination. La rapidité de son mécanisme de pensée n'en émoussait d'ailleurs pas la rigueur. Il ne détestait pas de surprendre, il aimait aussi démontrer. Ce qu'il n'aimait pas, c'était se heurter à des frontières, à des barrières. La connaissance était pour lui un champ d'exercices où toutes les sorties valaient d'être tentées. Il se plaisait aussi à rapprocher la création mathématique de la création artistique.

De l'analyse des particules élémentaires et des forces qui régissent le comportement de la matière, des considérations fondamentales sur la symétrie et l'invariance, il savait s'échapper pour aller vers les sciences sociales, l'étude du risque, l'organisation des assurances.

La part de l'œuvre de Moshé Flato qui est considérée par les physiciens théoriciens comme tout-à-fait essentielle est ce que l'on appelle «quantification par déformations». Les meilleurs spécialistes vont, au cours de cette conférence, nous servir de guides dans ce domaine en voie d'essartage.

S'il fallait en deux traits, en deux mots, résumer la personnalité de Moshé Flato, je choisirais intensité et générosité. C'est dans cet esprit aussi que le programme de cette conférence a été composé.

* * *

Allocution de Madame Yvette Chassagne

Ancien Préfet, Président Honoraire de l'Union des Assurances de Paris

lue par Madame Chantalle Reimeringer

Pendant cette conférence, vous parlerez de questions scientifiques en pensant à Moshé. Auparavant, il m'a été demandé de parler de l'ami, de l'homme – parce que je ne saurais pas parler de science et surtout parce qu'il a été ces dernières années, un ami, peut-être plus un frère qu'un ami.

Qui était-il derrière cette façade volontairement humoristique, que cachait cette ironie et cet humour ? D'abord, je crois, une réelle timidité, j'ai failli dire humilité parce que ce savant était humble devant ses semblables, pour autant que ceux-ci répondaient à des critères de valeur morale et de qualités humaines, essentielles à ses yeux. Ce qui l'autorisait à être intraitable avec les autres, inconscient de sa valeur exceptionnelle, inconscient de sa supériorité, gêné lorsqu'elle se révélait malgré lui.

Ensuite un profond respect de l'autre quel qu'il fût, même qualifié amicalement de 'débile', un mot qui contenait beaucoup d'affection. Ce respect est, je crois, la clé de son extraordinaire faculté d'attention et d'écoute, de sa totale disponibilité. Plus encore de sa capacité de répondre à toutes les attentes, même celles qu'on n'osait formuler. Les mois ont passé et je n'ai pu encore admettre qu'il suffisait de téléphoner n'importe où, à n'importe quelle heure pour être compris, conseillé, aidé dans tous les domaines.

Rigoureux, intransigeant, il ne recherchait d'abord chez l'autre que ce qui pouvait le rapprocher de lui. Ses extraordinaires capacités d'analyse et de synthèse, appliquées aux comportements d'autrui, lui ont, pratiquement, toujours donné raison.

Parce que nous partagions la même espérance, les organisateurs m'ont demandé de parler de cette part de nous qui a disparu avec lui. En venant à Dijon aujourd'hui, nous ne pouvons espérer combler le Vide qu'il a laissé derrière lui en chacun d'entre nous. Mais nous pouvons ensemble, quelles que soient nos convictions propres, communier encore une fois dans ce socialisme humain auquel il croyait tant que j'oserai dire qu'il l'incarnait.

Beaucoup de choses importantes seront dites ici, des voix plus qualifiées certes que la mienne parleront de la valeur de ses travaux, des amis plus proches, évoqueront mieux ces avancées et ces perspectives qui définissent si bien l'orientation de sa pensée en tous les domaines.

Si j'ai accepté de dire ces quelques mots, c'est parce que jamais au cours des années passées, pendant lesquelles nous étions si proches, je n'ai osé lui parler de lui, lui dire combien il m'était exceptionnellement précieux. Si je le fais aujourd'hui, c'est pour être un modeste écho de toutes les amitiés silencieuses qu'il avait su nouer autour de lui, de cette affection qu'il portait à tous.

Moshé ne peut pas nous manquer, ce serait la première fois. Si nous sommes réunis, c'est pour être avec lui.

* * *

Jérôme Lichnerowicz

Bonjour.

Ô que c'est difficile! Ce qu'on me demande aujourd'hui est très difficile. Devant des auditeurs que, pour la plupart, je ne connais pas, je dois livrer mes sentiments sur des gens et des moments qui furent d'exception. A la demande des organisateurs de cet hommage, je dois parler d'un absent au nom d'un autre absent.

C'est un exercice de style impossible auquel je me livre avec émotion au nom de l'amitié et de la mémoire. Il n'est pas dans mes habitudes, et ceux qui me connaissent le savent, de livrer mes sentiments. Mais si je fais une exception aujourd'hui, je la fais pour Moshé, André, Daniel, Susa, ma mère.

Je ne vais pas vous parler de science, je n'y connais rien et je n'ai aucune légitimité pour cela. Mais postulant arbitrairement que je suis crédible sur ce sujet, je vais vous parler d'amitié. Je vais vous parler de l'amitié qui a lié pendant plus de trente-cinq ans Moshé Flato et mon père, André Lichnerowicz. Je vais vous parler de cette relation exceptionnelle entre ces deux hommes, de cette relation qui a pris fin lorsqu'ils moururent l'automne dernier, à quelques jours d'intervalle, d'un accident tellement comparable que c'en était troublant, à quelques mètres l'un de l'autre, dans le même hôpital, proches et associés jusqu'au bout peut-être par hasard...

Quand j'évoque ces deux hommes ma mémoire m'envoie des images nombreuses : J'ai vu Moshé, jeune homme mince (eh oui!), venir consulter un homme déjà mûr et j'ai vu la séduction naître entre eux. Adolescent, j'ai vu Moshé que je trouvais trop vieux pour être de ma génération et trop jeune à mes yeux pour être un ami de mon père, trouver sa place chez nous, s'y sentir bien, accepté, reconnu, ayant toujours l'attention et la phrase adaptée à son interlocuteur. J'ai vu mon père, anéanti par un drame familial, se demander s'il avait le droit de guider et de conseiller des plus jeunes. J'ai vu Moshé soutenir et encourager André dans ces moments de doute.

J'ai senti (plus que je ne l'ai vue) la modification de leur rapport dans la recherche. Il n'y avait plus de maître ni d'élève mais une synergie incroyable entre deux amis. J'ai vu Moshé soutenir André quand, les années passant, il doutait de ses capacités créatrices. Et j'imagine sans difficulté que l'inverse eut lieu si ce fut nécessaire. J'ai vu André soutenir à chaque occasion Moshé quand celui-ci a pu être en difficulté, mal jugé par des imbéciles.

J'ai vu mon père, vieux monsieur fatigué, parfois malade et imprudent, traverser la moitié de la ville pour ne pas rater un seul rendez-vous rue de Tournon. J'ai entendu Moshé me dire : «C'est incroyable, il a eu une période à vide, mais là il refait des maths comme avant !»

J'ai entendu André me dire, le jour de la mort de Moshé : «C'était une intelligence fulgurante!». Quelques jours plus tôt, le 16 novembre 1998, j'avais vu Moshé chez mon père. André était malade et j'ai lu une inquiétude immense dans les yeux de Moshé. Je ne les ai plus jamais revus ensemble après ce jour.

Mais je n'ai pas vu que des choses tristes ... Je me souviens aussi de ces repas gastronomiques, car Moshé et moi communiions dans le plaisir et l'admiration pour les grands chefs. Je me souviens de nos discussions politiques où la modération, la diplomatie et la bonne foi légendaires de Moshé (rappelez-vous donc de ses jugements si nuancés !) n'avaient d'égales que mes arguments sereins et ma pondération tout aussi légendaires ! Je me souviens aussi très bien de Daniel dans ces moments-là, attentif et amusé.

J'ai vu la générosité de Moshé s'exprimer si souvent vis-à-vis de mon père et de toute notre famille que s'il ne doit me rester qu'un seul mot pour parler de lui, c'est celui-ci : *générosité*. Je laisse les autres qualificatifs aux scientifiques, qu'ils m'accordent celui-là.

Il a semblé, à Daniel et aux initiateurs de l'hommage d'aujourd'hui à Moshé Flato, qu'il fallait y associer le souvenir d'André Lichnerowicz. Je dis à tous, avec beaucoup d'émotion, combien cette attention nous a touchée, Susa, Jacques et moi. Et si ma mère n'est pas présente aujourd'hui c'est que sa santé le lui interdit.

Si la rencontre de Moshé et d'André a eu lieu, si elle a débouché sur cette amitié exceptionnelle, s'ils furent si indispensable l'un pour l'autre – la paraphrase est facile – mais c'est «parce que c'était lui et parce que c'était lui».

De mon adolescence à aujourd'hui presque cinquante ans, j'ai avancé avec l'intuition intime qu'ils étaient indissociables dans la science comme dans la vie : ils le furent jusqu'au bout. Ils nous manquent dans leurs ressemblances, leurs différences, leur complémentarité, leur absence.

Merci.

Bernard Loiseau

Jamais Moshé ne serait venu en Bourgogne sans passer par la Côte d'Or. C'était devenu un ami. Il me contait toutes ses aventures épicuriennes de-ci, de-là. Il aimait vraiment la bonne nourriture et les bons vins, et le faisait partager à ses amis. Il fêtait souvent son anniversaire à Saulieu.

Il se délectait de la poularde Dumaine, cuite à la vapeur et farcie de truffes et de foie gras. Il appréciait les écrevisses sautées qu'il décortiquait à la main.

C'était vraiment un épicurien. Comme le disait François Mitterand, il faut toujours avoir une pensée pour ses morts : et bien voilà Moshé je pense à toi et te dis que la poularde et les écrevisses sont toujours aussi bonnes.

A bientôt Moshé !

* * *

Kay and Leon Cooper
read by Noriko Sakurai

A mathematician friend once said, "You Physicists are cowboys – you ride wildly over the range, jumping over fences and herding cattle – then we mathematicians have to come in to clean up the mess. But", he added, "You Physicists have all the fun".

Moshé was one of those rare physicist mathematicians who not only had fun but also didn't leave a mess. His many significant contributions to science are marked by their originality as well as their precision. He was a serious, dedicated and generous scientist, serious about his work, dedicated to his colleagues and students and sharing and generous with his ideas.

We remember Moshé for his penetrating intelligence, his ability to pose the one question that was particulary apt and most difficult to answer, for his gusto, and his 'joie de vivre'. We remember him as a great host, generous and caring to his friends and family. We remember the many times we ate and drank together, the camaraderie, laughter and above all the talk – the lively witty intelligent talk.

There is a silence now that he is gone – a silence that can only imperfectly be replaced by our memory of his lively mind and exuberant presence.

Moshé, your grieving family and your many friends thank you for the joy and laughter you shared with us. We bid you farewell.

Clarice Schwinger
read by Noriko Sakurai

Moshé was a man to be treasured, and we did. We admired his brilliance, respected his integrity, enjoyed his wit, were touched by his compassion and loved the joy he brought.

I trust it will be understood that "we" means Julian and me.

I know the meeting in Dijon will be moving. While I hadn't expected to be able to join, I am sorry that I can't.

* * *

Noriko Sakurai

I mourn the unexpected and untimely loss of an irreplaceable friend.

Moshé was a man amongst men.

* * *

Bengt Nagel
K.T.H., Stockholm; former President of the Nobel Committee for Physics

Ladies and gentlemen!

In the short time at my disposal I can give only some reminiscences of my meetings with Moshé, a person who has probably most affected my life and made the deepest impression on me after my immediate relatives and some of my teachers.

I first met Moshé at a seminar in connection with his first visit to Stockholm in 1967, arranged by his friend Gilbert Karpmann, then attaché scientifique at the French Embassy in Stockholm. The topic of Moshé's seminar was, I think, in a way typical for his attitude in science, and maybe in life: few things are really impossible. The seminar presented a counter-example to the then popular no-go theorems claiming the impossibility of combining in a non-trivial way relativistic spacetime symmetry with internal symmetry in particle and field theory.

After this first visit Moshé, with one or two of his collaborators Daniel Sternheimer and Jacques Simon, often returned to our department of theoretical physics at the Royal Institute of Technology; in particular he was a Nordita Guest Professor in 1970–71.

These visits were always highlights in the life of our department and influenced all members to a larger or smaller degree. The only small trouble we had was to cover up for Moshé's smoking in seminar and lecture rooms, which was prohibited. However, the last time he visited us, in October last year, he could refrain from smoking during his seminar.

The scientific exchange Paris–Dijon–Stockholm was very important for our group in Stockholm. My two best research students over the years, Göran Lindblad and Håkan Snellman, both owe much to Moshé's help and cooperation for their theses. Snellman wrote part of his dissertation in Paris together with Moshé, Daniel and Jacques, whereas Lindblad, who has always been somewhat of a lone ranger, at least got an important inspiration for his work in discussions with Moshé.

Primarily through his friendship with Lamek Hulthén, creator and then head of our department, and also a long time member and finally chairman of the Nobel Committee for Physics, Moshé got chosen for the special list of scientists who are invited every year to nominate candidates for the Nobel Prize in Physics. I should not pass by the fact that he was sometimes not too happy about the outcome of the committee work, and he certainly did not hide his opinion in such a case.

Moshé's way of doing and teaching science was a sort of socratic dialogue, generally treating also maybe not so bright ideas of his discussion partner in a constructive and understanding way. I think this very successful way of communicating science was related to the fact that Moshé's fantastic memory was largely auditive, related to speech and hearing.

This showed up also in seminars and lectures given by Moshé; they were great, but primarily through the oral impression. The visible remnants were often somewhat erratically scattered symbols, equations and words on the blackboard, and an array of broken pieces of chalk on the floor below. You could hardly imagine Moshé using a projector and overheads.

Moshé's auditive memory was important also for his interest and proficiency in musique and language. He could have become a prominent concert pianist; he could also have earned a handsome living as a professional imitator or impersonator, an ability he used to the delight of his audience, with verbal imitations of friends and foes alike.

Luckily for us mathematical physicists and for our profession Moshé chose to devote his professional life to mathematics and physics.

To sum up:

It has been a privilege and a pleasure for my wife and me to have been counted among Moshé's friends. The world seems grayer and colder without him.

* * *

Message from Ed Witten

read by Ruth Boulet-Kerins from Princeton

I am very sorry not to be here with you today, and I am grateful to Ruth Boulet for offering to convey my greetings. Moshé Flato's ideas and insights have a tremendous impact on physics. There has been a marvelous renewed interest in the last few years in many of Moshé Flato's favorite topics of deformation quantization, singletons, and Anti de Sitter space. Surely he must have enjoyed these developments in the last few years of his life. It is tragic that he is no longer here to help us in advancing the scientific frontiers that so animated him.

This informal letter about Moshé was received some time before the Conference from a very close friend, who authorized publication here but asked not to be formally identified. We include it because it expresses the feelings of many in a form which fits perfectly Mohsé's personality. DS

Cher Daniel,

Tu m'as demandé une contribution personnelle pour le volume en l'honneur de Moshé que vous préparez. J'aurais aimé, tu t'en doutes, savoir dire qui il était, comment il vivait au milieu de ses amis. Son intelligence, sa qualité de communication, son respect des autres jusque dans leurs insuffisances, ne nous ont jamais permis de voir combien il était différent de la plupart d'entre nous, combien il nous était tout simplement supérieur.

J'ai donc essayé de parler de lui, du peu que j'en connaissais réellement (alors que je croyais si bien le connaître). Pour cela il nous faut maintenant emprunter le mode passé, cela me paraît impossible. Il faisait tellement partie de nous que c'est accepter qu'une part de nous-même ait disparu. C'est sans doute vrai, mais si dur à dire.

Je ne peux imaginer qu'une telle réalité soit possible. Je ne peux penser qu'il n'y ait plus à n'importe quelle heure, au téléphone, au bout du monde quelqu'un à qui je puisse dire «le mur de Berlin est tombé», «Rabin a été assassiné» aussi bien que «l'un de mes enfants est malade». Il avait toujours cette surprenante attention, cette aptitude à se mettre à la place de l'autre qui supprimait les explications, les préliminaires. Tout était compris sans être dit, tout était partagé.

Comment pouvait-on ne pas être en colère comme lui, quand il explosait, levant un sourcil terrifié par la stupidité humaine ? Comment supporter le monde tel qu'il est, si Moshé ne l'explique pas, ne le comprend pas à sa manière, à la fois élémentaire et excessive, qui rendait si sûr notre propre jugement ?

Sans doute être quelquefois, le temps d'une pensée, d'une manifestation d'humour, d'une colère, un peu de Moshé, était un privilège rare qui enrichissait l'interlocuteur. En le quittant on se sentait meilleur, plus intelligent, plus apte à anticiper, qu'à l'arrivée. Il faudra sans lui, sans sa compréhension, sans ce qui n'était pas de la compassion mais de la participation, recevoir les peines futures et essayer de les accepter sans sa révolte qui était un tel réconfort.

Peut-on oublier comment tu le débarassais des médiocrités quotidiennes en restant présent à l'entretien. J'ai beaucoup pensé à toi ces derniers temps. On peut tout écrire comme on n'oserait pas parler, mais de tout cœur, tu sais je partage ta peine de cette absence à laquelle nous n'étions pas préparés. Avec Jacques et toi, il était seul à nous expliquer l'inéluctable, à nous accueillir et à nous le faire accepter et peut-être comprendre. Il nous manque tout simplement.

Excuse-moi Daniel, notre amitié m'a paru permettre de dire un peu ce que je sens beaucoup plus. Tout cela n'est pas en termes acceptables pour un ouvrage scientifique, fût-il en son honneur. Je ne pourrais faire un bon devoir de premier jour à son sujet, il l'admettrait et cela le ferait sans doute bien rire.

Je sais que je ne pourrai plus, depuis ce fatal après-midi à l'hôpital Cochin, comme chaque fois que nous nous rencontrions, reconstruire le monde, récrire la Bible ou le Discours de la Méthode en trois phrases.

Tu sais sans doute mieux que moi, car on a dû davantage te le dire, combien tous ceux qui l'avait simplement rencontré ont été blessés, je devrais dire frustrés, par sa disparition.

J'hésite à expédier ces quelques phrases «minables et débiles» aurait-il dit, parce qu'elles vont réveiller une peine, qu'avec une dignité extraordinaire tu as été capable de taire. Tous ses amis doivent être dignes de lui et de toi. C'est pour cela que je préfère t'avoir dit, très mal, trop peu, combien il ne peut vraiment être absent pour nous tous qui l'aimions.

Et pour une fois tu ne m'en voudras pas de te dire qu'on t'aime bien, qu'on t'aime beaucoup, toi le frère qui es encore là avec lui et nous.

A chaque fois que nous perdons quelqu'un, l'un des rares êtres que nous savons irremplaçables et uniques, nous sentons, interdits et heureux en même temps, combien notre cœur meurtri est encore capable de ressentir la douleur et de se révolter contre un sort qui nous ravit beaucoup trop tôt les meilleurs d'entre nous, ceux que nul ne pourra remplacer.

Stephan Zweig

* * *

Message received from Israel M. Gelfand

Dear Daniel,

The death of Moshé is a very great loss. I never make an overstatement. You can trust me that I have regarded Moshé as one of my intellectual closest friends. He was sociologically very important for me and I think that his approach to connection mathematics and physics I like the best.

The journal that he and you founded is one of the best in mathematical physics. For example, this is a main journal I read systematically and in every issue I find two or three very important for me papers. I am very happy that you decided to continue the journal in the same style. I will be more than happy if we will have now more personal contacts.

I start now several interesting works, and I'd like to dedicate the best of them to Moshé's memory.

With love,

I. Gelfand

V. S. Varadarajan

Department of Mathematics, University of California

Talk delivered on Dec 5, 1998, at the Lie group workshop held at UCLA, Dec. 5–6, 1998

Moshé Flato died on Friday, Nov 27, 1998, in Paris, of complications from a massive cerebral haemorrhage he suffered a couple of days earlier. His shocking and untimely death is a tremendous loss to the world of mathematical physics, and even more so, to his family and large circle of friends and admirers. He was buried in Tel-Aviv.

Moshé's work covers a large spectrum of high energy physics, straddling diverse aspects of quantum field theory and elementary particles. His interest was always in fundamental questions. He was constantly searching for a conceptual understanding of the strange world of elementary particles, and explored many new ideas and ways of thinking about them. He created a school in Dijon and Paris and shaped the intellectual developments of a huge number of young scientists, some of whom became his closest friends, such as Daniel Sternheimer and Jacques Simon.

I would like to make some very brief comments on his work. In quantum mechanics there is a famous uniqueness theorem due to Von Neumann and Stone, which says that any system of operators satisfying the Heisenberg–Weyl commutation rules is isomorphic to the standard system which is called the Schrödinger model. The content of this theorem is that there is only one way to quantize a classical system, namely through canonical quantization. In a famous paper published in the mid 1970's, Moshé, and his collaborators Bayen, Fronsdal, Lichnerowicz, and Sternheimer, proved a wonderful variant of this uniqueness in a very different context. Their idea was to think of quantization as a deformation of the Poisson algebra of classical observables. Canonical quantization furnishes such a deformation, the Moyal–Weyl deformation. They proved that this is essentially the only deformation of the classical Poisson algebra. In my opinion this uniqueness theorem lies at a much greater depth than the Von Neumann–Stone theorem and may be thought of as the definitive clarification of the questions that were raised first by Hermann Weyl in 1927 regarding the foundations of quantum physics. The ideas coming out of this paper revolutionized quantization theory, made it possible to think of quantization in contexts where canonical quantization does not make sense (curved space–time, constrained mechanics, etc), and led inevitably to the theory of quantum groups and quantum homogeneous spaces. Recent work on q–quantum mechanics by Finkelstein and others would not have been possible without this seminal achievement.

During the 1980's, Moshé, in collaboration with Frønsdal, made a deep analysis of the representation theory of the conformal group. Their theory of the so called singletons, which was very much ahead of its time, has recently come to the forefront in high energy physics.

Throughout his life he was deeply interested in formulating and understanding the nonlinear aspects of quantum field theory. He called this the "nonlinear program" and wrote many papers on this theme, with Simon and others. His recent work in collaboration with Simon and Taflin is a technical tour de force on the Cauchy problem for the coupled Maxwell and Dirac equations that arise in quantum electrodynamics.

He loved the subject of mathematical physics deeply and was very unhappy that (in those days) there was no forum where important and interesting discoveries in mathematical physics could be published quickly. The physicists had several such journals, but it was often very difficult to get papers published in them because of lack of sympathy of the physics community toward major discoveries that were more mathematical and did not lead to immediate physical interpretations. Moshé understood this situation and with characteristic energy and enthusiasm founded, in 1977, the Letters in Mathematical Physics precisely to correct this anamoly. The Letters has become the premier journal in mathematical physics. If you look through its pages you will find all the major developments in the last two decades represented there when they first came up. There is no question that he was the moving spirit behind the Letters. This is one achievement that will live long after him.

But bigger than all his achievements in science is the fact that he was a decent man. Behind his rough exterior lay a heart of extraordinary generosity. He was lavish in giving his time to others. I learnt a lot of physics by talking to him and reading his papers. He was especially generous to young people, understood their difficulties at the start of their scientific careers, and helped them in every possible way. He treated his friends royally, and influenced their lives in ways that cannot be enumerated. He was the only man I have known of whom it could be said that he was larger than life (a comment also made by one of the young scientists who was deeply influenced by Moshé).

As one gets older, one finds invariably that one cannot do all the things that one did when one was young, and certainly not at the depth that was reached in youth. The natural tendency to compensate for this is to reduce one's ambitions and do what one can to advance science. Moshé scorned this way of living. He pushed himself and others around him without any slackening and lived on the creative edge all the time, right up to the moment of his death. According to the Hindu view of life this is one of the noblest ways to live and die.

When I learnt of Moshé's death I was reminded of Isaac Stern's words on the occasion of David Oistrakh's death, when he referred to Oistrakh as a golden man. Moshé was a golden man.

Collection of addresses by session chairs

Marcel Guenin, former Rector of Geneva University

My first real interaction, not my first meeting, with Moshé was in 1972 in Moscow. As you already know, Moshé was very fluent in Russian. But he had not bothered with learning the cyrillic alphabet, which was about the only thing I knew in the Russian language. So we went through Moscow, I reading the street names or the menus in restaurants, and he understanding the meaning, a benign version of the fable of the blind and the lame.

I shall refrain from further anecdotes, but should like to insist on the lesson he taught us throughout his life. The lesson is that time is short, we have to make the best of it; enjoy life in all aspects, family, friends, sciences, applications, travel, and so on. We should remember that, as we are too often thinking in a sequential way, postponing so many things. But of course we are not Moshé... Finally, a regret for me, he did not become my brother in law – but maybe it was better for him!

* * *

Alan Weinstein, University of California, Berkeley

Before introducing the next lecture, I would like say a few words about an aspect of Moshé's talents which has not yet come up in the preceding tributes. I refer to something related to his acting skills, which have already been mentioned, namely has ability as an artists' model.

Visitors to Moshé's apartment were treated to, among many other delights, a collection of portraits of Moshé done by a variety of artists working in many media. When Moshé and Daniel visited Berkeley several years ago, Moshé saw some of my wife Margo's photographs taken through melting ice on sheets of glass and asked if he could be the subject for such photographs. You can see the results yourselves in the lobby of this conference center. Moshé took to the portrait sitting with gusto, going through a wide variety of poses, some of them worthy of the kabuki actors he so enjoyed. Margo found him to be one of the most interesting subjects she has ever photographed.

When, shortly after Moshé passed away, Daniel asked Margo to make a new set of prints of Moshé's portraits, she agreed of course, but found the printing job to be emotionally wrenching. As the portrait appeared in the developer tray (the reader can see it emerging at the website http://www.math.berkeley.edu/˜alanw/moshe.html), Moshé appeared to be waving goodbye from "the other side."

Our next speaker, Simone Gutt, has been closely associated with Moshé since her student days, when he was a member of her doctoral thesis committee. Over the years, she and her collaborators have made a series of fundamental contributions to the subject of deformation quantization. It is therefore most appropriate that she will speak at this Conference on "Variations on Deformation Quantization".

Yvette Kosmann-Schwarzbach, École Polytechnique

What comes first to mind, now that the initial shock of Moshé's sudden illness and death has passed, is his forceful character, directness, and the warmth of his friendship. As so many of his friends have mentioned already, he had a special affection for children, enquiring about the progress that each of ours was making in school, in music, and in life. I shall not forget my conversations with him during the intermissions of the CNU session of the Spring of 1997, when our daughter was hospitalized. He was able to encourage me because I felt that he truly cared for her, as he did for the children of all his colleagues and friends.

Such was the vigor of his character that his really important achievements as a scientist only come to mind in second place. I need not repeat what others have said and written about them. His œuvre will remain. The creation of *Letters in Mathematical Physics* focused the energies of the mathematical community and contributed in an essential way to the development of an important field of science. Deformation quantization was the brainchild of Moshé and his collaborators, François Bayen, Chris Frønsdal, André Lichnerowicz, and Daniel Sternheimer.

Daniel, Chris, and so many others are even now working on developments of deformation theory, while Daniel is keeping Moshé's institutional and scientific legacy alive with indefatigable zeal and success. But André Lichnerowicz died two weeks after Moshé. Moshé was my contemporary, but Lichnerowicz was my master, and I miss both of them, enormously. For me, the most moving moment of the opening ceremony of this Conference was the evocation by Dr. Jérôme Lichnerowicz of their mathematical collaboration and the deep friendship that united them.

What can I add to the tribute already paid to Moshé by mathematicians and physicists better qualified than myself? The fact that so many of them travelled from the ends of the world to attend this Conference, to make it a meeting that Moshé would have enjoyed, is proof that Moshé's spirit is still alive among us and as driving force in the scientific community.

* * *

Robert Finkelstein, UCLA

I first met Moshé in the early 70's when he, Daniel, and Jacques began to visit Chris Frønsdal at UCLA. In those days Moshé and his collaborators were engaged in their pioneering studies of deformation theory.

Although we were of different generations, Moshé and I had remotely similar academic backgrounds. He, as a student of Racah, and I as a student of van Vleck were naturally interested in the application of group theory to quantum mechanics. However, by the time we had met, Moshé had become a mathematical physicist with a capital M and I had remained a physicist. Still, our tastes and attitudes toward the problems of fundamental physics had much in common.

The principal connection between us did not, however, grow out of our common scientific interests. There was rather a long personal friendship. With his high spirits and straight-forward frank way of speaking, it was always a great pleasure for me to be with him, even when we disagreed — as we frequently did.

Moshé's grand style, his superb hospitality, his concern for others, and his capability to reach out to all manner of people were qualities that we all knew. He took a personal interest not just in the promising students. There seemed to be no one who crossed his path whom he would not try to help and no troubled social situation that he would not try to ameliorate. (As a result he knew more local gossip at UCLA than I knew myself.) He was a kind of universal entrepreneurial activist in the best sense of the word.

My wife, Norma, and I spent a week with Moshé and Daniel about two months before his fatal stroke. We had no knowledge of the medical episode in St. Petersburg or the reports of his subsequent depression. We were with him every evening and there was no indication to us that there was anything wrong. He was the same wonderful exuberant Moshé that we had always known and that is how we would like to remember him.

* * *

Murray Gerstenhaber, University of Pennsylvania

Sadly I knew Moshé for too short a time. Although I am ten years older, I felt that I was his disciple, for I quickly saw that I was in the presence of a great teacher. My name is really also Moshe, so I often signed my email messages "Moshe2". I hope that Moshe1 recognized the tribute.

Since our next speaker, Alain Connes, has referred to Moshé as a "mathematician rabbi", I want to tell a talmudic tale that illustrates one aspect of Moshé's intellectual make-up. I am sure that Moshé knew it. Rabbi Eliezer was arguing a point of law with his fellow rabbis and said, If I am right this carob tree will prove it. The tree was uprooted 150 feet, but the other rabbis said, We don't accept proofs from trees. So Rabbi Eliezer said, If the law is with me this stream will prove it. The water began to flow backwards but the others said, We don't bring proofs from streams of water. So Eliezer said, If I am right the walls will prove it. They lean to this day, but the others still rejected Rabbi Eliezer's position. Finally he said, Let the heavens prove it, and a voice said, Eliezer is right, why do you trouble him? But the other rabbis said that their reason outweighed heavenly proofs, and being in the majority, they excommunicated him. Like Eliezer, Moshé was usually stubbornly right, but never in the majority. Our next speaker, like Moshé, has had to overcome a hostile establishment which, however, ultimately recognized the power of his proofs.

Piotr Rączka, Warsaw University

I am speaking here for the whole Rączka family, and first of all for my father, Ryszard Rączka, who has known Moshé Flato for more than thirty years. My father met Moshé for the first time in mid-sixties in Trieste. At that time they were both deeply interested in mathematical aspects of theoretical physics. Their professional contacts quickly evolved into what became a lifelong friendship. It is fair to say that Moshé was one of the closest friends of my father. My father greatly appreciated Moshé's bold and sometimes highly critical scientific opinions, his energetic personality and a unique sense of humour. Looking back at these years one can also clearly see that interaction with Moshé played a vital role in my father's life. In these years the research in mathematical physics was not always popular in the physics community – encouragment and support from Moshé and his enthusiasm for beautiful mathematics were certainly very important for the development of my father's career as a mathematically minded theoretical physicist. The contacts acquired through Moshé and frequent visits to France on Moshé's invitation were also of great value to him.

Also for the rest of us – that is, for my mother, my siblings and myself – Moshé was someone special. He was always interested in our lives and remembered very well all the important facts about us. My mother recalls that talking to him about herself and her family was like talking to a close relative, who understands completely your way of life and your problems – something quite extraordinary, taking into account number of people with whom Moshé interacted and the fact that we did not meet very often. We all visited him in Paris one time or another and benefited from his warm hospitality there. Moshé's presence could be felt in so many moments of our lives that we treated him almost as a member of our family.

Moshé was always deeply moved by the problems of his friends and showed great compassion for people close to him. We could experience that ourselves. During the last couple years we went through some difficult times. What we received during that period from Moshé and Daniel went way beyond what we received from anyone else. Moshé and Daniel paid a special short visit to Poland right after the death of my father to comfort my mother and – despite the fact that Moshé's health was clearly deteriorating – they later came again twice to attend two conferences organized in Poland to honour the memory of my father. Moshé greatly contributed to the organization of these conferences.

We are grateful to Moshé for all he has done for us. We treated him as a member of our family, and we mourn for him as for a family member.

Hideki Omori, Science University of Tokyo, Noda

My colleagues and I came to know the name Moshé Flato from the famous paper [1] that first fixed the notion of deformation quantization. At that time in Japan, the notion of quantization was not a so familiar one for mathematicians.

We were working on groups generated by invertible Fourier integral operators on manifolds by which we wanted to treat wave equations on manifolds, just as in the history of early quantum theory in physics. We were interested in the asymptotic geometry of Maslov [2].

However, as every quantum physicist believes, maybe from a philosophical point of view, the Heisenberg picture and the Schrödinger picture are equivalent. Hence, the asymptotic geometry of Maslov and the notion of deformation quantization seem, for our simple mind, to give at least the same predictions.

Influenced by these great ideas, we have started in Japan to study the theory of deformation quantization, that is, star-products. The notion of deformation quantization does not require representation theory, we can merely start from a phase space: a symplectic manifold, or more generally, a Poisson manifold.

After that, Moshé and Daniel have visited our university several times to impulse and stimulate our study. During each visit, they encouraged us warmly just like elder brothers, but that was hard too sometimes.

Being free from representation theory looks like a significant advantage for the construction of the general theory of quantization. But from the very beginning of deformation theory, there was a deep question: Is every Poisson manifold deformation quantizable?

Suppose there were a Poisson manifold which is not deformation quantizable. How does a quantum physicist, who insists that the true nature of the world is given by quantum theory, and the Poisson structure is only its first approximation, explain such an object? If there were such an object, the theory of deformation quantization would meet a philosophical difficulty.

Now, it is my great pleasure, and probably Moshé would have the same thought, to introduce to the audience Professor Maxim Kontsevich, who proved that every Poisson manifold is formally (i.e, treating $\hbar$ as a formal parameter) deformation quantizable. By his theorem, we can avoid the above-mentioned philosophical crisis.

The theory of deformation quantization is celebrated and we are now at the beginning of a second generation. In this second stage, $\hbar$ will no more a formal parameter, but a positive small number.

[1] Bayen, F., Flato, M., Fronsdal, C., Lichnerowicz, A., and Sternheimer, D.: Deformation theory and quantization I, II, *Ann. Physics* **111** (1978), 61–110, 111–151.

[2] Karasëv, M. V. and Maslov, V. P.: Asymptotic and geometric quantization, *Russian Math. Surveys* **39** (1984), 133–205.

Nicolai Reshetikhin, University of California, Berkeley

The next speaker does not need an introduction. The name of Bert Kostant is now part of the Lie theory. He was also a great friend of Moshé.

So many good words have already been said about Moshé that it is hard to add something new. The first time I met Moshé goes back to the earlier 80's in Leningrad, when he was one of (relatively few) visitors from the other side of the Iron Curtain. He has impressed me as an independent, opinionated, but at the same time very kind, person. He was very friendly with young people and at that time this was very important for me and, I believe, to others of my age. This allowed us to make the connection with the older generation in a natural and friendly way. It filled the generation gap. This was the beginning of our friendship. It was very interesting to argue with Moshé. I could not always agree with him, but that was making the conversation even more interesting. I will miss Moshé, I will miss discussions with him, when everything was questioned from mathematics and physics to politics and history. I am very happy that I had such a friend.

* * *

Ivan Todorov, Sofia

Moshé at seminars

Moshé and I first saw each other over 30 years ago in Trieste. We kept meeting at conferences in various places in the subsequent decade but it was not until we got together in Istanbul (at the Bogazici University) in 1979 that we became friends. (My next visit to Istanbul – alas, without Moshé – took place only this summer: We arrived on August 16, just in time for the earthquake!) No need for an earthquake was felt 20 years ago: there was Moshé Flato to shake us at every session of the Conference. He stormed discussions in (what I later came to know as) his usual manner, attacking speakers and vocal participants alike, in particular those displaying an excessive self-confidence. We got a good fight there more than once. Moshé, of course, put me down when I ventured to intervene at a talk on deformation quantization but he seemed to appreciate my resistence in a few other cases of disagreement. After one of these public fights he apparently thought that I was ripe to visit Dijon for a month, so he invited me for next year.

Moshé's seminars in Dijon used to be rather picturesque. Dijon time in those days did not run according to clock; a loud voice was needed to announce the beginning: "Séminaire!". You could not fall asleep at a seminar. No matter who the speaker was, he would be interupted and challenged by Moshé before long. Sometimes this could be irritating and a new visitor could get upset; but he was soon to realize that the attacks were not malicious, they rather served to keep the audience alert. In case you were an habitué you were discouraged in advance to make unsubstantiated claims.

Moshé loved intellectual fights just as he liked good meals – and often combined the two. Given his generosity and compassion for people – that were stressed here repeatedly (and to which my family and I can testify) it is not surprising that Moshé has so many friends. It may appear less obvious, however, if you remember that he seemed to be looking for confrontations rather than diplomatically avoiding them. I believe that this may actually account for the quality of Moshé's friends. Is not it wonderful that a gathering of just friends of Moshé Flato as this one is becoming a most remarkable meeting on mathematical physics in years?

Well, it is time to give the word to Maxim Kontsevich who, for his second lecture here, will speak on the "Proof of Deligne conjecture". Maxim, I am sorry, I am not Moshé Flato to stop you during your talk. So, please, slow down yourself to make it easier for us to follow.

* * *

Boris Fedosov, Moscow and Potsdam

Moshé Flato as I remember him

I knew Moshé Flato not as long as most of the previous speakers but like all of them I was greatly impressed by his exceptional personality. I am very proud that in spite of a rather short period of our acquaintance he named me his friend.

Our first meeting really overwhelmed me. I vividly remember the day when I came to Dijon to talk at a seminar. It began at the restaurant at the Dijon station where I was invited for lunch by a telephone call from Paris. Moshé's energy and outstanding personality produced an unforgettable impression upon me from the moment I saw him. During the lunch he acquainted me with French cuisine and discussed many things: mathematics, politics, art and music. He proved to know the events in my country better than me! The lunch lasted rather long and it was an event. Meals according to Moshé were things to enjoy. I remember him saying: "I can't understand some American friends of mine who, looking at the watch say: 'I still have 5 minutes before my lecture to have a sandwich and a cup of coffee' ". Moshé said: "You better go to the lecture hungry but after it relax and have good dinner."

Then there was a seminar where I talked on deformation quantization. It sounded rather funny that I, a newcomer, who came to deformation quantization from partial differential equations and index theory, tried to "teach" an outstanding specialist in this field. Nothing surprising that at first I was treated by Moshé as an intruder. During my talk and after it there were many critical remarks from both Moshé and Daniel, which in the long run proved to be very valuable to me.

In the evening a famous Thursday dinner took place at a Dijon restaurant. We parted late at night at the hotel. I was completely exhausted, but not Moshé. I already felt that I parted from a good friend. And indeed, later on I felt not only his friendly attitude towards me and my wife but also a desire and readiness to help. Now, when he has left us, I still feel his friendly hand.

Moshé's Memoirs

as told, from time to time, to Daniel Sternheimer

Moshé was not reticent. He would speak about himself often. Since the colleagues who may read the scientific papers in this collection prepared in his honor might like to know something more about the man as well as the scientist, I shall try to retell his own stories, and those we shared. He spoke about himself – and about others – with amusement, pugnacity, profanity and a verve which are hard to reproduce, but I must try to imitate them because verve and enthusiasm for everything he did were part of Moshé's personality. A staid "life and times" of Moshé would necessarily miss the mark, widely. He was extraordinarily quick minded, had a unique intuition and a fantastic ability to interact strongly at the appropriate level with everyone, friend and foe alike. Quite naturally he occupied a special and often very large place in the lives of all those who ever met him. He was truly exceptional – mildly speaking.

Moshé was born in Tel Aviv, September 17, 1937, while it was still under the British Mandate. It was an auspicious date, the day after Yom Kippur and the anniversary of Bernhard Riemann's birth. His father, Shlomo, was a brilliant engineer who had been trained in the Dantzig Polytechnic School just before Hitler came to power. He was the son of a prominent Jewish family of Łodz and Warsaw and a Zionist practically from birth. He anticipated the fate of Poland and European Jewry so he refused a flattering invitation to join the faculty directly upon graduation; he went instead to Palestine. In a Tel Aviv café he met Anna Lunski, the youngest sister of one of the first beauty queens of Palestine – the model for Maroussia in Thomas Mann's *Der Zauberberg*! Anna was a psychology student and the daughter of a family of Russian industrialists. Her parents arrived in Palestine in 1922 with nothing but a collection of imperial 500 ruble bills but, through hard work, succeeded in rebuilding the family fortune. Shlomo followed his heart, settled in Tel Aviv and managed to bring over some members of his family before the war broke out. His talent and integrity brought him an appointment as chief engineer for public works under the mandate authority, so he moved to Jerusalem only to return to Tel Aviv after the end of the war. His official status did not inhibit him from giving technical advice to Menahem Begin's "Irgun", which was fighting and bombing the forces of the Mandate authority.

The members of Shlomo's family who remained in Poland were exterminated by the Nazis and their allies, except for a cousin named Stanisław who was a doctor and a communist. He had served in the International Brigade in Spain, then joined Chou En Lai in South China and spent the war years at his side. After the war he returned to Poland and in 1949 was named the first Polish ambassador to the People's Republic, but was imprisoned during Stalin's purge of the former members of the International Brigade. Chou's intervention saved him from trial

G. Dito and D. Sternheimer (eds.), Conférence Moshé Flato 1999, Vol. 1, 23 – 30.

and its inevitable verdict and sentence. He was rehabilitated under Gomułka and became Under-Secretary of Foreign Affairs, but was removed from that position in 1968 under Moczar, and not permitted to leave the country. Moshé met him in 1969, on the occasion of his first trip to Poland and acquired from him a taste for the Orient.

From infancy it was clear that Moshé was an unusual child. His mother hired a governess for him so that she could pursue her career, but Moshé would have none of her, so Anna had to renounce a professional life to take care of her son. She and Shlomo spoke to him in Hebrew, a grandmother, who always refused to learn that "language of the natives", spoke to him in Polish, and Anna's family spoke among themselves in Russian. Many of his father's friends spoke German. He learned English at the movies and in school, and Arabic in high school with a very fine teacher. Moshé, the enthusiastic polyglot, picked up French in France, with the predictable misadventures, like addressing a policeman as "Monsieur le flic", and rudiments of Swedish, Japanese, Italian, Spanish, Thai, Chinese and Turkish, all mixed with a heavy dose of slang which did not always emerge at the most opportune moments. Actually, Moshé caught the music of these languages and could imitate what he heard faultlessly and incomprehensibly.

Moshé's elementary and junior high school grades were unremarkable, except in mathematics for which his natural talents were already obvious. The rest was taught too slowly to interest him. This lasted until he got the highest grades in the "psychometric" tests since their institution, after which most of his teachers started to pay attention to their unusual pupil, and his marks leaped upwards. Nevertheless he loved to play hookey and pranks like writing a parody of a poem in the "Canaanite" style and passing it off on an unsuspecting teacher as a subject for a grave literary analysis. His teachers resented the pranks and sometimes punished him. As a result Moshé became an assiduous patron of the Tel Aviv cinemas.

At home, because of his father's position as a chief engineer for the Southern Command of the young Israeli Army, Moshé met Itzhak Rabin and he would later serve under Moshé Dayan and Ezer Weizmann. But he also took piano lessons with Daniel Barenboim's mother and acquired a solid technique. (A few years before his death he could still play the Hammerklavier sonata from memory, without too many dropped notes!) At one time he thought of making a career as a pianist or an actor, but ultimately chose mathematics and physics.

Towards the end of his high school years, Moshé's talent for mathematics and physics, nurtured by his father who was very gifted in these areas, became obvious. And in his turn Moshé gave private lessons in mathematics and physics to high school students, sometimes two at a time, for astronomical fees, but he was such a gifted teacher that his students got their money's worth and more. He once even bet his fee that he could coach a boy who was a disaster in math and get him through the "bagruth" (matriculation) exam. The lad passed, made a brilliant career after that, and Moshé won his fee.

Service in the Israeli Army used to be almost ineluctable and anyhow avoiding it would have been unthinkable for a child of Shlomo Flato, so incompatible as military discipline was with Moshé's quick-silver temperament, serve he did. In those days the most gifted high schoolers could go directly to the university, taking their basic training during their vacations and then serving three years as an officer. During one exercise in 1955 Moshé's platoon crossed the Jordanian border – by accident according to the official version – and ran into cross fire from the Jordanian troops. Moshé sized up the situation instantly and, in defiance of Israeli military rules and traditions, ordered "every man for himself", abandoned their equipment but saved himself and his two comrades. The bodies of the other seven soldiers were returned two days later at the Mandelbaum Gate in Jerusalem. Moshé's superior wanted to cast the blame for the bungled operation on the young lieutenant who was killed, but Moshé insisted on telling the truth, that the platoon fell into an ambush because of the poor preparation by the general staff of their "accidental" incursion into Jordanian territory. As a result of his impeccable honesty, Moshé suffered endless reprisals from this officer and was even sentenced to a month in prison for striking him during an inspection. Moshé being Moshé, he spent the month as the prison librarian and coached the commander in mathematics so that he could pass his "bagruth".

In the 1956 Sinai campaign he served in operations control, where his astuteness was generally appreciated (at some point, Dayan even told him "Sergeant, transmit your order"). But on one occasion, because the general in charge at that time insisted on following standard procedure instead of listening to the young sergeant that Moshé was then, forty soldiers were killed by "friendly fire", a fact Dayan would concede (without mentioning Moshé's advice) in memoirs published many years later.

Moshé passed the Hebrew University admission exams in physics brilliantly and studied with Giulio Racah and the several Nobel prize winning guest professors whom Racah had recruited. Moshé impressed them as much with his brilliance as with his awkwardness in the laboratory (he had "two left hands", which is what saved him from being sent to a technical high school). He never acquired the manual dexterity of an experimental physicist so he was condemned to be a theoretician and mathematician. It was at the University that I met Moshé who was a year ahead of me. (My mother's best friend was a cousin of Moshé's father, so when I needed course notes for Abraham Robinson's hydrodynamics course, it was natural to turn to Moshé.) Moshé hesitated between mathematics with Amitsur and theoretical physics, and finally chose the latter, asking Racah to give him a subject for his Master's essay. Racah posed a problem concerning symmetry breaking in the 3-dimensional space by crystalline material. Moshé solved the problem in three months by a method that Racah had difficulty in understanding. But since the problem was supposed to keep Moshé busy for a year, Racah added a second one, which Moshé nearly solved in a month – there remained only a very large matrix to diagonalize –, but he only submitted it to Racah at the last minute to escape a third assignment. This was a fine piece of work which, since it was not published at the time, Moshé decided to recycle as his "second thesis" for his Doctorat d'État (1965). Only then was it published, at the urging of the Sorbonne professor of spectroscopy to whom Moshé was referred then, who was ready to accept it for principal thesis: it had remained pertinent for five years, despite the rapidity of progress in crystallography! Moshé gave some additional notes on the subject, the debris of his thesis, to one of his students in Lyon, who has since made an honorable career developing and extending them.

In 1960, as part of his military service as Air Force lieutenant, Moshé was asked to prepare the first comprehensive theoretical program in mathematics and aerodynamics for Israeli pilots. When Moshé was responsible for that part of the course, a well-connected recruit failed; the decision was supported by the Air Force chief and by his brother-in-law, the chief-of-staff and father of the unfortunate recruit. After a year with the Air Force, Moshé was asked to become head of the theoretical group at an experimental nuclear reactor, without having to enlist in the army for five years as is the rule in Israel; in that capacity he sometimes refused politically motivated but scientifically hazardous instructions.

When Racah became rector of the Hebrew University, Moshé taught his course on group theory for a year, while still serving in the Air Force. The next year, while at the reactor, he was in addition asked to teach solid state physics at the religious Bar Ilan University. He accepted the appointment on two conditions that were typical of him: a taxi to drive him to the University, and an exemption from the obligation of wearing a skullcap in class. In 1962 Moshé obtained army permission to attend a summer school on theoretical physics in Italy, but an automobile accident on the way out of Dijon kept him in the local hospital for two weeks, followed by six months' treatment in Geneva and Tel Aviv. He used that time to finish work on his doctoral thesis on group theory in nuclear physics. His injury also earned him an honorable exemption from completing his military service.

In October 1963 Moshé came to France with a generous ASTEF grant. He had chosen Paris over CERN in Geneva or the USA because Racah had no power over him there, and because he wanted to add French to his other languages. He was assigned to Louis de Broglie's laboratory at IHP (Institut Henri Poincaré) and I helped him find a room nearby, in the rue Gay-Lussac. The room was as modest as the Bohemian student tradition required, but we – I was in Paris then, too and my studio was on the same floor – had income enough to live more comfortably than students' traditions permitted. Shortly afterwards I began working with Moshé, to the great displeasure of my official thesis advisor, Szolem Mandelbrojt, and I eventually passed my doctorate under Moshé's direction even though he was only nine months older than I. In 1963 we began a scientific collaboration that Norma Finkelstein would eventually call "a team" and another friend, "Moshé *et al.*", and which endured to his death. We were sufficiently similar in character and scientific profile to agree and cooperate on essentials, and sufficiently different to complement each other's work. Though I always remained in the shadow of

Moshé's vigorous personality, my only regret is that our collaboration terminated so abruptly, after only 35 years.

The arrival in Paris of the "sabra" bull-in-a-china-shop that Moshé was then, and remained to the very end, created quite a stir. He was recognized immediately by Louis de Broglie and André Lichnerowicz and became their close friend and, in the case of Lichnerowicz, scientific collaborator, until death. Lichnerowicz was devastated by Moshé's death and never recovered from the shock, himself dying two weeks later in the same hospital, also as a result of a stroke. Many of the ambitious physics students and scholars who were Moshé's competitors did not appreciate him and his vigorous demonstrations of the inadequacies in the theories proposed by the gurus in the field. This was the beginning of his often strained relations with the French physics establishment. Nevertheless, Moshé found an intellectual freedom in Paris that he could not have found elsewhere, and he flourished in it.

At the end of his scholarship year I persuaded Moshé to remain in France. It was not hard because I had the help from the fine Parisian restaurants and the chefs whom Moshé appreciated so much, as his expanding girth testified.

Jean Braconnier, then dean of sciences at Lyon, invited Moshé to join his faculty as an adjunct professor, even before he finished his doctorate. The offer was supported by Lichnerowicz and, in the CCU (national committee), by Alfred Kastler, who hoped to bring Moshé to Orsay to work with Chabbal in theoretical spectroscopy at Laboratoire Aimé Cotton. Moshé had the effrontery to refuse that offer, then the beaten path for ambitious young French physicists, to stray into the area of mathematical physics, which was then much less highly regarded. When he arrived in Lyon, Moshé discovered that the physics curriculum was out of date with "wave mechanics" introduced only after the Master's level, and quantum optics only sketchily treated. Nonetheless Moshé threw himself and some of his students into particle physics.

In 1965 Moshé defended his French doctoral thesis in Paris – that was before the University was broken up into the thirteen complementary and sometimes competing universities that we know today – before a jury chaired by Lichnerowicz. Since he was still a foreigner and had studied abroad, the difficulties of obtaining certificates and equivalences from the Sorbonne administration were epochal, and Moshé was more vigorous than patient. I leave the rest to your imagination. The CCU in physics refused to renew Moshé's appointment in 1967 because he had strayed from spectroscopy into a too original and mathematical approach to particle physics with which the establishment was uncomfortable, and that is an understatement. Also, his reputation as a scientific "trouble maker" was so well established that in 1966, during a very friendly conversation with a well-known theoretical physicist who happened to share a train compartment with him (and of course Moshé could not be in a train without talking to his neighbors), he was warned to avoid a guy named Flato, who would "shoot down" prominent physicists whenever they happened to say something incorrect in a seminar! But he was adopted by the mathematicians who were used to being hospitable to intellectual originality, and this was at a time when mathematical physics had not arrived at its current respectability.

Moshé was appointed to Dijon in January 1968. When he required the intervention of the union, the SNESup, to recover the three months' salary of the interval between the lapse of his appointment in Lyon and the assumption of his functions in Dijon, the union refused to help a "mandarin" and an Israeli to boot. This taught Moshé a lesson about the mentality of "apparatchiks" in all countries, that he never forgot. Despite the SNESup, he remained a leftist in the French scheme of politics because of his native generosity of character. He participated in the 1968 student revolution, joining the occupation of the IHP. When the revolution ended, several of the students who had joined Moshé in the IHP feared retaliation from their humourless professors, so Moshé brought them to Dijon. Despite his politics and the fact that some leading socialist politicians were personal friends, he generally got along well with centrists and even with right wing politicians; they appreciated his talent and good sense and named him twice to the national committee (in 1978 to the CCU in physics, and in 1996 to the CNU in mathematics).

Before Moshé's arrival in Dijon, the mathematics faculty was not too distinguished, though Maurice Gevrey, a fine mathematician, had taught there in the first half of the century. In Dijon Moshé brought distinction to his own field, of course, but also encouraged other domains. Not all his colleagues, even those he recruited, were always equally tolerant of competing specialties, but Moshé was an astute tactician and often prevailed in University committees to obtain decisions satisfactory to most of his colleagues.

Perhaps not so picturesque as Gulliver's, Moshé's travels and misadventures abroad would make an amusing book. Let me just mention the time he kept interrupting Murray Gell-Mann's Nobel Prize lecture with entirely pertinent questions that the new laureate was unable to answer, or the time he broke his foot in Araki's linoleum-covered bathroom, or the time he explained to the Taiwan immigration control that Moshé is spelled with an M, like Mao, which nearly landed us in prison! On the serious side, Moshé formed strong ties with the mathematics and physics communities in Sweden since 1967 – for 27 years he was a member of the advisory committee that proposes nominees for the Nobel prize –, in Poland since 1969, in China where he was named honorary professor of Fudan University in Shanghai, and in Japan. He always loved the Orient and its cuisine, but he distanced himself from China after the Tiananmen "event". Aside from bringing European developments in physics and mathematical physics in his baggage, he made friends with foreign colleagues all over the world and, when he created *Letters in Mathematical Physics*, he made a point of soliciting articles from all of them, helping to integrate Eastern scientists into the community of Occidental scientists. The International Association of Mathematical Physics, which he also initiated, had much the same mission in a period when the iron curtain was as much an intellectual barrier as a political one. Moshé was recruited by UNESCO in

1975 to evaluate graduate studies in mathematics, physics, chemistry, biology and economics in Thailand (at the request of the Thai government). His frank report, still known there as the "Flato Report", contained recommendations for various reforms, many of which have gradually been introduced.

Though Moshé was greatly appreciated abroad, he was frequently at odds with many French physicists. When Louis Leprince-Ringuet retired from Collège de France in 1971, the physics establishment proposed a single candidate, a then talented theoretician who had been groomed for three years to be named to that chair in experimental physics. Jean-Claude Pecker urged Moshé to draft a "project" for a chair in theoretical physics so that the professors of that venerable institution – no irony intended, it was created in 1531 and has included among its members some of France's greatest scientists and scholars, in areas ranging from ancient history to the most up-to-date mathematics – might have the courtesy of a choice. Moshé's candidacy enraged the physics establishment, frightened the union, provoked a demonstration, and I must admit he was not always a model of tact when he paid the obligatory courtesy calls to the professors who would vote on his project and candidacy. Despite all of that, and the unofficial quota for Jews in the Collège, which was already filled, Moshé obtained a respectable score and would probably have been elected to one of the next vacant chairs, but he found the affair disagreeable, and he was not sure that he was willing to undertake a Collège de France professor's traditional obligation of lecturing each year on his own new research. Being expected to come up with new results year after year, until retirement age (at 70 in the Collège), intimidated Moshé, strange as it may seem to all of us who knew his self-confidence, because he took all his obligations with the utmost seriousness, and made a point of always giving much more than was demanded of him.

Moshé had too much intellectual energy to be constrained by his chair at Dijon, his guest lecturing, LMP, IAMP and his participation in the intellectual life of Paris. He had to break out. When his friend, Yvette Chassagne, was named Chairperson of the Board of the Union des Assurances de Paris, he created for her a scientific council to advise the UAP on ways of anticipating advances in science (including economics) and scientific technology that would inevitably influence its business strategy. Moshé insisted that the members of the council be unpaid, to insure their intellectual independence, and he convinced many of France's intellectual stars and future stars to participate. Again, I shall resist the temptation of name dropping and of mentioning the science and technology that was in fact produced by what was supposed to be merely a consulting council.

The articles in this collection deal with areas in which Moshé worked and, in many cases, extend his ideas and results. I shall not anticipate here the many articles based on Moshé's scientific achievements to avoid redundancy, and because Moshé and I worked so closely. Distinguishing his work from mine in our joint publications may be a good exercise for a historian of science desperately searching for a thesis subject, but is not too important here.

Let me only add that Moshé was as generous a personality as he was a vigorous scientist and bon vivant. He was a devoted son and brother, and Uncle Moshé to the children of all his colleagues and friends, following and encouraging their childhood and adolescence and then their studies and careers. Frank and even tactless as he often tended to be, he could also offer sensitive moral support to younger and older colleagues when they suffered personal losses or losses of confidence. All of Moshé's friends who are so numerous and distinguished – mentioning them by name here would be presumptuous – will miss him terribly. His premature death leaves a void in the scientific community that will surely be filled, possibly by several of his friends and collaborators, and void in our lives that only time will blur.

PS. This in an English version, "verkurtzert und verbesert" by Bertram Schwarzbach, from the longer and less discreet French text written by Daniel Sternheimer for the Conference, which is available in postscript and pdf formats on the web site of CMF1999, `http://www.u-bourgogne.fr/monge/cmf/cmf1999/inaugural.html`.

In retrospect:
A personal view on Moshé Flato's scientific legacy

Daniel Sternheimer

When a talented sculptor is confronted with a block of marble, he is using tools that come from centuries of tradition and the training which he received from his masters. But from the first stroke he already has, in the back of his mind, the design of the sculpture which he wants to extract from that block. Creating a masterpiece is a long and painful process, interspersed with surprises due to the irregularities of the physical properties of the noble material he is using, and the design continues to evolve with the artist's maturing and subject to external influences. Only gradually does the shape emerge, at first visible to experts, then to the general public.

Beyond the technicalities, however important they may be, Moshé's scientific life was a perfect illustration of this idea. His masterpiece is not finished. Much polishing remains to be done, new elements still need to be added to the main themes which he conceived, and entirely new themes are now beginning to emerge from them. But a very partial view of what he had in mind can already be described. This is what I shall attempt here, based on my experience of having, for about 35 years, participated with others in the elaboration of his ideas. All those who have been close to him know that he used to call me affectionately an "idiot" more often than he used that expression for anybody else; this usually happened when my formulation of one of his ideas was not to his taste and needed to be corrected, either because he did not like some of the ingredients I was adding or when, while striving for too precise a formulation, I obscured an essential point which he had purposedly stated ambiguously. François Mitterrand used to say, paraphrasing the Cardinal de Retz, "On ne sort de l'ambiguïté qu'à son détriment". In his case, this was smart politics, in Moshé's, using seemingly ambiguous statements was the only way by which he could transmit the deep feeling which he had without the distortion of too precise, confining words. I hope that now, with him not around, my formulation will be faithful enough to the main ideas which he developed and often initiated. My purpose here is to show how, in retrospect, one can look at the genesis and evolution of ideas and concepts in some of the fields where Moshé made an impact that has only begun to be fully recognized. I shall refrain from entering into technicalities, because they are covered in part in the present volume and in recent review papers, and in order to make the ideas understandable to a general, scientifically educated audience.

Towards the end of the 19th century, physics seemed to have achieved a complete understanding of the world, with classical mechanics describing the motion of rigid bodies, electromagnetism waves and charged particles, and the Lorentz force their interactions. It was only a plateau because then the 'deformation dae-

G. Dito and D. Sternheimer (eds.), Conférence Moshé Flato 1999, Vol. 1, 31 – 41.

mon' struck. In 1881 an American physicist, Albert Michelson, came to Berlin and tried to observe the movement of the earth in the 'ether' by measuring the velocity of light in different directions, using the interferometer which he had developed. He found no differences, which did not conform to classical mechanics. In such a case, in Europe, the first reaction was and remains – as Charles Schulz wrote in a 'Peanuts' cartoon –, "my mind is made up, don't bother me with facts". That is how the first observation of neutral currents at the CERN, ten years before their theoretical prediction, was deemed a mistake and never published. However facts and some American scientists are stubborn. In 1887, with Edward Morley, Michelson performed a more refined experiment in Cleveland and the fact was proved. So there was a *paradox*, because the speed of light is a limit. The so-called black-body radiation provided another paradox until in 1900, as a last resort, Max Planck proposed the quantum hypothesis, the energy of light is not emitted continuously but in quanta, proportional to its frequency. He wrote h for the proportionality constant which bears his name.

At the same time the notion of *symmetry* started to appear in mathematics and in physics, developments in both sciences being very closely interrelated all through the twentieth century. In physics, equations and quantities are usually invariant under some transformation groups. The theory of the latter was systematized by Sophus Lie and developed further by many, among them Élie Cartan and Henri Poincaré, to remain within that time range and in mathematics. Emphasizing this invariance property has many advantages, but we shall not elaborate on that aspect here because it has been developed in many books. In any case, this is a concept that Moshé learned from his teacher Giulio Racah, and first applied to molecular spectroscopy in his celebrated M.Sc. thesis, completed in 1960 under Racah but only published in 1965 [13], and in his unpublished Ph.D. thesis on applications of group theory to nuclear physics, a problem involving the restriction of representations from unitary to symplectic groups, which Moshé also completed under Racah before coming to France in 1963 but never defended. As a consequence of the expertise which he gained in these applications of group theory to physics, he took with more than a grain of salt the quite naive applications that were fashionable among particle physicists in the sixties and, in a much more elaborate manner, still influence the present interpretation of experimental data. This can already be seen in his 1965 French D.Sc. thesis [14]. He never hesitated to express these views, which many considered blasphemous, in public, and eventually his outspokenness obliged him to seek "scientific asylum" in the French mathematical community while staying in contact with physicists around the world who dared and could afford to hold independent opinions.

The notion of symmetry guided Moshé throughout his scientific career, both as a mathematician and as a theoretical physicist, and led him to make a number of significant contributions. Even on the mathematical side, the motivation for the study of symmetry comes from physics and the mathematical theory has physical applications. In this connection I can mention two directions which his work, in collaboration with (among others) his D.Sc. student and close friend since 1968, Jacques Simon, took. The first is the theory of analytic vectors in Lie group representations [25], suggested by the problems that arise in passing from Lie algebras to Lie groups. The second is the cohomological study of nonlinear representations [24] of covariance groups of nonlinear partial differential equations, which leads to important mathematical developments with nontrivial physical consequences such as the "tour de force" on Maxwell-Dirac equations [26], the very basis of classical electrodynamics. A lesser known, third direction, is the notion of supersymmetry which one can find already in an embryonic form – close to the Wess-Zumino Poincaré supersymmetry – in his 1970 paper [18].

In 1905 Einstein solved the first paradox that we described above when he showed, in our terminology, that the Galilei invariance group $SO(3)\cdot\mathbb{R}^3\cdot\mathbb{R}^4$ (space rotations, addition of velocities and space-time translations) of Newtonian mechanics had to be *deformed* into the Poincaré group $SO(3,1)\cdot\mathbb{R}^4$, where velocities do not add linearly, of relativistic mechanics. That same year Einstein contributed, in addition to a third basic paper, on the Brownian motion, to the solution of the second paradox by his theory of the photoelectric effect. Incidentally, that theory was the main contribution recognized by the Nobel Prize in Physics which was awarded to him only in 1922. Revolutionary ideas take time to be recognized. Around 1920, an "agrégé d'histoire", Prince Louis de Broglie, was introduced to the photoelectric effect, together with the Planck-Einstein relations and the theory of relativity, in the laboratory of his much older brother, Maurice duc de Broglie. This led him, in 1923, to his discovery of the duality of waves and particles, which he described in his celebrated 1925 thesis. After Davisson and Germer's observation of the diffraction of electrons in 1927 confirmed his theory, Louis de Broglie was awarded the Nobel Prize in 1929.

Moshé had been sent to Louis de Broglie when he came to Paris in October 1963. There was immediate sympathy between these two very different but profound personalities. Moshé used to say that Louis de Broglie was an extraordinarily well-adjusted schizophrenic, and that without pejorative connotation, because he could see single physical phenomena as dualities. The wave-particle duality is one example, but there are others, like his theory of fusion, in particular, his short-lived neutrino theory of light. As de Broglie formulated that theory, with photons composed of two neutrinos, it was soon 'shot down', and justly so, but the general idea was vindicated, in a way, by Moshé when he showed in 1988 [19] that it was possible to formulate quantum electrodynamics with photons composed of two scalar singletons. The latter 'particles', in fact, massless particles in a 2+1 dimensional space-time at infinity in the anti-de Sitter universe, are just one step below neutrinos (at the limit of unitarity) in the diagram of representations of the anti-de Sitter group $SO(3,2)$, which is the last possible deformation of the Poincaré group in the category of Lie groups when, due to a tiny negative curvature *in the microworld*, space-time translations do not add linearly. The idea that singletons, with Higgs particles, should be considered as the ultimate constituents of matter at our level of understanding and experimental results, is an example of the interaction between Moshé's basic *deformation view of physical theories* [16] and some of the concepts which he learned when he arrived in France.

Moshé's deformation philosophy was inspired by Murray Gerstenhaber's mathematical theory of deformations of algebras [29], joined to Moshé's deep analysis of the evolution of physics. That analysis had epistemological consequences going beyond its mathematical formulation in terms of deformations in a well-chosen category. In particular, it tells us that there can be no "theory of everything". Our representation of physical reality is only a mathematical idealization of that reality in a given range, and there are several ranges without hermetic border between them. One has to be modest and realize that there are limitations to the capacities of the human brain, however brilliant and profound it may be. Only God, if one is willing to add to our representation of the universe what is, in my view, an axiom that is not contradictory to science but independent of it, may possibly go further. Moshé was a pantheist, à la Spinoza, and considered the universe itself to be the manifestation of what theologians call God.

Moshé used to discuss physical concepts and their philosophical implications with Louis de Broglie, Hideki Yukawa (who spent some time with de Broglie in Paris in 1964), David Bohm, Roger Penrose, and André Lichnerowicz. This led him in the mid-sixties to stress, sometimes provocatively [27], that one should not decouple *space-time* from the so-called 'internal symmetries' of elementary particles. Rather, the conformal group $SO(4,2)$ of (flat or de Sitter) space-time has a role to play [5]. Reality is probably much more complex than any of us could have foreseen at that time, but this gives even greater value to Moshé's strong warnings about the dangers of some of the simplistic and incorrectly formulated 'no-go' theorems that were then popular.

I am confident that the first half of the 21st century will show how fundamental is the singleton hypothesis. After formulating the singleton hypothesis around 1978, a hypothesis which he considered "too beautiful not to be true", Moshé never ceased trying to prove it. At this stage it is not yet a complete theory and, so far, it has not yielded a verified model or new predictions (see however [28]). But after all, in a different context, quarks are still only a hypothesis without a complete theory, and yet the quark hypothesis has given rise to a succession of models, the last of which is considered successful because a wide array of experimental data is consistent with it. Logically, a hypothesis is a sufficient condition, not a necessary one because data might have a different explanation. Moshé was fully aware of the delicate work remaining to be done and even discussed a few related mathematical questions with some of his students. With the confirmation of neutrino oscillations in 1998, Moshé saw the 'promised land' not too far away. He died before reaching it, but he has brought us through the desert to a point where we can cross the Jordan river, at least, and sound the trumpets around the walls of Jericho. The contribution of our long time friend and collaborator Christian Frønsdal in these *Proceedings*

[28], along the lines indicated in Moshé's last paper [20], is a further step in this direction.

More will come, ranging from *phenomenology* where a lot of work and possibly reinterpretation of data could be required, through subtle *theoretical physics* developments needed for the new theory (including considerations of flavor symmetry, *PC* violation, the number of generations of leptons, conformal covariance and the new light on renormalization which emerges from the latest works of Alain Connes and Dirk Kreimer [10]), to pure *mathematics*, including groups (their indecomposable representations and associated Gupta-Bleuler triplets), infinite-dimensional algebras, possibly with a notion of "square root of a superalgebra" (which would develop in a general context the construction introduced in [19], where the generators of the canonical commutation relations associated with photon fields are obtained from the commutators of singleton field creation and annihilation operators), deformation quantization in infinite-dimensional manifolds of initial data of nonlinear classical field equations (which will require a study of the latter, e.g., along the lines developed by Moshé in [26] for the Maxwell-Dirac equations of classical electrodynamics), all the way to the very abstract ideas and methods which are now being developed by Maxim Kontsevich [33]. This is a simplistic sketch of a program which will occupy a generation of mathematicians and physicists.

De Broglie had another basic idea, what he and Yukawa had called the 'bilocal model', which had been formulated – as an unintended caricature – by some of his collaborators before Moshé arrived in Paris. It did not take long for Moshé, already an expert in group theory, to realize how naive the formulation was. But the basic idea, that points in space-time are not one, but two, is not so far fetched. Just pull a string between those two points and you can get into string theory, a major trend in modern theoretical physics which, though so far lacking experimental verifications, has inspired some very beautiful mathematics. Or look at the $\mathbb{Z}_2$-doubling of space-time, as in the Connes and Lott interpretation of the standard model of elementary particles within noncommutative geometry [9].

The duality between waves and particles had brought de Broglie to what he called the 'mécanique ondulatoire'. When Moshé came to Lyon in 1965, that was still what quantum mechanics was called there, and it was only taught at the graduate level. It had taken some German and Austrian physicists, in particular, Hermann Weyl – less well-known to the physics public because he was mainly a mathematician –, Werner Heisenberg and Erwin Schrödinger, to transform it into quantum mechanics, based on operators in spaces of wave functions, endowed with a structure invented some years before by David Hilbert. Goethe used to say that "Mathematicians are like Frenchmen: they translate everything into their own language and henceforth it is something entirely different". Those German physicists did just that with de Broglie's approach! A consequence of the operatorial formulation is the 'Copenhagen' probabilistic interpretation systematized by Niels Bohr, which Moshé learned from Giulio Racah. It offers an efficient interpretation of the experimental results, e.g., in terms of the probability of presence of a particle given by some wave function and of expectation values of observables (operators in this case). But Einstein and de Broglie hated the idea.

Einstein would say that "God does not play dice with the universe" and formulated his celebrated paradox with Podolsky and Rosen [11]. De Broglie (and many others) looked for 'hidden variables', a classical mechanics underlying quantum mechanics. Eventually it was experimentally proven [1] that hidden variables do not exist, or rather – as Moshé pointed out [21] when the first experimental indications appeared – that if they exist they cannot be local. Deformation quantization, Moshé's main scientific contribution, the most striking example of his deformation philosophy and a seminal idea in mathematics, solves the mystery in a way that reconciles Einstein and de Broglie with Bohr. God does not play dice with the universe, 'it just feels like it' when you consider only the operatorial formulation.

Around 1960 there occurred two seemingly unrelated mathematical events to which Moshé and I were exposed. Some mathematicians (Kunihiko Kodaira and Donald Spencer [32] in 1958 for complex analytic structures, and Murray Gerstenhaber [29] in 1963-64 for algebras) developed a cohomological theory of deformations. At the symmetry group level it became immediately clear to us, and many others, that the passage from the Galilei to the Poincaré group (with deformation parameter c^{-1}, where c is the velocity of light) and then to the de Sitter group (with curvature as the deformation parameter) is a typical example of a deformation.

At about the same time other mathematicians introduced pseudodifferential operators (called singular integral operators in 1957 by Alberto Calderón and Antoni Zygmund [7] who defined their symbols), which provided a rigorous framework for notions such as square roots of partial differential operators (of the Laplacian for instance). They gained celebrity outside the world of partial differential equations by their use, in 1963, by Michael Atiyah and Isadore Singer [2] and subsequently by many other mathematicians, in the proof of index theorems, a problem that had been posed by Israel Gelfand, how to express the analytic index of an operator in topological terms. I had personally taken an active part in a weekly seminar [8], led by Henri Cartan and Laurent Schwartz in Paris, which studied index theorems for elliptic operators on a compact manifold from both the analytical and topological points of view, and shaped the future of many of the most promising young mathematicians of our generation. Moshé did not attend the seminar but was exposed to it by me.

Around 1974 Moshé came to the conclusion that physics evolves in stages, when it hits a paradox [17], the passage from one level of scales, e.g., velocities and distances, to another being mathematically described by a deformation, in an appropriate category. This led us in 1976 – 1978 [23, 3] to the formulation of quantum mechanics and quantum theories on the same observables as classical mechanics, functions on a phase space, but with a deformed composition law which we called a star-product. The first term in the deformation series beyond the usual product of functions was the Poisson bracket of classical mechanics and the deformation parameter $\frac{i\hbar}{2}$, the Planck constant being $h = 2\pi\hbar$. Quantization should therefore be understood as a deformation, what is now called *deformation quantization*. This in effect reconciles Einstein and de Broglie with Bohr because it shows that, while there cannot be an underlying classical level with hidden variables, classical and quantum theories can be formulated on the same algebras of observables, and physical results can be described by appropriate mathematical notions, equivalent to what is given by the operatorial formulation when both theories apply, but without need for the latter's explicit probabilistic interpretation. Moreover deformation quantization can be applied to cases not covered by the existing operatorial formulation [36].

The way things developed is interesting in itself. Since it is at present the most recognized of Moshé's contributions and mathematically the most seminal, I shall elaborate a little. Technically, it started with the renewal of interest, in the late sixties, in infinite-dimensional symmetries, which date back to the work of Élie Cartan at the beginning of the century. Among them were what are now called Kac-Moody algebras and algebras of vector fields (first order differential operators) on some manifolds, with their Gelfand-Fuks cohomology. Our friend André Lichnerowicz who, among his many talents, was a master of tensor calculus on manifolds, had computed what he called the 1-differentiable cohomologies of such algebras [34]. Since deformations, in Gerstenhaber's sense, are governed by second and third cohomologies, it was natural to apply Lichnerowicz's results to deformations, which we did in the 1-differentiable case [22] because we did not yet know how to handle the more complicated differentiable case. A remarkable example stood out, the algebra of differentiable functions over a symplectic manifold endowed with the Poisson bracket, which is at the base of the geometric, Hamiltonian formulation of classical mechanics. There we discovered that infinitesimal deformations of the Poisson bracket algebra are, subject to some natural technical conditions, classified by the deformations of the symplectic structure, which are themselves governed by the second de Rham cohomology of the manifold.

From the beginning, Moshé had in mind that the deformation approach should provide a clue to a true understanding of quantization. As evidence I shall translate here an excerpt from his introduction to our *Compositio Mathematica* paper, submitted November 1, 1974. The introductory and concluding remarks in papers by Moshé are very inspiring, often going beyond the strict purpose of the paper. In retrospect it is clear that, beyond the technical difficulty which we encountered at that time, he had already foreseen deformation quantization 'around the corner':

> The study of deformations of those algebras is in part motivated by theoretical physics which appears as a generalization of classical mechanics where those algebras are naturally introduced. For instance, in quantum mechanics, the Poisson brackets of functions are replaced by commutators of operators in Hilbert space; in classical and quantum field theory, the symplectic manifold becomes infinite dimensional; in various branches of statistical mechanics, phase space plays an important role. ...

Our work inspired Jacques Vey [37] to pass in 1975 from 1-differentiable deformations to general differentiable deformations of the Poisson bracket algebra, at least subject to a technical assumption on the topology of the manifold, which was waived much later. His work relied on the determination of the differentiable Hochschild cohomology of the algebra of functions, itself related to a previous important algebraic result, well hidden in a now famous paper [31] by Hochschild, Kostant and Rosenberg. In passing he derived, on flat space $\mathbb{R}^{2n}$, a bracket which in fact had been obtained in England in 1949, in a completely different perspective related to the hidden variables question, by a student of Maurice Bartlett named José Enrique Moyal [35], as the inverse image, in the sense of Wigner, of the commutator of operators in the Weyl quantization procedure. Vey also considered the corresponding associative deformed product, that we now call star-product. In the Spring of 1975, I lectured about our work on the 1-differentiable case in Austin, Texas. Our friend Jerzy Plebański, who had mentioned Moyal brackets in lectures given in 1969 in Poland – much later we learned that another friend, Ludwig Faddeev, had also done so in Leningrad around the same time – was in the audience and said that it reminded him of Moyal. I could not find the paper [35] in the library there and *Mathematical Reviews* made no mention of the aspect for which that paper is now widely quoted, but Moshé decided that we should order a photocopy from the CNRS. At the same time we had received Vey's preprint. The bracket was in both places, independently, and that is exactly what Moshé had intuited in the passage we just quoted! This led us to our first Note in 1976 [23] and, with Frønsdal and Bayen who joined us, to the papers [3] where we showed, among many other things, that it is possible to obtain the most fundamental results of quantum mechanics, such as the description of the harmonic oscillator and the determination of the spectrum of the hydrogen atom, by considering the same algebra of observables as classical mechanics, but with a deformed product.

It turned out that the composition of symbols of pseudodifferential operators, a key ingredient in the proofs of the index theorems, is exactly a star-product, related to what physicists called the standard ordering, which consists, for differential operators, in writing first the coefficients and then the differentiations. So the two 'unrelated' events from the early sixties were related after all, and we, like Molière's M. Jourdain, had been 'speaking prose' for over twelve years without knowing it! Looking at the footnote by Louis Boutet de Monvel in [38], p.18, one can see that neither Vey, because of his algebraic approach, nor most mathematicians working in the field of pseudodifferential operators, realized the connection for quite some time. Nevertheless it must have had some impact on Moshé and me at the subconscious level. One should mention, for completeness, what is now called 'geometric quantization', developed around 1970, independently, by Bertram Kostant and Jean-Marie Souriau, related to the orbit method of Alexandre Kirillov and to ideas on quantization formulated at that time by Felix "Alik" Berezin, of which we became aware much later. It proved successful in group representation theory but, due to a too conservative approach which omitted the deformation concept, not equally so in physical applications.

I shall not elaborate further on what is now well-known history. For a more detailed exposition and for references the reader may consult my review [36], the "promenade" by Simone Gutt [30] in these *Proceedings* and the contribution by Louis Boutet de Monvel [6]. Let me just add that star-products gave rise to an 'autonomous' representation theory of Lie groups on spaces of functions over coadjoint orbits with deformed products, and that quantum groups are an avatar of deformation quantization in the category of Hopf algebras, and can, in many interesting cases, be reduced to deformations of the product of functions over Lie groups, leaving the coproduct undeformed. The questions of existence and classification of star-products over symplectic manifolds have been advanced by the more geometrical and algorithmic approach of Boris Fedosov [12], who incidentally came to this branch of mathematics from index theorems. Essentially, one combines our 1-differentiable deformations, classified by formal series in the second cohomology classes of the manifold, with a Moyal product.

In 1997 Maxim Kontsevich put what seemed at first 'the frosting on the cake' that we had baked, by proving his formality conjecture and giving a complete solution to the problem of existence and classification of deformation quantizations on general Poisson manifolds. It now appears [33] that this has roots going very deep into modern mathematics, using in part methods, such as graphs, inspired by physics and extending, in particular via the notion of deformations of algebras over operads which he defines in his contribution to these *Proceedings*, far into seemingly unrelated notions like Feynman path integrals, periods, and Grothendieck's unfinished symphony of algebraic geometry. All this will develop well into the 21st century.

This is, in my view, part of the kernel of Moshé's scientific legacy. I have not attempted here to develop a final historical description or philosophical judgment on it. I am merely doing some phenomenological work, trying to bring a little order and indicate some guidelines in the scientific production of a brain that was as complex and in constant evolution as life itself. There is a lot which remains to be done, beautiful mathematics for sure, in the interplay with fundamental physical developments. It will require a variety of talents which no single scientist may possess. I shall be very happy if I live to see the beginning of some breakthrough in that direction.

I shall end this overview by some reflection on the man behind it. Moshé used to say that scientists, especially mathematicians, can be divided into two categories, "problem solvers" and "theory makers". But where does he stand in this dichotomy? He was a theory maker who could solve problems and inspire others to solve problems, certainly in mathematics and in theoretical physics, but also in epistemology, and he had many other facets as can be seen from these *Proceedings* and the personal recollections by some of his numerous friends [38]. As three friends say there, it was impossible to define him when he was alive, and it would be presumptuous to attempt now to fix his image: *Personnage inclassable de son vivant, il serait bien présomptueux de vouloir aujourd'hui figer son image.*

Acknowledgements. Thanks are due to many friends, especially Wilfried Schmid, who, by their comments, contributed directly or indirectly to polish my original formulation. A special mention is due to Bertram Schwarzbach for his editorial help.

References

1. Aspect, A., Dalibard, J., and Roger, G.: Experimental test of Bell's inequalities using time-varying analyzers, *Phys. Rev. Lett.* **49** (1982), no. 25, 1804–1807.
2. Atiyah, M. F. and Singer, I. M.: The index of elliptic operators on compact manifolds, *Bull. Amer. Math. Soc.* **69** (1963), 422–433; The index of elliptic operators I, III and IV, V, *Ann. of Math.* **87** (1968), 484–530, 546–604 and **93** (1971), 119–138, 139–149.
3. Bayen, F., Flato, M., Fronsdal, C., Lichnerowicz, A., and Sternheimer, D.: Deformation theory and quantization I, II, *Ann. Physics* **111** (1978), 61–110, 111–151.
4. Bell, J. S.: On the problem of hidden variables in quantum mechanics, *Rev. Modern Phys.* **38** (1966), 447–452.
5. Bohm, D., Flato, M., Sternheimer, D., and Vigier J.-P.: Conformal group symmetry of elementary particles, *Nuovo Cimento* **38** (1965), 1941–1944.
6. Boutet de Monvel, L.: Complex star-products, *Math. Phys. Analysis and Geometry* **2** (1999), 113–139; Star-produits et star-algèbres holomorphes, these proceedings, pp. 109–120.
7. Calderón, A. P. and Zygmund, A.: Singular integral operators and differential equations, *Amer. J. Math.* **77** (1957), 901–921.
8. Cartan, H. and Schwartz, L.: *Séminaire Henri Cartan 1963/64, "Théorème d'Atiyah-Singer sur l'indice d'un opérateur différentiel elliptique"*, fasc. 1 & 2, Secrétariat mathématique, Paris (1965).
9. Connes, A.: *Noncommutative geometry*, Academic Press, San Diego, 1994.
10. Connes, A. and Kreimer, D.: Renormalization in quantum field theory and the Riemann-Hilbert problem, *JHEP* 9909 (1999) 024, (hep-th/9909126).
11. Einstein, A., Podolsky, B., and Rosen, N.: Can quantum-mechanical description of physical reality be considered complete? *Phys. Rev.* (2) **47** (1935), 777–780.
12. Fedosov, B. V.: A simple geometrical construction of deformation quantization, *J. Diff. Geom.* **40** (1994), 213–238.
13. Flato, M.: Ionic energy levels in trigonal and tetragonal fields, *J. Mol. Spec.* **17** (1965), 300–324.
14. Flato, M.: *Symétries de type lorentzien et interactions fortes*, Gauthier-Villars, Paris, 1967.
15. Flato, M.: Quantum mechanics and determinism, in: *Quantum mechanics, Determinism, Causality and Particles*, Mathematical Physics and Applied Mathematics **1**, D. Reidel, Dordrecht, 1976, pp. 19–31.
16. Flato, M.: Deformation view of physical theories, *Czech. J. Phys.* **B32** (1982), 472–475.
17. Flato, M.: *Le pouvoir des mathématiques*, Hachette, Paris, 1990.
18. Flato, M. and Hillion, P.: Poincaré-like group associated with neutrino physics, and some applications, *Phys. Rev. D1* (1970), 1667–1673.
19. Flato, M. and Fronsdal, C.: Composite Electrodynamics, *J. Geom. Phys.* **5** (1988), 37–61; Flato, M., Fronsdal, C., and Sternheimer, D.: Singletons as a basis for composite conformal quantum electrodynamics, in: M. Flato and M. Cahen, (eds.), *Quantum Theories and Geometry*, Mathematical Physics Studies **10**, Kluwer Acad. Publ., Dordrecht, 1988, pp. 65–76.
20. Flato, M., Frønsdal, C., and Sternheimer, D.: Singletons, physics in AdS universe and oscillations of composite neutrinos, *Lett. Math. Phys.* **48** (1999), 109–119.

21. Flato, M., Piron, C., Gréa, J., Sternheimer, D., and Vigier, J.-P.: Are Bell's inequalities concerning hidden variables really conclusive? *Helv. Phys. Acta* **48** (1975), 219–225.
22. Flato, M., Lichnerowicz, A., and Sternheimer, D.: Déformations 1-différentiables d'algèbres de Lie attachées à une variété symplectique ou de contact, *C.R. Acad. Sci. Paris Sér. A* **279** (1974), 877–881; and *Compositio Mathematica* **31** (1975,) 47–82; Algèbres de Lie attachées à une variété canonique, *J. Math. Pures Appl.* **54** (1975), 445–480; Deformations of Poisson brackets, Dirac brackets and applications, *J. Math. Phys.* **17** (1976), 1754–1762.
23. Flato, M., Lichnerowicz, A., and Sternheimer, D.: Crochet de Moyal-Vey et quantification, *C. R. Acad. Sci. Paris Sér. A-B* **283** (1976), Aii, A19–A24.
24. Flato, M., Pinczon, G., and Simon J.: Non-linear representations of Lie groups, *Ann. Sci. Éc. Norm. Sup.* (4) **10** (1977), 405–418; Flato, M. and Simon, J.: Non-linear wave equations and covariance, *Lett. Math. Phys.* **2** (1977), 115–160.
25. Flato, M., Simon, J., Snellman, H., and Sternheimer, D.: Simple facts about analytic vectors and integrability, *Ann. Sci. Éc. Norm. Sup.* (4) **5** (1972), 432–434.
26. Flato, M., Simon, J. C. H., and Taflin, E.: *The Maxwell-Dirac equations: The Cauchy problem, asymptotic completeness and the infrared problem*, Memoirs of the American Mathematical Society **127** (no. 606, May 1997).
27. Flato, M. and Sternheimer, D.: Remarks on the connection between external and internal symmetries, *Phys. Rev. Lett.* **15** (1965), 934–936.
28. Frønsdal, C.: Singletons and Neutrinos, these proceedings, pp. 209–216; Frønsdal, C. and Sternheimer, D.: Multiple Higgs mesons: mass spectrum and interactions (to be published).
29. Gerstenhaber, M.: The cohomology structure of an associative ring, *Ann. of Math.* **78** (1963), 267–288; On the deformation of rings and algebras, *Ann. of Math.* **79** (1964); 59–103; and (IV), *ibid.* **99** (1974), 257–276.
30. Gutt, S.: Variations on deformation quantization, these proceedings, pp. 217–254.
31. Hochschild, G., Kostant, B., and Rosenberg, A.: Differential forms on regular affine algebras, *Trans. Am. Math. Soc.* **102** (1962), 383–406.
32. Kodaira, K. and Spencer, D. C.: On the variation of almost-complex structure, in: *Algebraic geometry and topology. A symposium in honor of S. Lefschetz*, Princeton University Press, Princeton, N. J., 1957, pp. 139–150; On deformations of complex analytic structures I, II, *Ann. of Math.* **67** (1958), 328–466; III, Stability theorems for complex structures *ibid.* **71** (1960), 43–76.
33. Kontsevich, M.: Formality conjecture, in: D. Sternheimer et al., (eds.), *Deformation Theory and Symplectic Geometry*, Math. Phys. Stud. **20**, Kluwer Acad. Publ., Dordrecht, 1997, pp. 139–156; Deformation quantization of Poisson manifolds, I, preprint q-alg/9709040; Operads and motives in deformation quantization, *Lett. Math. Phys.*, **48** (1999), 35–72; Kontsevich, M. and Soibelman, Y.: Deformations of algebras over operads and the Deligne conjecture, these proceedings, pp. 255-307.
34. Lichnerowicz, A.: Cohomologie 1-différentiable des algèbres de Lie attachées à une variété symplectique ou de contact, *J. Math. Pures Appl.* **53** (1974), 459–483.
35. Moyal, J. E.: Quantum mechanics as a statistical theory, *Proc. Cambridge Phil. Soc.* **45** (1949), 99–124.
36. Sternheimer, D.: Deformation quantization: Twenty years after, in: J. Rembieliński, (ed.), *Particles, Fields, and Gravitation (Lodz 1998)* AIP Press, New York, 1998, pp. 107–145 (math.QA/9809056).
37. Vey, J.: Déformation du crochet de Poisson sur une variété symplectique, *Comment. Math. Helv.* **50** (1975), 421–454.
38. Moshé Flato – Personal Recollections, *Lett. Math. Phys.* **48** (1999), 5–34.

Moshé at 18 (from his Israeli ID card).

Recent picture of Moshé (extracted from a group picture).

Moshé and his sister Irit (Moshé's 60th anniversary at a Parisian three star restaurant).
Additional pictures can be seen on http://www.u-bourgogne.fr/monge/cmf/cmf1999/moshe.html.

Curriculum vitae of Moshé Flato

Born September 17, 1937 in Tel Aviv.

Died November 27, 1998 in Paris.

Tel Aviv, 1955. Matriculation (scientific section), mark: excellent.

Jerusalem, 1960. Master of Science *(summa cum laude)* at the Hebrew University with physics (major), pure and applied mathematics (secondary subjects), after submitting an original Thesis.

Israel, 1960–1961. Military service as lieutenant in the Israeli Air Force. In charge of the Group theory course and seminar at the Hebrew University (substituting for Racah, then Rector).

Israel, 1961–1962. Head of the Theory Group in an experimental nuclear reactor (military service). Lecturer in Group Theory at the Hebrew University and in Solid State Physics at Bar-Ilan University.

Paris, 1963–1964. At Institut Henri Poincaré with an "ASTEF" high level grant from the French government.

Lyon, 1964–1967. Invited Professor of physics at the University.

Paris, June 1965. Doctorat-ès-Sciences Physiques.

March 10, 1967. French citizen.

Dijon, 1968–1977. Professor of mathematics at the University.

Stockholm, 1970–1971. Nordita guest professor at the Royal Institute of Technology (KTH).

1971–1998. Consulted by the Nobel Committee for Physics (under Chapter 6 of the regulations).

Geneva, 1974–1975. Visiting professor at the University.

Bangkok, fall 1975. UNESCO expert, reporting on the graduate programs in Thailand in Mathematics, Physics, Chemistry, Biology, and Economics.

Dijon, 1977–1987. Chair Professor (on Gevrey's Chair) of mathematics at the University.

Los Angeles, 1977–1987 (and other years). Visiting Professor at UCLA (fall quarter in general), Department of Physics.

France, 1978–1980. Member of the national *Comité Consultatif des Universités* (in theoretical physics).

Paris, 1980. Saintour Prize laureate (an interdisciplinary prize), Collège de France.

Paris, 1984–1992. President of the Scientific Council of Union des Assurances de Paris.

Athens, GA, 1986–1989. Adjunct Professor in Mathematics, Physics, and at the Graduate School, University of Georgia.

Shanghai, 1986. Honorary Professor at Fudan University.

France, 1988. Professeur de Classe Exceptionnelle.

Kyoto, 1988–1989 and 1995. Visiting Professor (from Monbusho) at the Research Institute for Mathematical Sciences.

France, 1996–1998. Member of the *Comité National des Universités*, in mathematics.

Dijon, 1968–1988 and 1996–1998. Founder and Director of the present *Laboratoire Gevrey de mathématique physique*, Université de Bourgogne.

1975–1998. Founding Editor of *Letters in Mathematical Physics* and of two book series at D. Reidel (presently Kluwer Academic Publishers).

Adviser of 15 junior Theses, 20 D.Sc. Theses and 6 Ph.D. Theses in Lyon and Dijon. Co-adviser for a number of Ph.D. in Universities worldwide (including 7 at UCLA).

The publications of Moshé Flato

1964

1. Sur les formules de masses fortes et électromagnétiques des particules élémentaires, *C. R. Acad. Sci. Paris* **249**, 3455–3458 (with J. Sternheimer).

1965

2. Ionic energy levels in trigonal and tetragonal fields, *J. Molecular Spectroscopy* **17**, 300–324.
3. Definition of *P.C.T.* operators for the relativistic-rotator model and the Bronzan-Low symmetry, *Nuclear Phys.* **61**, 250–256 (with G. Rideau and J.-P. Vigier).
4. On the "space-time character" of internal symmetries of elementary particles, *Nuovo Cimento* **36**, 672–675 (with D. Bohm, F. Halbwachs, P. Hillion, and J.-P. Vigier).
5. Algèbres unifiantes. Applications à l'algèbre de Lie du groupe de Poincaré, *C. R. Acad. Sci. Paris* **260**, 3532–3534 (with D. Sternheimer).
6. Le groupe conforme comme possibilité de symétrie unifiée en physique des interactions fortes, *C. R. Acad. Sci. Paris* **260**, 3869–3872 (with D. Sternheimer and J.-P. Vigier).
7. Conformal group symmetry of elementary particles, *Nuovo Cimento* **38**, 1941–1944 (with D. Bohm, D. Sternheimer, and J.-P. Vigier).
8. Remarks on the connection between external and internal symmetries, *Phys. Rev. Lett.* **15**, 934–936 (with D. Sternheimer).

1966

9. On the connection between external and internal symmetries of strongly interacting particles, *J. Math. Phys.* **7**, 1932–1958 (with D. Sternheimer).
10. On the masses of nonstrange pseudoscalar mesons and the generalized Klein-Gordon equation, *Nuovo Cimento A* **42**, 431–434 (with D. and J. Sternheimer, J.-P. Vigier, and G. Wataghin).
11. Local representations and mass-spectrum, *Phys. Rev. Lett.* **16**, 1185–1186 (with D. Sternheimer).
12. Remarques sur les automorphismes causals de l'espace-temps, *C. R. Acad. Sci. Paris Sér. A-B* **263**, A935–A938 (with D. Sternheimer).

1967

13. *Symétries de type lorentzien et interactions fortes,* Gauthier-Villars Éditeur, Paris. (Thèse de Doctorat-ès-Sciences, Juin 1965.)

14. Un groupe de type Poincaré associé à l'equation de Dirac-Weyl, *C. R. Acad. Sci. Paris* **264**, 85–88 (with P. Hillion and D. Sternheimer).

15. Nonuniqueness of mass-formulas from infinite-dimensional groups and algebras, *Phys. Rev. Lett.* **19**, 254–255 (with D. Sternheimer).

1968

16. Sur l'unification des symétries externe et interne, in: *L'extension du groupe de Poincaré aux symétries internes des particules élementaires*, Éditions du CNRS, Paris, pp. 47–60. (Colloque international No 159, Gif-sur-Yvette, April 1966.)

1969

17. Poincaré partially integrable local representations and mass-spectrum, *Comm. Math. Phys.* **12**, 296–303 (with D. Sternheimer).

18. On an infinite-dimensional group, *Comm. Math. Phys.* **14**, 5–12 (with D. Sternheimer).

19. Groupe conforme, covariance des équations des champs et analyse dimensionnelle, *C. R. Acad. Sci. Paris* **268**, 1065–1068 (with P. Hillion and J. Simon).

20. Sur la covariance conforme d'une équation de spin quelconque, *C. R. Acad. Sci. Paris Sér. A-B* **269**, A1051–A1054 (with J. Simon).

1970

21. Poincaré-like group associated with neutrino physics, and some applications, *Phys. Rev. D* **1**, 1667–1673 (with P. Hillion).

22. Conformal covariance of field equations, *Ann. Physics* **61**, 78–97 (with J. Simon and D. Sternheimer).

23. Sur les représentations de l'algèbre de Lie de Poincaré, *C. R. Acad. Sci. Paris Sér. A-B* **270**, A914–A917 (with J.-C. Guillot and D. Sternheimer).

1971

24. Field theory on curved space-time and the TCP theorem, *Phys. Scripta* **3**, 53–54 (with J. Simon).

25. Flots ergodiques et mesures complexes, *C. R. Acad. Sci. Paris Sér. A-B* **272**, A892–A895 (with B. Nagel and D. Sternheimer).

1972

26. Wightman formulation for quantization of the gravitational field, *Phys. Rev. D* **5**, 332–341 (with J. Simon).

27. Restrictions from geometry on quantization in curved space-time, *Phys. Rev. Lett.* **28**, 385–387 (with J. Simon and D. Sternheimer).

28. Comments on proposed two-component spin 1/2 wave equations, *Phys. Scripta* **5**, 153–4 (with G. Lindblad and B. Nagel).

29. Simple facts about analytic vectors and integrability, *Ann. Sci. École Norm. Sup. (4)* **5**, 432–434 (with J. Simon, H. Snellman, and D. Sternheimer).

1973

30. A short survey of modern statistical mechanics, in: A. O. Barut, (ed.), *Studies in Mathematical Physics (Istanbul 1970)*, D. Reidel, Dordrecht, pp. 113–134.

31. Separate and joint analyticity in Lie groups representations, *J. Funct. Anal.* **13**, 268–276 (with J. Simon).

32. Sur l'intégrabilité des représentations antisymétriques des algèbres de Lie compactes, *C. R. Acad. Sci. Paris Sér. A-B* **277**, A939–A942 (with J. Simon and D. Sternheimer).

1974

33. Regge Lattice and Lie algebra representations, *J. Math. Phys.* **15**, 857–860 (with H. Snellman).

34. Generalized covariance principles and neutrino physics, *J. Math. Phys.* **15**, 1288–1294 (with A. Piard).

35. Remarks on axiomatic approach to quantum gravity, *Helv. Phys. Acta* **47**, 114–118 (with J. Simon and D. Sternheimer).

36. On the localizability of massless particles. *Phys. Scripta* **9**, 173–183 (with E. Angelopoulos and F. Bayen).

37. Déformations 1-différentiables des algèbres de Lie attachées à une variété symplectique ou de contact, *C. R. Acad. Sci. Paris Sér. A-B* **279**, A877–A881 (with A. Lichnerowicz and D. Sternheimer).

1975

38. Remarks concerning Nambu's generalized mechanics, *Phys. Rev. D* **11**, 3049–3053 (with F. Bayen).

39. Are Bell's inequalities concerning hidden variables really conclusive? *Helv. Phys. Acta* **48**, 219–225 (with J. Gréa, C. Piron, D. Sternheimer, and J.-P. Vigier).

40. Déformations 1-différentiables d'algèbres de Lie attachées à une variéte symplectique ou de contact, *Compositio Math.* **31**, 47–82 (with A. Lichnerowicz and D. Sternheimer).

41. Algèbres de Lie attachées à une variété canonique, *J. Math. Pures Appl.* **54**, 445–480 (with A. Lichnerowicz and D. Sternheimer).

1976

42. Quantum mechanics and determinism. In: M. Flato et al., (eds.), *Quantum mechanics, Determinism, Causality and Particle*, Mathematical Physics and Applied Mathematics vol. 1, D. Reidel, Dordrecht, pp. 19–31.

43. Remarks on conformal space, *J. Math. Phys.* **17**, 1112–1114 (with F. Bayen).

44. Theory of analytic vectors and applications, in: K. Maurin and R. Rączka, (eds.), *Mathematical physics and physical mathematics*, Mathematical Physics and Applied Mathematics vol. 2, D. Reidel, Dordrecht, pp. 231–250.

45. Deformation of Poisson brackets, Dirac Brackets, and applications, *J. Math. Phys.* **17**, 1754–1762 (with A. Lichnerowicz and D. Sternheimer).

46. Crochet de Moyal-Vey et quantification *C. R. Acad. Sci. Paris Sér. A-B* **283**, A19–A24 (with A. Lichnerowicz and D. Sternheimer).

1977

47. On space-like fields, *Helv. Phys. Acta* **50**, 117–122 (with M. Guenin).

48. On parity, charge-conjugation and time-reversal violation in relativistic classical non-linear field theory, *Lett. Math. Phys.* **1**, 443–453 (with R. Rączka).

49. Quantum mechanics as a deformation of classical mechanics, *Lett. Math. Phys.* **1**, 521–530 (with F. Bayen, C. Fronsdal, A.Lichnerowicz, and D. Sternheimer).

50. Non-linear representations of Lie groups, *Ann. Sci. École Norm. Sup. (4)* **10**, 405–418 (with G. Pinczon and J. Simon).

51. Non-linear wave equations and covariance, Lett. Math. Phys. **2**, 115–160 (with J. Simon).

1978

52. Deformation theory and quantization I, Deformations of symplectic structures, *Ann. Physics* **111**, 61–110 (with F. Bayen, C. Fronsdal, A. Lichnerowicz, and D. Sternheimer).

53. Deformation theory and quantization II, Physical Applications, *Ann. Physics* **111**, 111–151 (with F. Bayen, C. Fronsdal, A. Lichnerowicz, and D. Sternheimer).

54. On unitary implementability of conformal transformations, *Lett. Math. Phys.* **2**, 405–412 (with E. Angeloupoulos).

55. One massless particle equals two Dirac singletons, *Lett. Math. Phys.* **2**, 421–426 (with C. Fronsdal).

1979

56. Yang-Mills equations are formally linearizable, *Lett. Math. Phys.* **3**, 279–283 (with J. Simon).

1980

57. Deformations of Poisson brackets, separate and joint analyticity in group representations, nonlinear group representations and physical applications, in: J. A. Wolf et al., (eds.), *Harmonic analysis and representations of semisimple Lie groups*, Mathematical Physics and Applied Mathematics vol. 5, D. Reidel, Dordrecht, pp. 385–448 (with D. Sternheimer).

58. Linearization of relativistic non-linear wave equations, *J. Math. Phys.* **21**, 913–917 (with J. Simon).

59. On parity, charge conjugation and time reversal violation in models of relativistic quantum field theory, *Phys. Rev. D* **21**, 341–451 (with R. Rączka).

60. On a linearization program of non-linear field equations, *Phys. Lett. B* **94**, 518–522 (with J. Simon).

61. Cohomologie des représentations définies par la dérivation de Lie et à valeurs dans les formes de l'algèbre de Lie des champs de vecteurs d'une variété différentiable. Premiers espaces de cohomologie. Applications, *C. R. Acad. Sci. Paris Sér. A-B* **291**, A331–A335 (with A. Lichnerowicz).

62. On Dis and Racs, *Phys. Lett. B* **97**, 236–240 (with C. Fronsdal).

1981

63. Star products. Quantization and representations without operators, in: E. Tirapegui, (ed.), *Field theory, Quantization and Statistical Physics*, Mathematical Physics and Applied Mathematics vol. 6, D. Reidel, Dordrecht, pp. 85–111 (with D. Arnal, J.-C. Cortet, and D. Sternheimer).

64. Massless particles, conformal group and De Sitter universe, *Phys. Rev. D* **23**, 1278–1289 (with E. Angelopoulos, C. Fronsdal, and D. Sternheimer).

65. Quantum field theory of singletons. The Rac. *J. Math. Phys.* **22**, 1100–1105 (with (C. Fronsdal).

66. Mass generation within conformal invariant theories, *Nuovo Cimento* **31**, 248–252 (with M. Guenin).

1982

67. Presentation of the contributions in group theory at this conference, in: R Schrader et al., (eds.), *Mathematical problems in theoretical physics,* Proceedings of the 1981 IAMP Conference on Mathematical Physics. Lectures Notes in Physics vol. 153, Springer Verlag, Berlin-New York, pp. 327–328.

68. Deformation view of physical theories, *Czech. J. Phys. B* **32**, 472–475.

69. De Sitter conformal field theories, *Czech. J. Phys. B* **32**, 439–471 (with B. Binegar, C. Fronsdal, and S. Salamo).

70. Mass-generation, symmetry breaking and conformal invariance, *Czech. J. Phys. B* **32**, 490–493 (with M. Guenin).

1983

71. Physique et Information. Exposé au séminaire interdisciplinaire du Collège de France, in: A. Lichnerowicz et al., (eds.), *Information et communication,* Maloine, Paris, pp. 27–40.

72. Difficulties with massless particles? *Comm. Math. Phys.* **90**, 563–573 (with C. Fronsdal and D. Sternheimer).

73. Particules de masse nulle revisitées, in: *Journées Relativistes*, Lyon 1982.

1984

74. Representations of conformal supersymmetry, *Lett. Math. Phys.* **8**, 159–162 (with C. Fronsdal).

75. Déformations et mécanique statistique, *C. R. Acad. Sci. Paris Sér. I Math.*, **299**, 405–410 (with H. Basart, A. Lichnerowicz, and D. Sternheimer).

76. Deformation theory applied to quantization and statistical mechanics, *Lett. Math. Phys.* **8**, 483–494 (with H. Basart, A. Lichnerowicz, and D. Sternheimer).

1985

77. Conformal invariance and gauge fixing in QED, *Phys. Rev. D* **32**, 2673–2682 (with F. Bayen, C. Fronsdal, and A. Haidari).

78. Do different deformations lead to the same spectrum? *J. Geom. Phys.* **2**, 35–48 (with M. Cahen, S. Gutt, and D. Sternheimer).

79. Small global strong solutions for the Navier-Stokes equations in R^4 (multigraphié, Dijon, with J. Simon and E. Taflin).

1986

80. Masslessness and light-cone propagation in 3+2 De Sitter and 2+1 Minkowski spaces, *Phys. Rev. D* **33**, 415–420 (with C. Fronsdal and J.-P. Gazeau).

81. Quarks or Singletons? *Phys. Lett. B* **172**, 412–416 (with C. Fronsdal).

82. Spontaneously generated field theories, zero-center modules, colored singletons and the virtues of $N = 6$ supergravity, in: C. Fronsdal, (ed.), *Essays in supersymmetry* , Mathematical Physics Studies vol. 8, D. Reidel, Dordrecht, pp. 123–162 (with C. Fronsdal).

83. Singletons: Fundamental Gauge Theory, in: J. Hietarinta and J. Westerholm, (eds.), *International Symposium on Topological and Geometrical Methods in Field Theory (Espoo, 1986),* World Scientific Publ., Teaneck, NJ, pp. 273–290 (with C. Fronsdal).

1987

84. The singleton dipole, *Comm. Math. Phys.* **108**, 469–482 (with C. Fronsdal).

85. How BRST invariance of QED is induced from the underlying field theory, *Phys. Lett. B* **189**, 145–148 (with C. Fronsdal).

86. On global solutions of the Maxwell-Dirac equations, *Comm. Math. Phys.* **112** (special issue for W. Thirring), 21–49 (with J. Simon and E. Taflin).

1988

87. Singletons as a basis for composite conformal quantum electrodynamics, in: M. Cahen and M. Flato, (eds.), *Quantum Theories and Geometry*, Mathematical Physics Studies vol. 10, Kluwer Acad. Publ., Dordrecht, pp. 65–76 (with C. Fronsdal and D. Sternheimer).

88. Composite electrodynamics, *J. Geom. Phys.* **5**, 37–61 (special issue for I. M. Gelfand), (with C. Fronsdal).

1989

89. Parastatistics, highest weight $osp(n,\infty)$ modules, singleton statistics and confinement, *J. Geom. Phys.* **6**, 293–309 (with C. Fronsdal).

90. Three-dimensional singletons, *Lett. Math. Phys.* **20**, 65–74 (with C. Fronsdal).

1990

91. Physique et Vérité, in: Lothaire, M., (ed.), *Mots, volume in honor of M. Schutzenberger*, Hermès, Paris, pp. 70–80.

92. *Le Pouvoir des Mathématiques,* Hachette, Paris. (Translated into several languages, including in 1992 in English at McGraw-Hill.)

1991

93. Non abelian singletons, *J. Math. Phys.* **32**, 524–531 (with C. Fronsdal).

94. Remarks on quantum groups, *Lett. Math. Phys.* **21**, 85–88 (with Z. C. Lu).

95. On a possible origin of quantum groups, *Lett. Math. Phys.* **22**, 155–160 (with D. Sternheimer).

1992

96. Three D Singletons and 2-D C.F.T., *Internat. J. Modern Phys. A* **7**, 2193–2206 (with A. M. Harun Ar.Rashid and C. Fronsdal).

97. Closed star-products and cyclic cohomology, *Lett. Math. Phys.* **24**, 1–12 (with A. Connes and D. Sternheimer).

98. A natural and rigid model of quantum groups, *Lett. Math. Phys.* **25**, 75–84 (with P. Bonneau and G. Pinczon).

1993

99. From where do quantum groups come? *Found. Phys.* **23**, 587–598 (with D. Sternheimer and Z. C. Lu).

100. Quantum deformations of singletons and of free zero-mass fields, *Found. Phys.* **23**, 571–586 (with L. K. Hadjiivanov and I. T. Todorov).

101. Quantum groups, star-products and cyclic cohomology, in: H. Araki, K. Ito, A. Kishimoto, and I. Ojima, (eds.), *Quantum and Non-Commutative Analysis*, Mathematical Physics Studies vol. 16, Kluwer Acad. Publ., Dordrecht, pp. 221–337 (with D. Sternheimer).

1994

102. Star-products, quantum groups, cyclic cohomology and pseudo-differential calculus, in: P. Sally et al., (eds.), *Mathematical aspects of conformal and topological field theories and quantum groups (South Hadley, MA, 1992)*, Contemp. Math., 175, Amer. Math. Soc., Providence, RI, pp. 53–72 (with D.Sternheimer).

103. The hidden group structure of quantum groups: strong duality, rigidity and preferred deformations, *Comm. Math. Phys.* **161**, 125–156 (with P. Bonneau, M. Gerstenhaber, and G. Pinczon).

104. The Maxwell-Dirac equations: asymptotic completeness and the infrared problem, *Rev. Math. Phys.* **6** (special issue in honor of E. Lieb), 1071–1083 (with J. Simon and E. Taflin).

105. Closedness of star products and cohomologies, in: J.-L. Brylinski et al., (eds.), *Lie Theory and Geometry: In Honor of B. Kostant*, Progress in Mathematics, Birkhäuser, Boston, pp. 241–259 (with D. Sternheimer).

106. Topological quantum groups, star products and their relations, *St. Petersburg Math. J.* (Algebra and Analysis) **6**, 242–251 (with D. Sternheimer).

1995

107. Cohomology and deformation of Leibniz pairs, *Lett. Math. Phys.* **34**, 77–90 (with M. Gerstenhaber and A. Voronov).

1996

108. Nambu mechanics: classical and quantum aspects, Notes for the symposium *Symplectic and Poisson Geometry* (Ogaki, 27 November 1996).

1997

109. Deformation Quantization and Nambu Mechanics, *Comm. Math. Phys.* **183**, 1–22 (with G. Dito, D. Sternheimer, and L. Takhtajan).

110. Nonlinear Relativistic Wave Equations in General Dimensions, in: Li Ta-Tsien, (ed.), *Collected Papers on Geometry, Analysis and Mathematical Physics, in Honour of Professor Gu ChaoHao*, World Scientific, pp. 53–70 (with J. C. H. Simon, D. Sternheimer, and E. Taflin).

111. *The Maxwell-Dirac equations: the Cauchy problem, asymptotic completeness and the infrared problem*, Memoirs of the American Mathematical Society, **127**, no. 606 (x + 312 pages), (with J. C. H. Simon and E. Taflin).

112. Generalized Abelian Deformations: Application to Nambu Mechanics, *Lett. Math. Phys.* **39**, 107–125 (with G. Dito).

113. Nambu mechanics, *n*-ary operations and their quantization, in: D. sternheimer et al., (eds.), *Deformation theory and symplectic geometry (Ascona 1996)*, Math. Physics Studies vol. 20, Kluwer Acad. Publ., Dordrecht, pp. 43–66 (with G. Dito and D. Sternheimer).

1998

114. Some infinite dimensional algebras arising in spin systems and in particle physics and their grand algebra, *Lett. Math. Phys.* **43**, 155–171 (with H. Araki, S. Michéa, and D. Sternheimer).

115. Interacting singletons, *Lett. Math. Phys.* **44**, 249–259 (with C. Frønsdal).

116. Two disjoint aspects of the deformation programme: quantizing Nambu mechanics; singleton physics, in: J. Rembieliński, (ed.), *Particles, Fields and Gravitation* (AIP Conference Proceedings **453**), American Institute of Physics, Woodbury, NY, pp. 49–52.

117. The deformation programme in physical theories: usual and unusual deformation quantization; singleton physics, in: Yu. Novozhilov, (ed.), *Quantum Theory, in honour of Vladimir A. Fock*, Proceedings of the VIII Unesco International School of Physics, St. Petersburg, 25 May - 6 June 1998, 10–37 (with D. Sternheimer).

1999

118. Singleton physics, *Proceedings of the Steklov Mathematical Institute* **226**, 185–192 (with C. Frønsdal and D. Sternheimer).

119. Singletons, physics in AdS universe and oscillations of composite neutrinos, *Lett. Math. Phys.* **48**, 109–119 (with C. Frønsdal and D. Sternheimer).

Applications of sciences to insurance and finance

Program of the special session

Under the auspices of Denis Kessler
President of the Fédération Française des Sociétés d'Assurances

Chaired by Dominique Perrin, President of the Université de Marne-la-Vallée

Monday September 6, In the University, Bâtiment Sciences Mirande

9:15 – 9:30 Opening of the special session and of the scientific conference by Arthur Jaffe and Daniel Sternheimer

9:30 – 10:10 Introduction:
Yvette Chassagne, Honorary Chairperson of UAP
(presented by Roland Lejart, former General Director of UAP)

Le Conseil scientifique de l'UAP et ses réalisations

Erik Taflin (AXA)

A new method for optimal equity allocation and portfolio selection

10:10 – 10:50 Jean-Claude Simon†
PDG, A2iA and Emeritus professor, Université Paris VI

La reconnaissance automatique de l'écriture manuscrite

10:50 – 11:30 Hélyette Geman
Professor of Finance, Université Paris IX and ESSEC

From Bachelier and Lundberg to insurance derivatives

12:00 – 14:00 Lunch at *Clos du Roy* in Chenôve

Le Conseil scientifique de l'UAP et ses réalisations *

Yvette Chassagne
Président d'honneur de l'Union des Assurances de Paris

Pour beaucoup, l'idée d'un conseil scientifique auprès du dirigeant d'une compagnie d'assurances paraît aller de soi. Cependant, quand j'ai commencé à y penser et à en parler, et d'abord avec mes collaborateurs les plus proches, ce n'était pas une évidence. Pourtant ayant à faire des choix qui engageaient l'avenir de la mission qui m'était confiée, j'ai eu une conscience plus grande de l'insuffisance de mes connaissances, Moshé aurait dit de ma nullité. Ce n'est qu'avec lui que je pouvais me permettre de dire «je ne sais pas » sachant qu'il aurait la patience de m'écouter et de me répondre en termes énergiques, mais parfaitement clairs.

Parce que l'assurance s'applique à tout ce qui concerne la vie des *êtres* et l'existence des *biens*, ses points de rencontre avec la science sont multiples. Comment apprécier la valeur objective d'un risque si l'on ignore «l'état de l'art » dans les disciplines principales de la science, fondamentale ou appliquée ? Pour vous, ici présents, comme pour Moshé il y a quinze ans, c'est une «trivialité », un mot que j'extrais du vocabulaire courant de Moshé.

C'est précisément pour l'appréciation de ces risques que le concours de scientifiques de haut niveau était et reste indispensable. D'où, sous la conduite de Moshé, la composition de notre Conseil scientifique à l'UAP : des mathématiciens, des physiciens, des médecins, des linguistes, des économistes, des statisticiens, des informaticiens. Tout ceci donnait lieu à des recherches et à des réalisations dont nous pouvons, ensemble, être fiers, notamment dans les domaines de l'acceptation des risques en assurance vie, de la gestion électronique des documents, du pilotage à distance des ordinateurs, de la reconnaissance automatique de l'écriture, des agrégats statistiques et des «tables d'expérience ». On en parlera sans nul doute aujourd'hui et l'on appréciera tous les progrès réalisés depuis.

Pour le chef d'Entreprise, si l'on se reporte au début des années 80, il était indispensable de maîtriser tout ce que les avancées scientifiques pouvaient avoir de conséquences sur les méthodes et les équipements techniques. Mais surtout cela entraînait une mutation culturelle pour des personnels qui n'étaient absolument pas préparés à des changements fondamentaux des formes de pensée et des outils, au quotidien.

Cette mutation, il fallait la préparer, la contrôler, surtout s'en donner les moyens. Nous avons dû passer entre 1984 et 1987 au tout début de l'année, d'environ 300 terminaux, à une dernière commande qui nous doterait de 10 000. Des dépenses importantes qui ne nous ont pas empêchés (ou, plus exactement, qui nous ont permis) d'améliorer de plus de 20%, sur même période, notre productivité

* Allocution lue et commentée par Monsieur Roland Lejart, ancien Directeur Général de l'UAP.

G. Dito and D. Sternheimer (eds.), Conférence Moshé Flato 1999, Vol. 1, 57 – 58.

administrative. L'aspect financier n'est donc pas négligeable et il était indispensable que la plupart des responsables bénéficient d'une remise à jour de leurs connaissances et d'une appréciation des perspectives qui leur permettent de n'être pas trop vite dépassés.

C'était encore le début de l'application des modifications fondamentales que les systèmes de gestion allaient subir. L'allégement des tâches qui en résulterait, aurait naturellement pour conséquence l'obligation de diminuer les effectifs employés, sauf à développer le volume de nos affaires avec des effectifs constants. Ce fut notre politique, tant il me semblait évident que mon devoir de Chef d'Entreprise n'était pas seulement d'ordre économique mais aussi d'ordre social. Le bon équilibre n'est pas chose simple, mais c'est une faute lourde que de ne pas le rechercher en permanence. C'est avec cette préoccupation que nous avons déterminé que seule la mise à jour continue des connaissances pour tous permettrait d'organiser les transitions et d'éviter autant que faire se peut les drames sociaux. Nous les avons évités.

Aucun d'entre-nous n'était capable d'apprécier totalement les progrès à venir dans la gestion de l'image et la reconnaissance des formes, l'analyse linguistique, pour le moins. Il nous fallait obtenir l'avis préalable, la collaboration permanente de fondamentalistes capables de se mettre à la portée des applications les plus élémentaires, et je sais que ce n'est pas facile. Moshé s'est passionné pour cette forme nouvelle du travail scientifique. À chacune de mes questions dont il négligeait superbement la maladresse, il trouvait une réponse qui était généralement un nom.

Jamais sans lui, je n'aurais pu obtenir le concours de certains qui sont ici, d'autres hélas prématurément disparus mais si vivants dans notre souvenir. C'est lui qui, avec l'aide de Daniel et de bon nombre d'entre vous, a constitué ce Conseil, avec le concours de Roland Lejart. C'est lui qui a orienté puis défini le programme des travaux, c'est lui qui a eu l'idée de ce «prix » destiné à encourager des scientifiques intéressés par nos recherches. Pour un Chef d'Entreprise, le confort moral que peut donner une telle sécurité est inappréciable.

Vous ne m'en voudrez pas de penser encore au volume important de dépenses inutiles qu'aurait pu provoquer la méconnaissance de l'état le plus récent des connaissances, des travaux universitaires et de vos travaux personnels à cette époque. Or, il faut en témoigner, la participation au conseil scientifique a été en permanence totalement bénévole.

D'autres que moi pourront vous dire ce qu'a pu apporter une telle nouvelle méthode de travail, et combien chacun des participants y contribuait. Je n'ai jamais pu dire comme il l'aurait fallu à tous ceux qui y ont travaillé et surtout à Moshé lui-même, tout ce que tous m'ont donné et appris. Aujourd'hui que cette opportunité m'est accordée, je sais combien tout cela m'aurait été difficile sans Moshé et sans vous, et je sais combien je vous suis redevable.

Vous me permettrez, avant de vous laisser une fois de plus travailler ensemble, de dire le seul mot qui convienne : *Merci !*

A new method for optimal equity allocation and portfolio selection[‡]

Erik Taflin
AXA, 23 avenue Matignon, 75008 Paris, France
e-mail: `erik.taflin@ceremade.dauphine.fr`, `erik.taflin@u-bourgogne.fr`

Abstract. We review, for an audience of mathematically minded listeners, but non specialists on insurance and finance, some recent research results on optimal equity allocation and portfolio selection. These results, first developed in the domain of reinsurance, are also applicable to insurance, banking and more generally to corporate finance. A multi-time stochastic model, is presented within the context of a company with several portfolios (or subsidiaries), representing both liabilities and assets. The model has solutions respecting constraints on ROE's, ruin probabilities, non-solvency probabilities and market shares currently in practical use. The solutions, which give global and optimal risk management strategies of the company, also generalize VaR (Value at Risk) conditions.

Keywords: equity allocation, portfolio selection, risk management, value at risk

Mathematics Subject Classifications (2000): 91Bxx, 49xx, 60Gxx

1. Introduction and mathematical model

To fix the ideas and for simplicity, in this talk, we shall think of the equity of a company, at a given time, as the value of its accumulated wealth at that time. We consider an idealized situation, where the variation of the equity over a given time period is the difference of the result of the company and the dividends, paid to the shareholders, during that period. The result is the difference of all the incomes and expenses (dividends excluded). By basic economic and financial hypotheses, it is expected that the ROE (Return On Equity), i.e., the result per equity, is sufficiently high. For example it should, in some sense, be higher than the return on equally liquid values, with lower risks. The problem we shall consider in this talk is to find a satisfactory global solution on the balance between Low Equity Risk and High ROE. We formulate it as an optimization problem involving essentially three different kind of variables: portfolios, equities and dividends, which are discrete time stochastic processes. Acceptable upper bounds of risks, lower bounds of annual ROE's and positivity of portfolios are introduced as constraints and we optimize the expected final utility, being the wealth produced, until no more financial flows are generated. The solution of this problem consists of those stochastic processes, satisfying the constraints, for which the optimum is reached. A large class of *Equity Allocation and Portfolio Selection Models*, for which solutions exist in a certain sense, was introduced in references [5] and [6], where also a constructive approach to the solutions was given. Portfolios representing both assets and liabilities

[‡] The content of this research and development paper represents only the opinions of the author and does not engage AXA.

G. Dito and D. Sternheimer (eds.), Conférence Moshé Flato 1999, Vol. 1, 59 – 71.

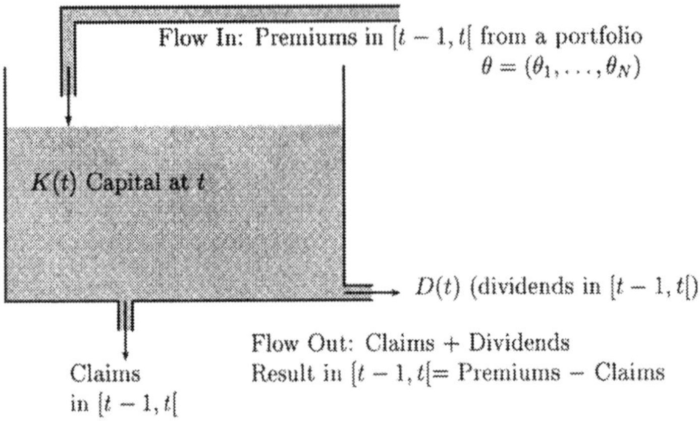

Figure 1. Ordinary reservoir model of an insurance company. All flows are stochastic.

are admitted in these references. The very popular VAR conditions are particular cases of constraints in [6]. These models therefore also give, in the context under consideration, a framework for *Optimal ALM* (Asset and Liability Management) and *Optimal Risk Management* of a company. In this talk, I will mainly review these results. The mathematical model as well as results are outlined in the sequel of this introduction. Somewhat more detailed results, an idea of their proofs and some supplementary mathematical points are given in §2.

We shall consider companies of two different levels of complexity. Let S be a company with only one portfolio and let H be a company having several subsidiaries. Each subsidiary has one portfolio and can, in certain situations, ceases its activity. For S we shall study the problem of a optimal portfolio selection. Its result is a basic building block for the more complex problem of an optimal equity distribution (between different subsidiaries and shareholders) and portfolio selection of H.

Let us first consider the company S, which for the purpose of this introduction, for example can be a reinsurance company. S has a portfolio $\theta = (\theta_1, \dots, \theta_N)$ of N different kind of contracts and with an initial capital $K(0) > 0$, at $t = 0$. It can be thought of, which is frequent, as a reservoir (see Figure 1), filled with capital $K(t)$ at time t, with premiums as in flows and claims and dividends $D(t)$ as out flows during the time interval $[t-1, t[$.

The in flow is a (discrete time) stochastic process. In fact in this model, the amount $\theta_i(t)$ of contract i written at time t, is known for past times, $t < 0$, is determined from past information at the present, $t = 0$, and the amount written at future times $t > 0$ will depend on information collected until that instant. Similarly, the dividends are given by a stochastic process. The claims are of course stochastic. It is supposed that the amounts $\theta_i(t)$ vanish after a sufficiently large time $\bar{T}$ (as well as before a sufficiently past time). Similarly, it is also supposed that the flows vanish after a sufficiently large time $\bar{T}+T$. The portfolio can be decomposed into $\theta = \xi + \eta$, where ξ is the run-off, at time $t=0$, concluded at a finite number of past times $t<0$ and where η is the (present and future) underwriting portfolio, to be concluded at a given finite number of times $0,\dots,\bar{T}$, $\bar{T}\geq 1$.

We introduce a Cramér–Lundberg like *utility function* U of the portfolio θ (cf. [2]), whose value at time t is

$$U(t,\theta) = \textit{accumulated premiums in } [0,t[- \textit{accumulated claims in } [0,t[. \quad (1.1)$$

The *final utility* $U(\infty,\theta) = \lim_{t\to\infty} U(t,\theta)$ is simply the wealth produced in $[0,\bar{T}+T[$. The *capital* is then given by

$$K(t) = K(0) + U(t,\theta) - \sum_{1\leq k\leq t} D(k). \quad (1.2)$$

It is constant for $t \geq \bar{T}+T$. A negative value of the capital corresponds to ruin. The ROE at time t is given by $(U(t,\theta)-U(t-1,\theta))/K(t-1)$. According to what has been said above we want, for given initial capital $K(0)$ and run-off ξ, to find a underwriting portfolio η, such that the capital is maintained non-negative with a high probability, such that the annual ROE's are bounded below by given positive constants and such that the final utility $U(\infty,\theta)$ is optimal. Generically such an optimal portfolio does not exist, so we shall give a weaker formulation useful in practical applications and where non-existence is not the rule. Tentatively we shall try, for given $K(0)$ and ξ, to select η as to achieve the following ($E(X)$ and $V(X)$ are respectively the expected value and the variance of the random variable X):

- A) Optimize the final expected utility $E(U(\infty,\theta))$, i.e., the expected wealth produced in $[0,\bar{T}+T[$;

- B) The expected result satisfies $E(U(t,\theta)-U(t-1,\theta)) \geq c(t)E(K(t-1))$, for $0\leq t\leq \bar{T}+T$, where $c(t)$ are pre-determined positive lower limits of ROE's;

- C) Sufficiently high probability that $K(t)\geq 0$, for $t\geq 0$, so the company will reasonably not make ruin and insurance claims can be paid;

- D) Present and future underwriting levels are positive, $0\leq \theta_i(t)$, where $1\leq i\leq N$ and $0\leq t\leq \bar{T}$.

We have here also introduced a constraint, corresponding to the more difficult case of positive underwriting levels.

In order to have a well-defined problem, we at least need to know the function $\theta \mapsto U(t,\theta)$. To this end, let a *unit contract* be an insurance contract whose total premium is one currency unit. The utility $u_i(t,t')$, at $t'\in\mathbb{N}$ of the unit contract i,

$1 \leq i \leq N$, concluded at $t \in \mathbb{Z}$, is by definition:

$$u_i(t,t') = \begin{cases} \textit{accumulated result in the time interval } [0,t'[, \textit{ if } t < 0, \\ \textit{accumulated result in the time interval } [t,t'[, \textit{ if } 0 \leq t \leq t', \\ 0, \textit{ if } 0 \leq t' \leq t. \end{cases} \quad (1.3)$$

The utility function of the portfolio θ at time t is then, according to (1.1) given by

$$U(t,\theta) = \sum_{k \leq t} \theta(k) \cdot u(k,t), \quad (1.4)$$

where the dot denotes the scalar product in $\mathbb{R}^N$. The final utility of θ is

$$U(\infty,\theta) = \sum_{k \in \mathbb{Z}} \theta(k) \cdot u^\infty(k), \quad (1.5)$$

where only a finite number of terms are non vanishing and $u^\infty(k) = u(k,\infty)$ exists since all flows vanish from time $\bar{T} + T$.

The random variables $u_i(t,t')$, are essential inputs of the model. They determine its probability space $(\Omega, P, \mathscr{F})$ and its information structure by a filtration $\mathscr{A} = \{\mathscr{F}_t\}_{t \in \mathbb{N}}$, of sub σ-algebras of the σ-algebra $\mathscr{F}$, i.e., $\mathscr{F}_0 = \{\Omega, \emptyset\}$ and $\mathscr{F}_s \subset \mathscr{F}_t \subset \mathscr{F}$ for $0 \leq s \leq t$. For given $k \in \mathbb{N}$ and $0 \leq i \leq N$, the utility $u_i(k,t)$ is the difference of accumulated premiums and claims in the time interval $[k,t[$, $k \leq t$. The accumulated premiums are usually a deterministic function of t, equal to 1 for sufficiently large times. The probability space $(\Omega, P, \mathscr{F})$ and the filtration $\mathscr{A}$ are then given by claims processes of unit contracts. We here suppose that $u_i(k,t)$ is $\mathscr{F}_t$-measurable (i.e., its value is known at time t), so the processes $(u(k,t))_{t \geq 0}$ is $\mathscr{A}$-adapted. Loosely speaking, since we have not defined what is meant by a claim, this amounts to neglect IBNR (Incurred But Not Registered claims) effects. The amount $\theta_i(t)$ of contract i written at time t, is known at t, so the process $(\theta(t))_{t \geq 0}$ is $\mathscr{A}$-adapted.

Secondly, let us consider the company H, which as already said is more complex than S. We shall study the capital structure and dividend flows for H, in addition to the portfolio selection. H is organized as a holding with several subsidiaries $S^{(1)}, \ldots, S^{(\aleph)}$, where $\aleph \geq 1$ is an integer (see Figure 2). The companies $S^{(i)}$ here only correspond to a division of the activities of H into parts, whose equity, profitability, portfolio selection and certain other properties need to be considered individually. This allows localization of capital flows and results. In general $S^{(i)}$ does not have to have a legal existence. Each company $S^{(i)}$ is similar to S, with the exception that its activity can cease. We introduce a supplementary value τ_f, which is a final state of the process $(\theta^{(i)}(t))_{t \in \mathbb{Z}}$, reached when the activity of the company $S^{(i)}$ ceases. $\theta^{(i)}(t)$ is then an $\mathbb{R}^{N^{(i)}} \cup \{\tau_f\}$-valued random-variable. We modify the definition of the utility function for $S^{(i)}$, such that the already written contracts, but no new, continue to be honored. The utility of $\theta^{(i)}$ at time t can then

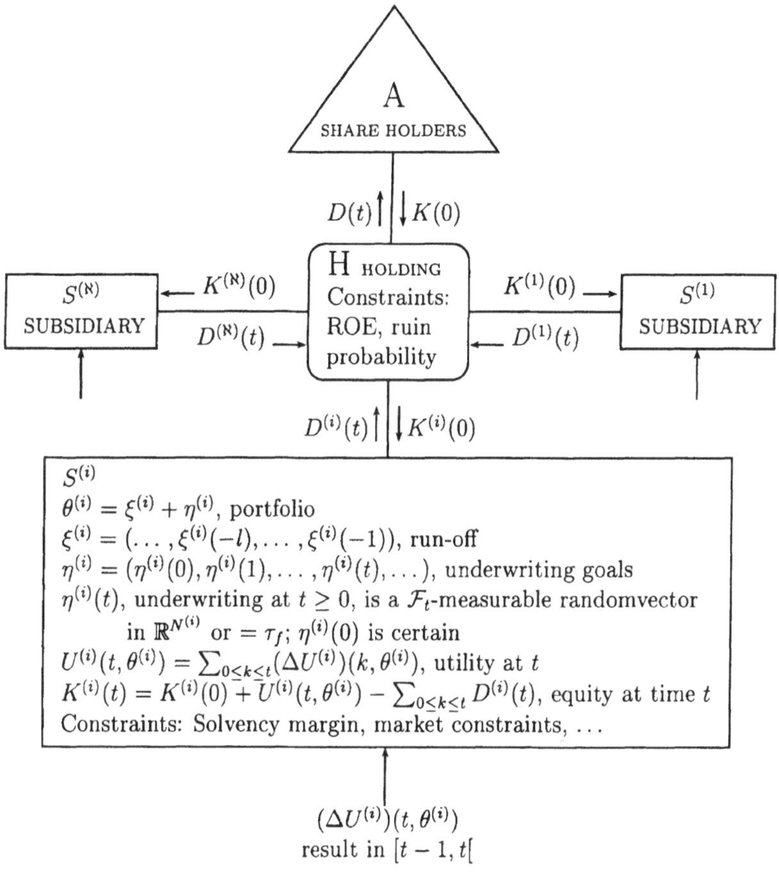

Figure 2. Flows, stocks and constraints.

be written in the following form:

$$U^{(i)}(t,\theta^{(i)}) = U^{(i)}(t, p\circ\theta^{(i)}) = \sum_{k\leq t}(p\circ\theta^{(i)}(k))\cdot u^{(i)}(k,t), \tag{1.6}$$

where $p : \mathbb{R}^N \cup \{\tau_f\} \to \mathbb{R}^N$ is defined by $p(\tau_f) = 0$ and $p(x) = x$, for $x \in \mathbb{R}^N$, where $U^{(i)}$ in the second member is defined as for S in (1.4) and where the second equality follows from (1.4). The result in the time interval $[t, t+1[$ is

$$(\Delta U^{(i)})(t+1,\theta^{(i)}) = U^{(i)}(t+1,\theta^{(i)}) - U(t,\theta^{(i)}), \tag{1.7}$$

for $t \geq 0$.

The utility of an aggregate portfolio

$$\theta = (\theta^{(1)}, \ldots, \theta^{(\aleph)}) \tag{1.8}$$

is defined by

$$U(t,\theta) = \sum_{1 \leq i \leq \aleph} U^{(i)}(t,\theta^{(i)}). \tag{1.9}$$

The dividends D paid to the shareholders by the company H and the equity of H are now given by $D = \sum_{1 \leq i \leq \aleph} D^{(i)}$ and $K = \sum_{1 \leq i \leq \aleph} K^{(i)}$ respectively.

It is supposed that the dividend paid to the shareholders by the company H at time t only depends of the aggregate portfolio θ. As matter of fact, it is usually a function of the aggregate results $(\Delta U)(1,\theta), \ldots, (\Delta U)(t,\theta)$ and often we simply have $(D(\theta))(t) = ((\Delta U)(t,\theta) - c)_+$, for some $c \geq 0$, where $(x)_+ = 0$ if $x < 0$ and $(x)_+ = x$ if $x \geq 0$.

To formulate the equity allocation and portfolio selection problem in the context of the company H, let us use the decomposition of $\theta = \xi + \eta$, into its run-off ξ (at time $t = 0$) and into its underwriting portfolio η and let us introduce the notation $\vec{K}(0) = (K^{(1)}(0), \ldots, K^{(\aleph)}(0))$ and $\vec{D} = (D^{(1)}, \ldots, D^{(\aleph)})$.

We shall try, for given $K(0)$ and ξ, to determine η, $\vec{K}(0)$ and $\vec{D}$ as to achieve the items (A)–(D) (with (D) slightly modified as follows) and certain supplementary constraints:

- D) Present and future underwriting levels are positive, $0 \leq \theta_k^{(i)}(t)$, where $1 \leq k \leq N$ and $0 \leq t \leq \bar{T}$ and there are upper and lower limits on $\theta_k^{(i)}(t)$;
- E) $K(0) = \sum_{1 \leq i \leq \aleph} K^{(i)}(0)$;
- F) $D = \sum_{1 \leq i \leq \aleph} D^{(i)}$;
- G) Sufficiently high probability that $K^{(i)}(t) \geq m^{(i)}(t)$, for $t \geq 0$, where $m^{(i)}$ is a given solvency margin depending of $\theta^{(i)}$;
- H) The activity of $S^{(i)}$ ceases just after that the solvency margin is not satisfied, i.e., the final state τ_f is reached.

The modification of item (D) corresponds to the introduction of upper and lower market limits. Items (E) and (F) are just budget constraints. In item (G), the solvency margin $m^{(i)}$ is often determined by stronger conditions than the usual ruin condition (i.e., $m^{(i)} = 0$). For example, it can be the case that a company $S^{(i)}$ has to be highly quoted (AAA, ...) on the market. Item (H) is a constraint defining the dynamics of how the final state τ_f of ceased activity of $S^{(i)}$ is reached. We remind that the portfolio $\theta^{(i)}$ is managed as a run-off from the instant when τ_f is reached.

We next sum up the results, which are given in more detail in §2. The optimization problem for the company S (resp. H) given by (A)–(D) (resp. (A)–(H)) gives rise to highly non-linear equations, due to the non-solvability probabilities in the constraints. In order to construct approximate solutions, *but satisfying the original constraints*, the constraints with non-solvability probabilities can be replaced by stronger quadratic variance constraints (cf. Theorem 2.4 of reference

[6]). We note that the portfolio of S is then an extension of Markowitz portfolio [4] to a multiperiod stochastic portfolio, as suggested by [3] (cf. also [1]). In this situation and under certain mild conditions, (H_1), (H_2) and (H_3) of §2, on the result processes of the unit-contracts, there is a unique solution of the problem for S, in the simplest case when the run-off vanishes, $\xi = 0$ (Theorem 2.2). Moreover the solution, is derived from a Lagrangian formalism (2.9), and is given by formula (2.12). The inverse of the integral operator, defined by the quadratic part of the Lagrangian, can be obtained explicitly (Corollary 2.2; for details see Proposition A.3 and Proposition A.4 of [5]). Condition (H_1) says that the final utility (sum of all results) of a unit contract, written at time k, is independent of events occurred before k. In practice, this is generally not true, among other things, because of feed-forward phenomena in the prizing. Condition (H_2) is equivalent to the statement that no non-trivial linear combinations of final utilities, of contracts written at a certain time, is a certain random-variable. This can also be coined, in more financial terms: An underwriting portfolio $\eta(t)$, constituted at time t, cannot be risk-free. Condition (H_3) says that the final utility of unit contracts, written at different times are independent. The conditions (H_1), (H_2) and (H_3), which exclude many interesting situations, like cyclic markets, have here been chosen for mathematical simplicity. They also constitute an approximation of real situations and can largely be weakened, without altering the results of this paper. They reflect to some extent how reinsurance underwriters tempt to build portfolios. In the case of the company H a solution exists (Theorem 2.5), under similar conditions. However it is usually not unique (see Theorem 2.8 of [6]). There is also a Lagrangian formalism, obtained to the price of some complications. An important point, in both cases S and H, is that no particular distributions (statistical laws) are required.

2. Results and mathematical complements

We will here first give results (§2.1) in the simplest case, i.e., the basic model of S with vanishing run-off, $\xi = 0$, and dividends $D = 0$ and then (§2.2) in the general case of H. But first let us introduce some notations.

We define spaces $\mathscr{E}^q(\mathbb{R}^N)$ of discrete and $\mathscr{E}^q(\mathbb{R}^N, \mathscr{A})$ of discrete adapted processes, for $1 \le q \le \infty$. Let $1 \le q < \infty$. Then $(X_i)_{0 \le i} \in \mathscr{E}^q(\mathbb{R}^N)$ if and only if $X_i : \Omega \to \mathbb{R}^N$ is $\mathscr{F}$-measurable and $\|X_i\|_{L^q(\Omega,\mathbb{R}^N)} = (E(|X_i(\omega)|^q_{\mathbb{R}^N}))^{1/q} < \infty$ for $i \ge 0$, where $|\ |_{\mathbb{R}^N}$ is the norm in $\mathbb{R}^N$. Let $\mathscr{E}^q(\mathbb{R}^N, \mathscr{A})$ be the subspace of $\mathscr{A}$-adapted processes in $\mathscr{E}^q(\mathbb{R}^N)$. Similarly $\mathscr{E}^\infty(\mathbb{R}^N)$ and $\mathscr{E}^\infty(\mathbb{R}^N, \mathscr{A})$ are defined by replacing the L^q-norm by the L^∞-norm. We define $\mathscr{E}(\mathbb{R}^N) = \cap_{q \ge 1} \mathscr{E}^q(\mathbb{R}^N)$ and $\mathscr{E}(\mathbb{R}^N, \mathscr{A}) = \cap_{q \ge 1} \mathscr{E}^q(\mathbb{R}^N, \mathscr{A})$ (where of course the intersection is over only finitely many q's). Let $\mathscr{E}^q_{\bar{T}}(\mathbb{R}^N, \mathscr{A})$ and $\mathscr{E}_{\bar{T}}(\mathbb{R}^N, \mathscr{A})$ be the subspace of elements $\eta \in \mathscr{E}^q(\mathbb{R}^N, \mathscr{A})$ and $\mathscr{E}(\mathbb{R}^N, \mathscr{A})$ respectively, with $\eta(t) = 0$ for $t > \bar{T}$.

2.1. Basic model

We shall here sum up certain results obtained in reference [5] in the case of the company S with vanishing run-off and dividends. The portfolio η is an element of the Hilbert space $\mathscr{H} = \mathscr{E}^2_T(\mathbb{R}^N, \mathscr{A})$. We remind that, in this situation, the equity is according to formula (1.2) given by

$$K(t) = K(0) + U(t,\eta), \tag{2.1}$$

where $K(0) \geq 0$ is the initial equity at $t = 0$.

In the sequel of this paragraph, we closely follow reference [5]. The constraints (B), (C) and (D) on the portfolio η take, after the introduction of variance constraints in §1, the following form:

- C_3) $E((\Delta U)(t+1,\eta)) \geq c(t)E(K(t))$, $c(t) \in \mathbb{R}^+$ is given (constraint on profitability);
- C_4) $E((U(\infty,\eta) - E(U(\infty,\eta)))^2) \leq \sigma^2$, where $\sigma^2 > 0$ is given (acceptable level of the variance of the final utility);
- C_6) $0 \leq \eta_i(t)$, where $1 \leq i \leq N$ (only positive subscription levels).

Let $\mathscr{C}_0$ be the set of portfolios $\eta \in \mathscr{H}$ such that constraints (C_3), (C_4), and (C_6) are satisfied. This is well-defined. In fact the quadratic form

$$\eta \mapsto \mathfrak{a}(\eta) = E((U(\infty,\eta))^2), \tag{2.2}$$

in $\mathscr{H}$, has a maximal domain $\mathscr{D}(\mathfrak{a})$, since for each $\eta \in \mathscr{H}$, the stochastic process $(U(t,\theta))_{t\geq 0}$ is an element of the space $\mathscr{E}^p(\mathbb{R},\mathscr{A})$, for $1 \leq p < 2$ (which follows directly from the Schwarz inequality). The optimization problem is now, to find all $\hat{\eta} \in \mathscr{C}_0$, such that

$$E(U(\infty,\hat{\eta})) = \sup_{\eta\in\mathscr{C}_0} E(U(\infty,\eta)). \tag{2.3}$$

The solution of this optimization problem is largely based on the study of the quadratic form

$$\eta \mapsto \mathfrak{b}(\eta) = E((U(\infty,\eta) - E(U(\infty,\eta)))^2), \tag{2.4}$$

in $\mathscr{H}$, with (maximal) domain $\mathscr{D}(\mathfrak{b}) = \mathscr{D}(\mathfrak{a})$.

We make certain (technical) hypotheses on the claim processes:

- H_1) $u^\infty(k)$ is independent of $\mathscr{F}_k$ for $k \in \mathbb{N}$;
- H_2) for $k \in \mathbb{N}$ the $N \times N$ (positive) matrix $c(k)$ with elements $c_{ij}(k) = E((u^\infty_i(k) - E(u^\infty_i(k))(u^\infty_j(k) - E(u^\infty_j(k)))$ is strictly positive;
- H_3) $u^\infty_i(k)$ and $u^\infty_j(l)$ are independent for $k \neq l$.

The next crucial result (Lemma 2.2 and Theorem 2.3 of [5]) show that (the square root of) each one of the quadratic forms $\mathfrak{b}$ and $\mathfrak{a}$ is equivalent to the norm in $\mathscr{H}$.

Theorem 2.1 *If the hypotheses* (H_1), (H_2), *and* (H_3) *are satisfied, then the quadratic forms* $\mathfrak{b}$ *and* $\mathfrak{a}$ *are bounded from below and from above, by strictly positive numbers* c *and* C *respectively, where* $0 < c \le C$, *i.e.,*

$$c\|\eta\|^2_{\mathscr{H}} \le \mathfrak{b}(\eta) \le \mathfrak{a}(\eta) \le C\|\eta\|^2_{\mathscr{H}}, \tag{2.5}$$

for $\eta \in \mathscr{H}$.

Since there is no risk for confusion, the bilinear form corresponding to the quadratic form $\mathfrak{b}$ (resp. $\mathfrak{a}$) will also be denoted $\mathfrak{b}$ (resp. $\mathfrak{a}$). The operators B (resp. A) in $\mathscr{H}$, associated with $\mathfrak{b}$ (resp. $\mathfrak{a}$), by the representation theorem, i.e.,

$$\mathfrak{b}(\xi,\eta) = (\xi, B\eta)_{\mathscr{H}} \quad (\text{resp.} \quad \mathfrak{a}(\xi,\eta) = (\xi, A\eta)_{\mathscr{H}}), \tag{2.6}$$

for $\xi \in \mathscr{H}$ and $\eta \in \mathscr{H}$, are strictly positive, bounded, self-adjoint operators onto $\mathscr{H}$ with bounded inverses. There exist $c \in \mathbb{R}$, such that $0 < cI \le B \le A$, where I is the identity operator. It follows from formula (2.6), that an explicit expression of A is given by

$$(A\eta)(k) = E(U(\infty,\eta)u^{\infty}(k)|\mathscr{F}_k) \tag{2.7}$$

and that an explicit expression of B is given by

$$(B\eta)(k) = E((U(\infty,\eta) - E(U(\infty,\eta)))u^{\infty}(k)|\mathscr{F}_k), \tag{2.8}$$

for $\eta \in \mathscr{H}$, where $0 \le k \le \bar{T}$.

Theorem 2.1 leads to the solution of the optimization problem of this paragraph (see Corollary 2.6 of [5]). In fact, using a simple scaling argument, optimization problem (2.3) can be reformulated (see Lemma 2.5 of [5]) as an elementary Hilbert space problem. Namely, to determine the points closest to the origin, in a closed convex set defined by conditions (C_3), (C_6), and $E(U(\infty,\eta)) \le e$, for some $e > 0$. As it is well-known, this problem has a unique solution.

Theorem 2.2 *Let hypotheses* (H_1), (H_2), *and* (H_3) *be satisfied. If* $\mathscr{C}_0$ *is non-empty, then the optimization problem (2.3) has a unique solution* $\hat{\eta} \in \mathscr{C}_0$.

The solution $\hat{\eta}$ is given by a constructive approach in [5]. In fact, in that reference a Lagrangian formalism, an algorithm to invert the operators A and B and approximation methods for determining the multipliers are given.

To sum up the Lagrangian formalism, let $\lambda_0, \lambda_1, \dots \lambda_{\bar{T}+T-1}$, and μ be real numbers and let $v \in \mathscr{H}$. These are the multipliers. A Lagrangian (slightly different from the Lagrangian (2.13) if [5]) is defined by

$$\begin{aligned} &h_{\lambda,\mu,v}(\eta) \\ &= \frac{1}{2}\mathfrak{b}(\eta) - \sum_{0\le t\le \bar{T}+T-1} \lambda_t (E((\Delta U)(t+1,\eta)) - c(t)E(U(t,\eta))) \\ &- \mu E(U(\infty,\eta)) - (v,\eta)_{\mathscr{H}}, \end{aligned} \tag{2.9}$$

where $\eta \in \mathscr{H}$. We now have to find the critical points in $\mathscr{H}$, for fixed λ, μ and ν and determine the multipliers such that the critical point in fact is an element of $\mathscr{C}_0$. The multipliers shall satisfy

$$\lambda_t \geq 0,\ \lambda_t(E((\Delta U)(t+1,\hat{\eta})) - c(t)E(K(t,\hat{\eta}))) = 0, \tag{2.10}$$

for $0 \leq t \leq \bar{T} + T - 1$ and

$$\nu_i(k) \geq 0,\ (\nu_i(k))(\omega) = 0 \quad \text{a.e. for} \quad \omega \in supp\,\hat{\eta}_i(k), \tag{2.11}$$

$0 \leq k \leq \bar{T}$ and $1 \leq i \leq N$. The solution to the optimization problem is given by the Lagrangian formalism (Theorem 2.7 of [5]):

Theorem 2.3 *Let $\mathscr{C}_0$ be non-empty. Then there exist multipliers, satisfying (2.10) and (2.11), such that the solution $\hat{\eta}$, of the optimization problem (2.3) is given by the unique solution $\hat{\eta}$ of the equation $(Dh_{\lambda,\mu,\nu})(\eta) = 0$. Moreover,*

$$\hat{\eta} = B^{-1}(\mu m + \sum_{0 \leq t \leq \bar{T}+T-1} \lambda_t l_t + \nu), \tag{2.12}$$

where $m \in \mathscr{H}$ is given by the linear functional $(m,\eta)_{\mathscr{H}} = E(U(\infty,\eta))$ and $l_t \in \mathscr{H}$ is given by the linear functional $(l_t,\eta)_{\mathscr{H}} = E((\Delta U)(t+1,\eta)) - c(t)E(U(t,\eta))$, for $0 \leq t \leq \bar{T} + T - 1$. Explicitly

$$m(k) = E(u^{\infty}(k)), \tag{2.13}$$

where $0 \leq k \leq \bar{T}$ and

$$l_t(k) = E(u(k,t+1) - (1+c(t))u(k,t)|\mathscr{F}_k), \tag{2.14}$$

if $0 \leq k \leq t$ and $l_t(k) = 0$, if $t < k$, where $0 \leq t \leq \bar{T} + T - 1$ and $0 \leq k \leq \bar{T}$.

Approximation methods to determine the multipliers (only ν is difficult) are discussed in [5]. A crucial point for the application of formula (2.12) is the inversion of the operators A and B. We developed (Appendix A of [5]) an algorithm giving the inverses in a small number of steps, only involving simple matrix algebra and conditional expectation. In particular we proved (see conclusion after Proposition A.4 of [5]):

Theorem 2.4 *The spectrum of each one of the operators A and B is a finite set of strictly positive real numbers.*

2.2. General case: *H*

In this section we outline results concerning the company H obtained in [6]. The (aggregate) underwriting portfolio $\eta = (\eta^{(1)}, \ldots, \eta^{(\aleph)})$ is supposed to be an element of a space of square-integrable portfolios $\mathscr{P}_{u,\bar{T}}$. This space is defined by the conditions that $p \circ \theta^{(i)} \in \mathscr{E}^2_{\bar{T}}(\mathbb{R}^{N^{(i)}}, \mathscr{A})$ and $\lambda \circ \theta^{(i)} \in \mathscr{E}^2_{\bar{T}}(\mathbb{R}, \mathscr{A})$, for $1 \leq i \leq \aleph$, where the function p is given before formula (1.6) and where the function λ is given by $\lambda(\tau_f) = 1$ and $\lambda(x) = 0$ for $x \in \mathbb{R}^N$ and $N \in \mathbb{N}$ (for details see [6]). We also suppose that the dividend allocation process $\vec{D}$ is square-integrable, i.e., $\vec{D} \in \mathscr{E}^2_{\bar{T}+T}(\mathbb{R}^{\aleph}, \mathscr{A})$. We remind that the equity allocation $\vec{K}(0) \in \mathbb{R}^{\aleph}$ and that $D^{(j)}(0) = 0$, for $1 \leq j \leq \aleph$.

Next we shall formulate more precisely the constraints (B)–(H). Reordering them and introducing the variance constraints in §1 (and see Theorem 2.4 of [6]), we obtain the following constraints on $(\eta, \vec{K}(0), \vec{D}) \in \mathscr{P}_{u,\bar{T}} \times \mathbb{R}^{\aleph} \times \mathscr{E}^2_{\bar{T}+T}(\mathbb{R}^{\aleph}, \mathscr{A})$ (We remind that $V(X)$ denotes the variance of the random variable X):

- $c'_1)$ $K(0) = \sum_{1 \leq j \leq \aleph} K^{(j)}(0)$ (*budget constraint*);
- $c'_2)$ $D = \sum_{1 \leq j \leq \aleph} D^{(j)}$ and $D^{(j)}(0) = 0$ (*budget constraint*);
- $c'_3)$ $E((\Delta U)(t+1, \xi + \eta)) \geq c(t) E(\sum_{1 \leq j \leq \aleph} K^{(j)}(t))$, where $c(t) \in \mathbb{R}^+$ is given for $t \in \mathbb{N}$, (*constraint on ROE*);
- $c'_4)$ $V(\sum_{1 \leq j \leq \aleph} K^{(j)}(t)) \leq \varepsilon'(t)(\delta(t)K(0))^2$, $E(\sum_{1 \leq j \leq \aleph} K^{(j)}(t)) \geq \delta(t)K(0)$, $t \in \mathbb{N}$, where $\varepsilon'(t) \geq 0$ and $\delta(t) > 0$ (*modified ruin constraint on H*);
- $c'_5)$ supplementary constraints, to be specified, on $D^{(j)}$, (e.g. to increase the equity of $S^{(l)}$, one can set $D^{(l)} = 0$). To be general, we only suppose that there are real valued functions F_α, $\alpha \in I$, an index set, such that $F_\alpha(t, \eta, \vec{K}(0), \vec{D}) \leq C_\alpha(t, K(0), \xi)$, where $C_\alpha(t, K(0), \xi)$ are constants only depending of the initial equity and the run-off;
- $c'_6)$ if $(\eta_i^{(j)}(t))(\omega) \neq \tau_f$, then $((\underline{c}_i^{(j)}(\eta))(t))(\omega) \leq (\eta_i^{(j)}(t))(\omega) < \infty$ and $(\eta_i^{(j)}(t))(\omega) \leq ((\bar{c}_i^{(j)}(\eta)))(t))(\omega)$, for $\omega \in \Omega$ (a.e.), $1 \leq j \leq \aleph$, $1 \leq i \leq N^{(j)}$ and $0 \leq t \leq \bar{T}$. Here $\underline{c}_i^{(j)}(\eta) \in \mathscr{E}^2(\mathbb{R}, \mathscr{A})$ and $\bar{c}_i^{(j)}(\eta)$ are given $\mathscr{A}$-adapted processes, which are causal functions of η and which satisfy $((\underline{c}_i^{(j)}(\eta))(t))(\omega) \in [0, \infty[$ and $((\bar{c}_i^{(j)}(\eta))(t))(\omega) \in [0, \infty]$ (*positivity and market constraints*);
- $c'_7)$ $V(K^{(j)}(t) - m^{(j)}(t, \xi^{(j)} + \eta^{(j)})) \leq \varepsilon'^{(j)}(t)(\delta^{(j)}(t)K(0))^2$ and $E(K^{(j)}(t) - m^{(j)}(t, \xi^{(j)} + \eta^{(j)})) \geq \delta^{(j)}(t)K(0)$, for $1 \leq j \leq \aleph$ and $t \in \mathbb{N}$, where $m^{(j)}$ are as in (c_7) and where $\varepsilon'^{(j)}(t) \geq 0$ and $\delta^{(j)}(t) > 0$ (*modified non-solvability constraint*);
- $c'_8)$ if $t > (t_f^{(j)})(\omega)$, then $(\eta^{(j)}(t))(\omega) = \tau_f$, where $(t_f^{(j)})(\omega)$ is the smallest time in $\mathbb{N}$ such that $K^{(j)}(t_f^{(j)}) < m^{(j)}(t_f^{(j)}, \xi^{(j)} + \eta^{(j)})$, and if $t \leq (t_f^{(j)})(\omega)$ and $(K^{(j)}(t-1))(\omega) > (m^{(j)}(t-1, \xi^{(j)} + \eta^{(j)}))(\omega)$, then $(\eta^{(j)}(t))(\omega) \in \mathbb{R}^{N^{(j)}}$,

for $\omega \in \Omega$ (a.e.), $1 \le j \le \aleph$ and $t \ge 1$ (*the activity of* $S^{(j)}$ *ceases just after that the solvency margin is not satisfied*).

Let $\mathscr{C}_{c'}$ be the set of all $(\eta, \vec{K}(0), \vec{D})$ satisfying the constraints (c'_1)–(c'_8). Thus we sum up the constraints (c'_1)–(c'_8) in the form:

- c') $(\eta, \vec{K}(0), \vec{D}) \in \mathscr{C}_{c'}$.

The original constraints with non-solvency probabilities are satisfied, when $(\eta, \vec{K}(0), \vec{D}) \in \mathscr{C}_{c'}$, under conditions on the constants $\varepsilon'(t)$ and $\varepsilon'^{(j)}(k)$ in (c'_4) and (c'_7) (see Theorem 2.4 of [6]).

The equity allocation and portfolio selection problem for H is then, for given run-off ξ, initial equity $K(0)$ and functional form $\eta \mapsto D(\xi+\eta)$ of the (total) dividend process for H, to find the solutions $\widehat{\eta} \in \mathscr{C}_{c'}$ of the equation

$$E(U(\infty, \widehat{\eta}+\xi)) = \sup_{(\eta,\vec{K}(0),\vec{D}) \in \mathscr{C}_{c'}} E(U(\infty, \eta+\xi)). \tag{2.15}$$

We need to make some hypotheses on the unit contract result processes. To be as simple as possible, suppose that

$$u^{(j)\infty} \textit{ satisfies hypotheses } (H_1), (H_2), \textit{and } (H_3) \textit{ of §2.1 for } 1 \le j \le \aleph, \tag{2.16}$$

that

$$\|E((u^{(j)}(k,t))^2 | \mathscr{F}_k)\|_{L^\infty} < \infty \textit{for } k < t \tag{2.17}$$

and that

$$u_i^{(p)\infty}(k) \textit{ and } u_j^{(r)\infty}(l) \textit{ are independent for } p \neq r \tag{2.18}$$

(see hypotheses (h_1)–(h_4) of [5]). For reference we sum up these hypotheses

- *h*) Hypotheses (2.16), (2.17) and (2.18) on unit contract result processes.

We can now state, the result on the solutions of optimization problem (2.15) (for details see Theorem 2.5 of [6]). We denote by d-topology, the topology of convergence for distributions.

Theorem 2.5 *Let the utilities $u^{(p)}(k,t)$ and $u^{(p)\infty}(k)$, of unit contracts, satisfy (h) and let the functions $\eta^{(j)} \mapsto m^{(j)}(t, \xi^{(j)} + \eta^{(j)})$, $\eta \mapsto D(\xi+\eta)$ and $\eta \mapsto \underline{c}_i^{(j)}(\eta)$ to $\mathscr{E}^2(\mathbb{R}, \mathscr{A})$ map bounded sets into bounded sets. In the d-topology, let $\eta^{(j)} \mapsto m^{(j)}(t, \xi^{(j)} + \eta^{(j)})$ and $\eta \mapsto D(\xi+\eta)$ be continuous, let $(\eta, \vec{K}(0), \vec{D}) \mapsto F_\alpha(t, \eta, \vec{K}(0), \vec{D})$, $\alpha \in I$, be lower semi-continuous and let $\eta \mapsto (\underline{c}_i^{(j)}(\eta))(t,\omega)$ and $\eta \mapsto -(\overline{c}_i^{(j)}(\eta))(t,\omega)$ be lower semi-continuous (a.e.). If*

$$V\Big(\sum_{1 \le k \le \bar{T}+T} D(k,\theta)\Big) \le c^2 V(U(\infty,\theta)),$$

where $0 \le c < 1$, and if $\mathscr{C}_{c'}$ is non-empty, then the optimization problem (2.15) has a solution $\widehat{x} = (\widehat{\eta}, \widehat{\vec{K}}(0), \widehat{\vec{D}}) \in \mathscr{C}_{c'}$.

The variance hypothesis of the theorem just states that the total accumulated dividend is less volatile than the accumulated final result (see Remark 2.6 of [6]). The supposed regularity properties are usually satisfied in applications.

The proof of this result in [6] goes along the following lines. Beginning with the fact that $\mathfrak{b}^{1/2}$ is equivalent to the norm in $\mathscr{H}$, one proves that $\mathscr{C}_{c^J}$ is d-compact. The function $\eta \mapsto E(U(\infty, \eta + \xi))$ is then proved to be d-continuous, so it takes its maximum in $\mathscr{C}_{c^J}$.

References

1. Dana, R. A. and Jeanblanc-Picqué, M.: *Marchés Financiers en Temps Continue: Valorisation et Équilibre*, Economica, 1994.
2. Embrechts, P., Klüppelberg, C., and Mikosh, T.: *Modelling Extremal Events*, Applications of Mathematics, Vol. 33, Springer-Verlag, 1997.
3. Harrison, M. and Pliska, S.: Martingales and Arbitrage in Multiperiod Securities Markets, *J. Econom. Theory* **20** (1979), 381–408.
4. Markowitz, H.: Portfolio Selection, *Jour. Finance* **7** (1952), 77–91.
5. Taflin, E.: Equity Allocation and Portfolio Selection in Insurance: A simplified Portfolio Model, Preprint AXA, March 1998, http://xxx.lanl.gov/abs/math/9907142.
6. Taflin, E.: Equity Allocation and Portfolio Selection in Insurance, Preprint AXA, March 1999, http://xxx.lanl.gov/abs/math/9907160. To appear in *Insurance: Mathematics and Economics*.

The recognition of handwriting

How a successful model of vision can be obtained by algorithmic information processing

J.-C. Simon†
A2iA, Paris France

Abstract. Only 15 years ago, the recognition of handwriting was considered as Science Fiction. Today ICR[1] Softwares executed on simple PCs allow to identify the amount of 70% of handwritten checks with the same rate of confusion that human operators, this not only on figures but also on words. Today 60 sites in France, UK, USA are using these process in production. But to reach this state several difficulties had to be solved, a company had to be started to finance research and application. The scientific interest of this endeavor is several: Show how a model of vision could be obtained by algorithmic information processing; introduce experimental method in a formerly philosophical interpretation; set up new scientific concepts; show that algorithmic models may not be only mere applications of mathematical models.

Résumé. Il y a 15 ans seulement la Reconnaissance Automatique de l'écriture manuscrite se limitait à des essais de laboratoire, à la limite de la Science Fiction. Notre recherche a permis de développer un logiciel qui est aujourd'hui installé dans 60 sites industriels de production et reconnaît journellement les montants de 20% des chèques manuscrits français, non seulement en chiffres mais aussi en lettres. Il a fallu résoudre d'innombrables problèmes scientifiques et techniques, et pour cela développer dans des conditions difficiles une entreprise en France et à l'Etranger. L'intérêt scientifique de cette étude est multiple : dégager des concepts scientifiques nouveaux; modéliser un domaine jusque là purement spéculatif; introduire la méthode expérimentale; montrer la validité d'une modélisation informatique autre qu'une simple mise en œuvre d'un modèle mathématique.

Keywords: handwriting recognition, information processing

Mathematics Subject Classifications (2000): 68T35, G8T10, 68Q30

A new type of models in science

Since Galileo, the generally agreed paradigm is that Mathematics will model the World. The success of this principle has been the success of Science, in so many domains, Mechanics, Astronomy, Relativity, Quantum Physics. The predictive power of the mathematical models is sometimes astounding as it is witnessed in Particle Physics. The modern powerful computers allow to apply the mathematical algebraic or differential equations, i.e., to compute solutions. But this approach does not seem productive for information processing, and many new domains are indeed relevant to information processing: Economy, Biology in particular Molecular Biology, and all of the brain processes such as vision, hearing, language, thinking. Certainly a knowledge of mathematics is useful, but no one has ever found the equations that may explain brain processing.

[1] ICR is an acronym for Intelligent Character Recognition.

G. Dito and D. Sternheimer (eds.), Conférence Moshé Flato 1999, Vol. 1, 73 – 80.

Our study of handwriting recognition demonstrates that new type of models may be build up without the Platonic approach: Conceive a mathematical model, then compute solutions. We call **models of 2nd type** these new algorithmic models, implemented directly by programs.Then "the program is the model".

This new approach does not imply that any hacker will program successfully such 2nd type models, by cut and try. One has to be guided by concepts, ideas and experiments. An important scientific community deals with the Pattern Recognition – **PR** – problems. IAPR, International Association for Pattern Recognition, is an umbrella association which covers 38 national PR associations, the members of which are in Universities, Engineering Schools, Industries. Applications are coming on the market in speech recognition, robotics, document understanding and recognition (cf. [1, 2, 3, 4, 5]). Let us show what are the main scientific concepts which are our guidelines in handwriting recognition.

Important concepts in handwriting recognition

Principles common to all algorithmic Pattern Recognitions

Basically PR goal is to interpret data. Let us limit to images. An image data is obtained through an optical sensor, which interprets the outside world, the result of which is a set of pixels[2], which may have several values, or only two, then they are said binary. To be processed by computers, a pixel is represented by a number of bits, one if it is a binary pixel. The image of a A4 page has currently $n = 3.6$ million pixels. The number of such possible binary images is 2^n, an unthinkable number. One could describe such an image as a point in a space of 2^n dimensions. The PR problem is to obtain projections of this point in sets of small cardinality, for example the set of names of the objects in the image. It is clear that this cannot be done in one step.

A chicken and egg problem: Segmentation and recognition

To find the sub images of interest, i.e., representing an object, we have to identify or at least determine the class of the object. But for this we have to segment into sub images. Clearly a vicious circle. The way out is to find simple *primitive features*, and as we see later regular features.

Finding primitive features

Primitive features in a digital image are textures, edges, lines, loops, any elementary features, which are found by the corresponding *operators*. An operator is an algorithm implemented in a program executed by a computer. One may consider it as a black box, the input data of which is a sub image, the output being a "soft decision": A score of credibility – similar to a probability – and parameters of the feature. In fact it is not possible to separate feature and operator: *An operator is the constructive definition of a feature*, whether or not it corresponds to what a human observer estimates the feature to be.

2 A pixel is the result of the measure of the illumination amplitude in a small solid angle.

Regular versus singular features

Experience has shown that some features are more easy to find. Taking the example of speech, 95% of vowels are correctly found, instead of 40% of consonants. The vowel signals are (pseudo) periodic, a consonant being an accident of a vowel. The operators finding the vowels are *invariant to translation.* This is true also for texture detection in images. We give the attribute of *regular* to periodic features and to the corresponding operators invariant to translation. Features which are not regular are *singular.* The interest is to find first the easy regular features. What is left is singular features, more easy to detect as they are then segmented. This approach has been used successfully in handwritten word recognition.

Operator complexity

The price to pay for constructive process is computing complexity. The execution of an algorithm requires a number of elementary steps function of the size n of the input data. When n is large enough, the complexity is measured by an order of magnitude $O(f(n))$. Most of the PR operators imply an exponential complexity $O(e^{\alpha n})$. As soon as n is large, though the number of steps is finite, it is considered infinite in practice, and not practicable no matter what the computing power is. Thus, operators of lower complexity have to be used.

Exchanging precision for speed

Very much faster polynomial operators of $O(n^{\alpha})$ are in use, preferably linear. But their precision is lower; in other words they introduce errors. The price to pay to come out of the time deadlock is to lower precision. At this point, we may remark that regular operators are of low complexity, which facilitate the approach suggested above.

Combination of operators

Experience has shown that by appropriate combination of several operator outputs, precision could be restored. Many research efforts were done to find best combinations. It is now admitted that a linear combination of the logarithms of the credibility scores using a Neural Net learning is close to the optimum. It was found also that such combinations of several low complexity operators reduce the generally *chaotic character of algorithms*, i.e., a small change in the input data would cause a large change in the output results.

Multiple levels of decision

The recognition process is performed in several levels of *interpretation.* At each, operators (algorithmic programs) give an *interpretation* of the input *representation.* The first level is the optical sensor level which interpret the outside world as a digital image. This data is the representation to be interpreted as features at the next level. These features will be the representation of the next level being interpreted as objects; and so on.

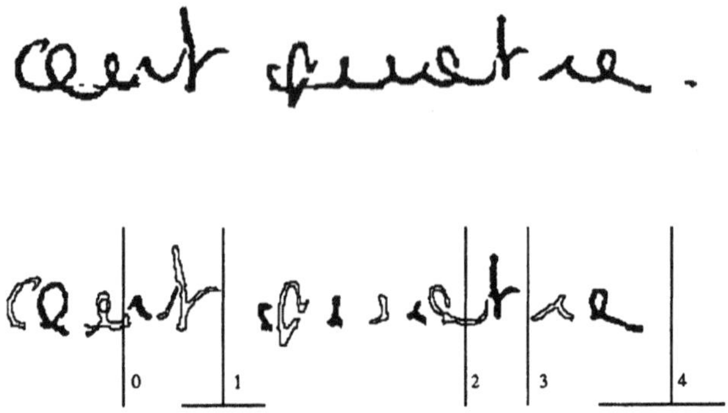

Figure 1. Words segmented in graphemes.

Soft decisions
Instead of "hard" decision by Yes or No, the result of an operator or of a combination of operators is a "soft " decision, i.e., a *list* of possible features or objects each with a credibility score. The next recognition level uses the set of lists to set up a new set of lists of possible objects with credibility scores

Increased abstraction
At each level abstraction is increased, and localization is decreased. For example if a letter 'e' is found in a sub image of $p \times p$ pixels, the information of the letter does not give the precise sub image. A huge number of sub images $p \times p$ can represent an 'e'. What is gained in abstraction is lost in localization (extension).

Handwriting recognition
Using letters to recognize a words
Figure 1 shows on the top the binary image two French words "cent quatre". Using the approach to find first the regular features, the words can be segmented in singular features, the *graphemes*, which are shown alternatively in black and white. A letter is made of one grapheme – like the 'c', the 'e' – or two graphemes – like the 'n', the 'u', or more like an 'm'. So the problem of segmentation in letters is well advanced. On Figure 1, the possible segmentations in words is also shown by vertical bars. The correct ones having an horizontal bar. But we do not know them a priori. So all combinations of segmentation have to be examined. Inside of each sub image the likelihood of a word is examined. For this all combination of one or two graphemes generates a possible list of couples (letter, score). Finding the optimal paths among these list generate a new list of couples (word, score). And so on up to a list of phrases- in the considered application a check amount. At each step, except the grapheme generation, all decisions are "soft". To have to consider many segmentations may generate a "combinatorial explosion" and again an impractibility. The credibility scores are of course essential to keep down the number of cases considered.

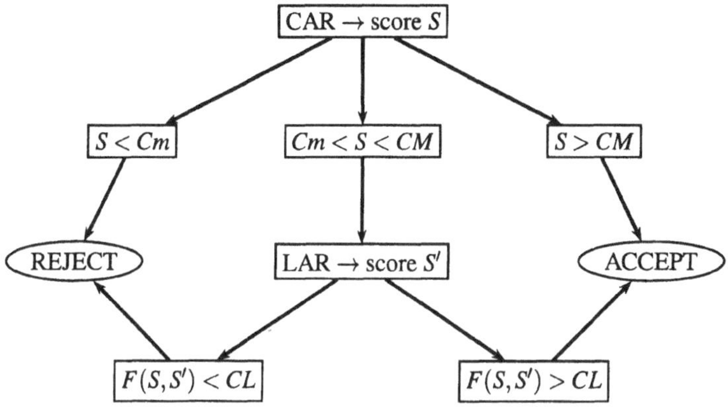

Figure 2. The use of CAR and LAR.

A global recognition
Another approach is to identify a word without finding the letters. Then the features are ascenders, descenders, loops, horizontal bars, etc.

Application to check amount recognition

5 billion checks in France, 60 billion in the USA are issued each year. Their amounts have to be coded to be processed in payment systems. Up to now it is done "by hand". It is a costly and unrewarding activity, which should be automatized. The amount of a check is coded twice, by numbers – the "courtesy amount" –, by words – the "legal amount". **CAR** is the acronym for Courtesy Amount Recognition, **LAR** for Legal Amount Recognition. The ICR Software "A2iA Check Reader" recognize both CAR and LAR, according to the scheme of Figure 2. CAR delivers a score S. If $S < Cm$ the image is rejected. If $S > CM$, the amount is accepted. Cm and CM are two thresholds which may be adjusted. In the grey zone, $Cm < S < CM$, LAR is performed giving a score S'. Let CL be another threshold; if $S' < CL$ the check is rejected. Otherwise the two CAR and LAR lists of couples – amount and score – are combined to give a final amount with a score, which is accepted of rejected according to a third threshold CL.

Applications in production
At the beginning of Year 2000, sixty payment system sites are using "A2iA Check Reader", mostly in France, but already also in the USA and UK. These ICR softwares are reading now of the order of five million check/day. Data entry – reading a check amount – by software is done by an ordinary Pentium 400MHz PC under Windows NT, at the speed of two amounts per second, 7000 per hour. If a higher speed is necessary, images are distributed among N PCs; the speed is multiplied by N. The ICR Software itself, written in 'C', has 85 000 lines of code. It is constantly improved and has also other added possibilities. A good human operator is coding

approximately 2 000 check amounts per hour, with a substitution error of 1%. By adjusting the thresholds *Cm*, *CM*, *CL*, the substitution rate of the ICR software can be kept down statistically at 1%. Then the ICR acceptance rate is in the range 65% to 75%, depending of the type of check, the quality of the image. The 35% to 25% rejected checks are read by human operators, who still beats the machine on difficult checks.

Concluding remarks

The A2iA team was quite lucky to be able to bring a basic research to the state of a massive application in the banking industry. Other applications will follow in the Postal address reading and more generally in document processing and recognition. This allows us to carry on research in the important field of *information processing and recognition.* Certainly we are not alone: An important scientific research community [6] is working on these topics, and many high tech companies are created.

Is information processing a science?
If Information Processing (I.P.) does not follow the Galilean paradigm: "Mathematical Models will explain the World ", – eventually may allow constructive predictions –, and if this is the necessary condition of a science, then I.P. is *not a science.* On the other hand, if I.P. gives constructive models of natural phenomena, which cannot be explained, modeled, or predicted by Mathematics, it should be *deemed to be a Science.* In fact a new scientific paradigm, to which we give the name of Models of 2nd type. It is easy to see that these models will predominate in fields where Information Processing, more precisely representations and interpretations, are basic: Economy, Molecular Biology, and all of the brain processes such as vision, hearing, language, thinking. Then the performances will be the proof of validity of the concepts which guide the build up of such 2nd type models.

Brain processes
The Nobel laureate Francis Crick wrote in *Scientific American 79*: "If most of philosophical thought is barren since 2000 years, and will likely remain so, it is because philosophers did not learn to understand the language of information processing". Brain processing, vision, language, understanding have been mostly a philosophical field, and *we know very little of the brain processing as an information processing system.* Knowing a lot about the neurons, does not help much to understand how 100 billion neurons are achieving the extraordinary brain processes, seeing, hearing, understanding, thinking, speaking.Building up constructive models of a brain process, even partial, is like introducing the experimental method in these domains, which are in a state similar to Medicine 200 years ago, when discourse was more important than practice.

Interest of vision models

A third of human cortex is devoted to vision, which is our most important sense. Vision is a lower level process in the sense that we have no control on our interpretation of visual perception: We cannot help to recognize objects. Our visual system is an *automatism*, which is not under our control, as are higher level processes like muscle control or thinking. Primitive animals like insects are visual automata. Vision appeared very early in the evolution of Life. It is likely that the brain processes of vision, which now are so refined in human beings, were the basis for most of the upper processes. Physiologists have found that a cortical zone, used for vision in the macaque monkey, has been confiscated for speech in human beings. Logical thinking appeared very recently in human beings. Instead of explaining brain process by logic, which is the strategy of Artificial Intelligence, it seems more appropriate to follow natural evolution and to try first to understand vision process. Though recognition of handwriting is a very small part of human vision, the algorithmic model, which has been developped, has similarities with the discoveries of brain physiologists: Neurons finding elementary features such as segments of different directions; layers of neurons to be compared to the several levels of recognition in our model; the many connections of neurons of a layer, to neurons of the next layer, similar to the model multiple combination of operators; at each layer, abstraction increases, etc. But a computer is not a brain. Our model uses the the most general and most efficient information processing machine available today to simulate vision. Our research has shown that the computer should not be used as a logical machine, but more and more as an analogue machine, to obtain good results. Thirty years ago, analogue computers were competitors of digital computers and could be started again with different technologies. But one point remains clear: We have to try to understand the brain processes to reproduce them, and use PR techniques rather that try to force some mathematical model. On the other hand, physiologists should be interested by constructive models. But more than ever, scientific interdisciplinarity seems more difficult to achieve, though many official declarations put it in the forefront of their preoccupations.

References

1. Duda, R. O. and Hart, P.: *Pattern Classification and Scene Analysis*, Wiley Interscience 1974.
2. Fukunaga, K.: *Pattern Recognition*, Academic Press, 1972.
3. Simon, J.-C.: *Patterns and Operators*, North Oxford Academic, 1986.
4. Bishop, C. M.: *Neural Network for Pattern Recognition*, Oxford University Press, 1995
5. Proceedings of ICPR, IEEE documents of 13 IAPR International Symposium,taking place every two years.
6. ICDAR '99, International Document Analysis and Recognition had 300 participants of 23 different nationalities, in September 99 at Bangalore India.

EDITORS' NOTE

Jean-Claude Simon, Professor Emeritus at Université Pierre et Marie Curie (Paris VI), Chairman and CEO of A2iA, died suddenly in Paris on February 16, 2000.

Jean-Claude Simon was brought to Paris VI from the private sector and always had an original approach to computer science problems. He has been a member of the Scientific Council of UAP, from its creation in 1984 by Yvette Chassagne and Moshé Flato, to its dissolution in 1992 (a few years before UAP itself disappeared).

The recognition of handwriting has been one of the projects initiated as early as 1985 by Yvette Chassagne together with the Scientific Council. In cooperation with the French Ministry of Industry when Madame Cresson was Minister, UAP and a 'startup' founded by Physics Nobel Prize laureate Leon Cooper developed a prototype which was presented in 1987 at the Journées Scientifiques of UAP. It was based on neural networks, a subject on which Leon Cooper worked for many years, and operational on PCs of that time, in a context and with an aim different from the A2iA approach. Difficulties in finding industrial partners and a change of policy from the new UAP leadership did not permit further development of the prototype and its production for the professional use in insurance marketing for which it had been designed.

We are happy that Jean-Claude Simon could see the success of his approach in a project which continued work started at the Scientific Council. Together with his family and colleagues we mourn the too early disappearance of a lively and original figure in the scientific community.

From Bachelier and Lundberg to insurance and weather derivatives

Hélyette Geman
ESSEC, avenue Bernard Hirsch, 95021 Cergy-Pontoise Cedex, France
and
Université Paris IX Dauphine
Place du Maréchal de Lattre-de-Tassigny, 75775 Paris Cedex 16 France
e-mail: geman@essec.fr

Keywords: catastrophic risk, jump-diffusion process, option pricing, actuarial valuation

Mathematics Subject Classifications (2000): 91B30, 91B02, 91B70

I. Introduction

Among other advances, the twentieth century has witnessed a number of spectacular achievements in three fields which are today closely inter-related, namely probability theory, finance and insurance. Building on the successive visionary intuitions of the botanist Brown observing the moves of pollen particles at the surface of a liquid, of the mathematician Bachelier modeling in 1900 the stock price dynamics on the Paris Bourse and of the physicist Einstein studying in 1905 the trajectories of molecules in a gas, Wiener provided in 1915 the precise definition of a remarkable stochastic process called Brownian motion. This process, whose mathematical properties are beautiful, has been playing for more than a century a key role in the representation of random moves when the assumption of *continuity* of the trajectories appears appropriate.

In a noteworthy coincidence of dates, the actuary-mathematician Lundberg introduced in 1905 the Poisson process to model the random arrival of claims in the portfolio of contracts of an insurance company. Today, the Poisson process remains the central instrument to represent the occurrence of natural catastrophic events as well as financial market crashes or firm bankruptcies (see [4]). The "simple" Poisson process has become compound and its intensity stochastic but in all cases, the key departing point in the modeling of discontinuous trajectories belongs to Lundberg.

More than twenty five years ago, the celebrated work by Black and Scholes [1], and Merton [11] on the valuation of equity options paved the road for the explosion of derivatives markets around the world. In the following decade, these results were extended to options on futures contracts, options on currencies and so forth. All these new formulas, quite useful up to the present time, have the property of being of the same vein as the Black-Scholes-Merton formula. This paper will try to explain why options on insurance, credit or weather present in several manners new challenges.

G. Dito and D. Sternheimer (eds.), Conférence Moshé Flato 1999, Vol. 1, 81 – 95.

To start with a straightforward but crucial observation, we first need to note that in the so-called first generation of options, the quantity which defines the option payout at maturity T is the price of a traded stock or asset S. This has a number of key consequences. Firstly, the number $S(T)$, hence the terminal value $\max(0, S(T) - k)$ of the call option is fully observable and so is the stock price $S(t)$ at any date t of the lifetime of the option. This property obviously does not hold in the case of options on insurance, credit, weather for which a clear *definition* of the underlying instrument or an *index* is necessary in order to avoid litigation on a contract which was improperly underwritten. Secondly, the seller of an equity or currency option is able to continuously hedge his/her position by holding the underlying instrument in a quantity equal to the delta of the option. Hence, except for rare occurrences of "liquidity holes", hedging plain-vanilla options on stocks does not pose any real difficulties. Delta-hedging a weather derivative through a long (or short) position in a temperature index is obviously not feasible. Other answers need to be found; they will definitely not have the accuracy nor the elegant simplicity of the Black-Scholes hedging portfolio. Thirdly, the mathematical modeling of the underlying state variable dynamics, namely the claim index for insurance derivatives or the firm rating for credit options, involve stochastic processes whose proper identification and calibration require an important database; obviously, data are readily available and numerous in the case of financial options. This is not necessarily the case for other types of underlyings.

In order to clarify these different points, in particular the crucial discussion of derivatives hedging, we are going to review the Black-Scholes-Merton formula and identify the steps in the proof which may not easily extend to insurance and weather derivatives.

II. The Black-Scholes-Merton formula (1973)

Let us first review the classical graphs corresponding to the payouts and gains at maturity T of simple elementary or combinations of put and call options written on a stock S. Keeping in mind that the respective payouts of call and put options

$$C(T) = \max(0, S(T) - k) \quad \text{and} \quad P(T) = \max(0, k - S(T)),$$

where k is the strike price of the option, we obtain the familiar graphs given in Figure 1.

If an investor has paid $C(t)$ (resp. $P(t)$) to buy the call (resp. put) at date t, his algebraic gain profile at maturity, expressed in dollars of date T, is shown in Figure 2.

Among the simple combinations of puts and calls that one can easily construct, let us mention one which will be particularly relevant for insurance derivatives, namely the so-called call spread. A call spread is the combination of a long position in a call with strike price k_1 and a short position in a call with strike price $k_2 > k_1$, both being written on the same underlying instrument S and having the

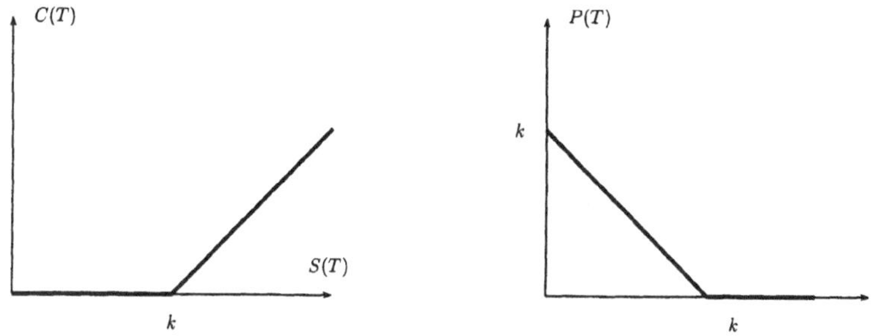

Figure 1.

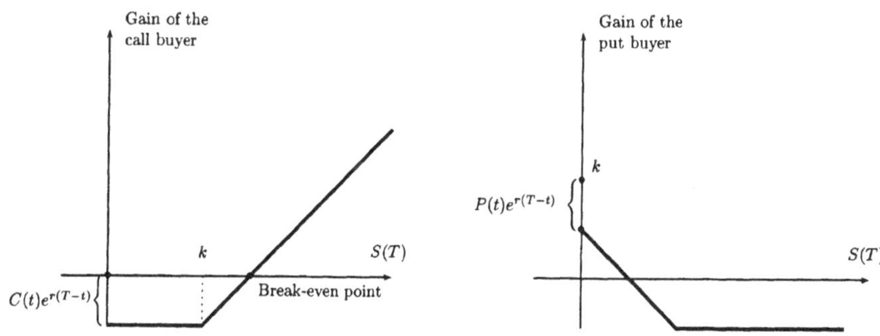

Figure 2.

same maturity T. As will be made precise below but is straight forward, the call price is higher for a lower strike price; hence the value $CS(t)$ at time t of a call spread is positive and the gain profile at maturity is represented in Figure 3.

Put-call parity

Before turning to the problem of valuing a call option, we are going to establish the relation prevailing at any time t between a call and a put written on the same underlying stock S, with the same strike price k and maturity T. It is worth noting that the result holds under fairly general assumptions, namely

A1 The market is supposed to be frictionless (no taxes, no transaction costs).

A2 The stock does not pay any dividend over the lifetime $[0,T]$ of the option (this assumption can easily be relaxed).

A3 The risk-free rate of borrowing or lending r is supposed to be constant over the period $[0,T]$.

A4 There are no arbitrage opportunities in the market, i.e., any riskless portfolio must have a return equal to the risk-free rate r.

Then, the following relationship holds at any date t:

$$S(t)+P(t)=C(t)+ke^{-r(T-t)},$$

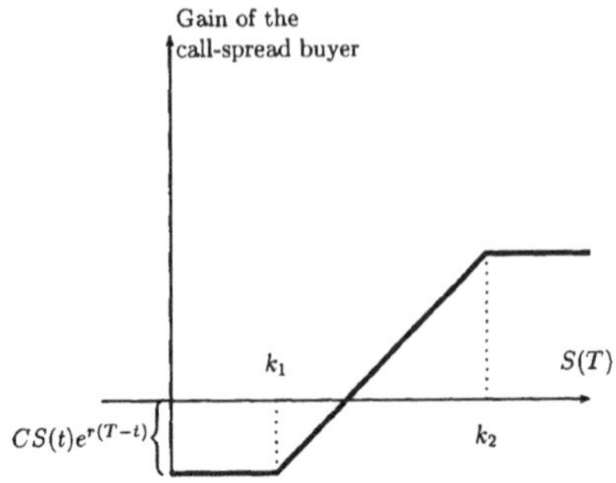

Figure 3.

where $C(t)$ is the call price, $P(t)$ the put price, $S(t)$ the stock price, and k the strike price.

The proof is straightforward and consists, as it is usual in finance, in considering an appropriate portfolio. Here, we build a position at date t which comprises a stock, a put and a short position in the call. The position is held and liquidated at date T.

	t	T	
		$S(T) < k$	$S(T) \geq k$
Buy a stock	$-S(t)$	$S(T)$	$S(T)$
Buy a put	$-P(t)$	$k - S(T)$	0
Sell a call	$C(t)$	0	$k - S(T)$
		k	k

We see that in all *states of the world*, the terminal value of the portfolio is equal to k, a fixed quantity *known* at date t. Hence, the portfolio is *riskless* since risk in finance (and insurance) is equivalent to randomness. The no arbitrage assumption then implies that the initial dollar value at date t of this portfolio is $+ke^{-r(T-t)}$, which needs to be invested in order to build the position. This proves the put-call parity relation. It is extremely important to observe that:

- This result does not depend on a particular choice for the spot price evolution between t and T. In particular the stock price process $S(t)$ may have continuous or discontinuous trajectories.

- A key point, however, is the possibility of buying the stock at date t and holding it up to time T.

When we turn to credit, insurance or weather derivatives, it will become difficult to buy a temperature or insurance index or a credit rating and keep it in a portfolio or in a position up to maturity. One has to keep the proofs of the Black-Scholes-Merton results in mind in order to identify the situations where their extension becomes hazardous.

To conclude on this put-call parity, let us make a final observation on a simple but fundamental use of the put option. Suppose that one buys at date t a stock $S(t)$ and a put written on the stock, with maturity T and strike price k. At date T, the value of this portfolio P is

$$\begin{aligned} V(T) &= S(T) + \max(0, k - S(T)) \\ &= \max(k, S(T)). \end{aligned}$$

This expression shows that the investor will benefit from a rise of the stock market since the portfolio value $V(T)$ at date T will be equal to $S(T)$; however, in the case of a collapse of the stock market, $V(T)$ will not vanish to zero but will be equal to k. The portfolio P is called an *insured portfolio* and k is called the floor of the portfolio insurance strategy. This use of put options plays a central role in the management of portfolios, both for small and institutional investors. It also illuminates the function of *insurance* structurally built in all financial options.

The Black-Scholes-Merton formula (1973)

We can observe first that if we only kept assumptions used to establish the put-call parity, we would not be able to find the price at date t (equally fair to the buyer and the seller) of the call option. One key assumption is missing, namely the one which describes the evolution of the stock price between t and T.

The Black-Scholes-Merton model

We make again the assumptions **A1** to **A4** and add the assumption

$$\textbf{A5} \quad \frac{dS_t}{S_t} = \mu dt + \sigma dW_t \tag{1}$$

where $(W_t)_{t \geq 0}$ denotes the Brownian motion and the positive constant σ denotes the so-called volatility of the stock price.

We can observe that, since the stock does not pay any dividend, the return over the interval $(t, t+dt)$ comes from the algebraic gain in capital represented by $S(t+dt) - S(t)$, hence is the left-hand side of Equation (1). Since, the right-hand side (an affine function of dWt) follows a normal distribution, it appears that the normality of returns is a key condition of applicability of the Black and Scholes formula and of its possible extension to other underlying instruments.

Moreover, if S_t is replaced by an index I_t of aggregate catastrophe claims when pricing an insurance derivative, the continuity of the trajectories implied by Equation (1) may not be desirable since it will poorly reflect the jumps in the amount insured losses following the occurrence of an earthquake or a hurricane. In the case of weather derivatives, S_t is replaced by a temperature index (or an arithmetic average of temperatures) and it is not clear that the relative change $\frac{dT_t}{I_t}$ over the time interval $(t, t+dt)$ is best represented by a normal distribution. Hence, in both cases, the mathematical modeling of the underlying source of uncertainly needs to rethought and an adequate option pricing formula established.

Besides the nature of the stochastic process itself, another key ingredient in the proof of the Black-Scholes-Merton is the *risk-neutrality* argument applied to a portfolio consisting of financial instruments, the underlying stock and the option. When the option is written on insurance, credit or weather, the portfolio to which one may apply this argument is more difficult to identify, if it exists at all.

Coming to the mathematics of the Black-Scholes-Merton formula, the well-know proof given by the authors is the following: The call price $C(t)$ at date t may be viewed as a function of the two state-variables t and S_t, namely $C(t) = f(t, S_t)$.

By application of Itô's lemma

$$dC_t = \frac{\partial C_t}{\partial t} + \frac{\partial C_t}{\partial S_t}\partial S_t + \frac{1}{2}\frac{\partial^2 C_t}{\partial S_t^2}(\partial S_t)^2,$$

where, from Equation (1),

$$dS_t = \mu S_t dt + \sigma S_t dW_t. \tag{1'}$$

Hence

$$dC_t = [\frac{\partial C_t}{\partial t} + \mu S_t \frac{\partial C_t}{\partial S_t} + \frac{1}{2}\sigma^2 S_t^2 \frac{\partial^2 C_t}{\partial S_t^2}]dt + \sigma S_t \frac{\partial C_t}{\partial S_t} dW_t. \tag{2}$$

In order to eliminate the risk expressed only by the presence of the Brownian motion (W_t), it suffices to consider the portfolio P built at time t with one call option and n stocks, where n is a real number held constant over the time interval $(t, t+dt)$.

The portfolio value at date t is

$$V_P(t) = C_t + nS_t.$$

Hence

$$\begin{aligned} dV_P(t) =& dC_t + ndS_t \\ =& dt[\frac{\partial C_t}{\partial t} + \mu S_t \frac{\partial C_t}{\partial S_t} + \frac{1}{2}\sigma^2 S_t^2 \frac{\partial^2 C_t}{\partial S_t^2} + nS_t\mu] + dW_t[\sigma S_t \frac{\partial C_t}{\partial S_t} + \mu n S_t]. \end{aligned} \tag{3}$$

By choosing $n = -\frac{\partial C_t}{\partial S_t}$, the coefficient of dW_t is nullified and the portfolio P becomes *riskless* over the interval $(t, t+dt)$.

Hence $V_P(t+dt)$, is non random at date t. By the *no-arbitrage* assumption **A4**, the return on this riskless portfolio has to be equal to the risk-free rate r and we obtain a second expression for $dV_P(t)$

$$dV_P(t) = rV_P(t)dt = r[C_t - \frac{\partial C_t}{\partial S_t} S_t]. \tag{4}$$

The comparison of Equations (3) and (4) leads to the fundamental *partial differential equation*

$$\frac{\partial C_t}{\partial t} + rS_t \frac{\partial C_t}{\partial S_t} + \frac{1}{2}\sigma^2 S_t^2 \frac{\partial^2 C_t}{\partial S_t^2} - rC_t = 0 \tag{5}$$

with the boundary condition $C(T) = \max(0, S(T) - k)$.

Up to a change of variables, Equation (5) may be reduced to the *heat equation* solved by Einstein. This leads to the call price at date t

$$C(t) = S(t)N(d_1) - ke^{r(T-t)}N(d_2), \tag{6}$$

where

- N denotes the cumulative function of a normal variable with mean 0 and variance 1
- $d_1 = \dfrac{\ln\left(\frac{S_t}{ke^{-r(T-t)}}\right) + \frac{1}{2}\sigma^2(T-t)}{\sigma\sqrt{T-t}}$
- $d_2 = d_1 - \sigma\sqrt{T-t}$

Before turning to the computation of the call price sensitivities to the different variables (the Greeks), some important observations are in order. Equation (4) establishes that the portfolio consisting of a long call and a number of stocks equal to $-\frac{\partial C_t}{\partial S_t}$ ($\frac{\partial C_t}{\partial S_t}$ will be defined below as the delta of the option) behaves over the interval $(t, t+dt)$ like a money market account; equivalently, the call is replicated on the interval $(t, t+dt)$ by Δ shares and some cash amount (in fact negative), the two quantities being functions of time (for instance, $\frac{\partial C_t}{\partial S_t}$ obviously varies with time), hence needing to be adjusted over time. This is *the dynamic replication* of the call the seller must perform over the interval $[0, T]$ in order to come up with the payout of the option $\max(0, S(T) - k)$ in all states of the world at time T.

This continuous replication of the option with the underlying instrument (and the cash, which is held by any institution) is obviously problematic for electricity derivatives, for insurance derivatives and for weather derivatives. In each case, an appropriate solution in terms of a hedging portfolio — if it exists — needs to be exhibited.

Sensitivities of the call price to the different variables

1. Delta of the call

The first quantity whose impact on the option price needs to be monitored is obviously the underlying source of risk S_t; hence the importance of the so-called delta of the call

$$\Delta = \frac{\partial C_t}{\partial S_t} = N(d_1),$$

as computed from Equation (6).

Hence, this sensitivity is positive, which is not surprising: The higher the stock price, the higher the option price. Moreover, Δ is smaller than 1, which shows that one uses less than one stock at any date t to replicate the call. Lastly, observing that over a time interval δt, one can write

$$\delta C_t \approx \Delta \cdot \delta S_t$$
$$\frac{\delta C_t}{C_t} \approx \frac{\delta S_t}{S_t} \frac{S_t}{C_t} \cdot \Delta.$$

From Equation (6), we have $S_t \cdot \Delta > C_t$, which leads to

$$\left|\frac{\delta C_t}{C_t}\right| > \left|\frac{\delta S_t}{S_t}\right|.$$

The return generated over $(t, t+dt)$ by investing in the call has the same sign as an investment in the stock but is always greater in absolute value, hence the *leverage effect* obtained by investing the same amount of money in the call rather than in the underlying stock when one has a view (upward or downward) on the stock moves. Lastly, we can observe that the delta of the put can be easily derived from the delta of the call which has the same strike price and maturity. Returning to the put-call parity relationship $S(t) + P(t) = C(t) + ke^{-r(T-t)}$, it suffices to take the partial derivative with respect to S_t on both sides to obtain $\Delta_{\text{put}} = \Delta_{\text{call}} - 1$.

It shows that the put is at any date t a decreasing function of the stock price (this is obvious at time T from the expression $\max(0, k - S(T))$ of the put price) and that the dynamic replication of the put will involve a short position in the underlying stock.

2. Gamma of the call

Given the importance of the delta of the call, a quantity of interest is the sensitivity of the delta itself to the stock price, namely

$$\Gamma = \frac{\partial \Delta_t}{\partial S_t} = \frac{\partial^2 C}{\partial S_t^2} = N'(d_1)\frac{\partial d_1}{\partial S_t} = \frac{1}{S_t}\frac{N'(d_1)}{\sigma\sqrt{T-t}}$$

which is obviously positive.

Hence at any date t, the call price is a convex function of the underlying stock price.

3. Sensitivity to time

$$\Theta = \frac{-dS_t}{dt} = -\left[re^{-r(T-t)}kN(d_2) + \frac{\sigma e^{-r(T-t)}}{2\sqrt{T-t}}kN'(d_1)\right] < 0.$$

As time goes by, the period during which the stock price may move to very high values before maturity shrinks (the downward moves below k remaining constantly translated into a zero payout of the call at date T) and the *time value* of the call price diminishes.

4. Sensitivity to the volatility: Vega of the option

$$\frac{\partial C}{\partial \sigma} = S\sqrt{T-t}N'(d_1) > 0.$$

A number of comments are in order at this point, particularly important from a practical standpoint. Firstly, let us mention that even though the volatility parameter σ is supposed to be constant in the Black-Scholes model, practitioners have been computed for more than twenty years the so-called *implied volatility* $\sigma_{\text{imp}}(t)$ embedded in the option market price $C(t)$ at date t (in a unique manner since $\frac{\partial C}{\partial \sigma} > 0$). In contrast with the model, $\sigma_{\text{imp}}(t)$ as reconstructed daily from prices of options written on a given stock, varies with time. Among the number of questions on this issue, the first one — which has been widely documented in the finance literature without any decisive conclusion — is the following: If one writes an option at time t for the maturity T, what is the best quantity to incorporate for σ in the Black-Scholes formula in order to obtain a quasi-perfect replication of the option by delta hedging? The main quantities which have been and are tested are the historical volatility, computed on a time series of past prices of the underlying stock, and the implied volatility derived from current option prices. As far as the latter is concerned, another difficulty arises from the fact that it also only varies with the strike price k of the option, generating the famous volatility smile $\sigma_{\text{impl}}(k)$ prevailing at date t.

III. Extensions of the Black-Scholes-Merton formula

A number of deviations from the original Black-Scholes-Merton model were documented both by academics and practitioners over the last twenty five years. We noted in the previous paragraph that the implied volatility rapidly became a time-varying parameter $\sigma_{\text{imp}}(t)$, depending also not only on the strike price of the option under analysis but also on its maturity T (for instance, options with a short time to maturity have a much higher volatility than the one predicted by the Black-Scholes formula). This dependence in t and T defines the so-called term structure of volatilities at date t, $\sigma_{\text{impl}}(t,T)$, whose dependence in k as well translates into the volatility surface $\sigma_{\text{impl}}(k,T)$ at date t.

The celebrated formula has been extended in several manners which can be roughly classified in three categories:

- Replacement of the classical geometric Brownian motion to represent the underlying price process by a more complicated stochastic process.
- Replacement of the classical call option payoff $\max(0, S(T)-k)$ by an "exotic" payoff involving not only $S(T)$ but the trajectory of the process $S(t)$ over the lifetime $[0,T]$ of the option.
- Replacement of the stock price as the underlying source of risk in the option by interest rates, currencies, and very recently by credit, insurance, weather,

electricity. As this article will try to show all these extensions apply to the case of derivatives on insurance, electricity, and weather.

A) Alternative representations of the underlying source of risk dynamics

We observed in the previous section that Equation (1), representing the return of the stock price over the interval $(t, t+dt)$ as the quantity $\mu dt + \sigma dW_t$, hence a normal distribution, was a key assumption in the Black-Scholes model. On the other hand, a huge literature in finance has been dedicated to the analysis of the normality of asset returns since this property play a central role in the construction of the Markowitz frontier of efficient portfolios and Capital Asset Pricing Model. The property of normality also has an impact on the nature of the *tails* of the distribution, which represent extreme events and are conveyed by the $1-(\alpha)$-quantile ($\alpha = 0.95$ for instance) called Value at Risk at the level of confidence α and playing a major role as a risk indicator. Deviations of returns from normality have been consistently exhibited in the finance literature and this is all the more true as the time period is shorter [5] — like a day or a week, time length which has become for practical purposes the horizon in portfolio management. Besides the fat tails or kurtosis, the distribution of observed returns is more skewed than the normal distribution. One way of addressing the issue of skewness and kurtosis is to introduce jumps in the representation of the stock price dynamics.

1. Jump diffusion processes for the underlying asset price

Back in 1976, Merton had proposed to extend the model proposed in [1] and [11] for the stock price dynamics by adding a jump component, namely

$$\frac{dS_t}{S_t} = \mu dt + \sigma dW_t + U dN_t, \tag{7}$$

where:
- (N_t) is a Poisson process with intensity λ;
- U is a real-valued random variable;
- The other quantities have the same meaning as before.

This assumption has a number of consequences:

a) From a mathematical viewpoint, the solution of (7) is obviously more complicated than the solution of (1) (in particular, it is not explicit any more but involves the so-called Doléans-Dade exponential).

b) From an economic viewpoint, the number of sources of randomness driving the stock price has been drastically increased, not only because of the presence of the Poisson process (N_t) but also because the magnitude of the jumps U is a random quantity. Hence, it is not any more possible to duplicate the call option with the underlying stock and the money-market account and in fact, an *infinite number* of hedging instruments are necessary and do not exist in practice.

In order to eliminate this market incompleteness, Merton had made the simplifying assumption that the jump risk could be diversified away, hence the economic difficulty — the most important one — was eliminated. And some mathematical considerations allowed to exhibit the option price in this new setting. This assumption of a null *market price of jump risk* is clearly unacceptable for *catastrophic insurance risk*. It is not valid either in the case of electricity derivatives where after the spike of June 1998 in the Illinois region — the spot price of electricity went from 25 dollars to 7500 dollars — options with a strike price of 1000 dollars got their market prices multiplied by 100, clearly indicating the existence of risk aversion and risk premium associated with the jump.

2. Time-varying volatility for the underlying asset price

Another way of explaining the volatility smile mentioned in Section II as well as the fact that the implied volatility $\sigma_{\text{impl}}(t)$ derived from a call option written on a given stock varies with time is to explicitly introduce in Equation (1) a volatility which varies with time.

a) In a first approach, $\sigma(t)$ may be expressed as a deterministic function of time

$$\frac{dS_t}{S_t} = \mu dt + \sigma(t)dW_t.$$

It is then easy to show that all Black-Scholes results hold, including the hedging strategy; it suffices to replace everywhere, including in the valuation formula (6), σ with Σ, where Σ is defined by

$$\Sigma^2(T-t) = \int_t^T \sigma^2(u)dw.$$

b) Going one step further, $\sigma(t)$ becomes stochastic but has the property of being a function of $S(t)$. In this case, we still have only one source of randomness, namely the Brownian motion involved in the dynamics of $(S(t))$. Hence, the option price (as well as the hedging portfolio) is unique. It can be obtained as a closed-form expression if $\sigma(S_t)$ is a simple function; otherwise, $C(t)$ may be computed using Monte-Carlo simulations.

c) In the most general *stochastic volatility model*, $\sigma(t)$ is a stochastic process; hence, assuming diffusion processes for the stock and the volatility, the dynamics of $(S(t))$ are defined by the two stochastic differential equations

$$\begin{cases} dS_t = \mu_1(t,S_t)dt + \sigma(t,S_t)dW_t^1 \\ dy(t) = \mu_2(t,y_t)dt + \eta(t,y(t))dW_t^2 \end{cases}$$

where $y(t) = [\sigma(t)]^2$, $dW_t^1 \cdot dW_t^2 = \rho(t)dt$.

The coefficient $\rho(t)$ allows for dependence of the Brownian motions $W_1(t)$ and $W_2(t)$; the functions μ_1 and μ_2 may eventually translate a property of mean reversion for the stock price or volatility. In this case, not only do the closed-form solutions for option prices become problematic but the unicity of the call value and of the hedging portfolio is lost since we still have only two hedging instruments and an extra-source of risk represented by the Brownian motion (W_t^2). A possible answer to this market incompleteness is to consider that the at-the-money call C^m (i.e., with a strike price equal to the value of the stock today) is a *basic security* whose price is agreed upon by all market players.

If then becomes possible to hedge all other calls and puts using the stock, the money-market account and the call C^m. The existence of this very liquid option completing the market is quite legitimate for options on stocks, on stock indexes or on currencies; it is certainly problematic for insurance or weather derivatives, where the generality of a stochastic volatility model is paid for by the loss of a unique price for the options.

B) Exotic pay-offs

As of the beginning of the 80's, so called exotic options were introduced to answer the hedging needs of financial and industrial firms as well as the search for new opportunities on the part of investors. We will describe the most popular exotic options, namely the Asian and the barrier options

a) Asian or average-rate options

Instead of the Black-Scholes pay-off $\max(0, ST - k)$, the buyer of an Asian call option receives at maturity T $\max(0, A_T - k)$, where A_T represents the arithmetic average of the stock price computed over the lifetime $[0, T]$ or over a subinterval of $[0, T]$.

Asian options are very much used for thinly traded assets such as gold since they obviously prevent a possible manipulation of prices on the maturity date of standard options by institutions holding large positions in the underlying asset. They represent a large percentage of the options traded on oil or oil spreads since oil indexes are themselves defined as averages of spot prices over a time period. In the FX markets, they allow corporate treasurers to hedge a series of cash-flows by buying a single average-rate option, as an alternative to a portfolio of standard options associated with each payment date, hence reducing the cost of FX risk management.

Other examples of Asian options will be presented in this article and explain why their valuation is such an important issue. Unfortunately, it is not a straightforward problem, the difficulties stemming from the fact that an average of prices S_t satisfying Equation (1) does not satisfy an equation of the same type or, in other terms, is not a geometric Brownian motion. The main methods used today to value Asian options are the following:

(i) The approximation of the true distribution of the average A_T by the log-normal distribution of S_T, with adjustment of the first two moments.

(ii) The use of Monte-Carlo simulations: Different paths of the stock price are generated, providing for each of them the values $S(t_1), \ldots, S(t_n)$ which are averaged to exhibit one outcome O_i of $\max(0, A_T - k)$, where k is the strike price of the Asian call.

From the law of large numbers, we know that the quantity $\frac{1}{N}\sum_{i=1}^{i=N} O_i$ will converge to the expectation of the pay-off which, discounted, gives the price of the Asian call at time zero.

(iii) An exact approach to the Asian option valuation and hedging: Geman and Yor [7] used sophisticated mathematical tools to compute the Laplace transform of the call price with respect to maturity.

Geman and Eydeland [6] on one hand, Fu, Madan, and Wang [3] on the other hand, apply different algorithms to invert the Geman-Yor formula but come up with remarkably close results. The former authors exhibit the superiority of their results over those obtained through Monte Carlo simulations, in particular in terms of the accuracy of delta hedging.

b) Barrier and double — barrier options

They have become increasingly popular in equity markets, and even more so, in FX and commodity markets. A barrier call option provides at maturity the classical Black-Scholes pay-off $\max(0, S_T - k)$ only if (unless) the underlying asset S has touched a prespecified barrier L over the lifetime $[0, T]$ of the option. For instance, if a non-zero pay-off is contingent on the fact that the value L is reached at some date t, the call is said to be "up and in" or "down and in".

It appears immediately that:

- The price of a barrier call at time 0 is smaller than the standard call with the same strike and maturity (since it pays zero more often).
- Hence, a barrier option is an inexpensive hedge against extreme events. An "up and in" option will be activated only if an insurance index, a foreign currency or the spot price of electricity reaches a very high level (defined as the barrier in the option contract).

A vast body of literature has been recently dedicated to barrier options. The pricing of "single barrier" options is not extremely difficult in the classical Black-Scholes-Merton framework and explicit formulas have existed for some time (see [11, 9]). For instance, if the barrier L is less than or equal to the strike price, the value at date 0 of a down-and-in call written on a non-dividend paying stock is

$$C(0) = S(0)\left(\frac{L}{S(0)}\right)^{2\alpha} N(d_1) - ke^{-rT}\left(\frac{L}{S(0)}\right)^{2\alpha} N(d_2),$$

where $\alpha = \frac{r+\frac{\sigma^2}{2}}{\sigma^2}$, d_1 and d_2 have the same expressions as in Formula (6).

Double-barrier options yield the Black-Scholes payout $\max(0, S_T - k)$ at maturity under the condition that S_t has not touched either of two barriers, denoted

L (lower barrier) and U (upper barrier), during the lifetime of the option $[0, T]$; they are then called corridor options and allow traders or investors *to take a view* on the range of a stock, a commodity or an index price. Another type of double barrier option is activated by the fact that one (at least) of the two barriers has been crossed by the trajectory of S_t between 0 and T. All types of double barrier options are more difficult to price than simple barrier options. In both cases, Monte Carlo methods may be dangerous: The discrete trajectories which are simulated may misrepresent the continuous possibility of (des)activation of the barrier option over the interval $[0, T]$.

Geman and Yor [8] provide the Laplace transforms of the exact prices of double barrier options and of their deltas, gammas, and other sensitivities as well .

c) Options on exotic underlyings

Over the last six years, a variety of non financial assets have become the source of risk on which standard and path-dependent options are written. Insurance derivatives were launched in December 1993 by the Chicago Board of Trade and are traded today both on exchanges and in OTC markets. An index, precluding manipulations and opacity, needed to be constructed for its value at maturity T to determine the option payout; this index is currently the PCS (Property Claim Services) index. Since, it is defined as an aggregate amount of losses, i.e., an average (up to constant), the insurance derivatives give rise to Asian-type problems. At the same time, the nature of the underlying instrument — an index which spikes in the case of an earthquake or a hurricane — requires the presence of a jump process in its modeling.

More recently, with all the credit events which shook the financial markets, from the Asian crisis to the Russian turmoil, credit derivatives developed as hedging instruments not only for defaultable bonds but also swaps and transactions with counterparties bearing default, i.e., all OTC transactions. Again, credit not being a traded instrument, a precise definition of the option contract and of the *insured risk* is necessary to avoid possible litigation later on.

Lastly, with deregulation of electricity on its way worldwide and in the United States in particular, weather derivatives have been introduced to allow utilities to hedge their electricity risk since, in the U.S., spikes in electricity price coincide very much with extreme weather conditions. Very quickly, these weather derivatives have appeared as a great instrument to farmers, construction companies or even clothe manufacturers. Also weather derivatives raise two types of difficulties: The first one is technical, since the pay-off of the option is related to an average of temperature spreads (65° Fahrenheit representing the "ideal" temperature), hence again Asian-type. A much bigger difficulty however is of an economic nature and resides in the fact that the seller of a weather derivative cannot delta-hedge since of his position this procedure would involve *owning* the underlying source of risk, i.e., the temperature index.

Hence, the pricing and hedging of weather derivatives require new answers, much closer to actuarial and insurance techniques than to the *dynamic hedging* of options written on traded assets.

References

1. Black, F. and Scholes, M.: The pricing of options and corporate liabilities, *J. Political Economy* **81** (1973), 637–654.
2. Clark, P.: A subordinated stochastic process with finite variance for speculative prices, *Econometrica* **41**, no. 1 (1973), 135–155.
3. Fu, M., Madan, D., and Wang, T.: Pricing continuous time Asian options: A comparison of Monte Carlo and Laplace transform inversion methods, *J. Computational Finance* **2** (1999), 49–74.
4. Geman, H.: Learning about risk: Some lessons from insurance, *European Finance Review* **2** (1998), 113–124.
5. Geman, H. and Ané, T.: Stochastic subordination, *Risk* September (1996), 146–149.
6. Geman, H. and Eydeland, A.: Domino effect: Inverting the Laplace transform, *Risk* April (1995), 64–67.
7. Geman, H. and Yor, M.: Bessel processes, Asian options and perpetuities, *Math. Finance* **3**, no. 4 (1993), 349–375.
8. Geman, H. and Yor, M.: Pricing and hedging path-dependent options: A probabilistic approach, *Math. Finance* **6**, no. 4 (1996), 365–378.
9. Goldman, M., Sosin, H., and Gatto, M.: Path dependent options: Buy at the low, sell at the high, *J. Finance* **34** (1979), 111–127.
10. Kunitomo, N. and Ikeda, N.: Pricing options with curved boundaries, *Math. Finance* **2**, no. 4 (1992), 275–298.
11. Merton, R. C.: Theory of rational option pricing, *Bell. J. Econ. Manag. Sci.* **4** (1973), 141–183.

Quantization, deformations, and symmetries

On the Bohr-Rosenfeld measurability problem for non-linear systems

Yakir Aharonov
School of Physics and Astronomy, Tel Aviv University, Tel Aviv 69978, Israel
and
Department of Physics, University of South Carolina, Columbia, SC 29208 USA

Benni Reznik
School of Physics and Astronomy, Tel Aviv University, Tel Aviv 69978, Israel

Abstract. The singular nature of fields at a space-time point, forces us to consider space-time weighted (smeared) field distributions as elementary observables of field theory. As a consequence, measurement theory for fields, must handle the non-trivial task of eliminating the back-reaction disturbance caused by the measurement process itself. For the case of a free field, Bohr and Rosenfeld showed that an additional spring-like mechanism can compensates this disturbance. In this article we attempt to extend their results to non-linear systems. We construct a perturbative extension of the Bohr-Rosenfeld compensation procedure. We show however that this method breaks down for sufficiently accurate measurements and leads to a lower bound on the precision of the measurement. The failure of the perturbative approach raises again the question of measurability for non-linear systems.

Keywords: quantum measurements, foundations of quantum theory

Mathematics Subject Classifications (2000): 81P15, 81B10

It is indeed a very sad occasion when we realize that Moshé is not with us anymore. I first heard about Moshé many years ago from my neighbor when I was in high school. It turned out that this neighbor was adopted by Moshé's parents when she came as an orphan to Israel. She was full of praise for their generosity and she told me that they had a son Moshé who was a mathematical genius, and that I should try to meet with him. That meeting took place years later in New York. I was immediately immensely impressed by his enthusiasm and his brilliant ideas. Since then we met many times and became close friends. It is still difficult for me to believe that we will never meet again. His untimely death is a great loss to all of us.

Y. Aharonov

1. Introduction

The measurement of an observable **A** is ideally described by an impulsive coupling between the system and the measuring device at a certain instant of time [1]. However in realistic measurements the duration of the coupling between the systems and the measuring device is finite. In this case the interaction with the measuring device modifies the evolution of the system and consequently instead of observing

G. Dito and D. Sternheimer (eds.), Conférence Moshé Flato 1999, Vol. 1, 99 – 108.

the undisturbed time average

$$\bar{\mathbf{A}} = T^{-1}\int_0^T \mathbf{A}(t')dt' \tag{1}$$

we observe a disturbed time average which also depends on the initial state of the measuring device.

This problem becomes particularly acute in the context of quantum field theory. The singular nature of fields at a point forces us to consider instead space-time weighted (smeared) fields as elementary observables of quantum field theory [2]. Therefore, in this case we have to deal on a fundamental level with the problem of measuring space-time averaged observables. Measurement theory for fields, must hence handle the non-trivial task of eliminating the back-reaction disturbance caused by the measurement process itself. Indeed Bohr and Rosenfeld showed that in addition to the usual measurement interaction, one can introduce a spring-like mechanism which compensates this disturbance and allows a precise observation of the undisturbed space-time average [3] . However their work was restricted to the special case of a non-interacting linear system.

In this article we will attempt to extend their results to non-linear systems. As a toy model we shall consider the simpler case of non-relativistic non-linear harmonic oscillator. This model simplifies the problem because we do not have to deal with spatial averages, but nevertheless it still contains the essential features of the full problem. We then consider problem of measuring a time averaged observable by means of a perturbative approach in the non-linear coupling constant. As we will show the validity of a perturbative method breaks down for sufficient accurate measurement and gives rise to a lower bound on the precision of the measurement.

2. Non-linear harmonic oscillator as a toy problem

We shall consider as a toy model for a non-linear system a harmonic oscillator with a potential given by

$$H = \frac{1}{2}(\mathbf{p}^2 + \Omega^2\mathbf{x}^2) - \frac{\lambda}{n}\mathbf{x}^n, \tag{2}$$

where $n > 2$.

Our first aim will be to describe a measurement of an averaged observable such as

$$\bar{\mathbf{x}} = \frac{1}{T}\int_0^T \mathbf{x}(t')dt'. \tag{3}$$

In the limit of $T \to 0$ the prescription is know: The appropriate interaction term is in this case

$$H_I = -g(t)\mathbf{Q}\mathbf{x}, \tag{4}$$

where $\mathbf{Q}$ is canonically conjugate to the "output variable" $\mathbf{P}$, and $g(t) = g_0\delta(t)$. It is assumed that the effective mass of the device is infinitely large, thus the kinetic part of the measuring device Hamiltonian vanishes and $\mathbf{Q}$ is a constant of motion. The interaction Hamiltonian (4) yields:

$$\delta \mathbf{P} \equiv \mathbf{P}(+\varepsilon) - \mathbf{P}(-\varepsilon) = g_0 \mathbf{x}(0). \tag{5}$$

As a starting point let us modify this interaction by replacing the Dirac delta function by a smooth function $g(t)$ with a compact support in the time interval $0 < t < T$.

The solution to the equations of motion

$$\begin{aligned} \dot{\mathbf{x}} &= \mathbf{p} \\ \dot{\mathbf{p}} &= -\Omega^2 \mathbf{x} + \lambda \mathbf{x}^{n-1} + g(t)\mathbf{Q} \\ \dot{\mathbf{P}} &= g(t)\mathbf{x} \end{aligned} \tag{6}$$

can be written in the integral form:

$$\mathbf{x}(t,g) = \mathbf{x}_0(t) + \int_0^t \frac{F(t')}{\Omega} \sin[\Omega(t-t')]dt' \tag{7}$$

$$\mathbf{p}(t,g) = \mathbf{p}_0(t) + \int_0^t F(t') \cos[\Omega(t-t')]dt', \tag{8}$$

where

$$F(t) = \lambda \mathbf{x}^{n-1}(t,g) + g(t)\mathbf{Q}, \tag{9}$$

and $\mathbf{x}_0(t)$, $\mathbf{p}_0(t)$ are free solutions ($\lambda = g = 0$), which coincide with the non-linear solution $\mathbf{x}(t)$ at $t = 0$. The presence of the coupling parameter g in $\mathbf{x}(t,g)$ will be used in the following to denote an explicit dependence of the solution on variables of the measuring device.

The solutions $\mathbf{x}(t,g)$ and $\mathbf{p}(t,g)$ depend on the back-reaction via the term $g(t)\mathbf{Q}$. Contrary to classical mechanics, due to the uncertainty principle, we cannot make $\mathbf{Q}$ as small as we wish and still obtain an accurate measurement, i.e., $\Delta\mathbf{P}(0) \to 0$. Consequently, we finally observe

$$\mathbf{P}(T) - \mathbf{P}(0) = \int_0^T g(t)\mathbf{x}(t,g)dt, \tag{10}$$

which is not the undisturbed (g=0) value of $\bar{\mathbf{x}}$. Only in the limiting case $T = 0$ does the error vanish.

In the next section we show how the undisturbed $\bar{\mathbf{x}}$ can be observed in the linear case: $\lambda = 0$.

3. The linear case

We shall now proceed to examine the case of a measurement of an averaged observable and show how to eliminate the back reaction in this special case.

For simplicity we shall choose the weight function as $g(t) = g_0$ for $t \in (0,T)$ and zero otherwise. In this case, the solution of the equations of motion when $\lambda = 0$ is

$$\begin{aligned} \mathbf{x}(t) &= \mathbf{x}_0(t) + \frac{g_0(1-\cos\Omega t)}{\Omega^2}\mathbf{Q} \\ &\equiv \mathbf{x}_0(t) + \alpha(t)\mathbf{Q}. \end{aligned} \tag{11}$$

Thus, the $\mathbf{P}$ coordinate of the measuring device will be shifted by

$$\begin{aligned} \delta\mathbf{P} &= \mathbf{P}(T) - \mathbf{P}(0) \\ &= g_0\int_0^T \mathbf{x}(t')dt' = g_0\int_0^T \mathbf{x}_0(t')dt' + \mathbf{Q}\frac{g_0^2 T}{\Omega^2}\left(1 - \frac{\sin\Omega T}{\Omega T}\right). \end{aligned} \tag{12}$$

The undisturbed average $\bar{\mathbf{x}} = \bar{\mathbf{x}}_0 = T^{-1}\int_0^T \mathbf{x}_0(t')dt'$ is given by

$$\bar{\mathbf{x}}_0 = \frac{\delta\mathbf{P}}{g_0 T} - \frac{g_0}{\Omega^2}\left(1 - \frac{\sin\Omega T}{\Omega T}\right)\mathbf{Q}. \tag{13}$$

Since $\bar{\mathbf{x}}_0$ depends linearly on both $\mathbf{P}$ and $\mathbf{Q}$, a precise measurement of $\mathbf{P}$ causes a larger uncertainty in the second term above, i.e., it increases the back reaction of the measuring device on the oscillator. Since the uncertainty in $\bar{\mathbf{x}}$ is

$$\Delta\bar{\mathbf{x}}_0 \approx \frac{\Delta\mathbf{P}}{g_0 T} + \frac{g_0}{\omega^2\Delta\mathbf{P}}\left(1 - \frac{\sin\Omega T}{\Omega T}\right), \tag{14}$$

the minimal uncertainty is

$$\Delta\bar{x}_{min} = \frac{2}{\Omega}\sqrt{1 - \frac{\sin\Omega T}{\Omega T}} \overset{T\to 0}{\to} \sqrt{\frac{2}{3}}T, \tag{15}$$

which vanishes only in the impulsive limit. We also note that in the limit of $T \to \infty$ the disturbance does not average out but rather approaches the constant $\frac{2}{\Omega}$.

The direct approach therefore fails to measure $\bar{\mathbf{x}}_0$ precisely. There are however ways to correct or eliminate the error above, and in the following we present three different ways to achieve this goal.

The idea of Bohr and Rosenfeld was to correct the error by adding to the Hamiltonian (4) a new "compensating" term. The error in the shift of $\mathbf{P}$ in (12) appears as linear in $\mathbf{Q}$, very much like the effect of a spring in the $\mathbf{Q}$-coordinates. Since the coefficient which multiplies $\mathbf{Q}$ in (12) is known, it is straightforward to compensate for this error by adding to the interaction Hamiltonian (4) a spring term:

$$H_I = -g(t)\mathbf{Q}\mathbf{x} + \frac{1}{2}k\mathbf{Q}^2, \tag{16}$$

where k can be chosen for example as

$$k(t) = \frac{g^2(t)T}{\Omega^2}\left(1 - \frac{\sin\Omega T}{\Omega T}\right). \tag{17}$$

It is straightforward to see that the new spring-term in the equations of motion precisely eliminates the back-reaction of the device on the system. In the limit $T \to 0$, the compensation is of course not needed since the spring constant vanishes, and we obtain back the ordinary impulsive measurement.

Another approach due to Unruh [4], does not require modification of the interaction (4) but requires instead an additional measuring device. Inspecting Equation (13), we notice that $\bar{\mathbf{x}}_0$ is given on the right-hand side in terms of a linear combination of $\mathbf{P}$ and $\mathbf{Q}$. Therefore, after preparing the measuring device in an initial state with a well-defined $\mathbf{P}$ one can measure by means of *another* measuring device the linear combination

$$\frac{1}{g_0 T}\mathbf{P} - \frac{k(T)}{g_0}\mathbf{Q}(T), \tag{18}$$

where k is given in (17). As before this method requires that the back-reaction effect on the system depends only on variables of the measuring device. As we shall see in Section 4, both methods apparently fail when non-linearities introduce back-reaction terms which depend also on the system itself.

A third different method to measure $\bar{\mathbf{x}}$ could be the following. By integrating $\mathbf{x}(t)$ we can express $\bar{\mathbf{x}}$ as

$$\frac{1}{T}\int_0^T (\mathbf{x}_0 \cos\Omega t' + \frac{1}{\Omega}\mathbf{p}_0 \sin\Omega t')dt' = \frac{\mathbf{x}_0}{\Omega T}\sin\Omega T + \frac{\mathbf{p}_0}{\Omega^2 T}(1 - \cos\Omega T). \tag{19}$$

Note that $\mathbf{x}_0$ and $\mathbf{p}_0$ are constants of motion. By inverting the solutions for the equations of motion for $\mathbf{x} = \mathbf{x}(\mathbf{x}_0, \mathbf{p}_0)$ and $\mathbf{p} = \mathbf{p}(\mathbf{x}_0, \mathbf{p}_0)$ we obtain

$$T\bar{\mathbf{x}}(t) = \frac{2}{\Omega}\sin\Omega T/2\left[\mathbf{x}(t)\cos\Omega(T/2 - t) + \frac{1}{\Omega}\mathbf{p}(t)\sin\Omega(T/2 - t)\right]. \tag{20}$$

Thus, we have expressed $\bar{\mathbf{x}}$ in terms of a (time dependent) constant of motion. We can now set up a continuous measurement

$$H_I = -g(t)\mathbf{Q}\bar{\mathbf{x}}(t). \tag{21}$$

Clearly the observable $\bar{\mathbf{x}}(t)$ remains a constant of motion and is not disturbed by the interaction. We have thus managed to measure $\bar{\mathbf{x}}$ by a continuous measurement of a constant of motion whose value is identical to $\bar{\mathbf{x}}$ for any t.

4. Non-linear case

The presence of non-linear terms makes the back-reaction dependent on both system and the measuring device. Let us now show that the methods of Section 3, which are directly based the idea of constructing a compensation are likely to fail. Let us assume again the exact solution is given, and let us express it as

$$\mathbf{x}(t,g) = \xi(t,\lambda,g=0) + \Delta\xi(t,\lambda,g). \tag{22}$$

Here, $\Delta\xi(t,\lambda,g) = \xi(t,\lambda,g) - \xi(t,\lambda,g=0) = x(t,g) - x(t,g=0)$ is the non-linear error due to the measurement. Thus, using the naive coupling (4) we obtain:

$$\frac{\delta \mathbf{P}}{g_0 T} = \bar{\mathbf{x}} + \frac{1}{g_0 T}\int_0^T \Delta\xi(t,\lambda,g)dt'. \tag{23}$$

To proceed, we would like to construct a compensation to the last term. Thus, $\Delta\xi$ must be expressed in terms of the dynamical variables $\mathbf{x}(t,g)$, $\mathbf{p}(t,g)$ and $\mathbf{Q}$. However, $\Delta\xi$ depends on $\mathbf{x}(t,g=0)$ and we first need to find the non-linear relation between the disturbed non-linear $\mathbf{x}(t,g)$ and the undisturbed $\mathbf{x}(t,g=0)$ solutions:

$$\mathbf{x}(t,g) = f[\mathbf{x}(t,0),\mathbf{p}(t,0),\mathbf{Q}], \tag{24}$$

etc. The latter equation corresponds to a "canonical" transformation between two non-linear solutions of two different systems. It is not clear if such a well-defined relation indeed exist. Assuming however that the transformation above does exist, we may now attempt, as before, to add to the Hamiltonian the compensation term $\int[k\mathbf{Q} + \Delta\xi(\lambda,g)]dq$. However, the new equations of motion will give rise to another non-linear error $\Delta\xi' \neq \Delta\xi$. Therefore, a self-consistent scheme for constructing a compensation must be found, which deals with the non-perturbative effects of the compensation on the system.

The difficulties in finding an exact non-perturbative compensation scheme, leads us to examine a more modest approach. It can be hoped that at least for small averaging time T, the effect of the non-linearities is small. Therefore, in the next section, we shall examine a perturbative approach.

5. A perturbative approach

For a small non-linear constant, λ, we shall attempt a perturbative approach. As a starting point we shall assume that the measurement can be expressed by:

$$H_I = -g(t)\mathbf{Q}\mathbf{x} + H_C, \tag{25}$$

i.e., as a sum of the naive measurement and a compensation term. Clearly in the limit of $T \to 0$, H_I should reduce to an ordinary impulsive measurement, i.e., $H_C(T=0) = 0$. Our aim will be to construct a self-consistent procedure to compute the compensation term H_C to any order in λ:

$$H_C = \Delta H^{(0)} + \sum_k \lambda^k \Delta H^{(k)}. \tag{26}$$

The terms $\Delta H^{(k)}$ are the analogous compensations to the back-reaction up to the k'th order in λ.

To this end, let us expand the solution for $\mathbf{x}(t)$ as

$$\mathbf{x}(t,g) = \mathbf{x}^{(0)} + \lambda \mathbf{x}^{(1)}(t) + \lambda^2 \mathbf{x}^{(2)}(t) + \cdots \tag{27}$$

and at each order separate the undisturbed solution, denoted as $\mathbf{x}_0^{(k)}$, from the non-linear error, denoted as $\Delta x^{(k)}$:

$$\mathbf{x}^{(k)}(g) = \mathbf{x}_0^{(k)}(t) + \Delta\mathbf{x}^{(k)}(t,g), \tag{28}$$

i.e., for $g = 0$ we have $\Delta\mathbf{x}^{(k)}(t, g = 0) = 0$ and $\mathbf{x}^{(k)}(g = 0) = \mathbf{x}_0^{(k)}(t)$ is the solution without the coupling to a measuring device.

Likewise, the shift of the output register, $\delta\mathbf{P} = \mathbf{P}(T) - \mathbf{P}(0)$, will be expanded in powers of λ, and at each order we shall evaluate the terms of the decomposition:

$$\delta\mathbf{P}^{(k)} = \delta\mathbf{P}_0^{(k)} + \Delta\mathbf{P}^{(k)}. \tag{29}$$

Here, $\Delta\mathbf{P}^{(k)}$ is the error due to the back reaction to the k'th order. The knowledge of $\Delta\mathbf{P}^{(k)}$ will allow us to construct the appropriate compensation $\Delta H^{(k)}$, up to the same order k.

To proceed we shall now simplify the problem, and let $n = 3$ in (2). It can be shown however that the following will be valid for any general non-linear potential.

To evaluate $\Delta\mathbf{P}^{(k)}$ we first use (7) to obtain an integral solution to any order:

$$\mathbf{x}^{(0)}(t,g) = \mathbf{x}_0(t) + \mathbf{Q}\int_0^t g(t')D(t-t')dt', \tag{30}$$

$$\mathbf{x}^{(1)}(t,g) = \int_0^t \mathbf{x}^{(0)}(t')\mathbf{x}^{(0)}(t')D(t-t')dt', \tag{31}$$

$$\mathbf{x}^{(2)}(t,g) = 2\int_0^t \mathbf{x}^{(0)}(t')\mathbf{x}^{(1)}(t')D(t-t')dt', \tag{32}$$

etc., where $\mathbf{x}_0$ is the free solution ($g = \lambda = 0$), and

$$D(t) = \frac{\sin\Omega t}{\Omega}. \tag{33}$$

The result for the zero'th order was already obtained above, where we found

$$\Delta H_0 = \frac{1}{2}k\mathbf{Q}^2, \tag{34}$$

where k is the c-number given by (17). Here, and in the following we shall ignore any potential problems with ordering of non-commuting operators.

To the first order in λ we obtain

$$\delta\mathbf{P}^{(1)} = g_0\int_0^T dt\int_0^t [\mathbf{x}^{(0)}]^2 dt'. \tag{35}$$

Thus,

$$
\begin{aligned}
\Delta\mathbf{P}^{(1)} &= g_0 \int_0^T dt \int_0^t \left[2\alpha(t')\mathbf{Q}\mathbf{x}_0(t') + \alpha^2(t')\mathbf{Q}^2\right] dt' \\
&= g_0 \int_0^\infty \left[2\alpha(t')\mathbf{Q}\mathbf{x}_0(t') + \alpha^2(t')\mathbf{Q}^2\right] dt' \int_0^T \theta(t-t')dt \\
&= g_0 \int_0^\infty \Big((T-t')\theta(T-t') - (-t')\theta(-t')\Big) \left[2\alpha(t')\mathbf{Q}\mathbf{x}_0(t') + \alpha^2(t')\mathbf{Q}^2\right] dt' \\
&= g_0 \int_0^T (T-t') \left[2\alpha(t')\mathbf{Q}\Big(\mathbf{x}^{(0)}(t') - \mathbf{Q}\alpha(t')\Big) + \alpha^2(t')\mathbf{Q}^2\right] dt' \\
&= g_0 \int_0^T (T-t') \left[2\alpha(t')\mathbf{Q}\mathbf{x}^{(0)}(t')dt' - \alpha^2(t)(T)\mathbf{Q}^2\right] dt'. \qquad (36)
\end{aligned}
$$

where $\alpha(t)$ is defined in (11). As could have been anticipated, the new feature of the error terms up to the first order in λ, is their depends on the variable $\mathbf{x}$ of the system. The last term does not depend on system variables and therefore may be trivially compensated as a spring-like compensation. Alternatively, as discussed in Section 3, we may use a second measuring device to measure the combination $\frac{1}{g_0 T}\mathbf{P} - k\mathbf{Q} - k'(T)\mathbf{Q}^2$, where k' can be found by integrating the last term in (36).

Equation (36) suggests adding as a compensation the term

$$
\Delta H^{(1)} = g^2(t)(T-t)\left[\alpha(t)\mathbf{Q}^2\mathbf{x}(t) - \frac{1}{3}\alpha^2(t)\mathbf{Q}^3\right]. \qquad (37)
$$

With the compensation $\Delta H^{(0)} + \lambda\Delta H^{(1)}$ added to the Hamiltonian, the equations of motion become

$$
\dot{\mathbf{P}} = g(t)\mathbf{x} - k(t)\mathbf{Q} + \lambda g^2(t)(T-t)\left[2\alpha(t)\mathbf{Q}\mathbf{x}(t) - \alpha^2(t)\mathbf{Q}^2\right], \qquad (38)
$$

$$
\frac{d^2\mathbf{x}}{dt^2} + \Omega^2\mathbf{x} = \lambda\mathbf{x}^2 + g(t)\mathbf{Q} + \lambda g^2(t)(T-t)\alpha(t)\mathbf{Q}^2. \qquad (39)
$$

The new term on the right-hand side of (39) modifies the solution for $x(t)$ to

$$
\mathbf{x}(t) = \mathbf{x}_0(t) + \alpha(t)\mathbf{Q} + \lambda\beta(t)\mathbf{Q}^2 + \lambda\int_0^T \mathbf{x}^2(t')dt'. \qquad (40)
$$

The first order is modified by the $\beta(t)\mathbf{Q}^2$ which can trivially be compensated by adding also the term $\frac{1}{3}\lambda g(t)\beta(t)\mathbf{Q}^3$. This completes the compensation to the first order. It is important to note that the a first order compensation cannot modify lower order compensations but only higher orders. This allows us to go to higher orders in λ without effecting the corrections found at lower orders.

Let us see how one can proceed to higher orders. To the second order we obtain:

$$
\frac{d\Delta\mathbf{P}^{(2)}}{dt} = g(t)\mathbf{Q}\Delta\mathbf{x}^{(2)} + 2g^2(t)(T-t)\alpha(t)\mathbf{Q}\Delta\mathbf{x}^{(1)}(t), \qquad (41)
$$

where

$$\Delta \mathbf{x}^{(1)}(t) = \beta(t)\mathbf{Q}^2 + \int_0^t \Delta[\mathbf{x}^{(0)}(t')]^2 dt', \tag{42}$$

and

$$\Delta \mathbf{x}^{(2)}(t) = 2\int_0^t \Delta[\mathbf{x}^{(0)}(t')\mathbf{x}^{(1)}(t')]dt' \tag{43}$$

where the notation $\Delta[\cdots]$, on the right-hand side in the equations above, indicates that only $\mathbf{Q}$-dependent terms of $[\cdots]$ are included.

The computation can be carried out as before. The only technical subtlety is that now $\Delta\mathbf{P}^{(2)}$ contain also triple integrals over time like:

$$\int dt \int x_0(t')dt' \int p_f(t'')dt''. \tag{44}$$

To proceed one has to reduce this integral to a single integration over time, or eliminate the integrals completely. Otherwise, the compensations would have to be non-local in time. In our case of a first quantized theory, we can always use the property of linear operators: $\mathbf{p}_0(t)$ (or $\mathbf{x}_0(t)$) can always be expressed as a linear combination of $\mathbf{x}_0(t=0)$ and $\mathbf{p}_0(t=0)$. Therefore, we can integrate over free operators and reduce: $\int_0^{t'} \mathbf{p}_0(t'')dt'' = c_1(t')\mathbf{x}_0(t') + c_2(t')\mathbf{p}_0(t')$. The integral above can be hence reduced to a single integral of the form $\int (c_1'(t)\mathbf{x}_0^2 + c_2'(t)\mathbf{x}_0\mathbf{p}_0)dt$.

Finally we find that the compensation to the second order has the following form:

$$\Delta H^{(2)} = \mathbf{Q}^2(\gamma_1\{\mathbf{x},\mathbf{p}\} + \gamma_2\mathbf{x}^2) + \mathbf{Q}^3(\gamma_3\mathbf{x} + \gamma_4\mathbf{p}) + \gamma_5\mathbf{Q}^4. \tag{45}$$

where $\gamma_i(T)$ are generally time dependent c-numbers. In principle it seems that this procedure can be carried out to any order in λ.

6. Break-down of the perturbative approach

Although by using a perturbative approach we can formally evaluate self-consistently compensation terms to any order in the nonlinear coupling constant, this approach still does not allow precise measurements. Unlike classical measurements, in the limit of a precise measurement, the variable $\mathbf{Q}$, which is conjugate to the "output register" $\mathbf{P}$, becomes completely uncertain. Since the expansion for the compensation H_C generally depends on $\mathbf{Q}$, for a given T, the validity of the perturbative expansion must break-down beyond a certain precision $\Delta\mathbf{x}_{min}$.

To show this consider the first order compensation term. From (37) we have $\Delta H^{(1)} \approx T\alpha(T)\mathbf{Q}^2\mathbf{x}$. A necessary (but not sufficient) condition for a well-behaved expansion is therefore:

$$\lambda T\alpha(T)\mathbf{Q}^2\mathbf{x} \sim \lambda T^3\mathbf{Q}^2\mathbf{x} << 1. \tag{46}$$

Substituting a rough approximation $\mathbf{Q}^2 \sim (\Delta\mathbf{Q})^2$ and $\mathbf{x} \sim \Delta x$ and using $\Delta\mathbf{P} \sim \Delta\mathbf{x}/g_0 T$ we obtain the condition

$$\Delta\mathbf{x} > \Delta\mathbf{x}_{min} \approx \lambda g_0^2 T^5. \tag{47}$$

Only by letting $T \to 0$ or $\lambda \to 0$ we regain $\Delta \mathbf{x}_{min} \to 0$.

7. Discussion

In this article we have examined the measurability of observables by means of a continuous measurement for linear and non-linear theories.

We found that a perturbative approach for constructing compensations breaks down for sufficiently accurate measurements. The reason for this failure is that the disturbance at each order generally depends on increasing powers of $\mathbf{Q}$, (the conjugate to the output variable $\mathbf{P}$). For a sufficiently precise measurements, $\Delta\mathbf{Q} \sim 1/\Delta\mathbf{P}$ becomes arbitrarily large. Hence, for any λ, and for any given interaction time T, there exists a sufficiently large $\Delta\mathbf{Q}$ which invalidates the perturbative expansion. Therefore our method gives rise to a lower bound on the accuracy.

Acknowledgements. We thank W. G. Unruh for valuable discussions and L. Vaidman for helpful remarks. We acknowledge the support from grant 471/98, of the Israel Science Foundation, established by the Israel Academy of Sciences and Humanities. The work of Y. A. was also supported by NSF grant PHY-9971005.

References

1. Von Neumann, J.: *Mathematical Foundations of Quantum Theory*, Princeton University Press, Princeton NJ, 1996.
2. Haag, R.: *Local Quantum Physics. Fields, Particles, Algebras*, Second edition, Springer-Verlag, Berlin, 1996.
3. Bohr, N. and Rosenfeld, L.: On the Question of the Measurability of Electromagnetic Field Quantities, in: J. A. Wheeler and W. H. Zurek, (eds.), *Quantum Theory and Measurement*, Princeton University Press, Princeton NJ, 1983, pp. 479–522.
4. Unruh, W. G.: private communications.

Star-produits et star-algèbres holomorphes

Louis Boutet de Monvel
Analyse Algébrique, Institut de Mathématiques de Jussieu
Université Pierre et Marie Curie, 4 place Jussieu, 75252 Paris cedex 05, France
e-mail: `boutet@math.jussieu.fr`

Abstract. We describe a slightly enlarged definition of star-products and star-algebras which is better adapted to holomorphic objects and contains the calculus of pseudo-differential operators and Toeplitz operators. Using this we give a classification of holomorphic star-algebras on the cotangent bundle of a complex manifold equipped with its canonical Poisson bracket (i.e., locally isomorphic to the algebra of pseudo-differential operators). It turns out that in dimension ≥ 3 these are essentially trivial and correspond to algebras of twisted differential operators on X; in dimension 1 and 2 there are many more — exotic or monstruous in dimension 2 (i.e., having globally not much to do with algebras of twisted differential operators); more tractable but still not trivial in dimension 1. We describe in more detail the case of the projective line. Further details can be found in [3].

Résumé. Dans cet exposé nous rappelons la définition des star-produits et des star-algèbres. Notre définition est faisceautique pour s'adapter mieux aux objets holomorphes, et légèrement élargie (cf. §2) : les star-produits vivent sur des "cônes" de façon à inclure le cas important des opérateurs pseudo-différentiels et des opérateurs de Toeplitz. Ces questions sont développées plus en détail dans [3].

Keywords: star-product, holomorphy, deformation

Mathematics Subject Classifications (2000): 16S32, 16S80, 32L05, 58J40

Cet exposé est dedié à la mémoire de mon ami Moshé Flato

1. Introduction

Commençons par rappeler la définition d'un star-produit : soit X une variété réelle ou complexe. Un star-produit semi-classique sur X, ou déformation formelle du produit usuel, est une loi d'algèbre associative et unitaire $(f,g) \to B(f,g)$ sur l'ensemble des séries formelles $f = \sum f_n(x)h^n$ à coefficients fonctions sur X, où $B = \sum h^n B_n$ est un opérateur bidifférentiel formel : dans un système de coordonnées locales sur X on a $B_n(f,g) = \sum b_{\alpha\beta}(x)\partial^\alpha f \partial^\beta g$. Unitaire signifie qu'on a $B(1,f) = B(f,1) = f$ pour toute f; l'associativité signifie qu'on a $B(f,B(g,h)) = B(B(f,g),h)$ pour toutes f,g,h. Plus loin (§2) nous utiliserons une définition un peu élargie, où h ne commute plus avec le reste, de façon à inclure dans la définition le calcul symbolique des opérateurs pseudo-différentiels ou de Toeplitz (cf. [5, 6]).

Si B est un star-produit, on lui associe un crochet de Poisson : $\{f,g\} = B_1(f,g) - B_1(g,f)$, terme dominant de la loi des commutateurs est un crochet de Poisson sur X (voir §2). Un problème de la théorie est de classifier les star-produits correspondant à un crochet de Poisson donné, et en particulier de montrer qu'il

G. Dito and D. Sternheimer (eds.), Conférence Moshé Flato 1999, Vol. 1, 109 – 120.

en existe. Lorsque X est réelle, ce problème a été résolu par M. De Wilde et P. Lecomte [10, 11] dans le cas symplectique, et par M. Kontsevich [19] dans le cas général. Kontsevich montre plus précisément qu'il y a une correspondance biunivoque entre les classes d'équivalence de star-produits sur X et les classes d'équivalence de familles formelles de crochets de Poisson $C = \sum h^n C_n$ sur X.

Les démonstrations dans le cas réel ne marchent pas sur une variété holomorphe, où on ne dispose pas de partitions de l'unité ou de voisinages tubulaires, et je ne sais d'ailleurs pas si un star produit global, associé à un crochet de Poisson donné, existe toujours, même dans le cas symplectique.[1]

Je crois néanmoins important de faire cette étude dans le cas complexe, parce que le calcul symbolique des opérateurs différentiels vit toujours sur une variété complexe (les symboles sont des polynômes), de sorte que les problèmes qui les concernent conduisent naturellement à des problèmes de star-produits holomorphes. De même il est important d'étudier les faisceaux de modules sur ces algèbres. Un autre problème important et de même nature est de comprendre le lien entre les faisceaux de modules sur l'algèbre des opérateurs pseudo-différentiels fournis par des constructions microlocales sur le fibré cotangent $T^*Y - \{0\}$ d'une variété Y privé de sa section nulle, et les $\mathscr{D}_Y$-modules : ce problème n'a de sens que sur un fibré cotangent complexe.

Ici nous nous limitons à l'étude et la classification des star-algèbres sur un fibré cotangent muni de sa structure symplectique standard, avec notre définition élargie, i.e. des algèbres localement isomorphes à la star-algèbre des opérateurs pseudo-différentiels. Il est aussi naturel de comparer ces star-algèbres aux algèbres d'opérateurs différentiels tordus (au sens de [2]).

Dans ce cas, la question d'existence ne se pose pas puisqu'on dispose de l'algèbre des opérateurs pseudo-différentiels, à laquelle toute star-algèbre du type étudié est localement isomorphe, et on peut faire une étude assez complète de la classification. Celle-ci n'apporte rien de nouveau en dimension ≥ 3, où les star-algèbres correspondent exactement aux algèbres d'opérateurs différentiels tordus. En dimension 2 apparaissent des objets nouveaux et monstrueux ou exotiques. En dimension 1 la classification est de nouveau raisonable mais diffère de façon significative de celle des algèbres d'opérateurs différentiels tordus.

Cette classification est décrite en détail dans [3], et je me contenterai ici d'en montrer les traits les plus saillants. Il est intéressant de noter que cette classification est essentiellement triviale en dimension ≥ 3, et beaucoup moins en dimension 1 ou 2.

[1] Néanmoins M. Kashiwara dans le cas symplectique ([18]) et M. Kontsevich en général (communication orale) montrent qu'on peut toujours construire une gerbe équivalente à la catégorie des faisceaux des modules sur une telle algèbre.

2. Star-algèbres sur un cône

2.1. Cônes et symboles

Définition 1 *On appelle cône réel (resp. complexe) un fibré principal* C^∞ *(resp. holomorphe)* Σ *de groupe* $\mathbf{R}_+^\times$ *(resp.* $\mathbf{C}^\times$*). La base du cône est* $B\Sigma = \Sigma/\mathbf{R}_+^\times$ *(resp.* $\Sigma/\mathbf{C}^\times$*).* [2]

Un cône réel est isomorphe à un produit $B\Sigma \times \mathbf{R}_+^\times$. Un cône complexe est isomorphe à $L - \{0\}$ (L moins sa section nulle), où L est un fibré linéaire de rang 1 sur $B\Sigma$. Le fibré L n'a pas de raison d'être trivial.

Définition 2 *(i) On note* $\mathscr{O}(m)$ *le faisceau des fonctions homogènes de degré* m *sur* Σ *(holomorphes dans le cas complexe); c'est un faisceau sur* $B\Sigma$*.*

(ii) On note $\widehat{\mathscr{O}}$ *le faisceau sur* $B\Sigma$ *des séries formelles (symboles) :*

$$f = \sum_{m \leq m_0} f_m \quad \text{avec} \quad f_m \in \mathscr{O}(m) \ (m \text{ entier}, \ m \to -\infty)$$

Définition 3 *On note* $\widehat{\mathscr{D}}_k$ *(*$k \geq 1$ *entier), le faisceau sur* $B\Sigma$ *des opérateurs k-différentiels formels :*

$$P(f_1, \ldots, f_k) = \sum_{m \leq m_0} P_m(f_1, \ldots, f_k)$$

où pour chaque entier $m \leq m_0$*,* P_m *est un opérateur k-différentiel, homogène de degré m pour les homothéties. Pour* $k = 1$ *on écrira simplement* $\widehat{\mathscr{D}}$*.*

Localement Σ est un cône produit, on peut choisir des coordonnées homogènes x_j (réelles ou complexes, de degré 0) sur la base et r de degré 1 sur la fibre, et $P_m(f_1, \ldots, f_k)$ est somme de monômes

$$a(x) r^m \, \partial_x^{\alpha_1} (r\partial_r)^{m_1} (f_1) \ldots \partial_x^{\alpha_k} (r\partial_r)^{m_k} (f_k).$$

Il n'y a pas de restriction sur l'ordre de P_m. Dans la suite, pour éviter les confusions, "degré" signifiera toujours degré d'homogénéité par rapport aux homothéties, ou le degré du terme dominant, et "ordre" l'ordre comme opérateur différentiel; ainsi dans l'expression ci-dessus P_m est de degré m et d'ordre fini arbitraire; la série formelle $\sum P_m$ peut être d'ordre infini.

Un opérateur $P = \sum P_k \in \widehat{\mathscr{D}}$ est inversible si et seulement si son terme dominant $\sigma(P) = P_{m_0}$ est inversible (en particulier il doit être d'ordre 0). On note $\widehat{\mathscr{D}}_-^\times$ le faisceau des $P \in \widehat{\mathscr{D}}$ inversibles de degré ≤ 0 tels que $P(1) = 1$ (i.e. $P_0 = 1$, et P_m n'a pas de "terme constant" pour $m < 0$).

---

[2] Toutes les variétés considérées ici sont supposées paracompactes.

Remarque 1 Les faisceaux sont tout à fait inutiles pour la théorie réelle, mais indispensables dans la théorie complexe où il peut y avoir trop peu de sections globales.

Remarque 2 Sur un cône analytique il y a aussi une notion de symbole convergent (introduite par l'auteur dans [7] pour définir les opérateurs pseudo-différentiels analytiques). Ce sont en fait les plus importants et pour beaucoup de questions il est essentiel d'utiliser des symboles "convergents" plutôt que formels. Cependant pour les problèmes de classification qui suivent, il n'y a pas de différence qualitative significative entre symboles formels et symboles convergents, et je me limiterai aux symboles formels en évitant les technicalités supplémentaires de la convergence.

2.2. Star-algèbres sur un cône

Définition 4 *Une star-algèbre sur Σ est un faisceau $\mathscr{A}$ sur la base $B\Sigma$, localement isomorphe à $\widehat{\mathscr{O}}$ (le faisceau structural est $\widehat{\mathscr{D}}_-^\times$), muni d'une loi d'algèbre associative unitaire (star-produit) $f * g = B(f,g)$, localement un opérateur bidifférentiel formel.*

Localement $f * g = \sum B_m(f,g)$ où B_m est un opérateur bidifférentiel homogène de degré m ($m \to -\infty$), $B_0 = 1$. Le faisceau structural de groupes qui sert à recoller les repères locaux est $\widehat{\mathscr{D}}_-^\times$, le faisceau des automorphismes de $\widehat{\mathscr{O}}$ qui préservent les opérateurs bidifférentiels formels, et l'unité 1.

Dans le cas semi-classique, Σ est un cône produit $\Sigma = B\Sigma \times L$ ($L = \mathbf{R}_+$ ou $\mathbf{C}^\times$), $\mathscr{A} = \widehat{\mathscr{O}}$, h est homogène de degré -1, et la loi de produit ne comporte pas de dérivation en h, donc h est central et joue le rôle d'une constante absolue. La définition ci-dessus contient le cas semi-classique, et aussi celui des algèbres d'opérateurs pseudo-différentiels ou des opérateurs de Toeplitz. Ce cadre pour décrire les star-produits a été introduit dans [4].

Dans le cas réel, $\mathscr{A}$ est toujours isomorphe à $\widehat{\mathscr{O}}$, comme faisceau de groupe structural $\widehat{\mathscr{D}}_-^\times$, i.e. on peut définir un "symbole total" global (cela se voit facilement en recollant au moyen d'une partition de l'unité). Il n'en est plus de même dans le cas holomorphe, même dans des cas très naturels (par exemple pour les opérateurs pseudodifférentiels sur la sphère de Riemann $P_1(\mathbf{C})$, il n'y a pas de symbole total — voir ci-dessous). Aussi on ne peut pas éviter cette présentation faisceautique dans le cas complexe.

2.3. Crochet de Poisson

Le groupe structural $\widehat{\mathscr{D}}_-^\times$ qui sert à recoller est constitué d'opérateurs de terme dominant 1, qui préservent la filtration canonique de $\widehat{\mathscr{O}}$ et induisent l'identité sur $\mathrm{gr}\,\widehat{\mathscr{O}}$. Donc une star-algèbre $\mathscr{A}$ a une filtration canonique, et il y a un isomorphisme canonique (calcul symbolique)

$$\mathrm{gr}\,\mathscr{A} \simeq \mathrm{gr}\,\widehat{\mathscr{O}} = \bigoplus \widehat{\mathscr{O}}(n)$$

$\operatorname{gr}\mathscr{A} \sim \operatorname{gr}\widehat{\mathscr{O}}$ est alors une algèbre de Poisson, dont le crochet de Poisson est donné par le terme dominant du commutateur :

$$\{f,g\} = B_1(f,g) - B_1(g,f).$$

Il s'agit d'un opérateur bidifférentiel alterné d'ordre 1, parce que c'est une dérivation par rapport au produit usuel, qui est le terme dominant de B : $B_0(f,g) = fg$; en outre il vérifie l'identité de Jacobi

$$\{f,\{g,h\}\} = \{\{f,g\},h\} + \{g,\{f,h\}\}$$

parce que le crochet $[a,b]$ de l'algèbre associative $\mathscr{A}$ est un crochet de Lie.

Remarque 3 L'existence d'un star-produit associé au crochet de Poisson d'un cône symplectique réel Σ a été prouvée par V. Guillemin et moi-même dans [6]. La preuve de M. Kontsevich pour un crochet de Poisson général s'adapte sans changement au cas conique ci-dessus (dans le cas réel) et montre que les classes de star-produits sont en correspondance biunivoque avec les classes d'équivalence de crochets de Poisson formels : $c = \sum_{k\leq -1} c_k$.

3. Star-algèbres sur un fibré cotangent

3.1. $\mathscr{E}$-algèbres

Soit X une variété complexe, et $\Sigma = T^*X - \{0\}$ le fibré cotangent, privé de sa section nulle, muni de sa structure symplectique canonique. La base est $B\Sigma = \Sigma/\mathbf{C}^\times = PX$, le fibré projectif associé à T^*X. Σ porte une star-algèbre canonique : l'algèbre $\mathscr{E}$ des opérateurs pseudo-différentiels, microlocalisation de l'algèbre des opérateurs différentiels de X : si on choisit des coordonnées locales $x = (x_1,\ldots,x_n)$ sur X et les coordonnées duales $\xi = (\xi_1,\ldots,\xi_n)$ sur les fibres, un opérateur a un symbole total $a(x,\xi)$ et le produit est donné par la règle de Leibniz : pour $f,g \in \widehat{\mathscr{O}}$ on a

$$f * g = \sum \frac{1}{\alpha!}\, \partial_\xi^\alpha f\, \partial_x^\alpha g. \tag{1}$$

Le cocycle de recollement est celui qui est défini par les changements de coordonnées; c'est un cocycle parce qu'il recolle effectivement les symboles totaux (polynomiaux) des opérateurs différentiels pour donner le faisceau $\mathscr{D}_X$ des opérateurs différentiels.

Nous dirons que $\mathscr{A}$ est une $\mathscr{E}$-algèbre si c'est une star-algèbre sur $\Sigma = T^*X - \{0\}$ associée au crochet de Poisson canonique. Il est élémentaire et connu qu'une $\mathscr{E}$-algèbre est localement isomorphe à $\mathscr{E}$; plus précisément si U est isomorphe à un sous-cône de $T^*\mathbf{C}^n$ dont la base est contractile et de Stein (par exemple l'ensemble $\{\xi_i \neq 0\} \subset T^*B$, où B est une boule de $\mathbf{C}^n$), toute $\mathscr{E}$-algèbre sur U est isomorphe à $\mathscr{E}$, et toute section de $\widehat{\mathscr{O}}(m)$ est le symbole d'une section de $\mathscr{A}_m$.

Par suite on obtient les $\mathscr{E}$-algèbres en recollant des modèles de $\mathscr{E}$. L'ensemble des classes d'isomorphismes de telles algèbres est

$$\mathrm{Alg}_{\mathscr{E}} \sim H^1(PX, \mathrm{Aut}\,\mathscr{E}), \tag{2}$$

l'ensemble des classes de cocycles à coefficients dans le faisceau des automorphismes de $\mathscr{E}$ (familles $(a_{ij}) \in \Gamma(X_{ij}, \mathrm{Aut}\,\mathscr{E})$, telles que $a_{ik} = a_{ij}a_{jk}$, indexées par les intersections $X_{ij} = X_i \cap X_j$ d'un recouvrement assez fin $X = \bigcup X_i$; deux cocycles a, a' étant équivalents si, après raffinement des recouvrements, on a $a'_{ij} = b_i a_{ij} b_j^{-1}$ pour une famille $b_i \in \Gamma(X_i, \mathrm{Aut}\,\mathscr{E})$ convenable).

3.2. $\mathscr{D}$-algèbres

Soit X une variété complexe, $\mathscr{D}_X$ le faisceau des opérateurs différentiels. Localement un automorphisme de $\mathscr{D}$ qui préserve les symboles est un automorphisme intérieur $P \to fPf^{-1}$ où f est une fonction holomorphe; on a $\mathrm{Int}\, e^f = \mathrm{Id}$ ssi f est constante, de sorte que le faisceau d'automorphismes de $\mathscr{D}$ est

$$\mathrm{Aut}\,\mathscr{D}_X \simeq \mathscr{O}_X^\times / \mathbf{C}^\times \simeq \mathscr{O}_X / \mathbf{C}. \tag{3}$$

Nous appellerons $\mathscr{D}$-algèbre (algèbre d'opérateurs différentiels tordus dans [2]) un faisceau d'algèbres sur X localement isomorphe à $\mathscr{D}_X$; l'ensemble $\mathrm{Alg}_{\mathscr{D}}$ des classes d'isomorphisme de telles algèbres est canoniquement isomorphe à l'ensemble $H^1(X, \mathscr{O}_X/\mathbf{C})$.

Une $\mathscr{D}$-algèbre définit bien sûr une star-algèbre sur PX, et il est naturel de comparer les ensembles $\mathrm{Alg}_{\mathscr{D}}$ et $\mathrm{Alg}_{\mathscr{E}}$.

3.3. Automorphismes

Pour comprendre comment on recolle des $\mathscr{E}$-algèbres pour former des objets globaux, il faut d'abord comprendre comment sont fait les automorphismes de $\mathscr{E}$.

Soit U un tel automorphisme : U préserve les symboles et l'unité, donc $U - 1$ est de degré -1 (i.e. abaisse le degré de 1). Alors le logarithme $D = \mathrm{Log}\, U$ est bien défini; c'est une dérivation de degré ≤ -1 de $\mathscr{E}$.

Si D est une dérivation homogène de degré $\leq k$, le symbole $\mathrm{gr}\, D = \sigma_k(D) := \delta$ est une dérivation homogène de degré k de l'algèbre de Poisson $\mathrm{gr}\,\mathscr{E} = \mathrm{gr}\,\widehat{\mathscr{O}}$, autrement dit un champ de vecteurs symplectique sur Σ, homogène de degré k. Il correspond à une 1-forme différentielle fermée α, homogène de degré $k+1$.

Il est élémentaire qu'une telle forme α est exacte, i.e. dérivée d'une fonction homogène, si $k+1 = 0$; et si $k+1 = 0$ elle est localement de la forme $c\frac{dr}{r} + df$, où f de degré 0, r est une fonction non nulle homogène de degré 1 fixée, $c = \langle \rho, \alpha \rangle = constante$ (ρ désigne le champ de vecteurs radial, générateur infinitésimal de l'action de $\mathbf{C}^\times$: $\rho = \sum \xi_j \partial_{\xi_j}$ si on a choisi des coordonnées locales).

Par approximations successives, on déduit de ceci que toute dérivation D de $\mathscr{E}$ est de la forme $s\,\mathrm{ad}\,(\mathrm{Log}\, P_1) + \mathrm{ad}\, Q$ où P_1 est elliptique de degré 1, $Q \in \mathscr{E}$, et que tout automorphisme de $\mathscr{E}$ est localement de la forme

$$U = (\mathrm{Int}\, P_1)^s \mathrm{Int}\, Q_0 \tag{4}$$

où P_1 est elliptique de degré 1, Q_0 elliptique de degré 0.[3] $\operatorname{Int} P$ désigne l'automorphisme intérieur $a \to PaP^{-1}$.

Si U est un automorphisme, nous noterons $\sigma(U)$ (symbole de U) la forme différentielle fermée homogène de degré 0 correspondant comme ci-dessus au terme dominant (de degré -1) de $\operatorname{Log} U$. Si $\sigma(U) = 0$, $U - 1$ est de degré ≤ -2, et il existe alors un unique P de degré 0 et de symbole 1 tel que $U = \operatorname{Int} P$:

Proposition 1 *Il y a une suite exacte de faisceaux de groupes sur PX :*

$$0 \to \mathscr{E}_-^\times \to \operatorname{Aut}\mathscr{E} \to \omega \to 0 \tag{5}$$

où $\mathscr{E}_-^\times$ est le faisceau des sections de $\mathscr{E}$ de symbole 1 ($\mathscr{E}_-^\times = 1 + \mathscr{E}_{-1}$) et ω est le faisceau (sur PX) des 1-formes fermées, homogènes de degré 0 sur Σ.

Si $\mathscr{A} \in \operatorname{Alg}_\mathscr{E} \simeq H^1(PX, \operatorname{Aut}\mathscr{E})$ le symbole $\sigma(\mathscr{A}) \in H^1(PX, \omega)$ est par définition l'élément déduit du cocycle qui définit $\mathscr{A}$.

3.4. Méthodes de calcul

Les calculs nécessaires pour étudier les star-algèbres dans le cadre ci-dessus utilisent une mixture de cohomologie non commutative (cf. [16]) et de cohomologie holomorphe, et la relation entre cohomologie et filtration. Il n'y a pas d'idée fondamentalement nouvelle — mais ça peut être un peu embrouillant. Je ne donne ici qu'un aperçu succinct, et renvoie à [3] pour les détails.

Si $0 \to A \to B \to C \to 0$ est une suite exacte de faisceaux de groupes sur un espace Y (A normal dans B), il y a une suite "exacte" de cohomologie :

$$\begin{aligned} 0 \to H^0(Y,A) \to H^0(Y,B) \to H^0(Y,C) \to \\ \to H^1(Y,A) \to H^1(Y,B) \to H^1(Y,C). \end{aligned} \tag{6}$$

cette suite est exacte au sens suivant :

(i) les H^0 sont des groupes et la suite est exacte aux 3 premières places.

(ii) Les H^1 sont des ensembles pointés. Le groupe $H^0(Y,C)$ opère sur $H^1(Y,A)$, et ses orbites sont les fibres de l'application $H^1(Y,A) \to H^1(Y,B)$. Il y a encore une relation renseignant sur les fibres de la dernière application $H^1(Y,B) \to H^1(Y,C)$: je renvoie à [16] ou [3].

On utilise cette suite avec $A = \mathscr{E}_-^\times$, $B = \operatorname{Aut}\mathscr{E}$, $C = \omega$; on utilise aussi les suites exactes $0 \to \mathscr{O}/\mathbf{C} \to \omega \to \mathbf{C} \to 0$, $0 \to \mathbf{C} \to \mathscr{O} \to \mathscr{O}/\mathbf{C} \to 0$, qui servent à étudier ω, et qui engendrent des longues suites de cohomologie, puisqu'il s'agit de faisceaux commutatifs.

Dans le cas commutatif le lien entre filtration et cohomologie est décrit au moyen d'une suite spectrale, dont il ne reste que des bribes dans le cas non commutatif. Retenons cependant que si $H^1(X, \operatorname{gr}\operatorname{Aut}\mathscr{E}) = 0$ alors $H^1(X, \operatorname{Aut}\mathscr{E}) = 0$; (de même avec $\mathscr{E}_-^\times$); si $H^2(X, \mathscr{E}_-^\times) = 0$ (i.e. $H^2(X, \mathscr{O}(m)) = 0$ pour $m < 0$) alors l'application symbole : $H^1(X, \operatorname{Aut}\mathscr{E}) \to H^1(X, \omega)$ est surjective.

[3] elliptique = inversible.

4. Le cas $\dim X \geq 2$

Si $n \geq 2$, les fonctions homogènes de degré $m \in \mathbf{Z}$ sur $\mathbf{C}^n - \{0\}$ sont en fait polynomiales, en particulier elles sont nulles si $m < 0$ et constantes si $m = 0$, et par suite une $\mathscr{E}$-algèbre $\mathscr{A}$ n'a pas de section globale de degré < 0, et deux automorphismes qui ont même symbole sont égaux, et leur symbole commun "est indépendant de la variable cotangente" (i.e. coïncide localement avec un symbole d'automorphisme intérieur défini par une fonction $f(x)$ sur la base X).

Alors si $\mathscr{A}, \mathscr{A}'$ sont deux $\mathscr{D}$-algèbres, tout morphisme des $\mathscr{E}$-algèbres qu'elles définissent provient en fait d'un unique morphisme de $\mathscr{D}$-algèbres, de sorte que l'application $\mathrm{Alg}_{\mathscr{D}} \to \mathrm{Alg}_{\mathscr{E}}$ est un foncteur fidèle.

Si $n \geq 3$, c'est en fait une équivalence. En effet dans ce cas on montre facilement que si X est une boule (ou un domaine de Stein contractile), toute $\mathscr{E}$-algèbre sur PX est isomorphe à $\mathscr{E}_X$ (ceci est lié au fait qu'on a $H^1(\mathbf{C}^n - \{0\}, \mathscr{O}) = 0$ si $n \geq 3$). Donc si $\dim X \geq 3$ toute $\mathscr{E}$-algèbre $\mathscr{A}$ sur X s'obtient en recollant des $\mathscr{D}$-algèbres sur des morceaux de X, au moyen d'isomorphismes provenant d'isomorphismes de $\mathscr{D}$-algèbres : autrement dit $\mathscr{A}$ provient d'une $\mathscr{D}$-algèbre.

Pour $n = 2$ on a $H^1(PX, \mathrm{Aut}\,\mathscr{E}) = H^1(PX), \mathscr{E}_-^\times) \neq 0$ (il est même très gros), et il y a beaucoup de $\mathscr{E}$-algèbres (exotiques ou monstrueuses) qui ne proviennent pas de $\mathscr{D}$-algèbres. La classification peut se simplifier lorsque X est compacte : dans [3] on montre que si $X = P_2(\mathbf{C})$ les $\mathscr{E}$-algèbres proviennent de nouveau toutes de $\mathscr{D}$-algèbres. La situation est une peu plus compliquée lorsque X est un "tore" : $X = \mathbf{C}^2/\Gamma (\Gamma \sim \mathbf{Z}^4$ discret) (la dichotomie $n = 2, n \geq 3$ intervient de façon similaire dans l'étude des $\mathscr{E}$-modules localement libres de [9] et [8]).

5. Le cas $\dim X = 1$

Dans ce cas on a $PX = X$ ce qui simplifie l'étude. En particulier une $\mathscr{E}$-algèbre sur X s'obtient en recollant des exemplaires de $\mathscr{E}$ au moyen d'automorphismes de $\mathscr{E}$ (provenant de $\mathscr{D}$) sur un recouvrement de X, et est triviale (isomorphe à $\mathscr{E}_X$) si X est simplement connexe (isomorphe à un disque ou à $\mathbf{C}$). En revanche il y a beaucoup d'opérateurs pseudo-différentiels et d'automorphismes de degré < 0 de sorte que le lien entre $\mathscr{E}$-algèbres et $\mathscr{D}$-algèbres est plus compliqué. Je ne donnerai ici que quelques indications pour les cas où X est ouverte ou de genre ≥ 1 (je renvoie à [3] pour plus de détails), et me concentrerai sur le cas où X est la droite projective.

5.1. Courbes ouvertes et courbes de genre $g \geq 1$

Si X est ouverte (donc de Stein), elle possède un champ de vecteurs sans zéro ξ (le fibré tangent, comme tout fibré linéaire, est trivial d'après le théorème de Cousin). Si (s_{ij}) est un cocycle à coefficients dans $\mathbf{C}$ sur X, $((\mathrm{Int}\,\xi)^{s_{ij}})$ est un cocycle à coefficients dans $\mathrm{Aut}\,\mathscr{E}$ et on vérifie facilement que l'application $H^1(X, \mathbf{C}) \to \mathrm{Alg}_{\mathscr{E}}$ ainsi définie est bijective (on a $H^1(X, \mathscr{O}/\mathbf{C}) = 0$) : ainsi toutes les $\mathscr{E}$-algèbres non triviales sont "exotiques".

Si X est de genre ≥ 3, on a $H^1(X,\mathscr{E}_-^\times) = 0$ ($H^1(X,\mathrm{gr}\,\mathscr{E}_-^\times)$ est nul car $\mathscr{O}(-1) \sim T^*X$ est un fibré "positif" donc $H^1(X,\mathscr{O}(k)) = 0$ pour $k < 0$). On a aussi $H^2(X,\mathrm{gr}\,\mathrm{Aut}\,\mathscr{E}) = 0$ parce que X est de dimension complexe 1. Donc $\mathrm{Alg}_{\mathscr{E}} \to H^1(X,\omega)$ est bijectif. Dans la suite exacte (commutative)

$$\cdots \to H^0(X,\mathbf{C}) \to H^1(X,\mathscr{O}/\mathbf{C}) \to H^1(X,\omega) \to H^1(X,\mathbf{C}) \to 0$$

la première flèche est bijective, parce que l'image du générateur de $H^0(X,\mathbf{C}) \sim \mathbf{C}$ correspond à la classe de Chern de T^*X qui est un générateur de $H^1(X,\mathscr{O}/\mathbf{C})$. Finalement on a encore $\mathrm{Alg}_{\mathscr{E}} \sim H^1(X,\mathbf{C})$, mais de façon moins simple que précédemment : il y a des $\mathscr{D}$-algèbres non triviales, mais la $\mathscr{E}$-algèbre correspondante l'est; les $\mathscr{E}$-algèbres non triviales sont toutes "exotiques", i.e. l'obstruction vient uniquement des exposants des automorphismes de recollement, mais le cocycle correspondant ne s'écrit pas aussi simplement que ci-dessus.

Si X est de genre 1 ($X \sim \mathbf{C}/\Gamma$, $\Gamma \sim \mathbf{Z}^2$), il possède un champ de vecteurs sans zéro, donc des $\mathscr{E}$-algèbres exotiques comme ci-dessus. On a toujours une application surjective $\mathrm{Alg}_{\mathscr{E}} \to H^1(X,\omega) \sim \mathbf{C}^3$, mais certaines fibres sont réduites à un point, d'autres très grosses (en particulier celles de $\mathscr{E}$). Ce cas est le plus compliqué et je renvoie à [3] pour les détails.

5.2. La droite projective

Soit $X = P_1(\mathbf{C})$ la droite projective. X est réunion des deux ouverts $X_0 = \{z \neq \infty\}$, $X_\infty = \{z \neq 0\}$, sur lesquels toute $\mathscr{E}$-algèbre (resp. $\mathscr{D}$-algèbre) est triviale. Donc les $\mathscr{E}$-algèbres (resp. $\mathscr{D}$-algèbres) sont classifiées par des cocycles réduits à une seule fonction $X_0 \cap X_\infty = \mathbf{C} - \{0\} \to \mathrm{Aut}\,\mathscr{E}$.

Les $\mathscr{D}$-algèbres sont classifiées par $H^1(X,\mathscr{O}/\mathbf{C}) = H^2(X,\mathbf{C}) = \mathbf{C}$. On note $\mathscr{D}_s$ ($s \in \mathbf{C} = H^1(X,\mathscr{O}/\mathbf{C})$) la $\mathscr{D}$-algèbre définie par le cocycle $(\mathrm{Int}\, z)^s$.

En coordonnées homogènes x,y ($z = \frac{x}{y}$) on dispose du faisceau des opérateurs différentiels homogènes $\mathscr{D}^{hom}$, i.e. qui commutent avec le générateur des homothéties $\rho = x\partial_x + y\partial_y$; cette algèbre est engendrée par ρ et les opérateurs

$$e = x\partial_y, \quad h = x\partial_x - y\partial_y, \quad f = y\partial_x \tag{7}$$

qui vérifient les relations

$$[h,e] = 2e, \quad [h,f] = -2f, \quad [e,f] = h, \quad h^2 + 2(ef + fe) = \rho(\rho+2). \tag{8}$$

$\mathscr{D}_s$ est isomorphe au quotient $\mathscr{D}^{hom}/(\rho + s)$; on peut y penser comme au faisceau des opérateurs différentiels sur le faisceau virtuel $\mathscr{O}(s)$ sur X des fonctions homogènes de degré s de x,y (qui n'existe que si s est entier).

Le résultat suivant est facile :

Lemme 1 *Si $X = P_1$ est la droite projective :*

(i) on a $H^0(X,\omega) = 0$, $H^0(X,\mathscr{E}_-^\times) = 0$, donc $H^0(X,\mathrm{Aut}\,\mathscr{A}) = 0$ pour toute $\mathscr{E}$-algèbre $\mathscr{A}$.

(ii) On a $H^1(X,\omega) = 0$ et l'application $H^1(X,\mathscr{E}_-^\times) \to H^1(X,\mathrm{Aut}\,\mathscr{E})$ est bijective.

Dans la longue suite exacte de cohomologie

$$H^0(X,\mathcal{O}/\mathbf{C}) \to H^0(X,\omega) \to H^0(X,\mathbf{C}) \to H^1(X,\mathcal{O}/\mathbf{C}) \to$$

$$\to H^1(X,\omega) \to H^1(X,\mathbf{C}) \to \cdots$$

on a $H^0(X,\mathcal{O}/\mathbf{C}) = 0$ car X ne porte pas de 1-forme holomorphe fermée non nulle; l'application exposant $H^0(X,\mathbf{C}) \to H^1(X,\mathcal{O}/\mathbf{C})$ est bijective parce que la classe de Chern de $\mathcal{O}(1)$ est $\neq 0$; on a $H^1(X,\mathbf{C}) = 0$ puisque X est simplement connexe. D'où $H^0(X,\omega) = H^1(X,\omega) = 0$.

Remarquons qu'un cocycle $q \in H^1(X,\mathrm{Aut}\,\mathscr{E})$ a un unique représentant "normalisé" :

$$q_{0\infty} = \sum_{0>p>2q} a_{pq} z^p \zeta^q \in \widehat{\mathcal{O}}(X_0 \cap X_\infty). \tag{9}$$

Ceci vient du fait que les deux champs de vecteurs $\partial_0 = \partial_z, \partial_\infty = \partial_{1/z}$ sont globalement holomorphes, elliptiques sur X_0 (resp. X_∞), et ont pour symbole ζ (resp. $-z^2\zeta$). Aussi, comme pour $H^1(X,\mathrm{gr}\,\mathscr{E}_-^\times)$, tout cocycle peut être réduit de façon unique à la forme ci-dessus.

Pour comparer les $\mathscr{D}$-algèbres et les $\mathscr{E}$-algèbres il est commode d'utiliser la suite exacte intermédiaire suivante. Soit $\mathrm{Int}\,\mathscr{E}_0 \simeq \mathscr{E}_0^\times/\mathbf{C}^\times$ le groupe des automorphismes intérieurs de $\mathscr{E}_0$; on a une suite exacte :

$$0 \to \mathrm{Int}\,\mathscr{E}_0 \to \mathrm{Aut}\,\mathscr{E} \to \mathbf{C} \to 0$$

d'où une surjection

$$H^1(X,\mathrm{Int}\,\mathscr{E}_0) \to H^1(X : \mathrm{Aut}\,\mathscr{E}) \tag{10}$$

dont les fibres sont les orbites de $\mathbf{C} = H^0(X,\mathbf{C})$ sur $H^1(X,\mathrm{Int}\,\mathscr{E})$ (par l'action $q_{0\infty} \to (\mathrm{Int}\,\partial_0)^s q_{0\infty} (\mathrm{Int}\,\partial_\infty)^{-s}$).

Lemme 2 *On a l'identité suivante (pour $k \in \mathbf{C}$) :*

$$(\mathrm{Int}\,z)^{-k-2} = (\mathrm{Int}\,\partial_0)^{k+1} (\mathrm{Int}\,z)^k (\mathrm{Int}\,\partial_\infty)^{-k-1}. \tag{11}$$

Preuve. Si k est entier positif, on a

$$z^{-k-2}(z^2\partial)^{k+1} = \partial^{k+1} z^k.$$

En effet les deux membres de cette égalité sont des opérateurs différentiels ordinaires d'ordre $k+1$, de même terme dominant $z^k\partial^{k+1}$, qui tuent tous les monômes $z^{-j}, 0 \geq j \geq -k$.

Le cas général en résulte, car l'identité (11) est polynomiale en s mod. $\mathscr{E}_m$, pour tout $m < 0$. ■

On voit donc que $\mathscr{D}_s$ (resp. $\mathscr{D}_{-s-2}$) définissent des $\mathscr{E}$-algèbres $\mathscr{E}_s$, ($\mathscr{E}_{-s-2}$) isomorphes, bien qu'elles ne soient pas isomorphes comme $\mathscr{D}$-algèbres. Ceci est le seul cas où deux $\mathscr{D}$-algèbres non isomorphes sur $X = P_1$ donnent des $\mathscr{E}$-algèbres isomorphes : en effet l'algèbre des sections globales d'une $\mathscr{E}$-algèbre en est un invariant évident. Les sections globales e, h, f (avec les notations ci-dessus) aussi, si elles existent car elles sont bien déterminées à une constante près par leurs symboles, et les relations de commutation fixent complètement ces constantes. La relation $h^2 + 2(ef + fe) = s(s+2)$ montre alors que, comme il se doit, le nombre $s(s+2)$ est un invariant de la $\mathscr{E}$-algèbre.

En dernière remarque, examinons le problème suivant : existe-t-il un symbole total, i.e. le faisceau sous-jacent à une $\mathscr{E}$-algèbre est-il isomorphe à $\widehat{\mathscr{O}}$? Comme on a remarqué ceci est toujours vrai pour une $\mathscr{E}$-algèbre réelle, pour laquelle on peut recoller des symboles totaux locaux grâce à une partition de l'unité.

Dans le cas de $X = P_1(\mathbf{C})$: $T^*X - \{0\}$ a un revêtement à deux feuillets canonique par $\mathbf{C}^2 - \{0\}$: $(u,v) \to (z = u/v, \zeta = \frac{1}{2}v^2)$. Si $\mathscr{A}$ est une $\mathscr{E}$-algèbre sur $\Sigma = T^*X$ elle se relève à $\Sigma' = \mathbf{C}^2 - \{0\}$ en une star-algèbre associée au crochet de Poisson canonique ($\{v,u\} = 1$), équivariante au-dessus de l'involution $(u,v) \to (-u,-v)$; noter que sur Σ', u et v sont de degré $\frac{1}{2}$. Si $\mathscr{A}$ a un calcul symbolique global, le relèvement définit un star-produit sur $\widehat{\mathscr{O}}(\Sigma')$.

Or sur $\widehat{\mathscr{O}}(\Sigma')$ il n'y a, à isomorphisme près, qu'une seule loi de star-produit associée au crochet de Poisson canonique : elle est engendrée par u, v soumis à la relation $[v,u] = 1$. À isomorphisme près elle est définie par la représentation $u \to u, v \to \partial_u$. Pour cette loi il y a beaucoup de sections globales (tous les polynômes de u, v); les sections globales e, h, f sont nécéssairement données par

$$e = -\frac{1}{2}u^2, \quad h = 2u * v + \frac{1}{2}, \quad f = \frac{1}{2}v^2 \tag{12}$$

parce que les symboles respectifs sont

$$\sigma(e) = -z^2\zeta = -\left(\frac{u}{v}\right)^2 \times \frac{1}{2}v^2, \quad \sigma(h) = 2z\zeta = 2(\frac{u}{v}) \times \frac{1}{2}v^2, \quad \sigma(f) = \zeta = \frac{1}{2}v^2$$

et que comme signalé plus haut ils déterminent e, h, f à des constantes additives près, que les relations de commutation achèvent de déterminer.

Ceci étant on obtient

$$h^2 + 2(ef + fe) = -\frac{3}{4} \qquad \left(s = -\frac{1}{2} \text{ ou } s = -\frac{3}{2}\right) \tag{13}$$

Ainsi les seules $\mathscr{D}$-algèbres qui possèdent un symbole total sont $\mathscr{D}_{-1/2}$ et $\mathscr{D}_{-3/2}$. En particulier il n'y a pas de symbole total sur $\mathscr{E}$ ou sur $\mathscr{D}$.

Références

1. Bayen F., Flato, M., Fronsdal, C., Lichnerowicz, A., and Sternheimer, D.: Deformation theory and quantization I, II, *Ann. Phys.* **111** (1978), 61–151.
2. Beilinson, A. and Bernstein, J.: Localisation de g-modules, *C. R. Acad. Sci., Paris, Sér. I Math.* **292** (1981), 15–18.
3. Boutet de Monvel, L.: Complex star-products, *Math. Phys. Analysis and Geometry* **2** (1999), 113–139.
4. Boutet de Monvel, L.: Star-Products on Conic Poisson Manifolds of constant rank, *Mat. Fiz. Anal. Geom. (Kharkov)* **2** (1995), 143–151.
5. Boutet de Monvel, L.: On the index of Toeplitz operators of several complex variables, *Invent. Math.* **50** (1978/79), 249–272.
6. Boutet de Monvel, L. and Guillemin, V.: The Spectral Theory of Toeplitz Operators, Annals of Mathematics Studies **99**, Princeton University Press, Princeton, 1981.
7. Boutet de Monvel, L. and Krée, P.: Pseudodifferential operators and Gevrey classes, *Ann. Inst. Fourier (Grenoble)* **17** (1967), 295–323.
8. Carette, M.: Exotic $\mathscr{D}$-modules in dimension 2, Thèse Paris VI, 1999.
9. D'Agnolo, A. and Schapira, P.: The Radon-Penrose Correspondence II: Line Bundles and Simple $\mathscr{D}$-modules, *J. Funct. Anal.* **153** (1998), 343–356.
10. De Wilde, M. and Lecomte, P.: Existence of star-products and of formal deformations of the Poisson Lie algebras of arbitrary symplectic manifolds, *Lett. Math. Phys.* **7** (1983), 487–496.
11. De Wilde, M. and Lecomte, P.: Formal deformations of the Poisson Lie algebra of a symplectic manifold and star-products. Existence, equivalence, derivations, in: M. Hazewinkel and M. Gerstenhaber, (eds.), *Deformation Theory of Algebras and Structures and Applications (Il Ciocco 1986)*, Kluwer Acad. Pub., Dordrecht, 1988, pp. 897–960.
12. Fedosov, B. V.: A simple geometrical construction of deformation quantization, *J. Differential Geom.* **40** (1994), 213–238.
13. Frenkel, J.: Cohomologie à valeurs dans un faisceau non abélien, *C. R. Acad. Sci. Paris* **240** (1955), 2368–2370.
14. Frenkel, J.: Cohomologie non abélienne et espaces fibrés, *Bull. Soc. Math. France* **85** (1957), 135–220.
15. Gerstenhaber, M.: On the deformation of rings and algebras, *Ann. of Math.* **79** (1964), 59–103.
16. Giraud, J.: *Cohomologie non Abélienne*, Die Grundlehren der mathematischen Wissenschaften, Band 179. Springer-Verlag, Berlin-New York, 1971.
17. Gutt, S.: Equivalence of deformations and associated star-products, *Lett. Math. Phys.* **3** (1979), 297-309.
18. Kashiwara, M.: Quantization of Contact Manifolds, *Publ. Res. Inst. Math. Sci.* **32** (1996), 1–7.
19. Kontsevich, M.: Deformation Quantization of Poisson Manifolds I, preprint q-alg/9709040.
20. Moyal, J.: Quantum mechanics as a statistical theory, *Proc. Cambridge Philos. Soc.* **45** (1949), 99–124.
21. Vey, J.: Déformation du crochet de Poisson sur une variété symplectique, *Comment. Math. Helv.* **50** (1975), 421–454.
22. Weinstein, A.: Deformation quantization, Séminaire Bourbaki, Vol. 1993/94. Exp. No. 789, Astérisque **227** (1995), 389–409.

Cyclic cohomology and Hopf algebra symmetry

Alain Connes
Collège de France, 3 rue Ulm, 75005 Paris, France
and
I.H.E.S., 35 route de Chartres, 91440 Bures-sur-Yvette, France

Henri Moscovici*
Department of Mathematics, The Ohio State University
231 W. 18th Avenue, Columbus, OH 43210, U.S.A.

Abstract. Cyclic cohomology has been recently adapted to the treatment of Hopf symmetry in noncommutative geometry. The resulting theory of characteristic classes for Hopf algebras and their actions on algebras allows to expand the range of applications of cyclic cohomology. It is the goal of the present paper to illustrate these recent developments, with special emphasis on the application to transverse index theory, and point towards future directions. In particular, we highlight the remarkable accord between our framework for cyclic cohomology of Hopf algebras on one hand and both the algebraic as well as the analytic theory of quantum groups on the other, manifest in the construction of the modular square.

Keywords: noncommutative geometry, cyclic cohomology, Hopf algebras, quantum groups

Mathematics Subject Classifications (2000): 58B34, 58B32, 46L87, 81R60

Dedicated to the memory of our friend Moshé Flato

Introduction

Cyclic cohomology of noncommutative algebras is playing in noncommutative geometry a similar rôle to that of de Rham cohomology in differential topology [11]. In [14] and [15], cyclic cohomology has been adapted to Hopf algebras and their actions on algebras, which are analogous to the Lie group/algebra actions on manifolds and embody a natural notion of symmetry in noncommutative geometry. The resulting theory of characteristic classes for Hopf actions allows in turn to widen the scope of applications of cyclic cohomology to index theory. It is the goal of the present paper to review these recent developments and point towards future directions.

The contents of the paper are as follows. In §1 we recall the basic notation pertaining to the cyclic theory. The adaptation of cyclic cohomology to Hopf algebras and Hopf actions is reviewed in §2, where we also discuss the relationship with Lie group/algebra cohomology. §3 deals with the geometric Hopf algebras arising

* The second author has conducted this work during employment by the Clay Mathematics Institute, as a CMI Scholar in residence at Harvard University. This material is based in part upon research supported by the National Science Foundation under award no. DMS-9706886.

G. Dito and D. Sternheimer (eds.), Conférence Moshé Flato 1999, Vol. 1, 121-147.

in transverse differential geometry and their application to transverse index theory. Finally, §4 illustrates the remarkable agreement between our framework for cyclic cohomology of Hopf algebras and both the algebraic as well as the analytic theory of quantum groups.

1. Cyclic cohomology

Cyclic cohomology has first appeared as a cohomology theory for algebras ([5], [7], [26]). In its simplest form, the cyclic cohomology $HC^*(\mathscr{A})$ of an algebra $\mathscr{A}$ (over $\mathbb{R}$ or $\mathbb{C}$ in what follows) is the cohomology of the cochain complex $\{C^*_\lambda(\mathscr{A}), b\}$, where $C^n_\lambda(\mathscr{A})$, $n \geq 0$, consists of the $(n+1)$-linear forms φ on $\mathscr{A}$ satisfying the cyclicity condition

$$\varphi(a^0,a^1,...,a^n) = (-1)^n \varphi(a^1,a^2,...,a^0), \quad a^0,a^1,...,a^n \in \mathscr{A} \tag{1.1}$$

and the coboundary operator is given by

$$(b\varphi)(a^0,\dots,a^{n+1}) = \sum_{j=0}^{n} (-1)^j \varphi(a^0,\dots,a^j a^{j+1},\dots,a^{n+1}) + (-1)^{n+1} \varphi(a^{n+1}a^0,a^1,\dots,a^n). \tag{1.2}$$

When the algebra $\mathscr{A}$ comes equipped with a locally convex topology for which the product is continuous, the above complex is replaced by its topological version: $C^n_\lambda(\mathscr{A})$ then consists of all continuous $(n+1)$-linear form on $\mathscr{A}$ satisfying (1.1).

Cyclic cohomology provides numerical invariants of K-theory classes as follows ([8]). Given an n-dimensional cyclic cocycle φ on $\mathscr{A}$, n even, the scalar

$$\varphi \otimes \mathrm{Tr}(E,E,...,E) \tag{1.3}$$

is invariant under homotopy for idempotents

$$E^2 = E \in M_N(\mathscr{A}) = \mathscr{A} \otimes M_N(\mathbb{C}).$$

In the above formula, $\varphi \otimes \mathrm{Tr}$ is the extension of φ to $M_N(\mathscr{A})$, using the standard trace Tr on $M_N(\mathbb{C})$:

$$\varphi \otimes \mathrm{Tr}(a^0 \otimes \mu^0, a^1 \otimes \mu^1,\dots,a^n \otimes \mu^n) = \varphi(a^0,a^1,\dots,a^n)\,\mathrm{Tr}(\mu^0\mu^1\cdots\mu^n).$$

This defines a pairing $\langle[\varphi],[E]\rangle$ between cyclic cohomology and K-theory, which extends to the general noncommutative framework the Chern-Weil construction of characteristic classes of vector bundles.

Indeed, if $\mathscr{A} = C^\infty(M)$ for a closed manifold M and

$$\varphi(f^0,f^1,...,f^n) = \langle \Phi, f^0 df^1 \wedge df^2 \wedge \ldots \wedge df^n \rangle, \quad f^0,f^1,...,f^n \in \mathscr{A},$$

where Φ is an n-dimensional closed de Rham current on M, then up to normalization the invariant defined by (1.3) is equal to

$$\langle \Phi, ch^*(\mathscr{E}) \rangle,$$

here $ch^*(\mathscr{E})$ denotes the Chern character of the rank N vector bundle $\mathscr{E}$ on M whose fiber at $x \in M$ is the range of $E(x) \in M_N(\mathbb{C})$.

Note that, in the above example,

$$\varphi(f^{\sigma(0)}, f^{\sigma(1)}, \ldots, f^{\sigma(n)}) = \varepsilon(\sigma)\, \varphi(f^0, f^1, \ldots, f^n),$$

for any permutation σ of the set $[n] = \{0, 1, \ldots, n\}$, with signature $\varepsilon(\sigma)$. However, the extension of $\varphi \otimes \mathrm{Tr}$ to $M_N(\mathscr{A})$, used in the pairing formula (1.3), retains only the *cyclic* invariance.

A simple but very useful class of examples of cyclic cocycles on a noncommutative algebra is obtained from group cohomology ([10], [3], [12]), as follows. Let Γ be an arbitrary group and let $\mathscr{A} = \mathbb{C}\Gamma$ be its group ring. Then any *normalized group cocycle* $c \in Z^n(\Gamma, \mathbb{C})$, representing an arbitrary cohomology class $[c] \in H^*(B\Gamma) = H^*(\Gamma)$, gives rise to a cyclic cocycle φ_c on the algebra $\mathscr{A}$ by means of the formula

$$\varphi_c(g_0, g_1, \cdots, g_n) = \begin{cases} 0 & \text{if } g_0 \cdots g_n \neq 1 \\ c(g_1, \cdots, g_n) & \text{if } g_0 \cdots g_n = 1 \end{cases} \tag{1.4}$$

extended by linearity to $\mathbb{C}\Gamma$.

In a dual fashion, one defines the cyclic homology $HC_*(\mathscr{A})$ of an algebra $\mathscr{A}$ as the homology of the chain complex $\{C_*^{\lambda}(\mathscr{A}), b\}$ consisting of the coinvariants under cyclic permutations of the tensor powers of $\mathscr{A}$, and with the boundary operator b obtained by transposing the coboundary formula (1.2). Then the pairing between cyclic cohomology and K-theory (1.3) factors through the natural pairing between cohomology and homology, i.e.,

$$\varphi \otimes \mathrm{Tr}(E, E, \ldots, E) = \langle \varphi, ch(E) \rangle, \tag{1.5}$$

where, again up to normalization,

$$ch_n(E) = E \otimes \cdots \otimes E \quad (n+1 \quad \text{times}) \tag{1.6}$$

represents the *Chern character* in $HC_n(\mathscr{A})$, for n even, of the K-theory class $[E] \in K_0(\mathscr{A})$.

The cyclic cohomology of an (unital) algebra $\mathscr{A}$ has an equivalent description, in terms of the bicomplex $(CC^{*,*}(\mathscr{A}), b, B)$, defined as follows. With $C^n(\mathscr{A})$ denoting the linear space of $(n+1)$-linear forms on A, set

$$\begin{aligned} CC^{p,q}(\mathscr{A}) &= C^{q-p}(\mathscr{A}), \qquad q \geq p, \\ CC^{p,q}(\mathscr{A}) &= 0, \qquad q < p. \end{aligned} \tag{1.7}$$

The vertical operator $b : C^n(\mathscr{A}) \to C^{n+1}(\mathscr{A})$ is defined as

$$\begin{aligned} (b\varphi)(a^0, \ldots, a^{n+1}) = \sum_{j=0}^{n} (-1)^j\, \varphi(a^0, \ldots, a^j a^{j+1}, \ldots, a^{n+1}) \\ + (-1)^{n+1}\, \varphi(a^{n+1} a^0, a^1, \ldots, a^n). \end{aligned} \tag{1.8}$$

The horizontal operator $B : C^n(\mathscr{A}) \to C^{n-1}(\mathscr{A})$ is defined by the formula

$$B = NB_0,$$

where

$$\begin{aligned} B_0\,\varphi(a^0,\dots,a^{n-1}) &= \varphi(1,a^0,\dots,a^{n-1}) - (-1)^n\,\varphi(a^0,\dots,a^{n-1},1), \\ (N\psi)(a^0,\dots,a^n) &= \sum_0^n (-1)^{nj}\,\psi(a^j,a^{j+1},\dots,a^{j-1}). \end{aligned} \tag{1.9}$$

Then $HC^*(\mathscr{A})$ is the cohomology of the *first quadrant* total complex $(TC^*(\mathscr{A}), b+B)$, formed as follows:

$$TC^n(\mathscr{A}) = \sum_{p=0}^{n} CC^{p,n-p}(\mathscr{A}). \tag{1.10}$$

On the other hand, the cohomology of the *full direct sum* total complex $(TC_*^{\Sigma}(\mathscr{A}), b+B)$, formed by taking the direct sums

$$TC_n^{\Sigma}(\mathscr{A}) = \sum_{p\in\mathbb{Z}} CC^{p,n-p}(\mathscr{A}), \tag{1.11}$$

gives the ($\mathbb{Z}/2$-graded) *periodic cyclic cohomology* groups $HC^*_{\mathrm{per}}(\mathscr{A})$.

There is a dual description for the cyclic homology of $\mathscr{A}$, in terms of the dual bicomplex $(CC_{*,*}(\mathscr{A}), b, B)$, with $C_n(\mathscr{A}) = \mathscr{A}^{\otimes n+1}$ and the boundary operators b, B obtained by transposing the corresponding coboundaries. The periodic cyclic homology groups $HC_*^{\mathrm{per}}(\mathscr{A})$ are obtained from the *full product* total complex $(TC^*_{\Pi}(\mathscr{A}), b+B)$, formed by taking direct products as follows:

$$TC^n_{\Pi}(\mathscr{A}) = \Pi_p CC_{p,n-p}(\mathscr{A}). \tag{1.12}$$

The Chern character of an idempotent $e^2 = e \in \mathscr{A}$ is given in this picture by the periodic cycle $(ch_n(e))_{n=2,4,\dots}$, with components:

$$ch_0(e) = e, \quad ch_{2k}(e) = (-1)^k \frac{(2k)!}{k!}\left(e^{\otimes 2k+1} - \frac{1}{2} \otimes e^{\otimes 2k}\right), \quad k \geq 1. \tag{1.13}$$

The functors HC^0 and HC_0 from the category of algebras to the category of vector spaces have clear intrinsic meaning: The first assigns to an algebra $\mathscr{A}$ the vector space of traces on $\mathscr{A}$, while the second associates to $\mathscr{A}$ its abelianization $\mathscr{A}/[\mathscr{A},\mathscr{A}]$. From a conceptual viewpoint, it is important to realize the higher co/homologies HC^*, resp. HC_*, as derived functors. The obvious obstruction to such an interpretation is the non-additive nature of the category of algebras and algebra homomorphisms. This has been remedied in [9], by replacing it with the category of Λ-modules over the *cyclic category* Λ.

The cyclic category Λ is a small category, obtained by enriching with *cyclic morphisms* the familiar *simplicial category* Δ of totally ordered finite sets and increasing maps. We recall the presentation of Δ by generators and relations. It has one object $[n] = \{0 < 1 < \cdots < n\}$ for each integer $n \geq 0$, and is generated by faces $\delta_i : [n-1] \to [n]$ (the injection that misses i), and degeneracies $\sigma_j : [n+1] \to [n]$ (the surjection which identifies j with $j+1$), with the following relations:

$$
\begin{aligned}
&\delta_j \delta_i = \delta_i \delta_{j-1} \qquad \text{for } i < j, \\
&\sigma_j \sigma_i = \sigma_i \sigma_{j+1} \qquad \text{for } i \leq j, \\
&\sigma_j \delta_i = \begin{cases} \delta_i \sigma_{j-1} & i < j, \\ 1_n & \text{if } i = j \text{ or } i = j+1, \\ \delta_{i-1} \sigma_j & i > j+1. \end{cases}
\end{aligned}
\tag{1.14}
$$

To obtain Λ one adds for each n a new morphism $\tau_n : [n] \to [n]$ such that,

$$
\begin{aligned}
&\tau_n \delta_i = \delta_{i-1} \tau_{n-1}, \qquad 1 \leq i \leq n, \\
&\tau_n \sigma_i = \sigma_{i-1} \tau_{n+1}, \qquad 1 \leq i \leq n, \\
&\tau_n^{n+1} = 1_n.
\end{aligned}
\tag{1.15}
$$

Note that the above relations also imply:

$$
\tau_n \delta_0 = \delta_n, \qquad \tau_n \sigma_0 = \sigma_n \tau_{n+1}^2. \tag{1.16}
$$

Alternatively, Λ can be defined by means of its "cyclic covering", the category $E\Lambda$. The latter has one object $(\mathbb{Z}, n)$ for each $n \geq 0$ and the morphisms $f : (\mathbb{Z}, n) \to (\mathbb{Z}, m)$ are given by non decreasing maps $f : \mathbb{Z} \to \mathbb{Z}$, such that $f(x+n) = f(x) + m, \quad \forall x \in \mathbb{Z}$. One has $\Lambda = E\Lambda/\mathbb{Z}$, with respect to the obvious action of $\mathbb{Z}$ by translation.

To any algebra $\mathscr{A}$ one associates a module $\mathscr{A}^\natural$ over the category Λ by assigning to each integer $n \geq 0$ the vector space $C^n(\mathscr{A})$ of $(n+1)$-linear forms $\varphi(a^0, \ldots, a^n)$ on $\mathscr{A}$, and to the generating morphisms the operators $\delta_i : C^{n-1} \to C^n$, $\sigma_i : C^{n+1} \to C^n$ defined as follows:

$$
\begin{aligned}
&(\delta_i \varphi)(a^0, \ldots, a^n) = \varphi(a^0, \ldots, a^i a^{i+1}, \ldots, a^n), \quad i = 0, 1, \ldots, n-1, \\
&(\delta_n \varphi)(a^0, \ldots, a^n) = \varphi(a^n a^0, a^1, \ldots, a^{n-1}); \\
&(\sigma_0 \varphi)(a^0, \ldots, a^n) = \varphi(a^0, 1, a^1, \ldots, a^n), \\
&(\sigma_j \varphi)(a^0, \ldots, a^n) = \varphi(a^0, \ldots, a^j, 1, a^{j+1}, \ldots, a^n), \quad j = 1, \ldots, n-1, \\
&(\sigma_n \varphi)(a^0, \ldots, a^n) = \varphi(a^0, \ldots, a^n, 1); \\
&(\tau_n \varphi)(a^0, \ldots, a^n) = \varphi(a^n, a^0, \ldots, a^{n-1}).
\end{aligned}
\tag{1.17}
$$

These operations satisfy the relations (1.14) and (1.15), which shows that $\mathscr{A}^\natural$ is indeed a Λ-module.

One thus obtains the desired interpretation of the cyclic co/homology groups of a k-algebra $\mathscr{A}$ over a ground ring k in terms of derived functors over the cyclic category ([9]):

$$HC^n(\mathscr{A}) \simeq Ext^n_\Lambda(k^\natural, \mathscr{A}^\natural) \quad \text{and} \quad HC_n(\mathscr{A}) \simeq Tor^\Lambda_n(\mathscr{A}^\natural, k^\natural).$$

Moreover, all of the fundamental properties of the cyclic co/homology of algebras, such as the long exact sequence relating it to Hochschild co/homology ([8], [23]), are shared by the functors $Ext^*_\Lambda/Tor^\Lambda_*$-functors and, in this generality, can be attributed to the coincidence between the classifying space $B\Lambda$ of the small category Λ and the classifying space $BS^1 \simeq P_\infty(\mathbb{C})$ of the circle group.

Let us finally mention that, from the very definition of $Ext^*_\Lambda(k^\natural, F)$ and the existence of a canonical projective biresolution for $k^\natural$ ([9]), it follows that the cyclic cohomology groups $HC^*(F)$ of a Λ-module F, as well as the periodic ones $HC^*_{\text{per}}(F)$, can be computed by means of a bicomplex analogous to (1.7). A similar statement holds for the cyclic homology groups.

2. Cyclic theory for Hopf algebras

The familiar antiequivalence between suitable categories of *spaces* and matching categories of *associative algebras*, effected by the passage to coordinates, is of great significance in both the purely algebraic context (affine schemes versus commutative algebras) as well as the topological one (locally compact spaces versus commutative C^*-algebras). By extension, it has been adopted as a fundamental principle of noncommutative geometry. When applied to the realm of *symmetry*, it leads to promoting the notion of a *group*, whose coordinates form a commutative Hopf algebra, to that of a general *Hopf algebra*. The cyclic categorical formulation recalled above allows to adapt cyclic co/homology in a natural way to the treatment of symmetry in noncommutative geometry. This has been done in [14], [15] and will be reviewed below.

We consider a Hopf algebra $\mathscr{H}$ over $k = \mathbb{R}$ or $\mathbb{C}$, with unit $\eta : k \to \mathscr{H}$, counit $\varepsilon : \mathscr{H} \to k$ and antipode $S : \mathscr{H} \to \mathscr{H}$. We use the standard definitions ([25]) together with the usual convention for denoting the coproduct:

$$\Delta(h) = \sum h_{(1)} \otimes h_{(2)}, \qquad h \in \mathscr{H}. \tag{2.1}$$

Although we work in the algebraic context, we shall include a datum intended to play the rôle of the modular function of a locally compact group. For reasons of consistency with the Hopf algebra context, this datum has a self-dual nature: It comprises both a character $\delta \in \mathscr{H}^*$:

$$\delta(ab) = \delta(a)\delta(b), \quad \forall a, b \in \mathscr{H}, \tag{2.2}$$

and a group-like element $\sigma \in \mathscr{H}$:

$$\Delta(\sigma) = \sigma \otimes \sigma, \qquad \varepsilon(\sigma) = 1, \tag{2.3}$$

related by the condition

$$\delta(\sigma) = 1. \tag{2.4}$$

Such a pair (δ, σ) will be called a *modular pair*. The character δ gives rise to a δ-*twisted antipode* $\widetilde{S} = S_\delta : \mathcal{H} \to \mathcal{H}$, defined by

$$\widetilde{S}(h) = \sum_{(h)} \delta(h_{(1)})\, S(h_{(2)}), \quad h \in \mathcal{H}. \tag{2.5}$$

Like the untwisted antipode, $\widetilde{S}$ is an algebra antihomomorphism:

$$\begin{aligned} \widetilde{S}(h^1 h^2) &= \widetilde{S}(h^2)\widetilde{S}(h^1), \quad \forall h^1, h^2 \in \mathcal{H}, \\ \widetilde{S}(1) &= 1, \end{aligned} \tag{2.6}$$

a coalgebra twisted antimorphism:

$$\Delta\widetilde{S}(h) = \sum_{(h)} S(h_{(2)}) \otimes \widetilde{S}(h_{(1)}), \quad \forall h \in \mathcal{H}, \tag{2.7}$$

and it also satisfies the identities:

$$\varepsilon \circ \widetilde{S} = \delta, \qquad \delta \circ \widetilde{S} = \varepsilon. \tag{2.8}$$

We start by associating to $\mathcal{H}$, viewed only as a *coalgebra*, the standard cosimplicial module known as the *cobar resolution* ([1], [4]), twisted by the insertion of the group-like element $\sigma \in \mathcal{H}$. Specifically, we set $C^n(\mathcal{H}) = \mathcal{H}^{\otimes n}$, $\forall n \geq 1$, and $C^0(\mathcal{H}) = k$, then define the *face* operators $\delta_i : C^{n-1}(\mathcal{H}) \to C^n(\mathcal{H})$, $0 \leq i \leq n$, as follows, if $n > 1$:

$$\begin{aligned} \delta_0(h^1 \otimes \ldots \otimes h^{n-1}) &= 1 \otimes h^1 \otimes \ldots \otimes h^{n-1}, \\ \delta_j(h^1 \otimes \ldots \otimes h^{n-1}) &= h^1 \otimes \ldots \otimes \Delta h^j \otimes \ldots \otimes h^{n-1} \\ &= \sum_{(h_j)} h^1 \otimes \ldots \otimes h^j_{(1)} \otimes h^j_{(2)} \otimes \ldots \otimes h^{n-1}, \quad 1 \leq j \leq n-1, \\ \delta_n(h^1 \otimes \ldots \otimes h^{n-1}) &= h^1 \otimes \ldots \otimes h^{n-1} \otimes \sigma, \end{aligned} \tag{2.9}$$

while, if $n = 1$:

$$\delta_0(1) = 1, \qquad \delta_1(1) = \sigma.$$

Next, the *degeneracy* operators $\sigma_i : C^{n+1}(\mathcal{H}) \to C^n(\mathcal{H})$, $0 \leq i \leq n$, are defined by:

$$\begin{aligned} \sigma_i(h^1 \otimes \ldots \otimes h^{n+1}) &= h^1 \otimes \ldots \otimes \varepsilon(h^{i+1}) \otimes \ldots \otimes h^{n+1} \\ &= \varepsilon(h^{i+1})\, h^1 \otimes \ldots \otimes h^i \otimes h^{i+2} \otimes \ldots \otimes h^{n+1}, \end{aligned} \tag{2.10}$$

and, for $n=0$:

$$\sigma_0(h)=\varepsilon(h), \quad h\in\mathcal{H}.$$

The remaining features of the given data, namely the *product* and the *antipode* of $\mathcal{H}$ together with the character $\delta\in\mathcal{H}^*$, are used to define the candidate for the *cyclic operator*, $\tau_n: C^n(\mathcal{H})\to C^n(\mathcal{H})$, as follows:

$$\begin{aligned}\tau_n(h^1\otimes\ldots\otimes h^n)=&(\Delta^{n-1}\widetilde{S}(h^1))\cdot h^2\otimes\ldots\otimes h^n\otimes\sigma\\ =&\sum_{(h^1)} S(h^1_{(n)})h^2\otimes\ldots\otimes S(h^1_{(2)})h^n\otimes\widetilde{S}(h^1_{(1)})\sigma.\end{aligned} \tag{2.11}$$

Note that

$$\tau_1^2(h)=\tau_1(\widetilde{S}(h)\sigma)=\sigma^{-1}\widetilde{S}^2(h)\sigma,$$

therefore the following is a necessary condition for cyclicity:

$$(\sigma^{-1}\circ\widetilde{S})^2=I. \tag{2.12}$$

The remarkable fact is that this condition is also sufficient for the implementation of the sought-for Λ-module.

A modular pair (δ,σ) satisfying (2.12) is called a *modular pair in involution.*

Theorem 1. ([14, 15]) *Let $\mathcal{H}$ be a Hopf algebra endowed with a modular pair (δ,σ) in involution. Then $\mathcal{H}^{\natural}_{(\delta,\sigma)}=\{C^n(\mathcal{H})\}_{n\geq 0}$ equipped with the operators given by (2.9) – (2.11) is a module over the cyclic category Λ.*

The cyclic cohomology groups corresponding to the Λ-module $\mathcal{H}^{\natural}_{(\delta,\sigma)}$, denoted by $HC^*_{(\delta,\sigma)}(\mathcal{H})$, can be computed from the bicomplex $(CC^{*,*}(\mathcal{H}), b, B)$, analogous to (1.7), defined as follows:

$$\begin{aligned}&CC^{p,q}(\mathcal{H})=C^{q-p}(\mathcal{H}), \quad q\geq p,\\ &CC^{p,q}(\mathcal{H})=0, \quad q<p,\end{aligned} \tag{2.13}$$

the operator

$$b: C^{n-1}(\mathcal{H})\to C^n(\mathcal{H}), \qquad b=\sum_{i=0}^{n}(-1)^i\delta_i, \tag{2.14}$$

is explicitly given, if $n\geq 1$, by

$$\begin{aligned}b(h^1\otimes\ldots\otimes h^{n-1})=&1\otimes h^1\otimes\ldots\otimes h^{n-1}\\ &+\sum_{j=1}^{n-1}(-1)^j\sum_{(h_j)} h^1\otimes\ldots\otimes h^j_{(1)}\otimes h^j_{(2)}\otimes\ldots\otimes h^{n-1}\\ &+(-1)^n h^1\otimes\ldots\otimes h^{n-1}\otimes\sigma,\end{aligned}$$

and, if $n = 0$

$$b(c) = c \cdot (1 - \sigma), \quad c \in k,$$

the operator $B : C^{n+1}(\mathcal{H}) \to C^n(\mathcal{H})$ is defined by the formula

$$B = N_n \circ \widetilde{\sigma}_{-1} \circ (1 + (-1)^n \tau_{n+1}), \quad n \geq 0, \tag{2.15}$$

where $\widetilde{\sigma}_{-1} : C^{n+1}(\mathcal{H}) \to C^n(\mathcal{H})$, is the *extra degeneracy* operator

$$\begin{aligned} \widetilde{\sigma}_{-1}(h^1 \otimes \ldots \otimes h^{n+1}) =& (\Delta^{n-1}\widetilde{S}(h^1)) \cdot h^2 \otimes \ldots \otimes h^{n+1} \\ =& \sum_{(h^1)} S(h^1_{(n)})h^2 \otimes \ldots \otimes S(h^1_{(2)})h^n \otimes \widetilde{S}(h^1_{(1)})h^{n+1}, \end{aligned} \tag{2.16}$$

$$\widetilde{\sigma}_1(h) = \delta(h), \qquad h \in \mathcal{H},$$

and

$$N_n = 1 + (-1)^n \tau_n + \ldots + (-1)^{n^2} \tau_n{}^n. \tag{2.17}$$

Explicitly:

$$\begin{aligned} &N_n(h^1 \otimes \ldots \otimes h^n) \\ &= \sum_{j=0}^{n} (-1)^{nj} \Delta^{n-1}\widetilde{S}(h^j) \cdot h^{j+1} \otimes \ldots \otimes h^n \otimes \sigma \otimes \widetilde{S}^2(h^0)\sigma \otimes \ldots \otimes \widetilde{S}^2(h^{j-1})\sigma \\ &= \sum_{j=0}^{n} (-1)^{nj} \Delta^{n-1}\widetilde{S}(h^j) \cdot h^{j+1} \otimes \ldots \otimes h^n \otimes \sigma \otimes \sigma h^0 \otimes \ldots \otimes \sigma h^{j-1} \\ &= \sum_{j=0}^{n} (-1)^{nj} \sum_{(h_j)} S(h^j_{(n)}) \otimes \ldots \otimes S(h^j_{(2)}) \otimes \widetilde{S}(h^j_{(1)}) \cdot \\ &\quad \cdot h^{j+1} \otimes \ldots \otimes h^n \otimes \sigma \otimes \sigma h^0 \otimes \ldots \otimes \sigma h^{j-1}. \end{aligned} \tag{2.18}$$

In particular, for $n = 0$:

$$B(h) = \delta(h) + \varepsilon(h), \quad \forall\, h \in \mathcal{H}. \tag{2.19}$$

The expression of the B-operator can be simplified by passing to the quasi-isomorphic *normalized bicomplex* $(C\bar{C}^{*,*}_{(\delta,\sigma)}(\mathcal{H}), b, \bar{B})$, defined as follows

$$\begin{aligned} C\bar{C}^{p,q}_{(\delta,\sigma)}(\mathcal{H}) =& \bar{C}^{q-p}(\mathcal{H}), \quad q \geq p, \\ C\bar{C}^{p,q}_{(\delta,\sigma)}(\mathcal{H}) =& 0, \qquad q < p, \end{aligned} \tag{2.20}$$

where

$$\bar{C}^n(\mathcal{H}) = (\mathrm{Ker}\ \varepsilon)^{\otimes n}, \quad \forall n \geq 1, \qquad \bar{C}^0(\mathcal{H}) = k,$$

while the formula for the b-operator remains unchanged, the new horizontal operator becomes

$$\bar{B} = N_n \circ \widetilde{\sigma}_{-1}, \quad n \geq 0, \tag{2.21}$$

in particular, for $n = 0$, one has

$$\bar{B}(h) = \delta(h), \quad \forall\, h \in \mathscr{H}. \tag{2.22}$$

An alternate description of the cyclic cohomology groups $HC^*_{(\delta,1)}(\mathscr{H})$, in terms of the Cuntz-Quillen formalism, is given by M. Crainic in [16].

We should also mention that the corresponding *cyclic homology* groups

$$HC_n^{(\delta,\sigma)}(\mathscr{H}) = Tor_n^{\Lambda}(\mathscr{H}^{\natural}_{(\delta,\sigma)}, k^{\natural}) \tag{2.23}$$

can be computed from the bicomplex $(CC_{*,*}(\mathscr{H}), b, B)$, obtained by dualizing (2.13) in the obvious fashion:

$$\begin{aligned} CC_{p,q}(\mathscr{H}) &= Hom(C^{q-p}(\mathscr{H}), k), \quad q \geq p, \\ CC_{p,q}(\mathscr{H}) &= 0, \quad q < p, \end{aligned} \tag{2.24}$$

with the boundary operators b and B the transposed of the corresponding coboundaries.

When applied to the usual notion of symmetry in differential geometry, the "Hopf algebraic" version of cyclic cohomology discussed above recovers both the Lie algebra co/homology and the differentiable cohomology of Lie groups, as illustrated by the following results.

Proposition 2. **([14])** *Let $\mathfrak{g}$ be a Lie algebra and let $\delta : \mathfrak{g} \to \mathbb{C}$ be a character of $\mathfrak{g}$. With $\mathscr{U}(\mathfrak{g})$ denoting the enveloping algebra of $\mathfrak{g}_{\mathbb{C}}$, viewed as a Hopf algebra with modular pair $(\delta, 1)$, one has*

$$HC^*_{\mathrm{per}(\delta,1)}(\mathscr{U}(\mathfrak{g})) \simeq \sum_{i \equiv *(2)}^{\oplus} H_i(\mathfrak{g}, \mathbb{C}_\delta),$$

where $\mathbb{C}_\delta$ is the 1-dimensional $\mathfrak{g}$-module associated to the character δ.

Remark 3 In a dual fashion, one can prove that

$$HC_*^{\mathrm{per}(\delta,1)}(\mathscr{U}(\mathfrak{g})) \simeq \sum_{i \equiv *(2)}^{\oplus} H^i(\mathfrak{g}, \mathbb{C}_\delta). \tag{2.25}$$

Proposition 4. **([14])** *Let $\mathscr{H}(G)$ be the Hopf algebra of polynomial functions on a simply connected affine algebraic nilpotent group G, with Lie algebra $\mathfrak{g}$. Then its periodic cyclic cohomology with respect to the trivial modular pair $(\varepsilon, 1)$ coincides with the Lie algebra cohomology:*

$$HC^*_{\mathrm{per}}(\mathscr{H}(G)) \simeq \sum_{i \equiv *(2)}^{\oplus} H^i(\mathfrak{g}, \mathbb{C}).$$

Remark 5 Since by Van Est's Theorem, the cohomology of the nilpotent Lie algebra $\mathfrak{g}$ is isomorphic to the differentiable group cohomology $H^*_{\mathrm{d}}(G)$, the above isomorphism can be reformulated as

$$HC^*_{\mathrm{per}}(\mathcal{H}(G)) \simeq \sum_{i\equiv *(2)}^{\oplus} H^i_{\mathrm{d}}(G,\mathbb{C}).$$

Under the latter form it continues to hold for any affine algebraic Lie group G, with the same proof as in [14].

Hopf algebras often arise implemented as endomorphisms of associative algebras. A *Hopf action* of a Hopf algebra $\mathcal{H}$ on an algebra $\mathcal{A}$ is given by a linear map, $\mathcal{H}\otimes\mathcal{A}\to\mathcal{A}$, $\quad h\otimes a\to h(a)$ satisfying the *action property*

$$h_1(h_2a) = (h_1h_2)(a), \qquad \forall h_i\in\mathcal{H}, a\in\mathcal{A}, \tag{2.26}$$

and the *Hopf-Leibniz rule*

$$h(ab) = \sum h_{(1)}(a)\, h_{(2)}(b), \qquad \forall a,b\in\mathcal{A}, h\in\mathcal{H}. \tag{2.27}$$

In practice, $\mathcal{H}$ first appears as a subalgebra of endomorphisms of an algebra $\mathcal{A}$ fulfilling (2.27), and it is precisely the Hopf-Leibniz rule which dictates the coproduct of $\mathcal{H}$. In turn, the modular pair of $\mathcal{H}$ arises in connection with the existence of a twisted trace on $\mathcal{A}$.

Given a Hopf action $\mathcal{H}\otimes\mathcal{A}\to\mathcal{A}$ together with a modular pair (δ,σ), a linear form $\tau:\mathcal{A}\to\mathbb{C}$ is called a σ*-trace* under the action of $\mathcal{H}$ if

$$\tau(ab) = \tau(b\sigma(a)), \qquad \forall a,b\in\mathcal{A}. \tag{2.28}$$

The σ-trace τ is called δ*-invariant* under the action of $\mathcal{H}$ if

$$\tau(h(a)b) = \tau(a\widetilde{S}(h)(b)), \qquad \forall a,b\in\mathcal{A},\ h\in\mathcal{H}. \tag{2.29}$$

If $\mathcal{A}$ is unital, the "integration by parts" formula (2.29) is equivalent to the δ-invariance condition

$$\tau(h(a)) = \delta(h)\,\tau(a), \qquad \forall a\in\mathcal{A},\ h\in\mathcal{H}.$$

With the above assumptions, the very definition of the cyclic co/homology of $\mathcal{H}$, with respect to a modular pair in involution (δ,σ), is uniquely dictated such that the following *Hopf action principle* holds:

Theorem 6. **([14, 15])** *Let $\tau:\mathcal{A}\to\mathbb{C}$ be a δ-invariant σ-trace under the Hopf action of $\mathcal{H}$ on $\mathcal{A}$. Then the assignment*

$$\gamma(h^1\otimes\ldots\otimes h^n)(a^0,\ldots,a^n) = \tau(a^0h^1(a^1)\ldots h^n(a^n)), \tag{2.30}$$

$\forall a^0,\ldots,a^n \in \mathscr{A}$, $h^1,\ldots,h^n \in \mathscr{H}$, defines a map of Λ-modules $\gamma^\natural : \mathscr{H}^\natural_{(\delta,\sigma)} \to \mathscr{A}^\natural$, which in turn induces characteristic homomorphisms in cyclic co/homology:

$$\gamma_\tau^* : HC^*_{(\delta,\sigma)}(\mathscr{H}) \to HC^*(\mathscr{A}); \tag{2.31}$$

$$\gamma_*^\tau : HC_*(\mathscr{A}) \to HC_*^{(\delta,\sigma)}(\mathscr{H}). \tag{2.32}$$

As a quick illustration, let us assume that $\mathfrak{g}$ is a Lie algebra of derivations of an algebra $\mathscr{A}$ and τ is a δ-invariant trace on $\mathscr{A}$. Then (2.32), composed with the Chern character in cyclic homology (1.13) on one hand and with the isomorphism (2.25) on the other, recovers the additive map

$$ch_\tau^* : K_*(\mathscr{A}) \to H^*(\mathfrak{g}, \mathbb{C}_\delta), \tag{2.33}$$

previously introduced in [6], in terms of $\mathfrak{g}$-invariant curvature forms associated to an arbitrary $\mathfrak{g}$-connection. Before applying the isomorphism (2.25), the periodic cyclic class $\gamma_*(ch_*(e)) \in HC_*^{\mathrm{per}(\delta,1)}(\mathscr{U}(\mathfrak{g}))$, for $e^2 = e \in \mathscr{A}$, is given by the cycle with the following components:

$$\begin{aligned}
&\gamma_*(ch_0(e)) = \tau(e),\\
&\gamma_*(ch_{2k}(e))\,(h^1,\ldots,h^{2k})\\
&\quad = (-1)^k \frac{(2k)!}{k!}\left(\tau(eh^1(e)\cdots h^{2k}(e)) - \frac{1}{2}\tau(h^1(e)\cdots h^{2k}(e))\right),\\
&\qquad\qquad \forall k \geq 1,\ \forall h^1,\ldots,h^{2k} \in \mathscr{U}(\mathfrak{g}),
\end{aligned} \tag{2.34}$$

The Lie algebra cocycle representing the class $ch_\tau^*(e) \in H^*(\mathfrak{g}, \mathbb{C}_\delta)$ in terms of the Grassmannian connection is obtained by restricting $\gamma_*^\tau(ch_*(e))$ to $\wedge^\bullet \mathfrak{g}$ via antisymmetrization.

3. Transverse index theory on general foliations

The developments discussed in the preceding section have been largely motivated by a challenging computational problem concerning the index of transversely hypoelliptic differential operators on foliations [13]. In turn, they were instrumental in settling it [14]. The goal of this section is to highlight the main steps involved.

The transverse geometry of a foliation $(V, \mathscr{F})$, i.e., the geometry of the "space" of leaves $V/\mathscr{F}$, provides a prototypical example of noncommutative space, which already exhibits many of the distinctive features of the general theory. In noncommutative geometry, a geometric space is given by a *spectral triple* $(\mathscr{A}, \mathscr{H}, D)$, where $\mathscr{A}$ is an involutive algebra of operators in a Hilbert space $\mathscr{H}$, representing the "local coordinates" of the space, and D is an unbounded self-adjoint operator on $\mathscr{H}$. The operator $D^{-1} = ds$ corresponds to the infinitesimal line element in Riemannian geometry and, in addition to its metric significance, it also carries nontrivial homological meaning, representing a *K-homology* class of $\mathscr{A}$. The construction of such a spectral triple associated to a general foliation ([13]) comprises several steps and incorporates important ideas from [10] and [20].

To begin with, we recall that a codimension n foliation $\mathcal{F}$ of an N-dimensional manifold V can be given by means of a *defining cocycle* (U_i, f_i, g_{ij}), where $\{U_i\}$ is an open cover of V, $f_i : U_i \to T_i$ are submersions with connected fibers onto n-dimensional manifolds $\{T_i\}$ and

$$g_{ij} : f_j(U_i \cap U_j) \to f_i(U_i \cap U_j),$$

are diffeomorphisms such that

$$\forall\, (i,j), \quad f_i = g_{ij} \circ f_j \text{ on } U_i \cap U_j.$$

The disjoint union $M = \cup_i T_i \times \{i\}$ can be regarded as a complete transversal for the foliation, while the collection of local diffeomorphisms $\{g_{ij}\}$ of M generates the transverse *holonomy pseudogroup* Γ. We should note that the notion of *pseudogroup* used here is slightly different from the standard one, since we do not enforce the customary hereditary condition; in particular, any group of diffeomorphisms is such a pseudogroup.

We shall assume $\mathcal{F}$ *transversely oriented*, which amounts to stipulating that M is oriented and that Γ consists of orientation preserving local diffeomorphisms. From M we shall pass by a Diff^+-functorial construction ([10]) to a quotient bundle, $\pi : P \to M$, of the frame bundle, whose sections are the Riemannian metrics on M. Specifically, $P = F/SO(n)$, where F is the $GL^+(n,\mathbb{R})$-principal bundle of oriented frames on M. The total space P admits a canonical *para-Riemannian structure* as follows. The vertical subbundle $\mathcal{V} \subset TP$, $\mathcal{V} = \operatorname{Ker} \pi_*$, carries natural Euclidean structures on each of its fibers, determined solely by the choice of a $GL^+(n,\mathbb{R})$-invariant Riemannian metric on the symmetric space $GL^+(n,\mathbb{R})/SO(n)$. On the other hand, the quotient bundle $\mathcal{N} = (TP)/\mathcal{V}$ comes equipped with a tautologically defined Riemannian structure: Every $p \in P$ is an Euclidean structure on $T_{\pi(p)}(M)$ which is identified to $\mathcal{N}_p$ via π_*.

The naturality of the above construction with respect to Diff^+ ensures that the action of the holonomy pseudogroup Γ lifts to both bundles F and P. One can thus form for each the associated *smooth étale groupoid* $F \rtimes \Gamma$, resp. $P \rtimes \Gamma$. An element of $F \rtimes \Gamma$ or of $P \rtimes \Gamma$ is given by a pair

$$(x, \varphi)\,, \qquad x \in \operatorname{Range} \varphi\,,$$

while the composition law is

$$(x, \varphi) \circ (y, \psi) = (x, \varphi \circ \psi) \quad \text{if} \quad y \in \operatorname{Dom} \varphi \quad \text{and} \quad \varphi(y) = x.$$

We let

$$\mathcal{A}_F = C_c^\infty(F \rtimes \Gamma)\,, \quad \text{resp.} \quad \mathcal{A}_P = C_c^\infty(P \rtimes \Gamma)$$

denote the corresponding convolution algebras. The elements of $\mathcal{A} = \mathcal{A}_P$ will serve as "functions of local coordinates" for the noncommutative space $V/\mathcal{F}$. A generic element of $\mathcal{A}$ can be represented as a linear combination of monomials

$$a = f\,U_\psi^*\,, \quad f \in C_c^\infty(\operatorname{Dom} \psi)\,,$$

where the star indicates a contravariant notation. The multiplication rule is

$$f_1 U^*_{\psi_1} \cdot f_2 U^*_{\psi_2} = f_1 \cdot (f_2 \circ \widetilde{\psi}_1) U^*_{\psi_2 \psi_1},$$

where by hypothesis the support of $f_1(f_2 \circ \widetilde{\psi}_1)$ is a compact subset of

$$\mathrm{Dom}\,\psi_1 \cap \psi_1^{-1} \mathrm{Dom}\,\psi_2 \subset \mathrm{Dom}\,\psi_2\,\psi_1 .$$

The algebras $\mathscr{A} = \mathscr{A}_P$ and $\mathscr{A}_F$ admit canonical $*$-representations on the Hilbert spaces

$$L^2(P) = L^2(P, vol_P), \quad \text{resp.} \quad L^2(F) = L^2(F, vol_F),$$

where vol_P, resp. vol_F, denotes the canonical Diff^+-invariant volume form on P, resp. on F. Explicitly, for $\mathscr{A} = \mathscr{A}_P$,

$$((f U^*_\psi)\xi)(p) = f(p)\,\xi(\psi(p)) \quad \forall p \in P,\ \xi \in L^2(P),\ f U^*_\psi \in \mathscr{A}, \tag{3.1}$$

and similarly for $\mathscr{A}_F$. We shall denote by $A = \bar{\mathscr{A}}$ the norm closure of $\mathscr{A}$ in this representation.

Evidently, the algebra $\mathscr{A}$ depends on the choice of the defining cocycle (U_i, f_i, g_{ij}). However, if (U'_i, f'_i, g'_{ij}) is another cocycle defining the same foliation $\mathscr{F}$, with corresponding algebra $\mathscr{A}'$ (resp. A'), then $\mathscr{A}$ and $\mathscr{A}'$ are *Morita equivalent*, while the C^*-algebras A and A' are *strongly Morita equivalent.* We recall that Morita equivalence preserves the cyclic co/homology and the K-theory/K-homology. Also, in the commutative case it simply reduces to isomorphism; e.g., absent any nontrivial pseudogroup of diffeomorphisms, two manifolds N and N' are diffeomorphic iff the algebras $C^\infty_c(N)$ and $C^\infty_c(N')$ are Morita equivalent.

To complete the description of the spectral triple associated to $V/\mathscr{F}$, we need to define the operator D. In practice, it is more convenient to work with another representative of the same K-homology class, namely the *hypoelliptic signature operator* $Q = D|D|$. The latter is a second order differential operator, acting on the Hilbert space

$$\mathscr{H} = \mathscr{H}_P := L^2(\wedge^\bullet \mathscr{V}^* \otimes \wedge^\bullet \mathscr{N}^*, vol_P), \tag{3.2}$$

it is defined as a graded sum

$$Q = (d_V^* d_V - d_V d_V^*) \oplus (d_H + d_H^*), \tag{3.3}$$

where d_V denotes the vertical exterior derivative and d_H stands for the horizontal exterior differentiation with respect to a fixed connection on the frame bundle. When $n \equiv 1$ or $2 \pmod 4$, for the vertical component to make sense, one has to replace P with $P \times S^1$ so that the dimension of the vertical fiber stays even.

Proposition 7. **([13])** *For any $a \in \mathscr{A}$, $[D,a]$ is bounded. For any $f \in C^\infty_c(P)$ and $\lambda \notin \mathbb{R}$, $f(D-\lambda)^{-1}$ is p-summable, $\forall p > m = \frac{n(n+1)}{2} + 2n$.*

One now confronts a well-posed index problem. The operator D determines an *index pairing* map $\mathrm{Index}_D : K_*(\mathscr{A}) \to \mathbb{Z}$ as follows:

(0) In the *graded* (or *even*) case,

$$\mathrm{Index}_D([e]) = \mathrm{Index}\,(eD^+e)\,, \quad e^2 = e \in \mathscr{A}\,;$$

(1) In the *ungraded* (or *odd*) case,

$$\mathrm{Index}_D([u]) = \mathrm{Index}\,(P^+uP^+)\,, \quad u \in GL_1(\mathscr{A})\,,$$

where $P^+ = \frac{1+F}{2}$, with $F = \mathrm{Sign}\,(D)$.

One of the main functions of cyclic co/homology, its *raison d'être* in some sense, is to compute the index pairing via the equality

$$\mathrm{Index}_D(\kappa) = \langle ch_*(D), \mathrm{ch}^*(\kappa)\rangle, \qquad \forall \kappa \in K_*(\mathscr{A}). \tag{3.4}$$

The cyclic cohomology class $ch_*(D) \in HC^*_{\mathrm{per}}(\mathscr{A})$, i.e., its *Chern character in K-homology*, is defined in the *ungraded case* by means of the cyclic cocycle

$$\tau_F(a^0,\dots,a^n) = \mathrm{Trace}\,(a^0[F,a^1]\dots[F,a^n])\,, \qquad a^j \in \mathscr{A}\,, \tag{3.5}$$

where n is any odd integer exceeding the dimension of the spectral triple $(\mathscr{A},\mathscr{H},D)$; in the *graded case*, Trace is replaced with the graded trace Trace_s and n is even. Being defined by means of the *operator trace*, the cocycle (3.5) is inherently difficult to compute. The problem is therefore to *provide an explicit formula for the Chern character in K-homology*.

We should note at this point that, for smooth groupoids such as those associated to foliations, the answer to (3.4) is indeed known for all *K-theory classes in the range of the assembly map* from the corresponding geometric K-group to the analytic one (cf. [11]).

As mentioned before, the functors K-theory/K-homology and cyclic co/homology are Morita invariant. Moreover, the corresponding Chern characters are *Morita equivariant*, in such a way that both sides of (3.4) are preserved by the canonical isomorphisms associated with a Morita equivalence datum. Thus, one may as well take advantage of the Morita invariant nature of the problem and choose from the start a defining cocycle (U_i, f_i, g_{ij}) for $\mathscr{F}$ with all local transversals T_i *flat affine manifolds*. This renders M itself as a flat affine manifold, although it *does not* allow one to assume that the affine structure is preserved by Γ. One can however take the horizontal component in (3.3) with respect to a *flat connection* ∇. It is then readily seen that the operator Q belongs to the class of operators of the form

$$R = \pi_a(R_{\mathscr{U}}), \quad \text{with} \quad R_{\mathscr{U}} \in (\mathscr{U}(G_a(n)) \otimes \mathrm{End}(E))^{SO(n)}, \tag{3.6}$$

where $G_a(n) = \mathbb{R}^n \rtimes GL(n,\mathbb{R})$ is the affine group, π_a denotes its right regular representation and E is a unitary $SO(n)$-module.

A differential operator R of the form (3.6) will be called *affine*. If in addition the principal symbol of R, with respect to (3.7), is invertible then R will be called an *hypoelliptic affine operator*.

By an easy adaptation of a classical theorem of Nelson and Stinespring, one can show that *any hypoelliptic affine operator R which is formally self-adjoint is in fact essentially self-adjoint*, with core any dense $G_a(n)$-invariant subspace of C^∞-vectors for π_a.

The hypoelliptic calculus adapted to the para-Riemannian structure of the manifold P and to the treatment of the above operators is a particular case of the pseudodifferential calculus on Heisenberg manifolds ([2]). One simply modifies the notion of homogeneity of symbols $\sigma(p,\xi)$ by using the homotheties

$$\lambda\cdot\xi = (\lambda\xi_v,\lambda^2\xi_n), \quad \forall\lambda\in\mathbb{R}_+^*, \tag{3.7}$$

where ξ_v, ξ_n are the vertical, resp. normal, components of the covector ξ. The above formula depends on local coordinates (x_v,x_n) adapted to the vertical foliation, but the corresponding pseudodifferential calculus is independent of such choices. The principal symbol of a hypoelliptic operator of order k is a function on the fibers of $\mathcal{V}^*\oplus\mathcal{N}^*$, homogeneous of degree k in the sense of (3.7). The distributional kernel $k(x,y)$ of a pseudodifferential operator T in this hypoelliptic calculus has the following behavior near the diagonal:

$$k(x,y)\sim\sum a_j(x,x-y)-a(x)\log|x-y|'+O(1), \tag{3.8}$$

where a_j is homogeneous of degree $-j$ in $x-y$ in the sense of (3.7), and the metric $|x-y|'$ is locally of the form

$$|x-y|' = ((x_v-y_v)^4+(x_n-y_n)^2)^{1/4}. \tag{3.9}$$

The 1-density $a(x)$ is independent on the choice of metric $|\cdot|'$ and can be obtained from the symbol of order $-m$ of T, where

$$m=\frac{n(n+1)}{2}+2n$$

is the *Hausdorff dimension* of the metric space $(P,|\cdot|')$. Like in the ordinary pseudodifferential calculus, this allows to define a residue of Wodzicki-Guillemin-Manin type, extending the Dixmier trace to operators of all degrees, by the equality

$$⨍ T=\frac{1}{m}\int_P a(x). \tag{3.10}$$

One uses the hypoelliptic calculus to prove ([13]) that the spectral triple $(\mathcal{A},\mathcal{H},D)$, or more generally that obtained by replacing Q with any hypoelliptic affine operator R (in which case $D|D|=R$), fulfills the hypotheses of the operator theoretic *local index theorem* of [13]. Its application allows to express the corresponding Chern character

$$ch_*(R)=ch_*(D)\in HC^*_{\mathrm{per}}(\mathcal{A}_P)$$

in terms of the *locally computable* residue (3.10). In the odd case, it is given by the cocycle $\Phi_R = \{\varphi_n\}_{n=1,3,\dots}$ in the (b,B)-bicomplex of $\mathcal{A}$ defined as follows:

$$\varphi_n(a^0,\dots,a^n) = \sum_k c_{n,k} \int\!\!\!\!\!- a^0[R,a^1]^{(k_1)}\cdots[R,a^n]^{(k_n)}\,|R|^{-n-2|k|}, \quad a^j \in \mathcal{A}, \quad (3.11)$$

where we used the abbreviations

$$T^{(k)} = \nabla^k(T) \qquad \text{and} \qquad \nabla(T) = D^2T - TD^2,$$

k is a multi-index, $|k| = k_1 + \dots + k_n$, and

$$c_{n,k} = (-1)^{|k|}\sqrt{2i}(k_1!\dots k_n!)^{-1}\left((k_1+1)\dots(k_1+\dots+k_n+n)\right)^{-1}\Gamma(|k|+\frac{n}{2});$$

there are finitely many nonzero terms in the above sum and only finitely many components of Φ_R are nonzero. In the even case, the corresponding cocycle $\Phi_R = \{\varphi_n\}_{n=0,2,\dots}$ is defined in a similar fashion, except for φ_0 (see [13]).

The expression (3.11) is definitely explicitly computable, but its actual computation is exceedingly difficult to perform. Already in the case of codimension 1 foliations, where we did carry through its calculation by hand, it involves computing thousands of terms. On the other hand, in the absence of a guiding principle, computer calculations are unlikely to produce an illuminating answer. However, a simple inspection of (3.11) reveals some helpful general features. For the clarity of the exposition, we shall restrict our comments to the case $R = Q$, which is our main case of interest anyway.

First of all, since the passage from $(\mathcal{A}_F, \mathcal{H}_F)$ to $(\mathcal{A}_P, \mathcal{H}_P)$ involves the rather harmless operation of taking K-invariants with respect to the compact group $K = SO(n)$, we may work directly at the level of the frame bundle, equivariantly with respect to K. Secondly, since we are interested in the flat case, we may as well assume for starters that $M = \mathbb{R}^n$, with the trivial connection. This being the case, we shall identify F with the affine group $G_a(n)$. We may also replace Γ by the full group $\mathrm{Diff}^+(\mathbb{R}^n)$ and thus work for a while with the algebra

$$\mathcal{A}(n) = C_c^\infty(F(\mathbb{R}^n) \rtimes \mathrm{Diff}^+(\mathbb{R}^n)).$$

We recall that Q is built from the vertical vector fields $\{Y_i^j; i,j = 1,\dots,n\}$ which form the canonical basis of $\mathfrak{gl}(n,\mathbb{R})$ and the horizontal vector fields $\{X_k, k = 1,\dots,n\}$ coming from the canonical basis of $\mathbb{R}^n$. Therefore, the expression under the residue-integral in (3.11) involve iterated commutators of these vector fields with multiplication operators of the form $a = fU_\psi^*$, $f \in C_c^\infty(F,)$, $\psi \in \Gamma$. Now the canonical action of $GL^+(n,\mathbb{R})$ on F commutes with the action of Γ and hence extends canonically to the crossed product $\mathcal{A}_F$. At the Lie algebra level, this implies that the operators on $\mathcal{A}_F$ defined by

$$Y_i^j(fU_\psi^*) = (Y_i^j f)U_\psi^* \qquad (3.12)$$

are derivations:

$$Y_i^j(ab) = Y_i^j(a)\,b + a\,Y_i^j(b)\,. \tag{3.13}$$

The horizontal vector fields X_k on F can also be made to act on the crossed product algebra, according to the rule

$$X_k(f\,U_\psi^*) = X_k(f)\,U_\psi^*\,. \tag{3.14}$$

However, since the trivial connection is not preserved by the action of Γ, the operators X_k are no longer derivations of $\mathscr{A}_F$; they satisfy instead

$$X_i(ab) = X_i(a)\,b + a\,X_i(b) + \sum \delta_{ji}^k(a)\,Y_k^j(b), \qquad a,b \in \mathscr{A}. \tag{3.15}$$

The linear operations δ_{ij}^k are of the form

$$\delta_{ij}^k(f\,U_\psi^*) = \gamma_{ij}^k\,f\,U_\psi^*, \tag{3.16}$$

with $\gamma_{ij}^k \in C^\infty(F \rtimes \Gamma)$ characterized by the identity

$$\psi^*\,\omega_j^i - \omega_j^i = \sum_k \gamma_{jk}^i\,\theta^k, \tag{3.17}$$

where ω is the standard *flat connection form* and θ is the *fundamental form* on $F = F(\mathbb{R}^n)$. From (3.17) it easily follows that each δ_{ij}^k is a derivation:

$$\delta_{ij}^k(ab) = \delta_{ij}^k(a)\,b + a\,\delta_{ij}^k(b)\,. \tag{3.18}$$

The commutation of the Y_i^j with δ_{ab}^c preserves the linear span of the δ_{ab}^c. However, the successive commutators with X_k produces new operators:

$$\delta_{ab,i_1\ldots i_r}^c = [X_{i_r}, \ldots [X_{i_1}, \delta_{ab}^c] \ldots], \qquad r \geq 1,$$

symmetric in the indices $i_1, \ldots, i_r$. They are all of the form $T(f\,U_\psi^*) = h\,f\,U_\psi^*$, with $h \in C^\infty(F \rtimes \Gamma)$; in particular, they pairwise commute.

A first observation is that the linear space

$$\mathfrak{h}(n) = \sum \mathbb{C} \cdot Y_i^j \oplus \sum \mathbb{C} \cdot X_k \oplus \sum \mathbb{C} \cdot \delta_{ab,i_1\ldots i_r}^c$$

forms a Lie algebra and furthermore, if we let

$$\mathscr{H}(n) = \mathscr{U}(\mathfrak{h}(n))$$

denote the corresponding enveloping algebra, then $\mathscr{H}(n)$ acts on $\mathscr{A}$ satisfying a Leibniz rule of the form

$$h(ab) = \sum h_{(0)}(a)\,h_{(1)}(b), \qquad \forall a,b \in \mathscr{A}. \tag{3.19}$$

A second observation is that the product rules (3.13), (3.15) and (3.18) can be converted into *coproduct* rules

$$\begin{aligned} \Delta Y_i^j &= Y_i^j \otimes 1 + 1 \otimes Y_i^j, \\ \Delta X_i &= X_i \otimes 1 + 1 \otimes X_i + \sum_k \delta_{ji}^k \otimes Y_k^j, \\ \Delta \delta_{ij}^k &= \delta_{ij}^k \otimes 1 + 1 \otimes \delta_{ij}^k. \end{aligned} \tag{3.20}$$

Together with the requirement:

$$\Delta[Z_1, Z_2] = [\Delta Z_1, \Delta Z_2], \qquad \forall Z_1, Z_2 \in \mathfrak{h}(n),$$

which is satisfied on generators because of the flatness of the connection, they uniquely determine a multiplicative coproduct

$$\Delta : \mathcal{H}(n) \to \mathcal{H}(n) \otimes \mathcal{H}(n). \tag{3.21}$$

One can check that this coproduct is coassociative and also that it is compatible with the Leibniz rule (3.19), in the sense that

$$\Delta h = \sum h_{(0)} \otimes h_{(1)} \quad \text{iff} \quad \Delta(h_1 h_2) = \Delta h_1 \cdot \Delta h_2, \qquad \forall h_j \in \mathcal{H}(n), \ \forall a, b \in \mathcal{A}. \tag{3.22}$$

Simple computations show that there is a unique antiautomorphism S of $\mathcal{H}(n)$ such that

$$S(Y_i^j) = -Y_i^j, \qquad S(\delta_{ab}^c) = -\delta_{ab}^c, \qquad S(X_a) = -X_a + \delta_{ab}^c Y_c^b.$$

Moreover, S serves as antipode for the bialgebra $\mathcal{H}(n)$; we should note though that $S^2 \neq I$.

To summarize, one has:

Proposition 8. **([14])** *The enveloping algebra $\mathcal{H}(n)$ of the Lie algebra generated by the canonical action of $\mathbb{R}^n \rtimes \mathfrak{gl}(n, \mathbb{R})$ on $\mathcal{A}(n)$ has a unique coproduct which turns it into a Hopf algebra such that its tautological action on $\mathcal{A}(n)$ is a Hopf action.*

Remark 9 The Hopf algebra $\mathcal{H}(n)$ acts canonically on the crossed product algebra $\mathcal{A}_{F(M)} = C_c^\infty(F(M) \rtimes \Gamma)$, for any flat affine manifold M with a pseudogroup Γ of orientation preserving diffeomorphisms. Using Morita equivalence, one can always reduce the case of a general manifold M to the flat case. The obstruction one encounters in trying to transfer the action of $\mathcal{H}(n)$, via the Morita equivalence data, from the flattened version to a non-flat M is exactly the curvature of the manifold M. *The analysis of this obstruction in the general context of actions of Hopf algebras on algebras should provide the correct generalization of the notion of Riemannian curvature in the framework of noncommutative geometry.*

There is a more revealing definition ([14]) of the Hopf algebra $\mathcal{H}(n)$, in terms of a bicrossproduct construction (cf. e.g. [21]) whose origin, in the case of finite groups, can be traced to the work of G. I. Kac. In our case, it leads to the interpretation of $\mathcal{H}(n)$ as a bicrossproduct of two Hopf algebras, $\mathcal{U}_a(n)$ and $\mathcal{S}_u(n)$, canonically associated to the decomposition of the diffeomorphism group as a set-theoretic product

$$\mathrm{Diff}(\mathbb{R}^n) = G_a(n) \cdot G_u(n),$$

where $G_u(n)$ is the group of diffeomorphisms of the form $\psi(x) = x + o(x)$. $\mathcal{U}_a(n)$ is just the universal enveloping $\mathcal{U}(\mathfrak{g}_a(n))$ of the group $G_a(n)$ of affine motions of $\mathbb{R}^n$, with its natural Hopf structure. $\mathcal{S}_u(n)$ is the Hopf algebra of polynomial functions on the pro-nilpotent group of formal diffeomorphisms associated to $G_u(n)$.

The modular character of the affine group $\delta = \mathrm{Trace} : \mathfrak{g}_a(n) \to \mathbb{R}$ extends to a character $\delta \in \mathcal{H}(n)^*$. It can be readily verified that the corresponding twisted antipode satisfies the involution condition $\widetilde{S}^2 = I$. It follows that the pair $(\delta, 1)$ fulfills (2.12) and hence forms a modular pair in involution.

The preceding bicrossproduct interpretation allows to relate the cyclic cohomology of $\mathcal{H}(n)$ with respect to $(\delta, 1)$ to the Gelfand-Fuchs cohomology [18] of the infinite-dimensional Lie algebra $\mathfrak{a}_n$ of formal vector fields on $\mathbb{R}^n$.

Theorem 10. ([14]) *There is a canonical Van Est-type map of complexes which induces an isomorphism*

$$\sum_{i \equiv *(2)} H^i(\mathfrak{a}_n) \simeq HC^*_{\mathrm{per}(\delta,1)}(\mathcal{H}(n)). \tag{3.23}$$

We now return to the spectral triple $(\mathcal{A}_F, \mathcal{H}_F, D)$ associated to (M, Γ) with M flat. Then we have the canonical Hopf action $\mathcal{H}(n) \otimes \mathcal{A}_F \to \mathcal{A}_F$. In addition, the crossed product $\mathcal{A}_F$ inherits a canonical trace $\tau_F : \mathcal{A}_F \to \mathbb{C}$, dual to the volume form vol_F,

$$\tau_F(f U_\psi^*) = 0 \quad \text{if} \quad \psi \neq 1 \quad \text{and} \quad \tau_F(f) = \int_F f\, vol_F. \tag{3.24}$$

Using the Γ-invariance of vol_F, it is easy to check that the trace τ_F is δ-invariant under the action of $\mathcal{H}(n)$, i.e., the property (2.29) holds. We therefore obtain a characteristic map

$$\gamma_F^* : HC^*_{\mathrm{per}(\delta,1)}(\mathcal{H}(n)) \to HC^*_{\mathrm{per}}(\mathcal{A}_F),$$

which together with (3.23) gives rise to a new characteristic homomorphism:

$$\chi_F : H^*(\mathfrak{a}_n) \to HC^*_{\mathrm{per}}(\mathcal{A}_F). \tag{3.25}$$

Passing to $SO(n)$-invariants, one obtains an induced characteristic map from the relative Lie algebra cohomology:

$$\chi_P : \sum_{i+p \equiv *(2)} H^i(\mathfrak{a}_n, SO(n)) \to HC^*_{\mathrm{per}}(\mathcal{A}_P), \tag{3.26}$$

which instead of τ_F involves the analogous trace τ_P of $\mathscr{A}_P$ and where $p = \dim P$.

Let us assume that the action of Γ on M has no degenerate fixed point. Recall the local formula (3.11) for $ch_*(R)$. Using the built-in affine invariance of a hypoelliptic affine operator R, one can show that any cochain on $\mathscr{A}_P$ of the form,

$$\varphi(a^0,\ldots,a^n) = \int\!\!\!\!\!\!- \; a^0[R,a^1]^{(k_1)}\cdots[R,a^n]^{(k_n)}\,|R|^{-(n+|2k|)}, \qquad \forall a^j \in \mathscr{A}_P,$$

can be written as a finite linear combination

$$\varphi(a^0,\ldots,a^n) = \sum_\alpha \tau_P(a^0 h_1^\alpha(a^1)\cdots h_n^\alpha(a^n)), \quad \text{with} \quad h_i^\alpha \in \mathscr{H}(n),$$

and therefore belongs to the range of the characteristic map χ_P. The structure of the cohomology ring $H^*(\mathfrak{a}_n, SO(n))$ is well-known ([19]). It is computed by the cohomology of the finite-dimensional complex

$$\{E(h_1,h_3,\ldots,h_m)\otimes P(c_1,\ldots,c_n)\ ,\ d\},$$

where $E(h_1,h_3,\ldots,h_m)$ is the exterior algebra in the generators h_i of dimension $2i-1$, with m the largest odd integer less than n and i odd, while $P(c_1,\ldots,c_n)$ is the polynomial algebra in the generators c_i of degree $2i$ truncated by the ideal of elements of weight $> 2n$. The coboundary d is defined by

$$dh_i = c_i, \quad i \text{ odd}, \quad dc_i = 0 \text{ for all } i.$$

In particular, the Pontryagin classes $p_i = c_{2i}$ are non-trivial for all $2i \leq n$.

The final outcome of the preceding discussion is the following index theorem for transversely hypoelliptic operators on foliations:

Theorem 11. **([14])** *For any hypoelliptic affine operator R on P, there exists a characteristic class $\mathscr{L}(R) \in \sum_{i+n\equiv *(2)} H^i(\mathfrak{a}_n, SO(n))$ such that*

$$ch_*(R) = \chi_P(\mathscr{L}(R)) \in HC^*_{\text{per}}(\mathscr{A}_P).$$

Remark 12 We conclude with the remark that similar considerations, leading to analogous results, can be implemented for more specialized cases of transverse geometries, such as complex analytic and symplectic (or Hamiltonian). For example, in the *symplectic case*, which is the less obvious of the two, the transverse data consists of a symplectic manifold (M^{2n},ω) together with a pseudogroup Γ_{sp} of local symplectomorphisms. The corresponding frame bundle is the principal $Sp(n,\mathbb{R})$-bundle F_{sp} of symplectic frames. Its quotient mod K, where $K = Sp(n,\mathbb{R}) \cap O_{2n} \simeq U(n)$, is the bundle P_{sp} whose fiber at $x \in M$ consists of the almost complex structures on T_xM compatible with ω_x. It carries an intrinsic *para-Kählerian structure*, obtained as follows. The typical fiber of P_{sp} can be identified with the noncompact Hermitian symmetric space $Sp(n,\mathbb{R})/U(n)$ and, as such, it inherits a canonical Kähler structure. This gives rise to a natural Kähler structure on the vertical subbundle $\mathscr{V}$ of the tangent bundle TP_{sp}. On the other hand, the normal bundle $\mathscr{N} = TP_{sp}/\mathscr{V} \simeq TM$ possesses a tautological Kähler structure. Indeed, a point in P_{sp} is by definition an almost complex structure and thus, together with ω, determines a "moving" Kähler structure. The entire construction is functorial with respect to local symplectomorphisms. One can therefore define the symplectic analogue Q_{sp} of the hypoelliptic signature operator as a graded direct sum

$$Q_{sp} = (\bar{\partial}_V^* \bar{\partial}_V - \bar{\partial}_V \bar{\partial}_V^*) \oplus (\bar{\partial}_H + \bar{\partial}_H^*),$$

where the horizontal $\bar{\partial}$-operator $\bar{\partial}_H$ is associated to a symplectic connection. *The corresponding index theorem asserts that, with (M^{2n}, ω) flat, acted upon by an arbitrary pseudogroup of local symplectomorphisms Γ_{sp}, and with*

$$\chi_{sp} : H^*(\mathfrak{a}_n^{sp}, U(n)) \to HC^*_{\mathrm{per}}(\mathscr{A}_{P_{sp}}), \tag{3.27}$$

denoting the characteristic map corresponding to the Lie algebra $\mathfrak{a}_n^{sp}$ of formal Hamiltonian vector fields on $\mathbb{R}^{2n}$, there exists a characteristic class $\mathscr{L}_{sp} \in \sum_{i\equiv *(2)} H^i(\mathfrak{a}_n^{sp}, U(n))$ such that

$$ch_*(Q_{sp}) = \chi_{sp}(\mathscr{L}_{sp}) \in HC^*_{\mathrm{per}}(\mathscr{A}_{P_{sp}}).$$

As in the para-Riemannian case, the proof relies on the local Chern character formula (3.11) and on the cyclic cohomology of the Hopf algebra $\mathscr{H}_{sp}(n)$, associated to the group of symplectomorphism of $\mathbb{R}^{2n}$ in the same manner as $\mathscr{H}(n)$ was constructed from $\mathrm{Diff}(\mathbb{R}^n)$.

4. Quantum groups and the modular square

The definition of cyclic co/homology of Hopf algebras hinges on the existence of modular pairs in involution. The necessity of this condition may appear as artificial. In fact, quite the opposite is true and the examples given below serve to illustrate that most Hopf algebras arising in "nature", including quantum groups and their duals, do come equipped with intrinsic modular pairs.

1. We begin with the class of *quasitriangular Hopf algebras*, introduced by Drinfeld [17], in connection with the quantum inverse scattering method for constructing quantum integrable systems. Such a Hopf algebra comes endowed with an universal *$\mathscr{R}$-matrix*, inducing solutions of the Yang-Baxter equation on each of their modules. (For a lucid introduction into the subject, see [21]).

A *quasitriangular Hopf algebra* is a Hopf algebra $\mathscr{H}$ which admits an invertible element $R = \sum_i s_i \otimes t_i \in \mathscr{H} \otimes \mathscr{H}$, such that

$$\begin{aligned} \Delta^{\mathrm{op}}(x) &= R\Delta(x)R^{-1}, \quad \forall x \in \mathscr{H}, \\ (\Delta \otimes I)(R) &= R_{13}R_{23}, \\ (I \otimes \Delta)(R) &= R_{13}R_{12}, \end{aligned} \tag{4.1}$$

where we have used the customary "leg numbering" notation, e.g.,

$$R_{23} = \sum_i 1 \otimes s_i \otimes t_i.$$

The square of the antipode is then an inner automorphism $S^2(x) = uxu^{-1}$, with

$$u = \sum S(t_i)s_i, \qquad u^{-1} = \sum S^{-1}(t_i)S(s_i).$$

Furthermore, $uS(u) = S(u)u$ is central in $\mathscr{H}$ and one has

$$\varepsilon(u) = 1, \quad \Delta u = (R_{21}R)^{-1}(u \otimes u) = (u \otimes u)(R_{21}R)^{-1}.$$

A quasitriangular Hopf algebra $\mathscr{H}$ is called a *ribbon algebra* [24], if there exists a *central element* $\theta \in \mathscr{H}$ such that

$$\Delta(\theta) = (R_{21}R)^{-1}(\theta \otimes \theta), \quad \varepsilon(\theta) = 1, \quad S(\theta) = \theta. \tag{4.2}$$

Any quasitriangular Hopf $\mathscr{H}$ algebra has a "double cover" cover $\widetilde{\mathscr{H}}$ satisfying the ribbon condition (4.2). More precisely (cf.[24]):

$$\widetilde{\mathscr{H}} = \mathscr{H}[\theta]/(\theta^2 - uS(u))$$

has a unique Hopf algebra structure such that, under the natural inclusion, $\mathscr{H}$ is a Hopf subalgebra.

If $\mathscr{H}$ is a ribbon algebra, by setting $\sigma = \theta^{-1}u$, one gets a group-like element

$$\Delta\sigma = \sigma \otimes \sigma, \quad \varepsilon(\sigma) = 1, \quad S(\sigma) = \sigma^{-1},$$

such that, for any $x \in \widetilde{\mathscr{H}}$,

$$\begin{aligned}(\sigma^{-1} \circ S)^2(x) &= \sigma^{-1}S(\sigma^{-1}S(x)) = \sigma^{-1}S^2(x)\sigma \\ &= \sigma^{-1}uxu^{-1}\sigma = \theta x \theta^{-1} = x.\end{aligned}$$

Thus, (ε, σ) is a modular pair in involution for $\mathscr{H}$.

By dualizing the above definitions one obtains the notion of *coquasitriangular*, resp. *coribbon algebra.* Among the most prominent examples of coribbon algebras are the function algebras of the classical quantum groups ($GL_q(N)$, $SL_q(N)$, $SO_q(N)$, $O_q(N)$, and $Sp_q(N)$). The analogue of the above *ribbon group-like element* σ for a coribbon algebra $\mathscr{H}$, is the *ribbon character* $\delta \in \mathscr{H}^*$. The corresponding twisted antipode satisfies the condition $\widetilde{S}^2 = 1$, which renders $(\delta, 1)$ as a canonical modular pair in involution for $\mathscr{H}$.

We thus have:

Proposition 13. **([15])** *Coribbon algebras and compact quantum groups are each intrinsically endowed with a modular pair in involution $(\delta, 1)$. Dually, ribbon algebras and duals of compact quantum groups are each intrinsically endowed with a modular pair in involution $(1, \sigma)$.*

For a *compact quantum group* in the sense of Woronowicz, the stated property follows from Theorem 5.6 of [28], describing the modular properties of the analogue of Haar measure.

2. Evidently, one can produce modular pairs in involution with both δ and σ nontrivial by forming tensor products of dual classes of Hopf algebras as in the preceding statement. The fully non-unimodular situation arises naturally however, in the case of *locally compact quantum groups*, because of the existence, by fiat or otherwise, of left and right Haar weights. We refer to [22] for the most recent and concise formalization of this notion, which is in remarkable agreement with our framework for cyclic co/homology of Hopf algebras and of Hopf actions. This accord is manifest in the following *construction of a modular square* associated to a Hopf algebra $\mathcal{H}$ modeling a locally compact group. Since the inherent analytic intricacies are beyond the scope of the present exposition, we shall keep the illustration at a formal level (comp. [27] for an algebraic setting).

By analogy with the definition of a C^*-algebraic quantum group in [22], we assume the existence and uniqueness (up to a scalar) of a *left invariant weight* φ, satisfying a *KMS-like condition*. The invariance means that

$$(I\otimes\varphi)((I\otimes x)\Delta(y)) = S((I\otimes\varphi)\Delta(x)(I\otimes y)), \qquad \forall x,y\in\mathcal{H}, \tag{4.3}$$

while the KMS condition stipulates the existence of a *modular group of automorphisms* σ_t of $\mathcal{H}$, such that

$$\varphi\circ\sigma_t = \varphi, \qquad \varphi(xy) = \varphi(\sigma_i(y)x), \quad \forall x,y\in\mathcal{H}. \tag{4.4}$$

Taking $\psi = \varphi\circ S^{-1}$, one obtains a a *right invariant weight*,

$$(\psi\otimes I)(\Delta(x)(y\otimes 1)) = S((\psi\otimes I)(x\otimes 1)\Delta(y)), \tag{4.5}$$

which is also unique up to a scalar and has modular group $\sigma'_t = S\circ\sigma_t\circ S^{-1}$.

It will be convenient to express the above properties in terms of the natural left and right actions of the dual Hopf algebra $\hat{\mathcal{H}} = \mathcal{H}^*$ on $\mathcal{H}$. Given $\omega\in\hat{\mathcal{H}}$ and $x\in\mathcal{H}$, we denote

$$\begin{aligned} \omega\cdot x &= \sum\omega(x_{(1)})\,x_{(2)} = (\omega\otimes I)\Delta(x)\,, \\ x\cdot\omega &= \sum x_{(1)}\,\omega(x_{(2)}) = (I\otimes\omega)\Delta(x)\,. \end{aligned} \tag{4.6}$$

With respect to the natural product of $\hat{\mathcal{H}}$,

$$(\omega_1 * \omega_2)(x) = <\omega_1\otimes\omega_2, \Delta(x)> = \sum\omega_1(x_{(1)})\,\omega_2(x_{(2)}), \qquad \forall x\in\mathcal{H},$$

the left action in (4.6) is the transpose of the left regular representation of $\hat{\mathcal{H}}$, hence defines a representation of the opposite algebra $\hat{\mathcal{H}}^{\mathrm{op}}$ on $\mathcal{H}$, while the right action in (4.6), being the transpose of the right regular representation of $\hat{\mathcal{H}}$, gives a representation of the algebra $\hat{\mathcal{H}}$ on $\mathcal{H}$. On the other hand, it is easy to check that both actions satisfy the rule (2.27) and therefore (4.6) *defines a Hopf action of the tensor product Hopf algebra* $\widetilde{\mathscr{H}} := \mathscr{H}^{\mathrm{op}} \otimes \mathscr{H}$ *on* $\mathscr{H}$,

$$\widetilde{\mathscr{H}} \otimes \mathscr{H} \to \mathscr{H}, \quad (\omega_1 \otimes \omega_2, x) \to \omega_1 \cdot x \cdot \omega_2 . \tag{4.7}$$

The invariance conditions (4.3) and (4.5) can now be rewritten as follows:

$$\begin{aligned} \varphi((\omega \cdot x) y) &= \varphi(x(S^{-1}(\omega) \cdot y)), \\ \psi((x \cdot \omega) y) &= \psi(x(y \cdot S(\omega))), \end{aligned} \tag{4.8}$$

where S^{-1} occurs in the first identity as the antipode of $\mathscr{H}^{\mathrm{op}}$.

The left invariance property of ψ gives the analogue of the modular function, namely a *group-like element* $\delta \in \mathscr{H}$ such that

$$(I \otimes \psi)\Delta(x) = \psi(x)\delta , \qquad \forall x \in \mathscr{H},$$

or equivalently

$$\psi(\omega \cdot x) = \omega(\delta)\psi(x), \qquad \forall x \in \mathscr{H}. \tag{4.9}$$

The *modular element* δ also relates the left and right Haar weights:

$$\varphi(x) = \psi(\delta^{\frac{1}{2}} x \delta^{\frac{1}{2}}), \qquad \forall x \in \mathscr{H}. \tag{4.10}$$

In particular, the full invariance property of φ under the action of $\widetilde{\mathscr{H}}$ is given by

$$\varphi((\omega_1 \cdot x \cdot \omega_2) y) = \varphi(x(S^{-1}(\omega_1) \cdot y \cdot S_{\delta^{-1}}(\omega_2))), \quad \forall \omega_1, \omega_2 \in \widetilde{\mathscr{H}},\ x, y \in \mathscr{H}, \tag{4.11}$$

where $S_{\delta^{-1}}$ denotes the twisted antipode (2.5) corresponding to δ^{-1}.

Let us now form the *midweight* τ,

$$\tau(x) = \varphi(\delta^{-\frac{1}{4}} x \delta^{-\frac{1}{4}}), \qquad \forall x \in \mathscr{H}. \tag{4.12}$$

One checks that its behavior under the action of $\widetilde{\mathscr{H}}$ is as follows:

$$\tau((\omega_1 \cdot x \cdot \omega_2) y) = \tau(x(S^{-1}_{\delta^{1/2}}(\omega_1) \cdot y \cdot S_{\delta^{-1/2}}(\omega_2))), \tag{4.13}$$

for any $\omega_1, \omega_2 \in \widetilde{\mathscr{H}}$ and $x, y \in \mathscr{H}$. In other words, τ is $\widetilde{\delta}$*-invariant under the action (4.7), with respect to the character*

$$\widetilde{\delta} = \delta^{\frac{1}{2}} \otimes \delta^{-\frac{1}{2}} \in \widetilde{\mathscr{H}}^{*} = \mathscr{H}^{\mathrm{op}} \otimes \mathscr{H}. \tag{4.14}$$

On the other hand, it follows as in [22], but with a slightly different notation, that there exists a group-like element $\sigma \in \mathscr{H}$, such that the modular groups of φ, ψ can be expressed as follows:

$$\begin{aligned} \sigma_t(x) &= \delta^{it/2}(\sigma^{it/2} \cdot x \cdot \sigma^{it/2})\delta^{-it/2}, \\ \sigma'_t(x) &= \delta^{-it/2}(\sigma^{it/2} \cdot x \cdot \sigma^{it/2})\delta^{it/2}, \quad \forall x \in \mathscr{H}. \end{aligned} \tag{4.15}$$

In terms of the modular group of τ, to be denoted by σ_t^τ , (4.15) is equivalent to

$$\sigma_t^\tau(x) = \sigma^{it/2}\cdot x\cdot \sigma^{it/2}, \quad \forall x \in \mathscr{H}. \tag{4.16}$$

This shows that τ is a $\widetilde{\sigma}$*-trace for the action (4.7), with group-like element*

$$\widetilde{\sigma} = \sigma^{\frac{1}{2}} \otimes \sigma^{\frac{1}{2}} \in \widetilde{\mathscr{H}}^* = \mathscr{H}^{\mathrm{op}} \otimes \mathscr{H}. \tag{4.17}$$

It remains to compute the square of the corresponding doubly twisted antipode of $\widetilde{\mathscr{H}}$,

$$\widetilde{\sigma}^{-1} \circ S_{\widetilde{d}} = \sigma^{-1/2} \circ S^{-1}_{\delta^{1/2}} \otimes \sigma^{-1/2} \circ S_{\delta^{-1/2}} : \hat{\mathscr{H}}^{\mathrm{op}} \otimes \hat{\mathscr{H}} \to \hat{\mathscr{H}}^{\mathrm{op}} \otimes \hat{\mathscr{H}}. \tag{4.18}$$

It suffices to compute the square of $\sigma^{-1/2} \circ S_{\delta^{-1/2}} : \hat{\mathscr{H}} \to \hat{\mathscr{H}}$, or equivalently, the square of its transpose, for which a straightforward calculation gives:

$$(\sigma^{-1/2} \circ S_{\delta^{-1/2}})^2 = < \sigma^{-1/2}, \delta^{-1/2} > I_{\hat{\mathscr{H}}}.$$

Since the passage to the opposite algebra gives the reciprocal scalar, it follows that

$$(\widetilde{\sigma}^{-1} \circ S_{\widetilde{\delta}})^2 = I_{\widetilde{\mathscr{H}}}. \tag{4.19}$$

We summarize the conclusions of the preceding discussion in the following result:

Theorem 14 (i) *The Hopf algebra* $\widetilde{\mathscr{H}} = \hat{\mathscr{H}}^{\mathrm{op}} \otimes \hat{\mathscr{H}}$ *possesses a canonical modular pair in involution* $(\widetilde{\delta} = \delta^{\frac{1}{2}} \otimes \delta^{-\frac{1}{2}}, \widetilde{\sigma} = \sigma^{\frac{1}{2}} \otimes \sigma^{\frac{1}{2}})$.

(ii) *The Haar midweight* τ*, given by (4.12), is a* $\widetilde{\delta}$*-invariant* $\widetilde{\sigma}$*-trace for the canonical action of* $\widetilde{\mathscr{H}}$ *on* $\mathscr{H}$*.*

The first statement characterizes the construction we referred to as *the modular square* associated to a Hopf algebra $\mathscr{H}$ that models a locally compact quantum group. Together with the Haar midweight τ of the second statement, it determines in cyclic cohomology a *modular characteristic homomorphism*:

$$\gamma_\tau^* : HC^*_{(\widetilde{\delta},\widetilde{\sigma})}(\widetilde{\mathscr{H}}) \to HC^*(\mathscr{H}). \tag{4.20}$$

References

1. Adams, J. F.: On the cobar construction, *Proc. Nat. Acad. Sci. U.S.A.* **42** (1956), 409–412.
2. Beals, R. and Greiner, P.: *Calculus on Heisenberg manifolds*, Annals of Math. Studies **119**, Princeton Univ. Press, Princeton, N.J., 1988.
3. Burghelea, R.: The cyclic homology of the group rings, *Comment. Math. Helv.* **60** (1985), 354–365.
4. Cartier, P.: Cohomologie des coalgèbres (Exposé 5), in: *Séminaire Sophus Lie, 1955–56*, Faculté des Sciences de Paris, 1957.

5. Connes, A.: Spectral sequence and homology of currents for operator algebras, *Math. Forschungsinst. Oberwolfach Tagungsber.* 41/81; *Funktionalanalysis und C^*-Algebren*, 27-9/3-10, 1981.
6. Connes, A.: C^* algèbres et géométrie differentielle, *C. R. Acad. Sci. Paris, Sér. A-B* **290** (1980), 599–604.
7. A. Connes: Noncommutative differential geometry. Part I: The Chern character in K-homology, *Preprint IHES* (M/82/53), 1982; Part II: de Rham homology and noncommutative algebra, *Preprint IHES* (M/83/19), 1983.
8. Connes, A.: Noncommutative differential geometry, *Inst. Hautes Etudes Sci. Publ. Math.* **62** (1985), 257–360.
9. Connes, A.: Cohomologie cyclique et foncteur Ext^n, *C. R. Acad. Sci. Paris, Sér. I Math* **296** (1983), 953–958.
10. Connes, A.: Cyclic cohomology and the transverse fundamental class of a foliation, in: H. Araki and E. G. Effros, (eds.), *Geometric methods in operator algebras, (Kyoto, 1983), Pitman Res. Notes in Math.* **123** Longman Sci. Tech., Harlow, 1986, pp. 52–144.
11. Connes, A.: *Noncommutative geometry*, Academic Press, San Diego, 1994.
12. Connes, A. and Moscovici, H.: Cyclic cohomology, the Novikov conjecture and hyperbolic groups, *Topology* **29** (1990), 345–388.
13. Connes, A. and Moscovici, H.: The local index formula in noncommutative geometry, *GAFA* **5** (1995), 174–243.
14. Connes, A. and Moscovici, H.: Hopf Algebras, Cyclic Cohomology and the Transverse Index Theorem, *Comm. Math. Phys.* **198** (1998), 199–246.
15. Connes, A. and Moscovici, H.: Cyclic Cohomology and Hopf Algebras, *Lett. Math. Phys.* **48** (1999), 97–108.
16. Crainic, M.: Cyclic Cohomology of Hopf Algebras and a Noncommutative Chern-Weil Theory, preprint QA/9812113.
17. Drinfeld, V. G.: Quantum groups, in: *Proc. Int. Cong. Math. (Berkeley, 1986)*, Amer. Math. Soc., Providence, RI, 1987, pp. 798–820.
18. Gelfand, I. M. and Fuchs, D. B.: Cohomology of the Lie algebra of formal vector fields, *Izv. Akad. Nauk SSSR* **34** (1970), 322–337.
19. Godbillon, G.: Cohomologies d'algèbres de Lie de champs de vecteurs formels, in: *Séminaire Bourbaki (1971/1972), Exposé No.421*, Lecture Notes in Math. Vol. **383**, Springer, Berlin, 1974, pp. 69–87.
20. Hilsum, M. and Skandalis, G., : Morphismes K-orientés d'espaces de feuilles et fonctorialité en théorie de Kasparov, *Ann. Sci. Ecole Norm. Sup. (4)* **20** (1987), 325–390.
21. Kassel, C.: *Quantum Groups*, Springer-Verlag, New York, 1995.
22. Kustermans, J. and Vaes, S.: A simple definition for locally compact quantum groups, *C. R. Acad. Sci. Paris, Sèr. I Math* **328** (1999), 871–876.
23. Loday, J.L.: *Cyclic Homology*, Springer-Verlag, Berlin, 1992, 1998.
24. Reshetikhin, N. Yu. and Turaev, V. G.: Ribbon graphs and their invariants derived from quantum groups, *Comm. Math. Phys.* **127** (1990), 1–26 .
25. Sweedler, M.E.: *Hopf Algebras*, W. A. Benjamin Inc., New York, 1969.
26. Tsygan, B. L.: Homology of matrix Lie algebras over rings and the Hochschild homology, *Uspekhi Math. Nauk.* **38** (1983), 217–218.
27. Van Daele, A.: An algebraic framework for group duality, *Adv. Math.* **140** (1998), 323–366.
28. Woronowicz, S.L.: Compact matrix pseudogroups, *Comm. Math. Phys.* **111** (1987), 613–665.

Modular double of a Quantum Group

Ludwig Faddeev
St. Petersburg Branch of Steklov Mathematical Institute
Russian Academy of Sciences, Fontanka 27, St. Petersburg, Russia
e-mail: `faddeev@pdmi.ras.ru`
and
Helsinki Institute of Physics
P.O. Box 9, FIN-00014 University of Helsinki, Finland

Abstract. The well known technical difficulties in the theory of quantum groups with $q = i\pi\tau$ appear when τ is real. It is argued that adding a second quantum group with dual parameter $1/\tau$ cures these difficulties. Moreover the new object, which we propose to call a modular double, has natural applications in conformal field models.

Keywords: quantum group, modular duality, factor II_1, universal R-matrix

Mathematics Subject Classifications (2000): 17B37, 20G42, 46L35

As is clear from the title, I shall deal with some questions connected with the theory of Quantum Groups. If I remember right, Moshé did not like Quantum Groups after this notion was crystallized by Drinfeld [1] in a purely algebraic manner. However his own attraction to deformations (as well as the pressure of authors in LMP) made him change his mind. When I presented the subject described below at a St. Petersburg meeting in May 1998, he did not express any bad feelings. So I decided to publish it in this memorial volume.

There are several sources to my proposal. I shall give just two of them, one "mathematical" and another "physical", as is appropriate for a paper in Mathematical Physics.

1) In the definition of Quantum Group one uses the deformation of the Chevalley generators K, f, e, whereas for the construction of the universal R-matrix one needs nonpolynomial elements like $H = \ln K$. Explicit formulas will be reminded below. This unfortunate obstacle can be circumvented in several ways: one, *à la* Lusztig [2] is just not to use the explicit formulas of Drinfeld; another, followed in most texts on Quantum Groups (see e.g. [3]), is to employ formal series in $\ln q$. However the value of R-matrix as a genuine operator is too high and deserves a more friendly attitude.

2) In the applications of Quantum Groups to Conformal Field Theory one explicitly sees, that together with the attributes of Quantum Groups (i.e. eigenvalues of Laplacians) for $q = e^{i\pi\tau}$ there enter analogous objects, corresponding to $\tilde{q} = e^{-i\pi/\tau}$. This modular duality is a well-known "experimental" fact, which goes without satisfactory explanation.

G. Dito and D. Sternheimer (eds.), Conférence Moshé Flato 1999, Vol. 1, 149 – 156.

In the following an extension of Quantum Groups will be described, which will throw some light on both topics above. Roughly speaking I propose to unify the Quantum Groups for q and $\tilde{q}$ in one object having a modular structure (see e.g. [4]). Thus the combination of words "Modular double of a Quantum Group" seems to be quite relevant for it.

The element $\ln K$ will have a natural definition and the expression for the universal R-matrix will become much more meaningfull in this extension. I believe also, that it is this modular double which in fact defines the hidden symmetry in the Conformal Field Theory. There are some indications of this in the literature [5, 6, 7] and recently it was made more explicit in [8, 9]. At this moment I know well how my proposal works in all details in the case of rank 1, the $SL(2)$ group. This will be presented below. The tools for the $SL(N)$ generalizations are known [10, 11], but other series of simple groups are not treated yet.

The work was partially supported by INTAS, RFFR (grant RFFR-96-01-00851) and the University of Helsinki.

1. Reminder on the Quantum Group $SL(2)$

I shall use an extension of $SL_q(2)$ which is the Drinfeld double of its Borel part. There are four generators K, K', e, f with familiar relations:

$$
\begin{aligned}
Ke &= q^2 eK & K'e &= q^{-2}eK' \\
Kf &= q^{-2}fK & K'f &= q^2 fK' \\
ef - fe &= \frac{K-K'}{q-q^{-1}} & KK' &= K'K.
\end{aligned}
$$

The algebra $\mathscr{U}_q$ generated by K, K', e, f over the field $\mathbb{C}$ of complex numbers, (so that q is a complex number) has two central elements:

$$
J = KK', \quad C = \frac{K-K'}{q-q^{-1}} + (q-q^{-1})^2(ef-fe).
$$

Reduction to $SL_q(2)$ is achieved if we put $J = \mathrm{id}$; however, we shall not do it here.

The universal R-matrix is affiliated with the tensor square $\mathscr{U}_q \otimes \mathscr{U}_q$ and defines the property of the Hopf multiplication $\triangle$ in $\mathscr{U}_q$, $\sigma \circ \triangle = R \triangle R^{-1}$, where σ is a permutation $\sigma(a \otimes b) = b \otimes a$. Drinfeld has given a formal expression for R:

$$
R = q^{-\frac{H \otimes H'}{2}} s_q(-(q-q^{-1})^2 e \otimes f)
$$

where $K = q^H$, $K' = q^{H'}$, and $s_q(w)$ is a q-exponent which can be written in several forms:

$$s_q(w) = \prod_{n=0}^{\infty}(1+q^{2n+1}w)$$
$$= 1 + \sum_{k=0}^{\infty} \frac{(-1)^k q^{\frac{n(n-1)}{2}} w^k}{(q-q^{-1})\cdots(q^k-q^{-k})}$$
$$= \exp \sum_{k=1}^{\infty} \frac{(-1)^k w^k}{k(q^k-q^{-k})}.$$

The term q-exponent is the most appropriate for the second form. In the third form there enters the q-deformed dilogarithm, which was explored in particular in [12, 13]. The title q-exponent is strongly supported by the following property, first found in [14]: let u, v be a Weyl pair

$$uv = q^2 vu,$$

then

$$s_q(u)s_q(v) = s_q(u+v).$$

Let us note that in [12] it was found that one more property holds:

$$s_q(v)s_q(u) = s_q(u+v+q^{-1}uv) = s_q(u)s_q(q^{-1}uv)s_q(v),$$

which can be called a "pentagon identity" and is a quantum deformation of the corresponding property of the dilogarithm [13]. Now we see one more deficiency in the definition of the universal R-matrix, besides the necessity of using $\log K$. The function $s_q(w)$ behaves badly for q lying on the unit circle $|q| = 1$. Indeed, for example, in the third form of $s_q(w)$ we see small denominators. Thus the expression for the universal R-matrix needs some mending.

2. The main idea

The problem of defining the log of an operator appears already in a simpler example of Weyl pair u, v. It is easy to realize the defining relation for u, v via the Heisenberg pair P, Q with relation $[Q,P] = 2\pi i\hbar I$.

Indeed, the pair $u = e^{\alpha P}$, $v = e^{\beta Q}$ satisfies Weyl relations: $uv = q^2 vu$ with $\ln q = \frac{\pi\alpha\beta\hbar}{i}$. Thus the pair (P,Q) defines (u,v). However the inverse is not true, partly because $\log u$ or $\log v$ are badly defined. A more subtle fact is that the pair P, Q defines a second Weyl pair

$$\tilde{u} = e^{\tilde{\alpha}P}, \quad \tilde{v} = e^{\tilde{\beta}Q}$$

with a different phase $\tilde{q}$, $\ln\tilde{q} = \frac{\pi\tilde{\alpha}\tilde{\beta}\hbar}{i}$, which commutes with the first pair if $\alpha\tilde{\beta} = \hbar^{-1}$ and $\tilde{\alpha}\beta = \hbar^{-1}$, so that in particular

$$\hbar^2\alpha\beta = \frac{1}{\tilde{\alpha}\tilde{\beta}}.$$

What is less trivial is the fact that, together, the commuting pairs (u,v) and $(\tilde{u},\tilde{v})$ define naturally P and Q for generic q. This fact was discussed explicitly in [15], but of course could be traced to earlier literature, in particular to A. Connes' monograph on noncommutative geometry [16] and to the paper [17].

We see more concisely, that the algebra $\mathscr{B}$ generated by P, Q is factored into the product

$$\mathscr{B} = \mathscr{A}_q \otimes \mathscr{A}_{\tilde{q}}$$

of commuting factors, generated by (u,v) and $(\tilde{u},\tilde{v})$ correspondingly. I used the term "factor" with full algebraic meaning: Indeed, neither $\mathscr{A}_q$ nor $\mathscr{A}_{\tilde{q}}$ have non-trivial center. However, I did not introduce any $*$-structure, so the connection with von Neumann theory (see e.g. [16]) is incomplete. In particular, the last formula must contain some closure. In other words, the use of tensor product in this formula is somewhat loose. Indeed, $\mathscr{A}_q$ and $\mathscr{A}_{\tilde{q}}$ are factors of type II_1 for generic q because they are infinite dimensional but allow for trace

$$\mathrm{tr}\left(\sum a_{mn}u^m v^n\right) = a_{00},$$

which is equal to 1 for the unit operator. On the other hand, $\mathscr{B}$ is a factor I_∞ since it can be realized as the algebra of all operators in $L_2(\mathbb{R})$.

I hope that now the idea of what to do in the Quantum Group case is clear — to define the $\log K$ one has to extend the algebra $\mathscr{A}_q$, adding to it some dual generators. The definition of "dual" is clear for the Weyl type operators

$$\tilde{K} = (K)^{1/\tau}$$

if we put $q = e^{i\pi\tau}$. However not all generators of a Quantum Group are of Weyl type. So we have to seek a new set of generators which have this property. Fortunately, the theory of integrable models, which previously led to main relations of Quantum Groups [18], produces also this relevant set of generators. Indeed, the Quantum Group generators appeared first as the elements of the Lax operator of the XXZ model. The lattice Sine-Gordon model introduced in [19] belongs to this class (see e.g. [20]) and, in its turn, naturally uses the Weyl type generators. In the next section we shall give the corresponding formulas in somewhat purified form.

3. Explicit construction

Consider the algebra $\mathscr{C}$ generated by four generators w_1, w_2, w_3, w_4, which is convenient to label by an index $n \in \mathbb{Z}_4$, so that $w_{n+4} = w_n$. We impose Weyl type relations on the nearest neighbours

$$w_n w_{n+1} = q^2 w_{n+1} w_n,$$

whereas

$$w_n w_m = w_m w_n \quad |m-n| > 1.$$

The algebra $\mathcal{C}_q$ (supplied by label q) has two central elements:

$$Z_1 = w_1 w_3, \quad Z_2 = w_2 w_4.$$

It is a simple exercise to check that

$$e = i\frac{w_1 + w_2}{q - q^{-1}} \qquad f = i\frac{w_3 + w_4}{q - q^{-1}}$$
$$K = q^{-1} w_2 w_3 \qquad K' = q^{-1} w_4 w_1$$

satisfy the defining relations of a Quantum Group. The relation between the central elements is as follows:

$$J = Z_1 Z_2, \quad C = Z_1 + Z_2,$$

and so $\mathcal{C}_q$ is a double cover of $\mathcal{A}_q$.

Following the reasoning of Section 2, I introduce a second algebra $\mathcal{C}_{\tilde{q}}$ with generators $\tilde{w}_n = w_n^{1/\tau}$. Both can be described in terms of the Heisenberg type generators p_n with relations[1]

$$[p_n, p_{n+1}] = -2\pi i I,$$

if we put $w_n = e^{bp_n}$, $\tilde{w}_n = e^{p_n/b}$, where $q = e^{i\pi b^2}$, $\tau = b^2$. In particular, we see that

$$K = e^{b(p_2+p_3)} \qquad K' = e^{b(p_1+p_4)}$$
$$\tilde{K} = e^{\frac{(p_2+p_3)}{b}} \qquad \tilde{K}' = e^{\frac{(p_1+p_4)}{b}}$$

and the first factor in the universal R-matrix is expressed as

$$q^{-\frac{H\otimes H'}{2}} = e^{\frac{\pi}{2i}(p_2+p_3)\otimes(p_1+p_4)} = \tilde{q}^{-\frac{\tilde{H}\otimes\tilde{H}'}{2}},$$

and does not depend on q, serving both quantum groups $\mathcal{A}_q$ and $\mathcal{A}_{\tilde{q}}$.

Let us turn now to the second factor in the universal R-matrix. We have

$$\begin{aligned} s_q(-(q-q^{-1})^2 e\otimes f) &= s_q((w_1+w_2)\otimes(w_3+w_4)) \\ &= s_q(w_1\otimes w_3)s_q(w_1\otimes w_4)s_q(w_2\otimes w_3)s_q(w_2\otimes w_4), \end{aligned}$$

(the last equality follows from Schützenberger relation) so that only Weyl type combinations enter here. Now I use the observation from [21, 15]: Consider the function

$$\psi(p) = \exp\frac{1}{4}\oint_{-\infty}^{\infty} \frac{e^{ip\xi/\pi}}{\operatorname{sh} b\xi \operatorname{sh}\xi/b}\frac{d\xi}{\xi},$$

where the singularity at $\xi = 0$ in the integral is circled from above. It is easy to see that

$$\psi(p) = \frac{s_q(w)}{s_{\tilde{q}}(\tilde{w})},$$

[1] We put now the Planck constant $\hbar$ equal to 1.

where $w = e^{bp}$, $\tilde{w} = e^{p/b}$. Thus the integral $\psi(p)$ unifies both q-exponents $s_q(w)$ and $s_{\tilde{q}}(\tilde{w})$, dual to each other. On the other hand, it definitely does not have their deficiencies; in particular, it does not suffer from the problem of small denominators. The pentagon relation for it takes the form

$$\psi(P)\psi(Q) = \psi(Q)\psi(P+Q)\psi(P)$$

if $[P,Q] = -2\pi i I$.

The function $\psi(p)$ was used extensively in [15, 21] and later by Kashaev in his new proposal for the knot invariants [22] and quantizing of the Teichmüller space [23], which was done also independently by Chekhov and Fock [24].

The relation of $\psi(p)$ to q-exponents makes our proposal quite clear: Consider the algebra

$$\mathcal{D} = \mathcal{C}_q \otimes \mathcal{C}_{\tilde{q}},$$

generated by the generators w_n, $\tilde{w}_n$, $n = 1,2,3,4$. The quantum groups $\mathcal{U}_q$ and $\mathcal{U}_{\tilde{q}}$ are naturally imbedded into $\mathcal{D}$ by means of defining relations between Chevalley generators and w-s. The algebra $\mathcal{D}$ can also be considered as being generated by the generators p_n, $n = 1,2,3,4$.

The element $\mathcal{R}$ in $\mathcal{D} \otimes \mathcal{D}$:

$$\mathcal{R} = \exp\left\{\frac{\pi}{2i}(p_2+p_3)\otimes(p_1+p_4)\right\}\psi(p_{13})\psi(p_{14})\psi(p_{23})\psi(p_{24}),$$

where $p_{ik} = p_i \otimes I + I \otimes p_k$, plays the role of the universal R-matrix for both $\mathcal{U}_q$ and $\mathcal{U}_{\tilde{q}}$. The Yang-Baxter relation $\mathcal{R}_{12}\mathcal{R}_{13}\mathcal{R}_{23} = \mathcal{R}_{23}\mathcal{R}_{13}\mathcal{R}_{12}$ is an easy consequence of the pentagon relation, as was shown by R. Kashaev and A. Volkov [11]. They also have obtained corresponding construction for the $SL_q(N)$ case.

There is a natural way to introduce the $*$-structure into $\mathcal{D}$: one is to consider p_n to be selfadjoint $p_n^* = p_n$. The corresponding formula in terms of w looks as follows:

$$w^* = w^{\bar{b}/b}, \quad \tilde{w}^* = \tilde{w}^{b/\bar{b}},$$

and does not respect the tensor structure of $\mathcal{D}$ in general.

However there are several particular cases when a $*$-involution can be related to this structure:

1) $\tau > 0$, so that b is real. The $*$ takes the form: $w^* = w$, $\tilde{w}^* = \tilde{w}$, and corresponds to the $SL_q(2,\mathbb{R})$ reduction.

2) $\tau < 0$, so that b is imaginary and $w^* = w^{-1}$, $\tilde{w}^* = \tilde{w}^{-1}$, which corresponds to the $SU_q(2)$ reduction.

3) $\tau = e^{i\theta}$, so that $\tau = e^{i\theta/2} = \bar{\bar{b}}$. The involution takes the form $w^* = \tilde{w}$, and so interchanges the factors in the modular double.

For all three cases the parameter of the central extension of the class mapping group

$$C = 1 + 6(\tau + 1/\tau + 2) = 1 + 6(b + 1/b)^2$$

is real. The values of C are $C \geq 25$, $C \leq 1$ and $1 \leq C \leq 25$ correspondingly for the case 1, 2 and 3. The relevance of this $*$ operation to Conformal Field Theory is discussed in [9].

In view of the modular duality relation $\tau \to -1/\tau$, it is natural to call $\mathscr{D}$ modular double.

I conclude by proposing several problems:

1) Give the generalization of our construction to other series of Quantum Groups;

2) Describe the coproduct in $\mathscr{U}_q$ in terms of w-generators;

3) Find a natural definition of the closure entering in formal tensor products like those for algebras $\mathscr{B}$ or $\mathscr{D}$.

Acknowledgements. It is my pleasure to acknowledge elucidating discussions with A. Connes, A. Volkov, and A. Vershik.

References

1. Drinfeld, V. G.: Quantum Groups, in *Proceedings of the International Congress of Mathematicians (Berkeley, 1986)*, Amer. Math. Soc., Providence, RI, 1987, pp. 798–820.
2. Lusztig, G.: *Introduction to Quantum Groups*, Birkhäuser Boston Inc., Boston, MA, 1993.
3. Kassel, Ch.: *Quantum Groups*, Springer-Verlag, New-York, 1995.
4. Haag, R.: *Local Quantum Physics: Fields, Particles, Algebras*, Springer-Verlag, Berlin, 1992.
5. Gervais, J. L.: Solving the strongly coupled 2D gravity. I. Unitary truncation and quantum group structure, *Comm. Math. Phys.* **138** (1991), 301–338.
6. Babelon, O.: Exchange formula and lattice deformation of the Virasoro algebra, *Phys. Lett. B* **B238** (1990), 234–238.
7. Bazhanov, V. V., Lukyanov, S. L., and Zamolodchikov, A. B.: Integrable structure of conformal field theory, quantum KdV theory and thermodynamic Bethe ansatz, *Comm. Math. Phys.* **177** (1996), 381-398; Integrable Structure of Conformal Field Theory. II. Q-operator and DDV equation, *Comm. Math. Phys.* **190** (1997), 247–278; Integrable structure of conformal field theory. III. The Yang-Baxter relation, *Comm. Math. Phys.* **200** (1999), 297–324.
8. Faddeev, L. D. and Volkov, A. Yu.: Algebraic Quantization of Integrable Models in Discrete Space-Time, in: *Discrete Integrable Geometry and Physics (Vienna, 1996)*, Oxford Lecture Ser. in Math. Appl., **16**, Oxford Univ. Press, New York, 1999, pp. 301–319.
9. Faddeev, L., Volkov, A., and Kashaev, R.: Towards the strongly coupled Liouville theory, in preparation.
10. Tarasov, O. V.: Cyclic monodromy matrices for $sl(n)$ trigonometric R-matrices, *Comm. Math. Phys.* **158** (1993), 459–483.
11. Kashaev, R. M. and Volkov, A. Yu.: From the Tetrahedron Equation to Universal R-matrices, preprint math.QA/9812017.
12. Faddeev, L. D. and Volkov, A. Yu.: Abelian current algebra and the Virasoro algebra on the lattice, *Phys. Lett. B* **B315** (1993), 311–318.
13. Faddeev, L. D. and Kashaev, R. M.: Quantum Dilogarithm, *Modern Phys. Lett. A* **9** (1994), 427–434.
14. Schützenberger, M. P.: Une interprétation de certaines solutions de l'équation fonctionelle: $F(x+y) = F(x)F(y)$, *C. R. Acad. Sci. Paris* **236** (1953), 352–353.
15. Faddeev, L. D.: Discrete Heisenberg-Weyl group and modular group, *Lett. Math. Phys.* **34** (1995), 249–254.
16. Connes, A.: *Noncommutative Geometry*, Academic Press, San Diego, 1994.

17. Rieffel, M. A.: C^*-algebras associated with irrational rotations. *Pacific J. Math.* **93** (1981), 415–429.
18. Kulish, P. P., Reshetikhin, N. Y.: Quantum Linear Problem For The Sine-Gordon Equation And Higher Representation, *J. Soviet Math.* **23** (1983), 2435.
19. Izergin, A. G., Korepin, V. E.: Lattice versions of quantum field theory models in two dimensions, *Nucl. Phys. B* **205** (1982), 410–413.
20. Faddeev, L. D.: How Algebraic Bethe Ansatz works for integrable models, in *Symmétries quantiques (Les Houches 1995)*, North-Holland, Amsterdam, 1998, pp. 149–219.
21. Faddeev, L. D.: Currentlike variables in massive and massless integrable models, in *Quantum groups and their applications in physics (Varenna 1994)*, Proc. Internat. School Phys. Enrico Fermi **127**, IOS Press, Amsterdam, 1994, pp. 117–135.
22. Kashaev, R. M.: The hyperbolic volume of knots from the quantum dilogarithm, *Lett. Math. Phys.* **39** (1997), 269–275.
23. Kashaev, R. M.: Quantization of Teichmüller Spaces and the Quantum Dilogarithm, *Lett. Math. Phys.* **43** (1998), 105-115.
24. Chekhov, L. O., Fock, V. V.: Quantum Mapping Class Group, Pentagon Relation and Geodesics, in *Proceedings of the Steklov Institute of Math.* **226**, IAPC "Nauka/Interperiodika" 1999.

On G-trace and G-index in deformation quantization

Boris Fedosov
Institut für Mathematik, Universität of Potsdam
Postfach 60 15 53, 14415 Potsdam, Germany
e-mail: `fedosov@math.uni-potsdam.de`

Abstract. For a compact group G of symplectomorphisms we define a G-trace on the algebra of quantum observables by postulating its properties. We give an explicit construction of such a trace and prove a G-index theorem similar to the Atiyah-Segal-Singer equivariant index theorem for elliptic operators.

Keywords: star-product, symmetry group, fixed-point manifold, G-trace, G-index

Mathematics Subject Classifications (2000): 53D55, 37J10

Dedicated to the memory of Moshé Flato

1. Introduction

Let (M,ω) be a symplectic manifold, $\partial = \partial^s$ a symplectic connection on M, $E \to M$ a complex vector bundle with a connection ∂^E. These data give rise to a *canonical deformation quantization* [5, 6] with coefficients in the bundle $K = \mathrm{Hom}(E,E)$. The algebra $(A_h, *)$ of quantum observables is isomorphic to $C^\infty(M,K)[[h]]$ as a linear space and the associative product $*$ is a deformation of the pointwise product of sections of the coefficient bundle K.

An important quantity in quantization theory is a *trace*. This is a linear functional which maps the ideal $A_h^c \in A_h$ of compactly supported observables $a(x,h)$ to "constants". The latter are formal Laurent series in h with finite number of negative exponents (notation $\mathbb{C}[h^{-1},h]]$). Although, in general, there are no operator representations of the algebra A_h a condition

$$\mathrm{Tr}\, a * b = \mathrm{Tr}\, b * a \tag{1.1}$$

is imposed on the trace as if the observables a, b were operators in a Hilbert space. It turns out that the trace exists, is unique up to a normalization factor and has a *canonical local expression*: The algebra A_h is locally isomorphic to a standard Weyl algebra W on $\mathbb{R}^{2n}$ and for $a \in W^c$

$$\mathrm{Tr}\, a = \frac{1}{(2\pi h)^n} \int_{\mathbb{R}^{2n}} \mathrm{tr}\, a(x,h) \frac{\omega^n}{n!}, \tag{1.2}$$

where tr means the trace in the coefficient bundle $K = \mathrm{Hom}(E,E)$.

G. Dito and D. Sternheimer (eds.), Conférence Moshé Flato 1999, Vol. 1, 157 – 176.

Finally, having a trace on the algebra A_h one can introduce an *index* for this algebra. In the simplest case when the manifold M is compact we define

$$\operatorname{ind} A_h = \operatorname{Tr} 1. \tag{1.3}$$

The *index theorem* for deformation quantization [6, 8] gives an explicit formula for this quantity

$$\operatorname{ind} A_h = \int_M \operatorname{ch} E \widehat{A}(M) \exp \frac{\omega}{2\pi h}, \tag{1.4}$$

where $\operatorname{ch} E$ is the Chern character form of ∂^E and $\widehat{A}(M)$ is the Atiyah-Hirzebruch form for ∂^s.

In this paper we construct a trace functional and prove an index theorem in an equivariant situation. The letter G in the title (and everywhere in this paper) means a *compact group* of symplectomorphisms of M acting also equivariantly on the bundle E. In particular, G may be a finite group. The averaging over a compact group provides us with G-invariant geometric objects: Riemannian metric, almost complex structure, G-invariant connections ∂^E and ∂^s, G-invariant canonical deformation quantization. The latter means that the pullback map $s^* a(x,h) = s^{-1} a(s(x),h) s$ is an automorphism of the algebra A_h.

For a fixed element $s \in G$ we would like to define a linear functional

$$\operatorname{Tr}_s : A_h^c \to \mathbb{C}[h]]$$

satisfying some conditions generalizing (1.1) and prompted by the representations of the algebra A_h and the group G in a Hilbert space (which do not exist in general). Assuming that they do exist so that the following equalities hold:

$$\widehat{a * b} = \widehat{a}\widehat{b}, \tag{1.5}$$

$$\widehat{s_1 s_2} = \widehat{s}_1 \widehat{s}_2, \tag{1.6}$$

$$\widehat{s}^{-1} \widehat{a} \widehat{s} = \widehat{s^* a}, \tag{1.7}$$

where the hat means the corresponding operator, we would define the trace Tr_s by

$$\operatorname{Tr}_s a = \operatorname{Tr} \widehat{s}\widehat{a}, \tag{1.8}$$

where Tr on the right means the operator trace. Formal manipulations yield

$$\begin{aligned} \operatorname{Tr}_s a * b &= \operatorname{Tr} \widehat{s}\widehat{a}\widehat{b} = \operatorname{Tr} \widehat{s}\widehat{a}\widehat{s}^{-1}\widehat{s}\widehat{b} \\ &= \operatorname{Tr} \widehat{s}\widehat{b}\widehat{(s^{-1})^* a} = \operatorname{Tr}_s b * ((s^{-1})^* a), \\ \operatorname{Tr}_{s_1 s_2} a &= \operatorname{Tr} \widehat{s}_1 \widehat{s}_2 \widehat{a} = \operatorname{Tr} \widehat{s}_2 \widehat{s}_1 \widehat{s}_2 \widehat{a} \widehat{s}_2^{-1} = \operatorname{Tr}_{s_2 s_1} (s_2^{-1})^* a. \end{aligned} \tag{1.9}$$

Dropping the intermediate terms, we postulate these identities as the properties of the functional Tr_s. Here we will mostly consider the first of them

$$\operatorname{Tr}_s a * b = \operatorname{Tr}_s b * ((s^{-1})^* a) \tag{1.10}$$

for a fixed $s \in G$. Similar motivations were used in [4] to express the Kubo-Martin-Schwinger condition in terms of deformation quantization.

As an immediate consequence of (1.10) we obtain that the distributional support of the functional Tr_s is contained in the fixed-point set of s. Indeed, let $U = \operatorname{supp} a$ and $V = \operatorname{supp}(s^{-1})^* a = s^{-1}(U)$ be compact sets with empty intersection. Taking a function ρ which separates these sets ($\rho = 1$ on U and $\rho = 0$ on V), we have

$$\mathrm{Tr}_s a = \mathrm{Tr}_s a * \rho = \mathrm{Tr}_s \rho * ((s^{-1})^* a) = 0,$$

because $\rho * (s^{-1})^* a = 0$ since $\rho = 0$ on $V = \operatorname{supp}(s^{-1})^* a$. On the other hand, if $\operatorname{supp} a$ is contained in a small neighborhood of a fixed point x_0 we find a *canonical local expression* for $\mathrm{Tr}_s a$ generalizing (1.2). This is done in Section 2. In Section 3 we study further properties of this expression, in particular (1.9). It is closely related to the notion of the $*$-exponential first introduced on an heuristic level in the pioneer work [3]. Section 4 deals with a parameter-dependent version: We calculate equivariant Chern character for a G-vector bundle with symplectic structure.

Next, in Section 5 we construct the functional Tr_s globally on a smooth connected component M_0 of the fixed-point set of the map s. Our main observation consists in the fact that the algebra A_h in the vicinity of M_0 may be treated as an algebra of quantum observables *on* M_0 with a non-trivial coefficient bundle defined by the normal bundle N of M_0 in M. The map s acts on coefficients only. This allows us to define a trace tr_s on the coefficient bundle in terms of the local equivariant trace constructed in Section 2. Then the global trace Tr_s is defined by (1.2) with the replacement of the coefficient trace tr in (1.2) by the equivariant coefficient trace tr_s. Assuming that M_0 is a compact submanifold, we introduce the s-index

$$\mathrm{ind}_s A_h = \mathrm{Tr}_s 1$$

and prove an explicit index formula for this quantity in the spirit of the Atiyah-Segal-Singer equivariant index theorem. We obtain it as a simple consequence of the usual index theorem for deformation quantization with non-trivial coefficients and the properties of the trace tr_s on the coefficient bundle. This is done in Section 6.

We have not considered applications in this paper postponing them to subsequent publications. Here we only indicate two of them: An index theorem for orbifolds and a global $*$-exponential. It should be mentioned, however, that further extension of "constant coefficients" is required for applications to include rapidly oscillating or decaying exponentials $\exp(ic/h)$ where the "weights" c are complex numbers with $\operatorname{Im} c \geq 0$.

Acknowledgements. This work was done during the author's stay at the Université de Bourgogne. The author would like to thank the Laboratoire Gevrey de Mathématique physique and especially Professor D. Sternheimer for the kind invitation and hospitality.

2. The local s-trace

In this section we construct an equivariant trace assuming that all the structures in question have constant coefficients. Thus,

$$\omega = dq^i \wedge dp_i$$

(summation over $i = 1, 2, \dots, n$) is a constant symplectic form on $\mathbb{R}^{2n}$ and the group G acts on $\mathbb{R}^{2n}$ by linear symplectic transformations. In this case the trace Tr_s actually may be defined via an operator representation, that is using (1.5)–(1.8). The price for this is that h must be a positive numerical parameter. It turns out, however, that the final expression for $\mathrm{Tr}_s a$ admits a stationary phase asymptotic expansion which is meaningful as a formal series in powers of h, so we may return to deformation quantization framework and define $\mathrm{Tr}_s a$ by this formal series. We would like to emphasize that the properties (1.9), (1.10) are fulfilled automatically on the level of asymptotic series by the very definition of $\mathrm{Tr}_s a$, thus, they are fulfilled on the level of formal series. This is why we need the operator representation, otherwise it would be difficult to prove these properties.

We begin with the Weyl correspondence $a \mapsto \mathrm{Op}(a)$ between functions $a(x) = a(q,p) \in C^\infty(\mathbb{R}^{2n})$ (satisfying some conditions at infinity) and pseudodifferential operators on $L^2(\mathbb{R}^n_q)$ given by

$$(\mathrm{Op}(a)u)(q) = \frac{1}{(2\pi h)^n} \int_{\mathbb{R}^n_p} \int_{\mathbb{R}^n_{q'}} e^{\frac{i}{h}\langle p, q-q'\rangle} a((q+q')/2, p) u(q') dq' dp. \tag{2.1}$$

The function $a(x)$ is called the Weyl symbol of the operator $\mathrm{Op}(a)$ (notation $a = \sigma_W(\mathrm{Op}(a))$. Although the operator (2.1) depends on the separation of variables $x = (q,p)$ into coordinates q and momenta p, the product $\widehat{ab}$ is independent of the separation and is given by the following formula (see, e.g., [6])

$$\sigma(\mathrm{Op}(a)\mathrm{Op}(b)) = \frac{1}{(\pi h)^{2n}} \int_{\mathbb{R}^{4n}} \exp\left(\frac{2i}{h}\omega(t,\tau)\right) a(x+t) b(x+\tau) dt d\tau. \tag{2.2}$$

Here

$$\omega(t,\tau) = \omega_{ij} t^i \tau^j, \quad (i,j = 1,2,\dots,2n)$$

is the symplectic form ω evaluated on vectors t, τ, integral (2.2) is understood as an oscillatory one. Applying stationary phase method to this integral (practically it means that we expand $a(x+t)b(x+\tau)$ in Taylor series in t, τ and integrate termwise), we come to the asymptotic series

$$\sigma(\mathrm{Op}(a)\mathrm{Op}(b)) \sim a * b = \exp\left(-\frac{ih}{2}\omega^{ij}\frac{\partial}{\partial t^i}\frac{\partial}{\partial \tau^j}\right) a(x+t) b(x+\tau)\Big|_{t=\tau=0} \tag{2.3}$$

called the *Moyal $*$-product* of two functions a and b. Observe that if a or b is a polynomial, the series (2.3) is finite and coincides with the integral (2.2). The algebra of functions on $\mathbb{R}^{2n}$ with the Moyal product $*$ will be called the *Weyl algebra* (notation $W = W(\mathbb{R}^{2n})$).

Unfortunately, the action of the group $G \subset \mathrm{Sp}(2n, \mathbb{R})$ cannot be defined on $L^2(\mathbb{R}^n)$ in a single-valued manner but it is possible to define a single-valued action on the Fock space.

Let J be a positive G-invariant complex structure on $\mathbb{R}^{2n}$, so that our symplectic space may be considered as a complex space $\mathbb{C}^n$ with a G-invariant hermitian scalar product $(\cdot,\cdot)$. In an orthonormal basis we have

$$(z,w) = \sum_{i=1}^{n} \bar{z}_i w_i = z^* w, \quad \omega = \frac{1}{2i} \sum_{i=1}^{n} d\bar{z}_i \wedge dz_i = \frac{1}{2i} dz^* \wedge dz,$$

where z on the right-hand side is treated as a column and z^* is a row of complex conjugate elements. The group G acts by unitary transformations on $\mathbb{C}^n$. The Moyal $*$-product on functions $a = a(z,\bar{z})$ may be rewritten as

$$\begin{aligned} a*b = & \frac{1}{(\pi h)^{2n}} \int_{\mathbb{C}^{2n}} \exp\left(\frac{1}{h}(u^* v - v^* u)\right) a(z+u, \bar{z}+\bar{u}) b(z+v, \bar{z}+\bar{v}) du d\bar{u} dv d\bar{v} \\ & \sim \sum_{\alpha,\beta} (-1)^{|\beta|} \frac{h^{\alpha+\beta}}{\alpha!\beta!} \frac{\partial^{\alpha+\beta}}{\partial z^\alpha \partial \bar{z}^\beta} a \frac{\partial^{\alpha+\beta}}{\partial \bar{z}^\alpha \partial z^\beta} b. \end{aligned} \tag{2.4}$$

Here we assume that h is a positive small parameter. In (2.4) $dud\bar{u}$, $dvd\bar{v}$ mean the usual Lebesgue measure on $\mathbb{C}^n$.

Consider an operator $\mathrm{Op}(p)$ on $L^2(\mathbb{R}^n_q)$ whose Weyl symbol has the following form in complex coordinates

$$p(z,\bar{z}) = 2^n \exp\left(-\frac{1}{h}|z|^2\right). \tag{2.5}$$

Using the integral formula (2.4), it is easy to verify that $\mathrm{Op}(p)$ is a projector in $L^2(\mathbb{R}^n_q)$. Moreover, since

$$\mathrm{Tr}\,\mathrm{Op}(p) = \frac{1}{(2\pi h)^n} \int_{\mathbb{C}^n} p(z,\bar{z}) dz d\bar{z} = 1,$$

its range is a one-dimensional subspace. We call it a *vacuum projector*.

The Fock space F corresponding to the Weyl algebra W and the vacuum projector (2.5) consists of all symbols having the form $a * p$ where $a \in W$. In other words, F is the left ideal in W generated by p. It is easy to see that $z_i * p = 0$, so F is generated by vectors

$$e^\alpha = \frac{(\bar{z})^\alpha * p}{\sqrt{\alpha!(2h)^{|\alpha|}}}. \tag{2.6}$$

At $\alpha = 0$ we obtain $e^0 = p$, this vector will be called the *vacuum vector*. Define a scalar product taking for two vectors $u = a * p$ and $v = b * p$

$$(u,v) = \mathrm{Tr}\,\mathrm{Op}(u)^*\mathrm{Op}(v) = \frac{1}{(2\pi h)^n}\int_{\mathbb{C}^n} p * \bar{a} * b * p\, dz d\bar{z}. \tag{2.7}$$

We assume that F is completed with respect to this scalar product, so that F is the Hilbert space with an orthonormal basis (2.6). The action of the Weyl algebra on F is defined by the $*$-multiplication: For $u = b * p \in F$ and $a \in W$ we set

$$\widehat{a}u = a * b * p,$$

so that W is represented by operators on the space F. We denote by $\widehat{a}$ the operator on the Fock space reserving notation $\mathrm{Op}(a)$ for the operator on $L^2(\mathbb{R}^n_q)$ given by (2.1). The representation on the Fock space has an advantage comparing with the original representation (2.1) in $L^2(\mathbb{R}^n)$, it admits a well-defined action of the unitary group $\mathrm{U}(n)$. Indeed, the unitary group acts on W:

$$s:\ a(z,\bar{z}) \mapsto a(s^{-1}z, \overline{s^{-1}z}) \tag{2.8}$$

by automorphisms. Considering $u = a * p \in F$ as an element of W, we have an action (2.8) of $s \in \mathrm{U}(n)$ on u. Since p is invariant for this action and $*$-product is preserved, the ideal F is preserved by this action too. Thus, we have a representation of $\mathrm{U}(n)$ on F which we denote by $s \mapsto \widehat{s}$ where

$$\widehat{s}u = u(s^{-1}z, \overline{s^{-1}z}).$$

In particular,

$$\widehat{s}e^\alpha = \frac{(\overline{s^{-1}z})^\alpha * p}{\sqrt{\alpha!(2h)^{|\alpha|}}}. \tag{2.9}$$

On the algebra W considered as the algebra of operators on F we have an adjoint representation

$$s : \widehat{a} \mapsto \widehat{s}\widehat{a}\widehat{s}^{-1}. \tag{2.10}$$

One can see immediately that the action (2.10) coincides with the previous action (2.8).

We are ready now to construct a G-trace on W for $G = \mathrm{U}(n)$. As was explained in the introduction, we set

$$\mathrm{Tr}_s a = \mathrm{Tr}\,\widehat{s}\widehat{a} \tag{2.11}$$

for $s \in \mathrm{U}(n)$, $a \in W$ where $\widehat{a}$ means the corresponding operator on F. This definition seems rather complicated. Quite remarkably, it admits a simple explicit formula in the spirit of deformation quantization.

Theorem 2.1 *Assume that 1 is not an eigenvalue of the matrix $s \in U(n)$. Then*

$$\mathrm{Tr}_s a = \det{}^{-1}(1 - s^{-1}) \exp\left(h\frac{\partial}{\partial z}\frac{1+s^{-1}}{1-s^{-1}}\frac{\partial}{\partial z^*}\right) a\Big|_{z=\bar{z}=0} \tag{2.12}$$

where $\partial/\partial z$ means a row $(\partial/\partial z_1, \partial/\partial z_2, \dots, \partial/\partial z_n)$ and $\partial/\partial z^*$ is a column of complex conjugate elements.

We will need well-known properties of the Cayley transformation

$$\sigma = C(s) = (1-s)(1+s)^{-1}$$

and its inverse

$$s = C(\sigma) = (1-\sigma)(1+\sigma)^{-1}$$

summarized in the following lemma (see also [6, Appendix A]).

Lemma 2.2 *Let s be a contraction in $\mathbb{C}^n$, that is*

$$|sz|^2 < |z|^2$$

for any $z \in \mathbb{C}^n$. The Cayley transform $\sigma = C(s)$ exists and belongs to the "Siegel right half-plane", that is the matrix $\sigma + \sigma^*$ is positive definite. Vice versa, if σ belongs to the "right half-plane" then $s = C(\sigma)$ exists and is a contraction in $\mathbb{C}^n$.

Proof. Since s is a contraction, the number -1 cannot be its eigenvalue, so the Cayley transform exists. Consider the quadratic form

$$2\mathrm{Re}(\sigma z, z) = (\sigma z, z) + (z, \sigma z). \tag{2.13}$$

Taking $z = (1+s)u$ (it is possible since $1+s$ is non-degenerate), we obtain

$$\begin{aligned} 2\mathrm{Re}(\sigma z, z) &= ((1-s)u, (1+s)u) + ((1+s)u, (1-s)u) \\ &= 2|u|^2 - 2|su|^2 > 0. \end{aligned} \tag{2.14}$$

Conversely, if the form (2.13) is positive for any $z \in \mathbb{C}^n$, then -1 cannot be an eigenvalue of σ, so that $s = C(\sigma)$ exists. Further, taking $z = (1+\sigma)u$, we obtain

$$\begin{aligned} (z,z) - (\sigma z, \sigma z) &= ((1+\sigma)u, (1+\sigma)u) - (1-\sigma)u, (1-\sigma)u) \\ &= 2(\sigma u, u) + 2(u, \sigma u) > 0, \end{aligned}$$

proving the lemma. ∎

Remark 2.3 As a limit case, we see that the Cayley transform gives a one-to-one correspondence between anti-hermitian matrices σ and unitary matrices s, provided that -1 is not an eigenvalue of σ or s.

Proof of Theorem 2.1. Let us calculate the trace (2.11) as the sum of the diagonal matrix entries in the basis (2.6)

$$\begin{aligned} \mathrm{Tr}\widehat{s}\widehat{a} &= \sum_{|\alpha|=0}^{\infty} (e^\alpha, \widehat{s}\widehat{a} e^\alpha) = \sum_{|\alpha|=0}^{\infty} (\widehat{s}^{-1} e^\alpha, \widehat{a} e^\alpha) \\ &= \frac{1}{(2\pi h)^n} \sum_{|\alpha|=0}^{\infty} \int_{\mathbb{C}^n} \frac{p * (s^{-1}z)^\alpha * a * \bar{z}^\alpha * p}{\alpha!(2h)^{|\alpha|}} dz d\bar{z}. \end{aligned}$$

We have made use of the formula (2.7) for the scalar product. The integral may be interpreted as a trace of an operator on $L^2(\mathbb{R}^n_q)$ via Weyl correspondence (2.1). In particular, the cyclic permutations of $*$-factors are possible. Since $p*p=p$, we obtain

$$\mathrm{Tr}\,\widehat{s}\widehat{a}=\frac{1}{(2\pi h)^n}\sum_{|\alpha|=0}^{\infty}\int_{\mathbb{C}^n}a*\frac{\bar{z}^\alpha * p*(s^{-1}z)^\alpha}{\alpha!(2h)^{|\alpha|}}dzd\bar{z}.$$

An important feature of the Weyl correspondence is that

$$\mathrm{Tr}\,\mathrm{Op}(a)\mathrm{Op}(b)=\frac{1}{(2\pi h)^n}\int_{\mathbb{C}^n}ab\,dzd\bar{z}$$

with the usual product ab instead of $*$-product (see, e.g. [6]). This gives us the possibility to reduce the number of $*$-products, obtaining

$$\mathrm{Tr}\,\widehat{s}\widehat{a}=\frac{1}{(2\pi h)^n}\sum_{|\alpha|=0}^{\infty}\int_{\mathbb{C}^n}a\frac{\bar{z}^\alpha * p*(s^{-1}z)^\alpha}{\alpha!(2h)^{|\alpha|}}dzd\bar{z}.$$

Observe now that for a polynomial $f(z)$ we have by (2.4)

$$p*f(z)=\sum\frac{1}{\alpha!}\frac{\partial^\alpha f(z)}{\partial z^\alpha}z^\alpha=f(2z)p,$$

so that only one $*$-product is left

$$\mathrm{Tr}\,\widehat{s}\widehat{a}=\frac{1}{(2\pi h)^n}\sum_{|\alpha|=0}^{\infty}\int_{\mathbb{C}^n}a\frac{\bar{z}^\alpha *(p(s^{-1}z)^\alpha)}{\alpha!h^{|\alpha|}}dzd\bar{z}.$$

For this remaining $*$-product we need the integral formula (2.4). The function $a(z,\bar{z})$ is supposed to have compact support. To feel free with infinite summation, we will assume that the factor $\exp(-\varepsilon|u|^2-\varepsilon|v|^2)$ is inserted into the integrand of (2.4) and the limit for $\varepsilon\to 0$ is taken after integration (this is the usual trick for oscillating integrals). We also will assume that s^{-1} is a contraction, so that $(1\pm s^{-1})^{-1}$ exists.

Thus,

$$\frac{\bar{z}^\alpha *(p(s^{-1}z)^\alpha)}{\alpha!h^{|\alpha|}}=\frac{1}{(\pi h)^{2n}}\int_{\mathbb{C}^n}\left(\frac{1}{h}\right)^{|\alpha|}\frac{1}{\alpha!}\exp\left(\frac{1}{h}(u^*v-v^*u-|z+v|^2)\right)$$
$$(z^*+u^*)^\alpha(s^{-1}(z+v))^\alpha dud\bar{u}dvd\bar{v}.$$

The series may be summed giving an exponential function

$$\exp\left(\frac{1}{h}(z^*+u^*)s^{-1}(z+v)\right),$$

and we finally obtain an oscillating integral

$$\frac{1}{(\pi h)^{3n}}\int_{\mathbb{C}^{3n}} a(z,\bar{z})\exp\big(\frac{1}{h}(u^*v-v^*u-(z^*+v^*)(z+v) \\ +(z^*+u^*)s^{-1}(z+v))\big)dud\bar{u}dvd\bar{v}dzd\bar{z},$$

which may be calculated asymptotically for $h\to 0$ by the stationary phase method. Moreover, the integrals over v, $\bar{v}$, u, $\bar{u}$ are Gaussian, so the integration over v, $\bar{v}$, u, and $\bar{u}$ may be performed explicitly. The integration over v, $\bar{v}$ yields

$$\frac{1}{(\pi h)^{2n}}\int_{\mathbb{C}^{2n}} a(z,\bar{z})\exp\left(-\frac{1}{h}(u^*(1+s^{-1})u-z^*(1-s^{-1})u+u^*z)\right)dud\bar{u}dzd\bar{z}.$$

The next integration over u, $\bar{u}$ results in

$$\frac{\det^{-1}(1+s^{-1})}{(\pi h)^n}\int_{\mathbb{C}^n} a(z,\bar{z})\exp\left(-\frac{1}{h}z^*\frac{1-s^{-1}}{1+s^{-1}}z\right)dzd\bar{z}. \tag{2.15}$$

At this step it is essential that the quadratic form $\mathrm{Re}\,u^*(1+s^{-1})^{-1}(1-s^{-1})u$ is positive definite since s^{-1} is a contraction. The exponent in (2.15) also has a negative real part. The stationary phase method applied to (2.15) means that we integrate termwise the formal Taylor expansion for $a(z,\bar{z})$ at $z=\bar{z}=0$. The result, obviously, may be written in the form (2.12). ■

Remark 2.4 Passing to the limit $s\to 1$ in (2.15), we obtain the usual formula for the trace

$$\mathrm{Tr}\widehat{a}=\frac{1}{(2\pi h)^n}\int_{\mathbb{C}^n} a(z,\bar{z})dzd\bar{z}. \tag{2.16}$$

We now may forget about the origin of the formula (2.12) and take it as a definition of $\mathrm{Tr}_s a$ for the algebra W in the case when s has no eigenvalue 1. This expression is meaningful even in the framework of deformation quantization. Moreover, if s has an eigenvalue 1, we split $\mathbb{C}^n$ into orthogonal direct sum of two subspaces $\mathbb{C}_1\oplus\mathbb{C}_1^{\perp}$ of dimensions k and $n-k$. On $\mathbb{C}_1$, s is equal to the identity matrix and on $\mathbb{C}_1^{\perp}$ the eigenvalues of s are not equal to 1. Thus, denoting $z=(z_1,z_{\perp})$, $s=1\oplus s_{\perp}$, we would have

$$\mathrm{Tr}_s a= \\ \frac{1}{(2\pi h)^k}\int_{\mathbb{C}_1} dz_1 d\bar{z}_1 \det{}^{-1}(1-s_{\perp}^{-1})\exp\left(h\frac{\partial}{\partial z_{\perp}}\frac{1+s_{\perp}^{-1}}{1-s_{\perp}^{-1}}\frac{\partial}{\partial z_{\perp}^*}\right)a\bigg|_{z_{\perp}=\bar{z}_{\perp}=0}. \tag{2.17}$$

We will also consider a regularization of the trace (2.17) taking $a(z,\bar{z})$ in the form

$$a(z,\bar{z})=\lim_{\varepsilon\to 0} a_{\varepsilon}(z,\bar{z})=\lim_{\varepsilon\to 0} a(z,\bar{z})e^{-\frac{\varepsilon}{h}|z|^2}. \tag{2.18}$$

Substituting into (2.15), we get

$$\mathrm{Tr}_s a_\varepsilon = \frac{\det^{-1}(1+s^{-1})}{(\pi h)^n}\int_{\mathbb{C}^n} a(z,\bar z)\exp\left(-\frac{1}{h}\left(z^*\left(\varepsilon+\frac{1-s^{-1}}{1+s^{-1}}\right)z\right)\right)dzd\bar z,$$

whence by the stationary phase method

$$\mathrm{Tr}_s a_\varepsilon = \tag{2.19}$$
$$\det{}^{-1}(\varepsilon(1+s^{-1})+1-s^{-1})\exp\left(h\frac{\partial}{\partial z}\frac{1+s^{-1}}{\varepsilon(1+s^{-1})+1-s^{-1}}\frac{\partial}{\partial z^*}\right)a\Big|_{z=\bar z=0}.$$

For $\varepsilon > 0$ this expression is meaningful as a formal series in h for any unitary matrix s. Moreover, if s has no eigenvalue eual to 1 then we have

$$\mathrm{Tr}_s a = \lim_{\varepsilon\to+0}\mathrm{Tr}_s a_\varepsilon$$

with the termwise limit of the formal series.

Proposition 2.5 *The properties* (1.9) *and* (1.10) *hold.*

Proof. Consider the operators $\widehat{s}$, $\widehat{a}$, $\widehat{b}$ in the Fock space corresponding to the symbols $a(x)$, $b(x) \in W$ (with compact supports) and to the unitary matrix s. Then

$$\mathrm{Tr}\widehat{s}\widehat{a}\widehat{b} = \mathrm{Tr}\widehat{s}\widehat{b}\widehat{s}\widehat{a}\widehat{s}^{-1}.$$

But $\widehat{s}\widehat{a}\widehat{s}^{-1} = \widehat{a(sx)}$, so we come to an identity

$$\mathrm{Tr}\widehat{s}\widehat{a}\widehat{b} = \mathrm{Tr}\widehat{s}\widehat{b}\widehat{a(sx)}.$$

Now, on both sides of this identity we pass to their deformation quantization counterparts given by (2.17). They must coincide since they give the asymptotic expansions of the identical operator expressions. This proves (1.10).

The proof of (1.9) is similar. ∎

Finally, we show that $\mathrm{Tr}_s a$ is independent of the choice of a positive complex structure needed for its definition.

Proposition 2.6 *Let S be a symplectic matrix whose eigenvalues are not equal to* 1 *and let J be a positive complex structure on $\mathbb{R}^{2n}$ commuting with S. Let s be a unitary matrix corresponding to S in a given complex structure. Then* $\mathrm{Tr}_s a$ *given by* (2.12) *is independent of J.*

Proof. Taking

$$J = \begin{pmatrix} 0 & 1 \\ -1 & 0 \end{pmatrix},$$

we have a correspondence between complex $n \times n$ matrices $s = s_1 + is_2$ with s_1, s_2 real, and real $2n \times 2n$ matrices commuting with J

$$S = \begin{pmatrix} s_1 & s_2 \\ -s_2 & s_1 \end{pmatrix}.$$

Being rational, the Cayley transformation preserves this correspondence. It means that

$$\frac{\partial}{\partial z} \frac{1-s^{-1}}{1+s^{-1}} \frac{\partial}{\partial z^*} = \frac{\partial}{\partial x} \frac{1-S^{-1}}{1+S^{-1}} \frac{\partial}{\partial x'}$$

where $\partial/\partial x$ is a row $(\partial/\partial x^1, \partial/\partial x^2, \ldots, \partial/\partial x^{2n})$ and $\partial/\partial x'$ is a column obtained by transposition. Thus, the second factor in (2.12) admits an expression in terms of the symplectic matrix S only, so it is independent of J. Let us show that the first factor

$$\det{}^{-1}(1 - s^{-1}) \tag{2.20}$$

is also independent of J. To this end we show that the variation of (2.20) with respect to J is identically 0. This implies the desired independence since any two positive complex structures are homotopic. So, consider variations δJ of complex structures commuting with S. From the identity $J^2 = 1$ it follows that $J\delta J + \delta J J = 0$, thus δJ commutes with S and anticommutes with J. Let us reduce $J + \delta J$ to its canonical form J. The reducing matrix $C = 1 + \delta C$ satisfies the equation

$$J\delta C - \delta C J + \delta J = 0.$$

Its solutions have the form

$$\delta C = \frac{1}{2} J \delta J + X,$$

where

$$X = \begin{pmatrix} x_1 & x_2 \\ -x_2 & x_1 \end{pmatrix}$$

is an arbitrary matrix commuting with J. This yields

$$C^{-1} S C = [S, \delta C] = [S, X],$$

since S commutes with J and δJ. Thus, the variation $\delta s = \delta s_1 + i\delta s_2$ of the corresponding complex matrix s is

$$\delta s = [s, x],$$

where $x = x_1 + ix_2$. Now,

$$\delta \det(1 - s^{-1}) = \det(1 - s^{-1}) \operatorname{tr} (1 - s^{-1})^{-1} s^{-1} \delta s^{-1}.$$

But for a rational function $f(s)$

$$\operatorname{tr} f(s) \delta s = \operatorname{tr} f(s) [s, x] = 0,$$

proving the proposition. ■

Remark 2.7 In the general situation, when S has an eigenvalue equal to 1 the functional Tr_s given by (2.17) is also independent of the positive complex structure commuting with the symplectic matrix S. Indeed, the k-dimensional complex eigensubspace $\mathbb{C}_1$ of the unitary matrix s corresponds to a $2k$-dimensional real eigensubspace of the symplectic matrix S. The orthogonal complement $\mathbb{C}_1^{\perp}$ with respect to the hermitian scalar product on $\mathbb{C}^n$ corresponds to the orthogonal complement of the real eigensubspace of S with respect to the symplectic form. So, everything in (2.17) may by defined in terms of symplectic geometry on $\mathbb{R}^{2n}$.

Remark 2.8 An important quantity for index theory is the trace of the identity element $1 \in W$. From (2.12) we have

$$\mathrm{Tr}_s 1 = \det{}^{-1}(1 - s^{-1}) \tag{2.21}$$

if s has no eigenvalues 1. This result has a simple explanation. The right-hand side of (2.21) may be viewed as the character of the representation of the unitary group $\mathrm{U}(n)$ on an infinite direct sum of symmetric tensor products $\mathrm{Sym}(\mathbb{C}^n \otimes \mathbb{C}^n \otimes \cdots \otimes \mathbb{C}^n)$. But the bosonic Fock space clearly coincides with this direct sum and the action of the unitary group is the same, so the characters must coincide. What is a little bit surprising is that the quantum effects are inessential.

3. The local $*$-exponential

In this section we consider a link between the G-trace and the $*$-exponential introduced in the pioneer work [3] as a hypothetical solution of the Schrödinger evolution equation

$$\frac{dU}{dt} = \frac{i}{h} H * U, \quad U|_{t=0} = 1.$$

Unfortunately, in general, such a solution makes no sense in the framework of deformation quantization.

Proposition 3.1 *Let $s(t) \in U(n)$ be a smooth family. Then for any $a \in W$ with compact support and any $\varepsilon > 0$*

$$2h \frac{d}{dt} \mathrm{Tr}_{s(t)}\, a_\varepsilon = \mathrm{Tr}_{s(t)}\, H * a_\varepsilon \tag{3.1}$$

where

$$H(t) = z^* s^{-1}(t)\dot{s}(t)z - p * (z^* s^{-1} \dot{s} z) * p = z^* s^{-1}(t)\dot{s}(t) * z \tag{3.2}$$

is a normal-ordered quadratic Hamiltonian and a_ε is defined as in (2.18).

Proof. Consider the operator counterpart of the left-hand side of (3.1)

$$2h \frac{d}{dt} \mathrm{Tr}\, \widehat{s}(t)\, \widehat{a}_\varepsilon = 2h \mathrm{Tr}\, \widehat{s}\, \widehat{s}^{-1} \widehat{\dot{s}}\, \widehat{a}_\varepsilon.$$

Let us show that the operator $2h\widehat{s}^{-1}\dot{\widehat{s}}$ is equal to $\widehat{H}$ with $H \in W$ given by (3.2). Indeed,

$$\widehat{s}(t)(\bar{z}^\alpha * p) = \overline{s^{-1}z}^\alpha * p.$$

Thus, in the algebra W we have

$$\frac{d}{dt}(s^{-1}(t)z)^\alpha = -\frac{\partial z^\alpha}{\partial z_k}(s^{-1}\dot{s}z)_k\bigg|_{z\mapsto s^{-1}z} = \frac{1}{2h}[z^* s^{-1}\dot{s}z, z^\alpha]\bigg|_{z\mapsto s^{-1}z},$$

where the commutator is understood with respect to the $*$-product (2.4). Denoting $H_0 = z^* s^{-1}\dot{s}z \in W$, we see that

$$2h\frac{d}{dt}\widehat{s}(t)(\bar{z}^\alpha * p) = \widehat{s}(t)((H_0 * \bar{z}^\alpha - \bar{z}^\alpha * H_0) * p).$$

We may add any constant C to H_0 without changing the result. If we take $C = -p * H_0 * p$, then $(H_0 + C) * p = 0$, so we obtain

$$2h\frac{d}{dt}\widehat{s}(t)(\bar{z}^\alpha * p) = \widehat{s}(t)(\widehat{H_0 + C})(\bar{z}^\alpha * p),$$

which means that

$$2h\frac{d}{dt}\widehat{s}(t) = \widehat{s}(t)\widehat{H},$$

where $H = H_0 + C$ is the normal-ordered Hamiltonian. Multiplying both sides by $\widehat{a}_\varepsilon$ and taking traces, we get

$$2h\frac{d}{dt}\mathrm{Tr}\,\widehat{s}(t)\,\widehat{a}_\varepsilon = \mathrm{Tr}\,\widehat{s}(t)\widehat{H}\,\widehat{a}_\varepsilon.$$

It remains to pass to deformation quantization counterparts of both sides to obtain (3.1). ∎

4. Equivariant Chern character

In this section we continue to study G-trace on the Fock space but in a parameter-dependent situation. Let M be an hermitian vector bundle with an hermitian connection ∂ and let a compact group G act on E by unitary transformations of fibers. Thus, choosing local orthonormal frames of the bundle E, we represent an element $s \in G$ as a unitary matrix $s(x)$ acting on fibers E_x and depending on a parameter $x \in M$. We suppose that the matrix $s(x)$ has no eigenvalue equal to 1 for any x. For any fiber E_x we may construct a Fock space F_x as was described in Section 2 and obtain an associated Fock bundle F over M. Moreover, we have a trace $\mathrm{Tr}_{s(x)}$ on fibers. The problem is to calculate the Chern character of the Fock bundle with respect to this trace.

First of all let us describe an associated connection $\widehat{\partial}$ on the Fock bundle. In a local orthonormal frame of E the connection ∂ is given by a matrix-valued 1-form $\Gamma = \Gamma_{ij}$, $i,j = 1,2,\dots,n$, which is an anti-hermitian matrix ($\Gamma = -\Gamma^*$). Consider a local 1-form

$$\widehat{\Gamma} = \frac{1}{2h}\Gamma_{ij}\bar{z}_i * z_j = \frac{1}{2h}z^*\Gamma * z \tag{4.1}$$

whose values are normally ordered quadratic Hamiltonians in F. We set for the basis (2.6)

$$\widehat{\partial}(\bar{z}^\alpha * p) = [\widehat{\Gamma}, \bar{z}^\alpha * p] = \widehat{\Gamma} * \bar{z}^\alpha * p, \tag{4.2}$$

the latter identity follows since $p * \Gamma = 0$ because of the normal ordering. In particular, the vacuum section p is covariantly constant, that is $\widehat{\partial} p = 0$.

The result of this section consists in the following theorem.

Theorem 4.1 *Let $\widehat{R}$ be the curvature of $\widehat{\partial}$. Then*

$$\mathrm{Tr}_s \exp\left(\frac{1}{2\pi i}\widehat{R}\right) = \det{}^{-1}\left(1 - s^{-1}(x)\exp\left(-\frac{R}{2\pi i}\right)\right). \tag{4.3}$$

Proof. First let us calculate the curvature. We have

$$\widehat{\partial}^2 e^\alpha = \widehat{\partial}(\widehat{\Gamma} e^\alpha) = (d\widehat{\Gamma} + \widehat{\Gamma} * \widehat{\Gamma})e^\alpha$$

so that

$$\widehat{R} = d\widehat{\Gamma} + \widehat{\Gamma} * \widehat{\Gamma} = d\widehat{\Gamma} + \frac{1}{2}[\widehat{\Gamma}, \widehat{\Gamma}]$$
$$= \frac{1}{2h}d\Gamma_{ij}\bar{z}_i * z_j + \frac{1}{8h^2}\Gamma_{ij}\wedge\Gamma_{kl}[\bar{z}_i * z_j, \bar{z}_k * z_l].$$

But, using (2.4), we have

$$[\bar{z}_i * z_j, \bar{z}_k * z_l] = [\bar{z}_i z_j, \bar{z}_k z_l] = 2h\delta_{jk}\bar{z}_i z_l - 2h\delta_{il}\bar{z}_k z_j$$
$$= 2h(\delta_{jk}\bar{z}_i * z_l - \delta_{il}\bar{z} * z_j),$$

so, that

$$\widehat{R} = \frac{1}{2h}z^*(d\Gamma + \Gamma\wedge\Gamma) * z = \frac{1}{2h}z^* R * z, \tag{4.4}$$

where R is the curvature of the connection ∂.

Now, since M is a finite-dimensional manifold, the element $\widehat{R}$ is nilpotent, so the $*$-exponential $\exp(\widehat{R}/2\pi i)$ is defined correctly by its Taylor series or via differential equation, of course, both definitions coincide. Thus, we may apply Proposition 3.1 to calculate the $*$-exponential and its trace. According to this proposition the $*$-exponential acts on the Fock space as a unitary matrix $\sigma = \exp(-R/2\pi i)$. In other words,

$$\exp\left(\frac{\widehat{R}}{2\pi i}\right) = \widehat{\exp\left(-\frac{R}{2\pi i}\right)} = \widehat{\sigma}.$$

Then

$$\mathrm{Tr}_s\,\widehat{\sigma} = \mathrm{Tr}\,\widehat{s}\widehat{\sigma} = \mathrm{Tr}\,\widehat{s\sigma}\,1 = \mathrm{Tr}_{s\sigma}\,1 = \det{}^{-1}(1-(s\sigma)^{-1}),$$

proving the theorem. ∎

5. Quantization of the fixed-point manifold

Let $(M,\ \omega)$ be a symplectic manifold with a symplectic connection ∂^s, $E \to M$ a complex vector bundle with a connection ∂^E, s a symplectic diffeomorphism belonging to a compact group G of symplectomorphisms preserving the connection ∂^s. We assume also that G has an equivariant lifting to E preserving the connection ∂^E, that is E is a G-bundle.

We pick a positive almost complex structure J on M which is preserved by G, this means that $\omega(Ju,v)$ is a G-invariant Riemannian metric on M. Such a structure always exists. A standard way to construct it consists in taking first any Riemannian metric g. Averaging it over a compact group G, we make it G-invariant. Then goes a standard procedure which for a pair $g,\ \omega$ gives an automorphism J of the tangent bundle such that $g(u,v) = \omega(Ju,v)$ (see, e.g. [6, Chapter 2]). The tangent bundle TM with the almost complex structure J becomes a complex vector bundle $T^{\mathbb{C}}M$ with a G-invariant hermitian scalar product (u,v) such that

$$\omega(u,v) = \mathrm{Im}(u,v). \tag{5.1}$$

Let, further, M_0 be a connected component of the fixed-point set of s. We assume that M_0 is a smooth compact submanifold and let N be its normal bundle (with respect to the metric g). At any point $x \in M_0$, s acts on T_xM as an orthogonal matrix. The vectors $u \in T_xM$ tangent to M_0 are the eigenvectors of s with the eigenvalue 1. We assume also that s is *non-degenerate*, that is there are no more eigenvectors with the eigenvalue 1. So, we have an s-invariant decomposition $T_xM = T_xM_0 \oplus N_x$ and s acts on N without fixed vectors (except the origin). Since the complex structure J is preserved s acts on $T_x^{\mathbb{C}}M$ as a unitary matrix and the decomposition

$$T_x^{\mathbb{C}}M = T_x^{\mathbb{C}}M_0 \oplus N_x^{\mathbb{C}}$$

also takes place. In particular, this decomposition is orthogonal with respect to the symplectic form ω on T_xM, too. Thus, the restriction of ω to T_xM_0 and N_x gives the symplectic structure on both of these subspaces. We come to the following proposition.

Proposition 5.1 *The restriction $\omega_0 = \omega|_{M_0}$ is a symplectic form on M_0. The normal bundle N has an hermitian structure $(\cdot,\cdot)$, s acts on N as a unitary automorphism and, moreover $1 \notin \mathrm{spec}\, s|_N$.*

In the sequel we treat N as a complex bundle simplifying the notation $N^{\mathbb{C}}$ to N. Pick an s-invariant hermitian connection on N, let $\Gamma^N = (\Gamma^N_{ij})$ be a connection 1-form ($i, j = 1, 2, \ldots, k = \dim_{\mathbb{C}} N$) in a local orthonormal frame, this is an anti-hermitian matrix (cf. Section 4). The connection gives rise to a closed 2-form $\widetilde{\omega}$ on the total space N. Denoting by (x, z) local coordinates on N where

$$z = (z_1, z_2, \ldots, z_k)^t$$

is a column of coordinates of the vector $z \in N_x$, set

$$\widetilde{\omega} = \frac{1}{2i} d(z^* \partial^N z) + \omega_0 = \frac{1}{2i}(dz^* dz + d(z^* \Gamma^N z)) + \omega_0. \tag{5.2}$$

This form is non-degenerate for z small enough, so we obtain a symplectic form $\widetilde{\omega}$ on a neighborhood N_ε of the zero section. The exponential mapping (with respect to the metric g) maps this neighborhood onto a tubular neighborhood of M_0 in M. The symplectic form ω on M is pulled back to N_ε and the forms ω and $\widetilde{\omega}$ coincide being restricted to a tangent space $T_x M$ at any point $x \in M_0$. The Weinstein arguments (see e.g. [6, Section 2.3]) show that there exists an s-equivariant diffeomorphism of N_ε to the tubular neighborhood of M_0 in M transforming ω to $\widetilde{\omega}$. So, in the sequel we will assume that this neighborhood is identified with N_ε with the symplectic form (5.2) which from now on we will denote simply by ω. This gives us a *normal form* of M in the vicinity of the fixed-point set component M_0.

Consider now a G-invariant canonical deformation quantization on M. The word "canonical" means that the algebra A_h of quantum observables on M is realized as the algebra of flat sections of the Weyl algebra bundle $W(TM)$ over M with respect to an Abelian connection D whose Weyl curvature is $-i\omega/h$ (see [6, Chapter 5] for more details). Restriction to the tubular neighborhood gives a canonical deformation quantization on N_ε with the symplectic form (5.2). But the structure of the vector bundle allows us to construct another realization of the algebra A_h (see [6, Section 6.6]). This construction goes in two steps. The first step is simple: We quantize each fiber N_x of N. The result is the Weyl algebra $W_x(\mathbb{C}^k)$ on the complex space N_x described in Section 2. These algebras form a bundle $W(N)$ over M_0 with fibers $W_x(\mathbb{C}^k)$. We may think of it as the bundle of "operators on the Fock bundle" as in Section 4. The trouble is, however, that such a viewpoint requires h to be a number. So, only the bundle $W(N)$ is meaningful for deformation quantization. Nevertheless, it is very helpful to keep in mind such a "realization" of the bundle $W(N)$. The connection $\widehat{\partial}^N$ gives rise to an associated connection $\widehat{\partial}$ on $W(N)$: If $a = a(x, z, \bar{z})$ is a section of $W(N)$, then

$$\widehat{\partial}^N a = d_x a + \frac{1}{2h}[z^* \Gamma * z, a] \tag{5.3}$$

(cf. Section 4). Observe, that the normal ordering in (5.3) may be replaced by the symmetric one, that is we may take $z^*\Gamma z$ instead of $z^*\Gamma * z$. Indeed, by (2.4)

$$\bar{z}_i * z_j = \bar{z}_i z_j - h\delta_{ij},$$

so that

$$z^*\gamma * z = z^*\Gamma^N z - \operatorname{tr}\Gamma^N$$

and the scalar $\operatorname{tr}\Gamma$ disappears under the commutator.

Remark 5.2 We see once again that it is useful to keep in mind the hypothetical Fock bundle $F(N)$. The normal ordering gives a "connection" on the Fock bundle with the connection form

$$\widehat{\Gamma}^N = \frac{1}{2h} z^*\Gamma^N * z$$

and the change of the ordering changes the connection itself (cf. Section 4).

The curvature of the connection (5.3) is by definition

$$\widehat{R}^N = d_x\widehat{\Gamma}^N + \widehat{\Gamma}^N * \widehat{\Gamma}^N = \frac{1}{2h} z^* R^N * z = \frac{1}{2h} z^* R^N z - \frac{1}{2}\operatorname{tr} R^N.$$

The second step is to quantize M_0 taking $W(N)$ as a coefficient bundle. More precisely, we have a bundle E and the original coefficient bundle $K = \operatorname{Hom}(E,E)$. Now, having constructed a bundle

$$W(N) = \text{"}\operatorname{Hom}(F(N),F(N))\text{"},$$

let us take

$$L = K \otimes W(N) = \text{"}K \otimes \operatorname{Hom}(F(E),F(E))\text{"}$$

as a new coefficient bundle with the connection

$$\partial^L = \partial^E \otimes 1 + 1 \otimes \widehat{\partial}^N \tag{5.4}$$

whose curvature is

$$R^L = R^E \otimes 1 + 1 \otimes \widehat{R}^N.$$

We apply a quantization scheme with a non-trivial coefficient bundle [6, Section 6.5]. According to this scheme we need to consider the Weyl algebra bundle $W(TM_0) \otimes L$ on the tangent bundle to M_0 and a connection on it

$$D = d + [\gamma, \cdot]$$

with

$$\gamma = \frac{i}{h}\omega^0_{ij} y^i dx^j + \frac{i}{2h}\Gamma^s_{ij} y^i y^j + \Gamma^L + \frac{i}{h} r.$$

Here ω^0_{ij} are coefficients of the symplectic form ω_0 in local Darboux coordinates on M_0, Γ^s is a connection form of the symplectic connection on M_0, Γ^L a connection form of ∂^L, $y = (y_1, y_2, \ldots, y^{2m})$, $2m = \dim M_0$, is a tangent vector to M_0 and r is an unknown 1-form subject to the condition

$$d\gamma + \gamma^2 = -\frac{i}{h}\omega_0 + \frac{1}{2}\operatorname{tr} R^N. \tag{5.5}$$

On the right-hand side of (5.5) stands the desired scalar curvature form (called the *Weyl curvature*). The additional lower-order term $1/2 \operatorname{tr} R^N$ is included to compensate the change of the ordering (from normal to symmetric ones) needed in future. In [6, Chapter 6] it is proved that such an r exists. Moreover, the quantum algebra A_h on the total space N_ε with the Weyl curvature $-i/h\,\omega$ given by (5.2) and the quantum algebra on M_0 with coefficients in L and the Weyl curvature (5.5) are isomorphic.

Thus, we have realized A_h as an algebra of flat sections on M_0 with coefficient bundle $L = K \otimes W(N)$. It is very important that s acts on coefficients only but not on M_0 or the Weyl algebra bundle $W(TM_0)$.

6. *G*-index theorem

To define the global s-trace for a given fixed-point set component M_0 we follow general prescriptions of [6, Chapters 5,6]. The algebra A_h constructed in Section 5 consists of flat sections of the Weyl algebra bundle $W(TM_0)$ over M_0 with coefficients in the bundle $L = K \otimes W(N)$. Restricting this algebra to a small coordinate neighborhood $O \subset M_0$, we first trivialize the coefficient bundle taking local frames of E and N over O. Then, similarly to [6, Section 6.5], we construct a trivialization isomorphism

$$I : A_h \to W(\mathbb{R}^{2m}) \otimes L$$

to a standard Weyl algebra with coefficients in the trivial bundle L. The group G acts on coefficients only, thus, the local s-trace (Section 2) makes sense (we denote here this trace by tr_s). Finally, define the global s-trace by

$$\operatorname{Tr}_s a = (2\pi h)^{-m} \int_{\mathbb{R}^{2m}} \operatorname{tr}_s Ia \frac{\omega_0^m}{m!}. \tag{6.1}$$

The general case is reduced to the case of a small support by using partition of unity. As usual, this construction is independent of the choice of trivialization isomorphisms.

The properties of the global trace Tr_s follow directly from the corresponding properties of the local coefficient trace tr_s.

Having constructed the desired s-trace, we calculate now $\operatorname{Tr}_s 1$. The following theorem which we call the *equivariant index theorem for deformation quantization* gives an explicit formula for this quantity.

Theorem 6.1 *Let M_0 be a compact submanifold, TM_0 its real tangent bundle, N its complex normal bundle. Let*

$$\kappa^t = \frac{R^t}{2\pi i}, \quad \kappa^N = \frac{R^N}{2\pi i}, \quad \kappa^E = \frac{R^E}{2\pi i},$$

be curvature forms for TM_0, N, and E, respectively. Then

$$\mathrm{Tr}_s 1 = \tag{6.2}$$
$$\int_{M_0} \frac{\mathrm{tr}\, s^E \exp \kappa^E}{\det(1-(s^N)^{-1}\exp(-\kappa^N))} \det{}^{-1/2}\left(\frac{\sinh \kappa^t/2}{\kappa^t/2}\right) \exp\left(\frac{\omega_0}{2\pi h} - \frac{1}{2}\mathrm{tr}\,\kappa^N\right)$$

where s^E and s^N mean automorphisms of E and N defined by s.

Proof. We apply the usual index theorem for deformation quantization [6] with a non-trivial coefficient bundle [7]. In our case the symplectic manifold is M_0 and the coefficient bundle is $L = K \otimes W(N)$ with the coefficient trace equal to tr_s. We take the Weyl curvature to be

$$\Omega_W = -\frac{i}{h}\omega_0 + \frac{1}{2}\mathrm{tr}\, R^N$$

and the connection in $K \otimes W(N)$ defined by (5.4) with the curvature

$$R^L = R^E \otimes 1 + 1 \otimes \frac{1}{2h} z^* R^N * z. \tag{6.3}$$

The only problem is to calculate the Chern character form for (6.3), that is $\mathrm{tr}_s \exp \kappa^L$. But

$$\begin{aligned} \mathrm{tr}_s \exp \kappa^L &= \mathrm{tr}_s \left(\exp \kappa^E \otimes \exp\left(\frac{1}{2h} z^* R^N * z\right)\right) \\ &= \mathrm{tr}_s \exp \kappa^E \, \mathrm{tr}_s \exp\left(\frac{1}{2h} z^* R^N * z\right). \end{aligned} \tag{6.4}$$

On the bundle E, s acts as a homomorphism, so $\mathrm{tr}_s \exp \kappa^E = \mathrm{tr}\, s^E \exp \kappa^E$. The second factor in (6.4) was calculated in Section 4. The result is

$$\det{}^{-1}(1-(s^N)^{-1}\exp(-\kappa^N)).$$

Substituting these expressions into the index formula

$$\mathrm{Tr}_s 1 = \int_{M_0} \mathrm{ch} L \exp\left(\frac{-\Omega_W}{2\pi i}\right) \widehat{A}(M_0),$$

we come to (6.2). ∎

Let us examine the expression (6.2) in two extreme cases. The first one is when s is the identity. Then $M_0 = M$, N is a zero bundle, so $\kappa^N = 0$, and we come to the usual index formula (1.4). In the other case when M_0 is a fixed point x_0 we obtain a familiar fixed-point contribution

$$\frac{\operatorname{tr} s^E(x_0)}{\det(1-(s^N)^{-1}(x_0))}.$$

It should be mentioned that this expression is defined up to a phase factor which depends on how the group G acts on the vacuum. Our normalization in (6.2) means that the action is trivial.

References

1. Atiyah, M. F. and Segal, G. B.: The index of elliptic operators. II, *Ann. of Math.* **87** (1968), 531–545.
2. Atiyah, M. F. and Singer, I. M.: The index of elliptic operators. III, *Ann. of Math.* **87** (1968), 546–604.
3. Bayen, F., Flato, M., Fronsdal, C., Lichnerowicz, A., and Sternheimer, D.: Deformation theory and quantization. I and II, *Ann. Physics* **111** (1978), 61–151.
4. Basart, H., Flato, M., Lichnerowicz, A., and Sternheimer, D.: Deformation theory applied to quantization and statistical mechanics, *Lett. Math. Phys.* **8** (1984), 483–494.
5. Fedosov, B.: A simple geometrical construction of deformation quantization. *J. Diff. Geom.* **40** (1994), 213–238.
6. Fedosov, B.: *Deformation Quantization and Index Theory*, Mathematical Topics 9, Akademie Verlag, Berlin, 1996.
7. Fedosov, B.: The Atiyah-Bott-Patodi method in deformation quantization, preprint 99/7, Univ. Potsdam, 1999.
8. Nest, R. and Tsygan, B.: Algebraic index theorem, *Comm. Math. Phys.* **172** (1995), 223–262.

Superconformal field theories, multiplet shortening, and the $AdS_5/SCFT_4$ correspondence

Sergio Ferrara
CERN Geneva, Switzerland
e-mail: `sergio.ferrara@cern.ch`

Alberto Zaffaroni
CERN Geneva, Switzerland
e-mail: `alberto.zaffaroni@cern.ch`

Abstract. We review the unitarity bounds and the multiplet shortening of UIR's of 4 dimensional superconformal algebras $SU(2,2|N)$, $(N = 1,2,4)$ in view of their dual role in the AdS/SCFT correspondence. Some applications to KK spectra, non-perturbative states and stringy states are given.

Keywords: superconformal algebras, singletons, supersymmetry

Mathematics Subject Classifications (2000): 81T60, 81T20

1. Introduction

The classification of Super Poincaré and Super Anti-de-Sitter algebras [1, 2, 3] in diverse dimensions has played a major role since the early days of supersymmetry and supergravity.

Super Poincaré algebras have been widely used to classify supersymmetric field theories and in constructing generic supergravity theories in diverse dimensions and with different numbers of supersymmetries [4]. Their central extension (more precisely the central extension of the supertranslational algebras) had a major role with the inclusion of p-branes supergravity solutions; a particular case is related to the black hole classification based on the BPS properties of the solutions.

Supergravity solutions for BPS p-branes are considered to be limits of analogous solutions in quantum theories of gravity, such as string theory or M-theory. However, in certain cases, due to non renormalization theorems of N-extended supersymmetry, such solutions are exact or at least reliable approximations of stringy or M-theory solutions.

The recent interplay between Anti-de-Sitter supergravity and conformal field theories follows from the conjectured duality [5] between the string (or M-theory) background describing the AdS_{p+2} horizon geometry [6] of black p-branes and the conformal field theory living on their world-volume, which can be thought as the AdS-boundary. This correspondence claims that the *correlation functions* of composite operators in the CFT can be computed as *bulk* Green functions of AdS states of supergravity or string theory [7, 8].

G. Dito and D. Sternheimer (eds.), Conférence Moshé Flato 1999, Vol. 1, 177 – 188.

In the present discussion, we only consider the case of D3-branes in type IIB supergravity. The world-volume theories are $N = 1,2,4$ 4d superconformal field theories [8, 9] and the corresponding bulk theory is type IIB on $AdS_5 \times X_5$, for internal manifolds X_5 giving $N = 2,4,8$ supergravities [10, 11, 12, 13], respectively. Operators in the SCFT are associated with the KK excitations, coming from the harmonic expansion on X_5.

These theories and their UIR's are therefore related to the $SU(2,2|N)$ algebras for $N = 1,2,4$ [14, 15, 16, 17]. The case $N = 4$, as we will see, deserves a special treatment since the $U(1)$ factor of the $U(N) = U(1) \times SU(N)$ R-symmetry becomes an outer automorphism of the superalgebra [18, 19, 17, 20] (this is a particular case of the $SU(N|M)$ superalgebra with $N = M$).

This contribution reviews the unitary bounds and multiplet shortening for highest weight UIR's of 4d superconformal algebras in view of the AdS_5/CFT_4 correspondence [9, 21, 22]. Special emphasis will be given to different properties of the $N = 1,2,4$ cases, due to the different supersingleton representations [23, 14, 18, 19]. The latter are the basic degrees of freedom of the superconformal theories on the boundary. In this respect, the pioneering work of Moshé Flato, in collaboration with Christian Frønsdal, has a crucial role in this analysis. The above authors extended, in particular, the unitary bounds of UIR's of $SU(2,2)$ to $SU(2,2|1)$, and classified the multiplet shortenings, which just correspond to the *thresholds* of the unitary bounds [14]. Extension of this analysis came later [18, 19] and is the latter that we will use in the context of the AdS_5/CFT_4 correspondence.

In Sections II and III, the unitarity bounds for $N = 1,2,4$ will be reviewed. UIR's satisfying these bounds have an interpretation in terms of superconformal boundary fields [8, 9, 24]. Some examples will be discussed. More importantly, the case of multiplet shortening corresponds to CFT operators of protected dimensions. This happens either if a boundary operator is a *conserved current*, thus corresponding to a massless bulk field, or if it is a shortened multiplet of the *chiral* type [8, 25]. Intermediate multiplets which are not chiral also exist that have protected dimensions [21, 22]. They can be formally obtained by the product of a conserved supercurrent with a *chiral* operator.

For extended superconformal algebras with $N = 2,4$ a new (exceptional) series of shortened multiplets also exists which has no analogous in $N = 1$. This has to do with the fact that $N = 2$ and $N = 4$ algebras admit *self-conjugate* supersingletons representations [16, 18, 19] (unlike the $N = 1$ case).

In Section IV, the application of different type of shortened multiplets to different boundary conformal field theories will be given. Shortened multiplets allow to make a detailed comparison between the AdS_5 bulk theory and the boundary superconformal field theory. Particular examples are the $N = 4$ KK towers [11, 26, 8, 24], corresponding to an exceptional series of the $N = 4$ superalgebra as well as multitrace operators of the $N = 4$ Yang-Mills theory [27], some of them having protected dimensions [27, 28, 29]. A richer structure exists for $N = 1$ superconformal theories where shortened multiplets with anomalous dimensions can appear with precise associated supergravity states. This is the case for $X_5 = T^{1,1} = SU(2) \times SU(2)/U(1)$, unique example [12] of a smooth manifold where both the bulk and the boundary theories have been worked out in full detail [30, 31, 32, 22, 33].

2. The unitary representations and shortening for $SU(2,2|N), N = 1,2$

In this Section we consider the unitarity bounds for the highest weight representations of the $SU(2,2|N)$ superalgebra.

For the case of the $SU(2,2)$ algebra itself, a given UIR is denoted, following Flato and Frønsdal [23, 14, 34], as $D(E_0, J_1, J_2)$ where E_0, J_1, J_2 are the quantum number of the highest weight state [16], given by a finite UIR of the maximal compact subgroup $SU(2) \times SU(2) \times U(1)$. The UIR's fall in three series [35, 9, 36]:

$$\begin{array}{lll} a)\ J_1 J_2 \neq 0, & E_0 \geq 2 + J_1 + J_2; & \\ b)\ J_2 J_1 = 0, & E_0 \geq 1 + J; & \quad (2.1) \\ c)\ J_1 = J_2 = 0, & E_0 = 0. & \end{array}$$

In the bulk interpretation, the inequality corresponds to massive representations in AdS_5, the bound in a) corresponds to massless bulk particles of *spin* $J_1 + J_2$ while the bound in b) corresponds to singletons of spin J.

Note that in the AdS/CFT correspondence the bulk-boundary quantum numbers (E_0, J_1, J_2) refer to the *compact* basis for the AdS states, while they refer to the non-compact basis $SL(2,C) \times O(1,1)$ for the boundary conformal operators [8, 17]. The highest weight state in AdS corresponds to a conformal operator $O(x)$ at $x = 0$, so the AdS energy E_0 corresponds to the conformal dimension Δ_0 and the (J_1, J_2) quantum numbers correspond to the Lorentz spin of $O(x)$.

In the CFT, the bound a) corresponds to conformal conserved currents of spin $J = J_1 + J_2$:

$$E_0 = 2 + J_1 + J_2, \qquad \partial^{\alpha_1 \dot{\alpha}_1} J_{\alpha_1 \ldots \alpha_{2J_1}, \dot{\alpha}_1 \ldots \dot{\alpha}_{2J_2}}(x) = 0, \qquad J_1 J_2 \neq 0, \qquad (2.2)$$

while the bound b) corresponds to massless spin J conformal fields on the boundary:

$$\begin{array}{lll} E_0 = 1 + J, & \partial^{\alpha_1 \dot{\alpha}_1} O_{\alpha_1 \ldots \alpha_{2J}} = 0, & J \neq 0; \quad (2.3) \\ & \partial^2 O(x) = 0, & J = 0. \end{array}$$

The case c) corresponds to the identity representation.

Let us now consider the case of $SU(2,2|N)$ superalgebras [15, 18]. In this case, the highest weight state is denoted by $D(E_0, J_1, J_2; r, a_1, \ldots, a_{N-1})$ where the quantum numbers in the bracket denote a UIR of $SU(2,2) \times U(1) \times SU(N)$, r being the quantum number of the $U(1)$ R-symmetry and $a_1, \ldots, a_{N-1}$ the Dynkin labels of a UIR of the non-abelian symmetry $SU(N)$. We will denote by R the $U(1)$ inside $U(N)$.

Note that for $N \neq 4$, the $SU(2,2|N)$ algebra is both a subalgebra and a quotient algebra of $U(2,2|N)$, since the supertrace generator (which is a central charge) can be eliminated by a redefinition of the $U(1)$ generator R of $U(N)$ [18, 19]. This redefinition is not however possible for $N=4$ since R drops from the supersymmetry anti-commutators and it becomes an outer automorphism of the algebra [19, 37, 20]. In this case we have therefore two inequivalent algebras (which do not include the $U(1)$ R generator), $PSU(2,2|4)$ and $PU(2,2|4)$, depending whether $r=0$ or $r \neq 0$.

We will adopt the convention that r is always the quantum number of the $U(1)$ generator of $U(2,2|N)$, so it will be the $U(1)$ subgroup of $U(N)$ for $N=1,2$ while it will be a central $U(1)$ for $N=4$ [18].

In the boundary CFT language, UIR's can be realized as conformal superfields. The superhighest weight state corresponds to a superfield $\phi(x,\theta)$ at $x=\theta=0$ [18, 24, 9, 38].

The unitarity bounds for $SU(2,2|1)$ were found by Flato and Frønsdal [14]. They generalize the cases a), b) and c) of (2.1). They are:

$$\text{a)}\ E_0 \geq 2+2J_2+r \geq 2+2J_1-r \quad (\text{or } J_1 \to J_2, r \to -r), \quad J_1,J_2 \geq 0; \tag{2.4}$$

which implies

$$E_0 \geq 2+J_1+J_2, \quad r \geq J_1-J_2, \quad 2+2J_1-E_0 \leq r \leq E_0-2-2J_2; \tag{2.5}$$

$$\text{b)}\ E_0 = r \geq 2+2J-r \quad (J_2=0, J_1=J, \text{ or } J_1=0, J_2=J, r \to -r); \tag{2.6}$$

which implies $E_0 \geq 1+J$,

$$\text{c)}\ E_0 = J_1 = J_2 = r = 0; \tag{2.7}$$

which corresponds to the identity representation.

Shortening in the case a) corresponds to

$$E_0 = 2+2J_2+r, \quad r \geq J_1-J_2, \quad (\text{or } J_1 \to J_2, r \to -r). \tag{2.8}$$

This is a semi-long AdS_5 multiplet [21] or, in conformal language, a *semiconserved* superfield [39, 22]:

$$\bar{D}^{\dot\alpha_1} L_{\alpha_1\ldots\alpha_{2J_1},\dot\alpha_1\ldots\dot\alpha_{2J_2}}(x,\theta,\bar\theta) = 0, \quad (\bar{D}^2 L_{\alpha_1\ldots\alpha_{2J_1}} = 0 \text{ for } J_2=0), \tag{2.9}$$

(in our conventions θ carries $\Delta=-1/2, r=1, \bar\theta$ has $\Delta=-1/2, r=-1$).

Maximal shortening in case a) corresponds to $E_0=2+J_1+J_2$, $r=J_1-J_2$. This is a conserved superfield which satisfies both left and right constraints:

$$\bar{D}^{\dot\alpha_1} J_{\alpha_1\ldots\alpha_{2J_1},\dot\alpha_1\ldots\dot\alpha_{2J_2}} = D^{\alpha_1} J_{\alpha_1\ldots\alpha_{2J_1},\dot\alpha_1\ldots\dot\alpha_{2J_2}} = 0. \tag{2.10}$$

The shortening in b) corresponds to *chiral superfields*. Maximal shortening in b) to *massless* chiral superfields, i.e., chiral singleton representations: $E_0 = r = 1 + J$. The corresponding superfield, for $E_0 = r$ satisfies

$$\bar{D}^{\dot{\alpha}} S_{\alpha_1 \ldots \alpha_{2J}} = 0, \tag{2.11}$$

and, for $E_0 = 1 + J$, it also satisfies

$$D^{\alpha_1} S_{\alpha_1 \ldots \alpha_{2J}} = 0, \quad (D^2 S = 0, \text{ for } J = 0). \tag{2.12}$$

These equations are the supersymmetric version of (2.2) and (2.3).

We may call, with an abuse of language, off-shell singletons chiral superfields in the sense that in an interacting conformal field theory singletons may acquire anomalous dimension and fall in (2.11).

It is also evident, from superfield multiplication, that by suitable multiplication of several free supersingletons one may get any other superfield of type (2.9), (2.10) or (2.11).

Note that superfields obeying (2.9), (2.11) may have anomalous dimensions since the shortening condition just implies a relation between E_0 and r without fixing their value.

The basic singleton multiplets for $N = 1$ gauge theories correspond to $J = 0, 1/2$ in (2.11), i.e., chiral scalar superfields S (Wess-Zumino multiplets) and Yang-Mills field strength multiplets W_α. Any other conformal operator is obtained by suitable multiplication of these two sets of basic superfields.

In type IIB on $T^{1,1}$ long, semi-long and chiral multiplets do indeed occur [30, 31, 22]. Chiral WZ singleton multiplets have in this case an anomalous dimension $\gamma = -1/4$ ($\Delta = 1 + \gamma$) and R-symmetry $R = 3/4$.

If we adopt the concept of *bulk* masslessness as corresponding to an UIR that is contained in the product of two supersingletons there are other massless bulk representations which are obtained by chiral multiplication of two singleton representations:

$$D(1+J_1, J_1, 0; 1+J_1) \otimes D(1+J_2, J_2, 0; 1+J_2) = D(2+J_1+J_2, J, 0; 2+J_1+J_2), \tag{2.13}$$

$$|J_1 - J_2| \leq J \leq J_1 + J_2.$$

These representations do not occur in the AdS/SCFT correspondence unless $J_1 = 0$, $J_2 = 0, 1/2$ since they do not correspond to global symmetries of the SCFT. One can indeed show that the symmetry of $N = 1$ SYM theory, as in the dual of type IIB on $AdS_5 \times T^{1,1}$, do not allow such UIR's.

We now turn to the $SU(2,2|2)$ superalgebra. In this case the highest weight state is $D(E_0, J_1, J_2; r, l)$ where l is the spin label of the R-symmetry $SU(2)$. The series of UIR's are now:

$$\alpha)\ E_0 \geq 2 + 2J_2 + r + 2l \geq 2 + 2J_1 - r + 2l \ (\text{or } J_1 \to J_2, r \to -r, l \to l), \\ J_1, J_2 \geq 0; \tag{2.14}$$

which implies

$$E_0 \geq 2 + J_1 + J_2 + 2l,$$
$$r \geq J_1 - J_2, \qquad (2.15)$$
$$2 + 2J_1 + 2l - E_0 \leq r \leq E_0 - 2 - 2J_2 - 2l.$$

$$\beta)\ E_0 = r + 2l \geq 2 + 2J - r + 2l\ (J_2 = 0, J_1 = J, \text{ or } J_1 = 0, J_2 = J, r \to -r); \qquad (2.16)$$

which implies $r \geq 1 + J$, $E_0 \geq 1 + J + 2l$,

$$\gamma)\ E_0 = 2l, \quad J_1 = J_2 = r = 0. \qquad (2.17)$$

The new phenomenon in $N = 2$ is that supersingletons appear in two different series, β) and γ). This is related to the fact that the γ) series does not reduce to the identity representation only, as for $N = 1$, but contains both massless bulk ($l = 1$) and massless boundary (supersingletons) ($l = 1/2$) representations. Let us discuss in detail the cases of massless bulk and massless boundary representations.

Supersingletons in β) correspond to the shortening condition

$$E_0 = r = 1 + J, \qquad l = 0. \qquad (2.18)$$

This is the highest weight state $D(1+J, J, 0; 1+J, 0)$. The highest spin in these multiplets is $J+1$. For $J = 0$ we get the Yang-Mills multiplet.

The $l = 1/2$ supersingleton in γ) is the hypermultiplet ($D(1,0,0;0,1/2)$).

Since supermassless bulk representations are defined as the product of two supersingletons we may obtain bulk massless representations by multiplying either two supersingletons (in a chiral and anti-chiral manner) in β) or by a supersingleton in β) and the $l = 1/2$ supersingleton in γ), or two $l = 1/2$ supersingletons in γ).

The spin 2 massless bulk states are obtained by chiral anti-chiral multiplication of the $J = 0$ chiral supersingleton in β), giving the shortened representations in α) with $J_1 = J_2 = r = l = 0$. This is the massless graviton multiplet: $D(2,0,0;0,0)$.

The massless spin 1 multiplet is obtained by multiplying two hypermultiplets and corresponds to the $l = 1$ UIR of case γ), giving $D(2,0,0;0,1)$.

Finally, the massless spin 3/2 multiplet is obtained by multiplying the $J = 0$ supersingleton in β) with the hypermultiplet, giving a bulk massless multiplet corresponding to the shortening $r = 1, J = 0, l = 1/2$ in β), i.e., $D(2,0,0;1,1/2)$. The superconformal realization of these superfields as current multiplets was given in [40, 41, 42].

Obviously these examples can be extended by replacing the chiral vector multiplet by an arbitrary spin J supersingleton. Also one can get, as for $N = 1$, *chiral* bulk massless multiplets by chiral multiplication of two spin J supersingletons as in (2.13).

In the AdS_5/CFT_4 correspondence the relevant supersingletons are the *chiral* vector multiplets (in β) for $J = 0$) and the hypermultiplets (in γ) for $l = 1/2$). Multiplet shortening in the AdS/CFT correspondence will be further discussed in Section IV. Here we note that the series γ) contains short multiplets with $E_0 = 2l$ for each value of l. These states can be explicitly constructed by multiplying $2l$ hypermultiplet singletons and their dimension is not renormalized and coincide with the canonical one. On the other hand, short multiplets in β), constructed using also chiral vector singletons, may have anomalous dimension since, due to (2.16), E_0 is related to the arbitrarily valued $U(1)$ charge r by $E_0 = r + 2l$.

3. UIR's of $PSU(2,2|4)$ and $PU(2,2|4)$ and shortening conditions

The $N = 4$ superalgebra is of great interest because it corresponds to $N = 4$ superconformal Yang-Mills theory. In the dual description, it lives at the boundary of AdS_5 [5, 7, 8]. The supergravity theory emerges as the low energy limit of type IIB string theory compactified on $AdS_5 \times X_5$, where X_5 is a manifold preserving $N = 8$ such as S_5 or RP_5 [11, 43].

The UIR's classes of the previous Section enter in this case as follows.

Let us consider a highest weight representation $D(E_0, J_1, J_2; r, p, k, q)$, where r is the central charge eigenvalue and (p, k, q) are the $SU(4)$ Dynkin labels. Then the three unitary series show up as follow [18]:

$$A)\ E_0 \geq 2 + 2J_2 + r + \frac{1}{2}(p + 2k + 3q) \geq 2 + 2J_1 - r + \frac{1}{2}(3p + 2k + q) \quad (3.1)$$
$$(\text{or } J_1 \to J_2,\ r \to -r,\ (p,k,q) \to (q,k,p));$$

which implies

$$E_0 \geq 2 + J_1 + J_2 + p + k + q, \qquad r \geq \frac{1}{2}(p - q) + J_1 - J_2,$$
$$2 + 2J_1 + \frac{1}{2}(3p + 2k + q) - E_0 \leq r \leq E_0 - 2 - 2J_2 - \frac{1}{2}(p + 2k + 3q). \quad (3.2)$$

Shortening occurs when

$$E_0 = 2 + 2J_2 + r + \frac{1}{2}(p + 2k + 3q), \quad (3.3)$$

and maximal shortening when

$$E_0 = 2 + J_1 + J_2 + p + k + q, \qquad r = \frac{1}{2}(p - q) + J_1 - J_2. \quad (3.4)$$

Generic massless bulk multiplets correspond to $p = k = q = 0$. They correspond to Table 12 in [16].

When we have a strict inequality in (3.1), we obtain generic massive multiplets with highest spin $(J_1 + 2, J_2 + 2)$ in the (p, k, q) representation of $SU(4)$. Note that all these long multiplets (with $J_{MAX} \geq 4$) must necessarily be associated to stringy states, since in supergravity all the states have $J \leq 2$.

$$B)\ E_0 = r + \frac{1}{2}(p+2k+3q) \geq 2+2J-r+\frac{1}{2}(3p+2k+q) \qquad (3.5)$$
$$(\text{for } J_2=0,\ J_1=J \text{ or } J_1 \to J_2,\ r \to -r,\ (p,k,q) \to (q,k,p));$$

which implies

$$E_0 \geq 1+J+p+k+q, \qquad r \geq 1+J+\frac{1}{2}(p-q). \qquad (3.6)$$

The maximal shortening occurs when

$$E_0 = 1+J+p+k+q, \qquad (3.7)$$

and the chiral supersingletons occur for $p=k=q=0$. They correspond to Table 6 of [16]. When a strict inequality occurs in (3.6) and $p=k=q=0$, we have UIR's described by $N=4$ chiral (left-handed) superfields with highest spin $(J+2,0)$ in $SU(4)$ singlets with $\Delta = r-4$.

$$C)\ E_0 = p+k+q, \quad r = \frac{1}{2}(p-q), \quad J_1 = J_2 = 0; \qquad (3.8)$$

which gives supersingleton representations for p, (or q)$=1$, $k=0$ or $p=q=0$, $k=1$. The self-conjugate (Yang-Mills) supersingleton with spin 1 corresponds to the $D(1,0,0;0,0,1,0)$ highest weight and the spin 3/2 supersingleton to $D(1,0,0;1/2,1,0,0)$. They correspond to Tables 1 and 2 of [17].

The UIR's of the $PSU(2,2|4)$ superalgebra are obtained by setting $r=0$ in the previous shortening conditions. Correspondingly, we also abolish the entry corresponding to r in the symbol D denoting highest weight states. We get the three classes of UIR's:

$$A')\ E_0 \geq 2+J_1+J_2+p+k+q, \qquad J_2-J_1 \geq \frac{1}{2}(p-q). \qquad (3.9)$$

Maximal shortening occurs when

$$E_0 = 2+J_1+J_2+p+k+q, \qquad J_2-J_1 = \frac{1}{2}(p-q). \qquad (3.10)$$

Massless bulk multiplets correspond to $p=k=q=0$ and $J_1=J_2$.

$$B')\ E_0 = \frac{1}{2}(p+2k+3q) \geq 2+2J+\frac{1}{2}(3p+2k+q) \qquad (3.11)$$
$$(J_2=0,\ J_1=J \text{ or } J_1 \to J_2,\ (p,k,q) \to (q,k,p));$$

with

$$E_0 \geq 1+J+p+k+q, \qquad 1+J \leq \frac{1}{2}(q-p). \qquad (3.12)$$

Maximal shortening occurs when $1+J=\frac{1}{2}(q-p)$, with highest weight $D(3+3J+2p+k,J,0;p,k,p+2+2J)$. No supersingletons appear in this series.

$$C') \; E_0 = 2p+k, \qquad p=q, \qquad J_1=J_2=0. \tag{3.13}$$

The highest weight states are $D(2p+k,0,0;p,k,p)$. The $p=0,k\geq 2$ UIR's correspond to the KK states of type IIB on $AdS_5 \times S_5$, the $k=2$ case being associated with the bulk graviton multiplet. The $p=0,k=1$ UIR corresponds to the only supersingleton of the $PSU(2,2|4)$ algebra [19, 16]. The infinite sequence of UIR's with $p=0$, multiplets with $J_{MAX}=2$, have been obtained in [26] with the oscillator construction. They correspond to the harmonic [44] holomorphic superfields of [45]. The case $p\neq 0$ may be relevant for multiparticles supergravity states, as will be discussed in the next Section.

To summarize the structure of UIR's of $SU(2,2|N)$ algebras: Generic massive supermultiplets, occurring when a strict inequality in (2.4), (2.14), (3.1) holds, have multiplicity 2^{4N}, generic multiplets when the inequality in (2.5), (2.16), (3.5) occurs have multiplicity 2^{2N} (the same is true for the series c), γ), C)). At the threshold of the unitarity bounds, these multiplicities shrink. For example, massless bulk multiplets have multiplicity 2^{2N}, while massless boundary multiplets (supersingletons) 2^N.

4. Applications to the AdS/CFT correspondence

The previous analysis of shortening conditions finds applications in the AdS/CFT correspondence, where many shortened multiplets are realized.

For $N=1$ theories, the prototype is type IIB on $AdS_5 \times T^{1,1}$, where all type of shortening a), b) and c) have been shown to occur for $J_1,J_2 \leq 1/2$ [31, 22, 32].

For $N=2$ theories also all types of shortening occur. Shortening β) includes, for instance, $N=2$ tensor multiplet KK recurrences, while shortening γ) corresponds to $N=2$ vector multiplet KK recurrences [46, 30].

The $N=4$ case includes, for the $PSU(2,2|4)$ algebra, all the KK states of type IIB on $AdS_5 \times S^5$ [11, 26, 8, 25, 24]. It also contain other states corresponding to multiparticle supergravity states, which can in principle be analyzed [27]; they correspond to multitrace conformal operators in the Yang-Mills theory [47, 48, 28, 29].

Shortening in A$'$) typically occurs in free-field theory but not in the interacting $N=4$ theory. For instance, the $D(E_0,0,0;0,0,0)$ highest weight corresponds to the Konishi multiplet, which undergoes a shortening in a free theory ($E_0=2$) while is long in the interacting Yang-Mills theory since $E_0>2$ [49, 41, 50, 9]. In the AdS/CFT correspondence, it is indeed the simplest example of operator corresponding to a stringy state. Another example of operator which is short only in free theory is the highest weight $D(3,0,0;0,0,2)$ in B$'$), which contains the superpotential of the $N=4$ theory when it is written in $N=1$ notations. It is not a primary conformal superfield in the interacting Yang-Mills theory since it can be obtained by total antisymmetrization of the product of three self-conjugate supersingletons [9, 51].

The sequence C′) has an interesting application to multitrace operators, since it predicts that primary operators with highest weight $D(2p+k,0,0;p,k,p)$ are not renormalized. In a single trace operator, representations with $p \neq 0$ are not primary since they involve at least a partial antisymmetrization of the Yang-Mills supersingletons, which makes it a descendent via equations of motion [8, 25, 24]. The same argument does not apply to multitrace operators. Consider the simplest case. Out of two single-trace primary operators in the 20_R $(p=0, k=2)$ we can construct by multiplication several multitrace operators. They are in the symmetric product $(20_R \times 20_R)_S = 105 + 84 + 20_R + 1$. The previous discussion implies that the 105 and 84 $((0,4,0)$ and $(2,0,2))$ can fit into a shortening condition and are not renormalized, so that anomalous dimensions can only show up for the 20_R and singlet pieces. This result has been recently confirmed in [27, 28, 29] for the $(0,4,0)$ piece and in [28] for the 84 piece, by an explicit perturbative computation.

It is interesting to speculate at this point whether UIR's of the $PU(2,2|4)$ algebra with $r \neq 0$ can occur. It is obvious that, since the Yang-Mills theory is built in terms of the self-conjugate multiplet with $r = 0$, any local operator constructed out of such multiplet will have $r = 0$ and will be in some representation of $PSU(2,2|4)$. However, as speculated in [17], if some AdS_5 states are dual to dyonic non-perturbative states of the Yang-Mills theory, then necessarily a new sector with $r \neq 0$ will appear at the non-perturbative level. Natural candidates are the 1/4 BPS dyonic states of the Yang-Mills theory [52], which would correspond to the second exceptional supersingleton representation in C) (with $J_{MAX} = 3/2$). To strengthen this interpretation, it would be interesting to understand how the central $U(1)$ acts on space-time type IIB fields. We could conjecture that non-perturbative states with $r \neq 0$ are obtained by considering type IIB (p,q) five-branes and strings. The mentioned 1/4 BPS states of the Yang-Mills theory certainly fall into this category, since they are obtained as junctions of (p,q) strings [52]. The central charge of the $U(2,2|4)$ algebra is an $SU(4)$ singlet and may temptatively be interpreted as coming from a wrapped five-brane, in some generalization of the construction in [43]. The question whether such states are actually BPS saturated in AdS_5 is somewhat unclear.

Acknowledgements. S. F. is supported in part by the DOE under grant DE-FG03-91ER40662, Task C, the NSF grant PHY94-07194, and by the ECC Science Program SCI*-CI92-0789 (INFN-Frascati). We would like to thank M. Bianchi and D.Z. Freedman for helpful discussions.

References

1. Haag, R., Łopuszański, J. T., and Sohnius, M.: *Nuclear Phys. B* **88** (1975), 257–274.
2. Nahm, W.: *Nuclear Phys. B* **135** (1978), 149.
3. Townsend, P. K.: *p*-Brane Democracy, preprint hep-th/9507048.
4. Salam, A. and Sezgin, E. (eds.): *Supergravities in Diverse Dimensions*, North-Holland, 1989.

5. Maldacena, J. M.: *Adv. Theor. Math. Phys.* **2** (1998), 231.
6. Gibbons, C. W. and Townsend, P. K.: *Phys. Rev. Lett.* **71** (1993), 3754.
7. Gubser, S. S., Klebanov, I. R., and Polyakov, A. M.: *Phys. Lett. B* **428** (1998), 105.
8. Witten, E.: *Adv. Theor. Math. Phys.* **2** (1998), 253.
9. Ferrara, S. and Zaffaroni, A.: *Phys. Lett. B* **431** (1998), 49; Bulk Gauge Fields in AdS Supergravity and Supersingletons, preprint hep-th/9807090.
10. Gunaydin, M., Romans L. J., and Warner, N. P.: *Phys. Lett. B* **154** (1985), 268.
11. Kim, H. J., Romans, L. J., and van Nieuwenhuizen, P.: *Phys. Rev. D* **32** (1985), 389.
12. Romans, L. J.: *Phys. Lett. B* **153** (1985), 392.
13. Gunaydin, M., Romans L. J., and Warner, N. P.: *Nuclear Phys. B* **272** (1986), 598.
14. Flato, M. and Frønsdal, C.: *Lett. Math. Phys.* **8** (1984), 159.
15. Bars, I. and Gunaydin, M.: *Comm. Math. Phys.* **91** (1983), 31; Gunaydin, M.: *J. Math. Phys.* **29** (1988), 1275.
16. Gunaydin, M., Minic, D., and Zagermann, M.: *Nuclear Phys. B* **534** (1998), 96.
17. Gunaydin, M., Minic, D., and Zagermann, M.: *Nuclear Phys. B* **544** (1999), 737.
18. Dobrev, V. K. and Petkova, V. B.: *Phys. Lett. B* **162** (1985), 127; *Lett. Math. Phys.* **9** (1985), 287; *Fortschr. Phys.* **35** (1987) 7, 537.
19. Binegar, B.: *Phys. Rev. D* **34** (1986), 525.
20. Intriligator, K.: *Nuclear Phys. B* **551** (1999), 575; Intriligator, K. and W. Skiba, W.: Bonus Symmetry and the Operator Product Expansion of N=4 Super-Yang-Mills, preprint hep-th/9905020.
21. Freedman, D. Z., Gubser, S. S., Pilch K., and Warner, N. P.: Renormalization Group Flows from Holography–Supersymmetry and a c-Theorem, preprint hep-th/9904017.
22. Ceresole, A., Dall'Agata, G., D'Auria, R., and Ferrara, S.: *Spectrum of Type IIB Supergravity on* $AdS_5 \times T^{11}$*: Predictions on* $N = 1$ *SCFT's*, preprint hep-th/9905226.
23. Flato, M. and Frønsdal, C.: *Lett. Math. Phys.* **2** (1978), 421; *Phys. Lett. B* **97** (1980), 236; *J. Math. Phys.* **22** (1981), 1100; *Phys. Lett. B* **172** (1986), 412.
24. Andrianopoli, L. and S. Ferrara, S.: *Phys. Lett. B* **430** (1998), 248; *Lett. Math. Phys.* **46** (1998), 26.
25. Ferrara, S., Frønsdal C., and Zaffaroni, A.: *Nuclear Phys. B* **532** (1998), 153.
26. Gunaydin, M. and Marcus, N.: *Class. Quant. Grav.* **2** (1985), L11.
27. Andrianopoli, L. and Ferrara, S.: On short and long $SU(2,2/4)$ multiplets in the AdS/CFT correspondence, preprint hep-th/9812067.
28. Bianchi, M., Kovacs, S., Rossi, G., and Stanev, Y. S.: On the logarithmic behaviour in $N = 4$ SYM theory, preprint hep-th/9906188.
29. Skiba, W.: Correlators of Short Multi-Trace Operators in $N = 4$ Supersymmetric Yang-Mills, preprint hep-th/9907088.
30. Klebanov I. R. and Witten, E.: *Nuclear Phys. B* **536** (1998), 199; AdS/CFT Correspondence and Symmetry Breaking, preprint hep-th/9905104.
31. Gubser, S. S.: *Phys. Rev. D* **59** (1999), 5006; Gubser, S. S. and Klebanov, I. R.: *Phys. Rev. D* **58** (1998), 5025.
32. Jatkar, D. P. and Randjbar-Daemi, S.: Type IIB string theory on $AdS_5 \times T^{nn'}$, preprint hep-th/9904187.
33. Ceresole, A., Dall'Agata G., and D'Auria, R.: KK Spectroscopy of Type IIB Supergravity on $AdS_5 \times T^{11}$, preprint hep-th/9907216.
34. Flato, M. and Frønsdal, C.: *Essays in Supersymmetries*, Math. Phys. Stud. **8**, D. Reidel, Dordecht, 1986.
35. Ferrara, S., Grillo A. F., and Gatto, R.: *Phys. Rev. D* **9** (1974), 3564.
36. Binegar, B., Frønsdal, C., and Heidenreich, W.: *J. Math. Phys.* **24** (1983), 2828.
37. Ferrara, S., Porrati, M., and Zaffaroni, A.: *Lett. Math. Phys.* **47** (1999), 255.
38. Park, J.: *Int. J. Mod. Phys. A* **13** (1998), 1743; Superconformal Symmetry and Correlation Functions, preprint hep-th/9903230.

39. Osborn, H.: *Ann. of Phys.* **272** (1999), 243.
40. Howe, P., Stelle K. S., and Townsend, P. K.: *Nuclear Phys. B* **192** (1981), 332.
41. Anselmi, D., Freedman, D. Z., Grisaru, M.T., and Johansen, A. A.: *Phys. Lett. B* **394** (1997), 329; *Nuclear Phys. B* **256** (1998), 543.
42. Kuzenko, S. M. and Theisen, S.: Correlation Functions of Conserved Currents in $N = 2$ Superconformal Theory, preprint hep-th/9907107.
43. Witten, E.: *J. High Energy Phys.* 9807 (1998) 006.
44. Galperin, A., Ivanov, E. A., Kalitsyn, S., and Ogievetsky, V.: *Class. Quant. Grav.* **1** (19984), 469; *Class. Quant. Grav.* **2** (1985), 155.
45. Howe P. S. and West, P. C.: *Phys. Lett. B* **389** (1996), 273; *Phys. Lett. B* **400** (1997), 307.
46. Gukov, S.: *Phys. Lett. B* **439** (1998), 2.
47. Freedman, D. Z., Mathur, S. D., Matusis, A., and Rastelli, L.: *Nuclear Phys. B* **546** (1999), 96.
48. Chalmers, G. and Schalm, K.: Holographic Normal Ordering and Multi-particle States in the AdS/CFT Correspondence, preprint hep-th/9901144.
49. Konishi, K.: *Phys. Lett. B* **135** (1984), 439.
50. Anselmi, D., Erlich, J., Freedman D. Z., and Johansen, A.: *Phys. Rev. D* **57** (1998), 7570.
51. Girardello, L., Petrini, M., Porrati, M., and Zaffaroni, A.: *J. High Energy Phys.* 9812 (1998) 022.
52. Bergman, O.: *Nuclear Phys. B* **525** (1998), 104; Bergman, O. and Kol, B.: *Nuclear Phys. B* **536** (1998), 149.

On a classical limit of quantum theory and the non-linear Hartree equation ‡

Jürg Fröhlich
Theoretical Physics, ETH-Hönggerberg, CH–8093 Zürich, Switzerland

Tai-Peng Tsai and Horng-Tzer Yau
Courant Institute of Mathematical Sciences, New York University
251 Mercer Street, New York, NY 10471, USA

Keywords: quantum theory, classical limit, Hartree equation, solitons, point-particle limit

Mathematics Subject Classifications (2000): 81T10, 81Q20, 35Q40

Dedicated to the memory of Moshé Flato, whose contributions to deformation quantization are fundamental

1. General perspective, description of the problem

Historically, the description of Nature in theoretical physics began with an encoding of rather special, simple natural phenomena exhibiting high approximate symmetry in special, unstable mathematical structures. For example, the shape of the orbits of planets around the sun was conceived to be circular. As the ambition to describe more general classes of natural phenomena grew those special mathematical structures had to be deformed and enlarged into more general, more stable mathematical structures. Kepler discovered that the idea of elliptic orbits, with the sun in one of the foci, provides a better model for the motion of planets.

With the advent of the idea of *dynamical* Laws of Nature, one discovered that it is these laws that may exhibit high symmetry, rather than special consequences of these laws. Newtonian mechanics led to a first model of space and time. The orbits of planets and moons could be understood as trajectories in Newtonian space-time which are solutions of the Newtonian equations of motion. These equations of motion exhibit high symmetry: In all inertial frames of reference they take the same form, the transition from one to another inertial frame being given by a Galilei transformation. These symmetries are accompanied by conservation laws.

During the twentieth century, our understanding of Nature underwent revolutionary developments and changes which led to at least three fundamental *deformations* of the mathematical structure underlying physical theory. The first one

‡ Notes accompanying a lecture presented by J. F. This paper has first appeared in the proceedings of the conference "Visions in Mathematics", Tel Aviv, August 1999, published in the special volume of GAFA.

G. Dito and D. Sternheimer (eds.), Conférence Moshé Flato 1999, Vol. 1, 189 – 207.

originated from the study of the laws of black-body radiation and resulted in the discovery of *quantum theory*. As early as 1900, it uncovered new fundamental constants of Nature, the elementary electric charge e, the Boltzmann constant k_B, Planck's constant $\hbar$ — all related to the atomic constitution of matter — and the Planck length ℓ_P which is connected with gravity. Planck's law of black-body radiation, combined with previously discovered laws of thermodynamics and statistical physics, enabled him to determine the approximate values of e, k_B and $\hbar$, and from the speed of light c, and Newton's law of gravitational attraction he could then infer the approximate value of ℓ_P.

The second revolution arose from attempts to find a unified description of mechanical and electromagnetic phenomena and resulted in Einstein's special theory of relativity and in the deformation of Newtonian space-time to Minkowski space-time and of the Galilei group to the Poincaré group. The underlying deformation parameter is the speed of light c. The third one arose from trying to find a relativistic theory of gravitation and an invariant (coordinate-independent) formulation of fundamental dynamical laws and resulted in Einstein's general theory of relativity. In this theory, Minkowski space-time is deformed to a dynamical space-time, described as a Lorentzian manifold, whose topology and metric shape are determined, in part, by the distribution of matter. *Global* space-time symmetries are promoted to *local* symmetries (gauge symmetries of the second kind). The underlying deformation parameter is ℓ_P.

The deformation of the mathematical structure underlying Hamiltonian mechanics to the one underlying non-relativistic quantum mechanics was accomplished last and has remained, perhaps, the most mysterious one. The underlying deformation parameter is Planck's constant $\hbar$. It represents the first example of a deformation of classical geometry to noncommutative geometry, namely the deformation of symplectic geometry to the noncommutative geometry of quantized phase spaces.

The three deformations mentioned above and their diverse fundamental physical consequences have set much of the agenda for the theoretical physics of the twentieth century and are likely to occupy an important place in our agenda for many years to come.

Moshé Flato was very actively involved in research on deformation theory and made or stimulated fundamental contributions in this area. His insights in this and many other developments will be missed!

From the experience gained from the deformations of Newtonian to relativistic mechanics, of special to general relativity and of Hamiltonian to quantum mechanics, one may infer that *dimensionful, fundamental constants of Nature*, c, $\hbar$, ℓ_P, play the role of *deformation parameters*, (to be distinguished from scale hierarchy parameters, such as the Rydberg, the QCD scale or the mass of the Z boson, and from dimensionless coupling constants, such as the fine structure constant $\alpha \simeq 1/137$, or Yukawa couplings). To "turn on" $1/c$ or $\hbar$ or ℓ_P necessitates a fundamental change in the mathematical structure of physical theory, the passage

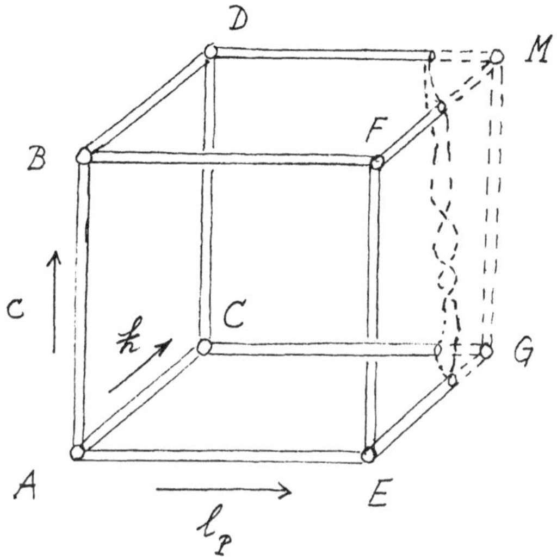

Figure 1.

from one family to a new family of theories, (while ratios of energy scales and dimensionless coupling constants can be viewed as *moduli* parametrizing a family of theories). Dimensionful, fundamental constants of nature permit us to introduce a natural system of units in which they take the value 1, i.e., $c = 1$, $\hbar = 1$ and $\ell_P = 1$, in natural units, (while ratios of energy scales and dimensionless coupling constants are pure numbers independent of our choice of units).

Deforming one family of theories to a new family of theories is the subject of *deformation theory*, while the passage from an effective description of natural phenomena at a high energy scale to an effective description of phenomena at a lower energy scale (within one family of effective theories) is the subject of *renormalization group theory*. Deformation theory is algebra, renormalization group theory is analysis. Progress in theoretical physics appears to rely on progress in deformation — and in renormalization group theory!

The sketch in Figure 1 may convey a very rough impression of where we stand, at present:

A. Newtonian mechanics, Euclidian geometry, Hamiltonian mechanics, symplectic geometry

B. Special relativistic classical field theory, Minkowski geometry, infinite-dimensional symplectic geometry

C. Non-relativistic quantum mechanics, noncommutative geometry of quantized phase spaces

D. Special relativistic quantum field theory, noncommutative geometry of infinite-dimensional quantized phase spaces, functional integration, BRST- and BV cohomology

E. Newtonian gravity and its reformulation as a geometrical theory, making use of Riemannian geometry

F. General Relativity, general Lorentzian geometry, ...

G. "Non-relativistic quantum gravity" (?), noncommutative Riemannian geometry

M. "Quantum gravity" or "M-theory" (?), infinite-dimensional noncommutative differential geometry (?), ...

The question marks indicate that the corresponding terms do not have a mathematically precise definition, yet. The "Escherian" aspect of the cube in Figure 1 is intended to indicate that we are still facing obstacles or, perhaps, obstructions in attempting to add the missing vertices G and M, and a link between them, to the cube.

Our discussion makes it clear that the following have been and will remain *important tasks* for theoretical physicists:

(1) Find suitable stable mathematical structures as foundations for physical theory by *deforming* more special, less stable mathematical structures known to underly physical theory in various asymptotic regimes.

For the future (ℓ_P): Find the deformations leading to vertex G and, most importantly, to vertex M, ("M-theory").

(2) Once a physical theory founded on a general, stable mathematical structure (e.g., non-relativistic quantum mechanics) has been found, one must understand in which asymptotic regime it reduces to one of its ancestor theories (e.g., Hamiltonian mechanics) founded on a less stable structure from which the former has arisen by deformation.

For the future (length scales $\gg \ell_P$):

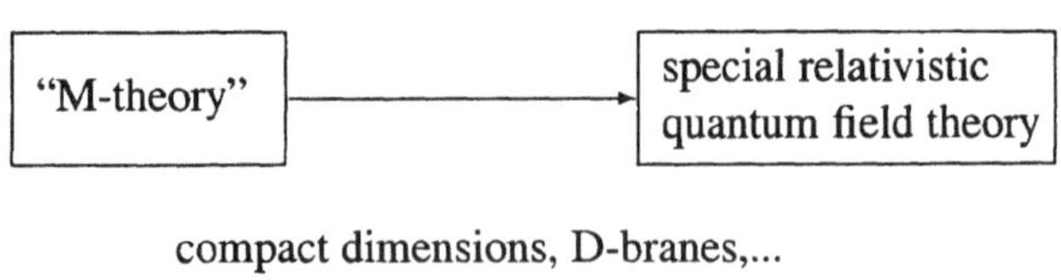

We do not really know precisely what "M-theory" is, and we do not understand how as yet unobserved symmetries (in particular, space-time supersymmetry) are broken and how a hierarchy of low-energy scales is generated.

For now ($c \gg$ typical velocities):

(i)

weak interaction strengths between compounds,
saturation of forces, $\alpha \simeq 1/137$ *small* $\Rightarrow v \ll c$.

For some results, see [1].

(ii)

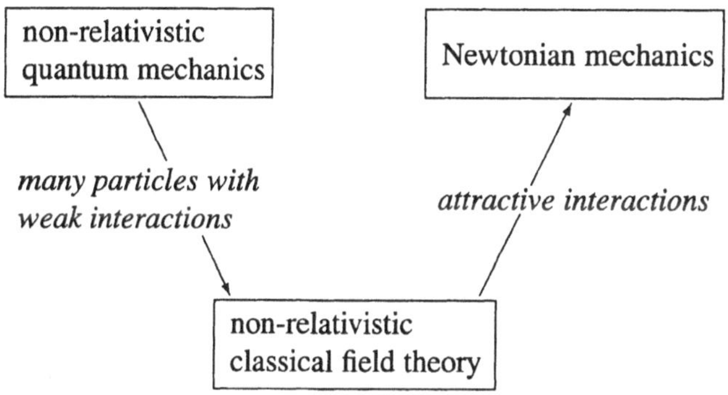

A simple example of a solution of problem (ii) is the *main subject* of these notes.
Warning: Since, in natural units, $c = 1$, $\hbar = 1$, $\ell_P = 1$, the problems formulated above *cannot* be solved by letting c^{-1} or $\hbar$ or ℓ_P tend to 0 !

(3) The passage from microscopic theories at short distance — and time scales to *effective theories* at large distance — and time scales (which, hopefully, are easier to analyze) necessitates a deeper and mathematically more precise understanding of *renormalization group methods*.

In these notes, we describe some recent results which provide an example of a solution of problem (2), (ii). We consider a system of very many non-relativistic bosons with very weak two-body interactions and exhibit a class of states with the property that the time evolution of expectation values of certain operators in these states is approximately described by classical Hamiltonian equations of motion, more specifically by the *non-linear Hartree equations*. We then sketch how certain solutions of the non-linear Hartree equations approach solutions of *Newtonian equations of motion* for point particles when the two-body forces are attractive at short distances. Details may be found in [2].

2. Large systems of non-relativistic bosons

In non-relativistic quantum mechanics, pure states are unit rays in a separable Hilbert space, $\mathscr{H}$, mixed states are density matrices on $\mathscr{H}$. A specific physical situation is encoded in the choice of some concrete *algebra $\mathscr{A}$ of bounded operators contained in (or equal to) the algebra $B(\mathscr{H})$ of all bounded operators on $\mathscr{H}$. The time-evolution of an operator $a \in \mathscr{A}$ in the Heisenberg picture is given by

$$\mathscr{A} \ni a \mapsto a(t) := e^{i\widehat{H}t/\hbar}\, a\, e^{-i\widehat{H}t/\hbar}\,, \tag{1}$$

where $\widehat{H}$ is a selfadjoint operator densely defined on $\mathscr{H}$ (whose spectrum is usually assumed to be bounded from below).

As an example, we consider a system of N spinless, non-relativistic, identical quantum mechanical (point) particles in physical space $\mathbb{E}^3$ moving under the influence of an external force with potential W and of two-body interactions with potential $-\Phi$. As an algebra $\mathscr{A}$ we may, e.g., choose the algebra of smooth, bounded functions $a(x_1,\ldots,x_N)$ on $\mathbb{R}^{3N}$ which are totally symmetric in their N arguments. Its spectrum corresponds to the space of configurations of N identical point particles.

The Hilbert space of state vectors $\mathscr{H} = \mathscr{H}_\sigma^{(N)}$ is chosen to be a subspace of the space $L^2(\mathbb{R}^{3N}, \Pi_{j=1}^N\, d^3x_j)$ of square-integrable functions on $\mathbb{R}^{3N}$ of a specific symmetry type σ under permutations of their N arguments ("quantum statistics"); i.e., a function $\psi \in \mathscr{H}_\sigma^{(N)}$ transforms under an irreducible representation σ of the group of permutations of $\{1,\ldots,N\}$ when its arguments are permuted. When σ is the trivial representation ψ is totally symmetric in its N arguments, and we speak of *Bose–Einstein statistics*; when σ is the alternating representation ψ is totally anti-symmetric in its N arguments, and we speak of *Fermi–Dirac statistics*. (Higher-dimensional representations σ are sometimes called parastatistics.)

The Hamiltonian of the system is given by

$$\widehat{H} \equiv H^{(N)} := \sum_{j=1}^{N}\Big(-\frac{\hbar^2}{2}\,\Delta_j + W(x_j)\Big) - \sum_{1\le i<j\le N} \Phi(x_i - x_j)\,, \tag{2}$$

and we have chosen units so that the mass of a particle is equal to 1.

In Equation (2), $\Delta_j = \sum_{a=1}^{3} \partial^2/(\partial x_j^a)^2$ and $x_j = (x_j^1, x_j^2, x_j^3)$ denotes the position of the jth particle in $\mathbb{E}^3$ (in a fixed system of Cartesian coordinates). If W is e.g. a smooth positive function on $\mathbb{E}^3$ and Φ is "small" relative to $-\Delta$, in the sense of Kato, then $H^{(N)}$ is essentially selfadjoint on a natural domain dense in $L^2(\mathbb{R}^{3N}, \Pi d^3x_j)$. Since $H^{(N)}$ is symmetric under permutations of the particle labels $1,\ldots,N$, the time evolution, $\exp(-iH^{(N)}t/\hbar)$, leaves the subspaces $\mathscr{H}_\sigma^{(N)}$ of $L^2(\mathbb{R}^{3N}, \Pi d^3x_j)$ invariant, for arbitrary σ.

There are straightforward generalizations of this class of systems to systems of several species of identical particles some of which may have non-zero quantum mechanical spin, e.g., to Coulomb system; see [3].

Two important problems in the analysis of such systems are (see [3]):

(A) *Stability* (i.e., the energy per particle is bounded from below by a constant $-\varepsilon$, uniformly in $N \Leftrightarrow H^{(N)} \ge -\varepsilon N, \forall N$); and

(B) *Existence of thermodynamic limit* (choose $W = W_\Lambda$ so as to confine the particles to a bounded region $\Lambda \subset \mathbb{E}^3$; keep the particle density $\rho := N/|\Lambda|$ fixed, where $|\Lambda|$

is the volume of Λ, and try to pass to the limit $\Lambda \nearrow \mathbb{E}^3$. — A necessary condition for the existence of a thermodynamic limit is stability.)

Much of *theoretical condensed matter physics* is the study of large systems of identical non-relativistic quantum mechanical point particles for which (A) and (B) hold. Such systems exhibit fascinating physical phenomena such as *Bose–Einstein condensation* (σ = trivial representation, two-body interactions repulsive); *superconductivity, Landau-* and *marginal Fermi liquids, Hall liquids, different forms of magnetism*, ... (σ = alternating representation). Precise mathematical understanding of some of these phenomena is only beginning to emerge; see e.g. [4] (and [5] for a review).

The topic of these notes is a study of dilute systems of a large number of neutral atoms with Bose–Einstein statistics (σ = trivial rep.) and with weak two-body interactions. If, in the initial state, the mean distance between atoms is sufficiently large and the average kinetic energy per atom is large then one may argue that, on sufficiently short time scales, short-range repulsion between the atoms can be neglected. Then the system can be described by a Hamiltonian $H^{(N)}$ of the form (2), where $-\Phi$ is e.g. the potential of a long-range attractive force. In this idealization, stability (A) is, however, usually violated.

We propose to identify a class of initial states for such systems with the property that the time evolution of the expectation value of certain operators in such states approaches one described by some classical, non-relativistic Hamiltonian PDE, in the limit where the strength κ of the two-body interaction tends to 0, with $N \propto \kappa^{-1}$

In order to analyze such systems, it is convenient (though dispensable) to make use of the formalism of "second quantization". We imagine that the number N of atoms may fluctuate (e.g., because the system is coupled to a reservoir of atoms not included in our description). It is then convenient to introduce the Hilbert space

$$\mathcal{F} = \bigoplus_{N=0}^{\infty} \mathcal{H}^{(N)} \quad \mathcal{H}^{(0)} = \mathbb{C}\,, \tag{3}$$

where $\mathcal{H}^{(N)} = \mathcal{H}^{(N)}_{\sigma=\text{trival rep.}}$, which is commonly called *Fock space*. A vector $\underline{\psi} \in \mathcal{F}$ is a sequence

$$\underline{\psi} = \left\{ \psi^{(N)} \mid \psi^{(N)} \in \mathcal{H}^{(N)} \right\}_{N=0}^{\infty} \tag{4}$$

with

$$\| \underline{\psi} \|^2 := \sum_{N=0}^{\infty} \| \psi^{(N)} \|_2^2 < \infty\,.$$

The scalar product of two vectors $\underline{\psi}$ and $\underline{\varphi}$ is given by

$$\langle \underline{\psi}, \underline{\varphi} \rangle := \sum_{N=0}^{\infty} \left(\psi^{(N)}, \varphi^{(N)} \right), \tag{5}$$

where $(\cdot,\cdot)$ is the usual L^2 scalar product. Let f be a test function on $\mathbb{E}^3$. We define an *"annihilation operator"* $a(f)$ on $\mathscr{F}$ by setting

$$a(f)\underline{\psi} = \underline{\psi}_f\,,$$

where

$$\psi_f^{(N)}(x_1,\ldots,x_N) := \sqrt{N+1}\int d^3x\,\overline{f(x)}\,\psi^{(N+1)}(x,x_1,\ldots,x_N)\,. \tag{6}$$

The operator $a(f)$ is unbounded, but closable. The *"creation operator"* $a^*(f)$ is defined to be the adjoint of $a(f)$ on $\mathscr{F}$. It is an unbounded closable operator. It is easy to verify that

$$\begin{aligned}[]
[a(f),a(g)] &= [a^*(f),a^*(g)] = 0\,,\\
[a(f),a^*(g)] &= (f,g)\,\mathbb{1}\,,
\end{aligned} \tag{7}$$

for arbitrary test functions f and g. Equations (7) are called *canonical commutation relations*. The Hamiltonian $\widehat{H}$ on Fock space is given by

$$\widehat{H} = \bigoplus_{N=0}^{\infty} H^{(N)}\,, \quad \text{with } H^{(0)} = 0\,.$$

In terms of annihilation- and creation operators, $\widehat{H}$ can be expressed as follows:

$$\begin{aligned}
\widehat{H} =& \frac{1}{2}\int d^3x\,[(\nabla a^*)(x)(\nabla a)(x) + W(x)\,a^*(x)\,a(x)]\\
&-\frac{1}{2}\int d^3x\int d^3y\,a^*(x)\,a^*(y)\,\Phi(x-y)\,a(y)\,a(x)\,,
\end{aligned} \tag{8}$$

and we now use units such that $\hbar = 1$. The two-body potential $-\Phi$ is chosen as follows:

$$\Phi = \kappa\Phi_c\,, \quad \kappa \ll 1\,, \tag{9}$$

where $\Phi_c \in L^{\frac{3}{2}} + L^\infty$ (with $L^p = L^p(\mathbb{R}^3, d^3x)$); (depending on the analytical methods employed, one may need to also impose some regularity properties on first and second derivatives of Φ; see e.g. [6]). An example for Φ_c is

$$\Phi_c(x) \simeq |x|^{-6} + g|x|^{-1}\,, \quad g \ll 1, \tag{10}$$

for $|x|$ large compared to the diameter of an atom, and $\Phi_c(x)$ is regular near $x = 0$. The first term in Φ_c describes van der Waals attraction, the second term gravitational attraction between atoms. The external potential W is assumed to be a smooth positive function.

It is easy to check that

$$H^{(N)} \geq -C\kappa N^2\,, \tag{11}$$

for some finite constant C. For $N \propto \kappa^{-1}$, kinetic and potential energy terms in $H^{(N)}$ both scale like κ^{-1}.

Next, we introduce rescaled annihilation- and creation operators:

$$A_\kappa(f,t) := \kappa^{1/2}\, e^{i\widehat{H}t}\, a(f)\, e^{-i\widehat{H}t} \tag{12}$$

($\hbar = 1$!), and

$$A^*_\kappa(f,t) := (A_\kappa(f,t))^* . \tag{13}$$

Let $\underline{\varphi}$ be a vector in Fock space $\mathscr{F}$ with a finite number of particles and smooth k-particle wave functions of compact support, for all k. We define a vector $\underline{\theta}_\kappa$ by setting

$$\underline{\theta}_\kappa := C_\kappa \left(a^*(\psi_0)\right)^{N_\kappa} \underline{\varphi} , \tag{14}$$

where ψ_0 is a smooth function of compact support which, for convenience, we choose to have $\| \psi_0 \|_2 = 1$,

$$N_\kappa = \left[\kappa^{-1}\right] , \tag{15}$$

where $[a]$ denotes the integer part of a positive number a, and the constant C_κ is chosen such that $\langle \underline{\theta}_\kappa, \underline{\theta}_\kappa \rangle = 1$.

Let $\psi(x,t)$ be a solution of the *non-linear Hartree equation*

$$i\partial_t \psi = -\frac{1}{2}\Delta\psi + W\psi - \left(\Phi_c * |\psi|^2\right)\psi \tag{16}$$

with initial condition

$$\psi(x,0) = \psi_0(x) . \tag{17}$$

We define

$$\psi(f,t) := \int d^3x\, \overline{f(x)}\, \psi(x,t) ,$$

and

$$\psi^*(f,t) := \overline{\psi(f,t)} ,$$

where f is a test function.

The following theorem is proven in [6].

Theorem 2.1 *Under the hypotheses described above,*

$$\lim_{\kappa\to 0} \langle \underline{\theta}_\kappa, \prod_{j=1}^{m} A^*_\kappa(f_j,t_j)\, A_\kappa(g_j,s_j)\, \underline{\theta}_\kappa \rangle = \prod_{j=1}^{m} \psi^*(f_j,t_j)\, \psi(g_j,s_j) , \tag{18}$$

for arbitrary $m = 1,2,\dots$, arbitrary test functions $f_1,\dots,f_m$, $g_1,\dots,g_m$ and finite times $t_1,\dots,t_m$, $s_1,\dots,s_m$.

Thus, a classical limit described by the non-linear Hartree equation (16) is approached in systems of $N \sim \kappa^{-1}$ bosons interacting through two-body forces of strength κ, as κ tends to 0, for an appropriate choice of the initial state of the system.

In the following, we show that the non-linear Hartree equation is an example of a Hamiltonian PDE and then explore a Newtonian point-particle limit of the non-linear Hartree equation, for two-body potentials which are *attractive* at short distances.

The theorem quoted above is but one example of a result on the approach of quantum dynamics towards some classical Hamiltonian dynamics in a suitable limit. Another example would be a quantum spin system whose dynamics is expected to approach the one described by the Landau–Lifshitz equation in the limit where the local magnetization becomes large.

An unsatisfactory aspect of results of this type is that one must choose rather special sequences of states, $\{\underline{\theta}_\kappa\}$, in order to approach a classical limit. States leading to a classical limit do *not* form a linear space (which may explain why there are no puzzles related to the superposition principle). We do not understand much, yet, about a general class of states leading to classical limits, and in which physical situations such states appear naturally.

The material in the following sections is motivated by the above theorem, Equation (18): We study the NL Hartree equation (16) and the question under what circumstances it reduces to the Newtonian mechanics of point particles. We begin with a brief review of the *Hamiltonian nature* of the *non-linear Hartree equations*.

3. NL Hartree equations as a Hamiltonian system

The phase space for the non-linear Hartree equation (16) is the energy (Sobolev) space

$$\Gamma = H^1\left(\mathbb{R}^3\right) \tag{19}$$

with complex coordinates $\psi(x)$, $\bar{\psi}(x)$. The symplectic two-form is given by $\frac{i}{2}\, d\psi \wedge d\bar{\psi}$. It leads to the following Poisson brackets

$$\begin{aligned} \{\psi(x),\, \psi(y)\} &= \{\bar{\psi}(x),\, \bar{\psi}(y)\} = 0 \\ \{\psi(x),\, \bar{\psi}(y)\} &= 2i\,\delta(x-y)\,. \end{aligned} \tag{20}$$

The Hamilton functional $\mathscr{H}$ leading to the NL Hartree equations (16) is given by

$$\mathscr{H}(\bar{\psi},\psi) = \frac{1}{4}\int\left[|\nabla\psi|^2 + 2W|\psi|^2 - \left(\Phi_c * |\psi|^2\right)|\psi|^2\right], \tag{21}$$

with $\psi = \psi(x)$, $\int(\cdot) = \int_{\mathbb{R}^3} d^3x(\cdot)$, and $*$ denotes convolution. For $\Phi_c \in L^{\frac{3}{2}} + L^\infty$, $\mathscr{H}$ is well-defined on Γ (Sobolev- and Hölder inequalities), and it is bounded below on the spheres

$$\mathscr{S}_N := \{\psi \mid \psi \in \Gamma,\, \mathscr{N}(\bar{\psi},\psi) = N < \infty\} \tag{22}$$

in phase space Γ, where

$$\mathscr{N}(\bar{\psi},\psi) := \int |\psi|^2\,; \tag{23}$$

see e.g. [7]. Hamilton's equations of motion for ψ are

$$\dot{\psi}(x,t) = \{\mathscr{H}(\bar{\psi},\psi),\, \psi\}\,(x,t) \tag{24}$$

and are easily seen to be equivalent to the NL Hartree equation (16). From the form of $\mathscr{H}$ we infer the following *symmetries* and corresponding *conservation laws*.

(1) Gauge invariance of the first kind:

$$\psi(x) \mapsto e^{i\theta}\,\psi(x), \quad \bar{\psi}(x) \mapsto e^{-i\theta}\,\bar{\psi}(x) \tag{25}$$

leaves $\mathscr{H}$ invariant; (25) is a symplectic flow generated by the Hamiltonian vector field associated with the functional $\frac{1}{2}\mathscr{N}(\bar{\psi},\psi)$. Thus $\mathscr{N}(\bar{\psi},\psi)$ must be conserved by the time evolution generated by (the Hamiltonian vector field determined by) $\mathscr{H}$, which means that

$$\{\mathscr{H},\mathscr{N}\} = 0\,. \tag{26}$$

It follows that the spheres $\mathscr{S}_N$ defined in (22) are *invariant* under the time evolution.

(2) If $W = 0$ and if Φ_c is rotation-invariant then arbitrary Galilei transformations are symmetries of $\mathscr{H}$. Space translations are generated by the momentum functional

$$\mathscr{P}(\bar{\psi},\psi) := \frac{i}{2}\int \bar{\psi}\nabla\psi\,, \tag{27}$$

and rotations by the angular momentum functional

$$\mathscr{L}(\bar{\psi},\psi) := \frac{i}{2}\int \bar{\psi}(x\wedge\nabla)\,\psi\,. \tag{28}$$

All components of $\mathscr{P}$ and $\mathscr{L}$ Poisson commute with $\mathscr{H}$. Velocity transformations (boosts), i.e. $x \mapsto x - vt$, are represented on time-dependent trajectories, $\psi(x,t)$, in Γ by

$$\psi(x,t) \mapsto \psi(v;x,t) := \psi(x-vt,t)\,e^{i(v\cdot x - \frac{v^2}{2}t)}\,. \tag{29}$$

These transformations map solutions of the equations of motion (24) to other solutions of (24). It follows that the "center of mass motion" of a solution $\psi(x,t)$ of (24) is inertial.

Instead of the Hamiltonian formalism, we may make use of the *Lagrangian formalism* to study the Hartree equations (16). These equations are then obtained as the Euler–Lagrange equations corresponding to the *action functional*

$$S(\bar{\psi},\psi) = \int_{t_1}^{t_2} dt\,\Big[\frac{i}{2}\int \bar{\psi}(\cdot,t)\,\frac{\partial}{\partial t}\,\psi(\cdot,t) - \mathscr{H}\,(\bar{\psi}(\cdot,t),\,\psi(\cdot,t))\Big]\,, \tag{30}$$

by variation with respect to $\bar{\psi}$, (with $\delta\bar{\psi}(\cdot,t_i) = 0$, for $i = 1,2$).

Existence and regularity of global solutions of the Cauchy problem associated with the non-linear Hartree equation, under the conditions on Φ_c described above, is known. If $-\Phi_c$ is radially symmetric, decreasing (*repulsive* two-body forces) and of sufficiently short range then the scattering map associated with the non-linear Hartree equations exists and satisfies asymptotic completeness. These results are reviewed in [7]. In the situation of interest in these notes ($-\Phi_c$ radially symmetric, but *increasing*, i.e., for *attractive* two-body forces) existence and regularity are known from the results reviewed in [7], but the scattering theory is much more subtle, and only very preliminary results are available. Our main interest is in the Newtonian point-particle limit of the NL Hartree equations, for two-body forces which, at small distances, are *attractive*.

4. The point-particle limit of the NL Hartree equations with attractive two-body forces

For reasons that will become apparent shortly, we start by trying to find (local and absolute) minima of the Hamilton functional $\mathscr{H}$ restricted to a sphere $\mathscr{S}_N$, as defined in (22), for a vanishing external potential, $W = 0$. In this situation, the momentum- and angular momentum functionals, $\mathscr{P}$ and $\mathscr{L}$, are conserved. We should therefore study the (generalized) energy functional

$$\begin{aligned}\mathscr{E}(\bar{\psi},\psi;\nu,\pi,\lambda) &:= \mathscr{H}(\bar{\psi},\psi) + \frac{\nu}{2}\,(N-\mathscr{N}(\bar{\psi},\psi)) \\ &\quad + \pi\cdot(P-\mathscr{P}(\bar{\psi},\psi)) + \lambda\cdot(L-\mathscr{L}(\bar{\psi},\psi))\,,\end{aligned} \tag{31}$$

where ν, π and λ are Lagrange multipliers. Variation with respect to $\psi, \bar{\psi}, \nu, \pi$ and λ yields the following variational equations:

$$-\frac{1}{2}\,\Delta\psi - (\Phi_c * |\psi|^2)\,\psi - \nu\psi - i\pi\cdot\nabla\psi - i\lambda\cdot(x\wedge\nabla)\,\psi = 0 \tag{32}$$

as well as the complex conjugate equation, and

$$\mathscr{N}(\bar{\psi},\psi) = N,\quad \mathscr{P}(\bar{\psi},\psi) = P,\quad \mathscr{L}(\bar{\psi},\psi) = L\,. \tag{33}$$

We do not know much about the general solutions to these equations. But, for the purposes of these notes, the following special case is particularly important: We search for an absolute minimum, Q_N, of $\mathscr{H}(\bar{\psi},\psi)\big|_{\mathscr{S}_N}$ which is rotation-invariant and has zero momentum. The function Q_N is the solution of the equations

$$\begin{aligned}-\frac{1}{2}\,\Delta\psi - (\Phi_c * |\psi|^2)\,\psi &= \nu\,\psi\,, \\ \mathscr{N}(\bar{\psi},\psi) &= N\,,\end{aligned} \tag{34}$$

and we henceforth set $\nu =: E$. It follows that

$$E = \frac{1}{2N}\int\left[|\nabla Q_N|^2 - (\Phi_c * |Q_N|^2)\,|Q_N|^2\right]\,, \tag{35}$$

which is strictly *negative*, for a non-trivial minimum Q_N.

Fairly standard methods of variational calculus show that there exists some finite $N_* = N_*(\Phi_c)$ such that, for $N > N_*$, Equations (34) have a non-trivial, smooth, positive solution $Q_N(x)$ which decays exponentially in $|x|$, with decay rate $\sqrt{-E}$. From the theory of binding in Schrödinger quantum mechanics one infers that $N_*(\Phi_c) = 0$, if $\Phi_c(x)$ is positive and decreases in $|x|$ more slowly than $|x|^{-2}$; but $N_*(\Phi_c) > 0$ if Φ_c is of short range. Covariance under translations and velocity transformations then implies that

$$\psi_N(v,a;x,t) \;:=\; Q_N(x-a-vt)\, e^{i\left(v\cdot x-\left[\frac{v^2}{2}+E\right]t\right)} \tag{36}$$

solves the non-linear Hartree equations (16), for an arbitrary velocity $v \in \mathbb{R}^3$. It describes a *"soliton"* traveling with velocity v.

Let $\mathscr{H}''$ denote the Hessian of $\mathscr{H}\big|_{\mathscr{S}_N}$ at Q_N. Then one finds that

$$\left(Q_N, \mathscr{H}''\, Q_N\right) \;=\; \varepsilon_0\, \left(Q_N, Q_N\right), \quad \varepsilon_0 < 0\,. \tag{37}$$

In fact, $\mathscr{H}''$ has precisely one eigenvector ϕ_0 corresponding to a negative eigenvalue. Furthermore,

$$\mathscr{H}''\, \frac{\partial}{\partial x^a}\, Q_N \;=\; 0\,, \;\text{ for } a = 1,2,3, \tag{38}$$

which follows from the translation invariance of $\mathscr{H}$, and the continuous spectrum of $\mathscr{H}''$ covers the interval $[-E,\infty)$, with $-E > 0$.

Assumption $*$ It will be assumed henceforth that Φ_c can be chosen such that the multiplicity of the eigenvalue $\varepsilon_1 = 0$ of $\mathscr{H}''$ is equal to 3.

The eigenspace of $\mathscr{H}''$ corresponding to the eigenvalue 0 is then spanned by the vectors $\frac{\partial}{\partial x^1}\, Q_N$, $\frac{\partial}{\partial x^2}\, Q_N$, $\frac{\partial}{\partial x^3}\, Q_N$, and it follows that

$$\left(\phi, \mathscr{H}''\phi\right) \;\geq\; \varepsilon_2\,(\phi,\phi)\,, \quad 0 < \varepsilon_2 \leq -E\,, \tag{39}$$

for arbitrary $\phi \in L^2(\mathbb{R}^3)$, with $\phi \perp Q_N$, $\frac{\partial}{\partial x^a}\, Q_N$, $a = 1,2,3$.

We are now prepared to discuss the Newtonian point-particle limit of the non-linear Hartree equation. Our discussion is heuristic; for mathematical results see [2].

We choose an external potential W of the form

$$W(x) \;\equiv\; W^{(\varepsilon)}(x) \;:=\; V(\varepsilon x)\,,$$

where V is a positive, smooth function on $\mathbb{R}^3$, and $\varepsilon \ll 1$ is a parameter. Furthermore, the two-body potential $-\Phi_c$ is chosen to be

$$\Phi_c(x) \;=\; \Phi_s(x) + \Phi_\ell(\varepsilon x)\,, \tag{40}$$

where $\Phi_s(x)$ is a rotation-invariant, smooth function decaying rapidly in $r := |x|$, as $r \to \infty$, and with the properties that

$$\frac{\partial \Phi_s(r)}{\partial r} < 0 \,, \quad \text{for } r > 0 \,, \tag{41}$$

and that Assumption $*$ above holds for $\Phi_c = \Phi_s$. The potential Φ_ℓ is rotationally invariant and may be of long range, e.g.,

$$|\Phi_\ell(x)| \sim r^{-1} \,, \quad \text{as } r \to \infty \,.$$

For technical convenience, we also assume that Φ_ℓ is once continuously differentiable, with $|\partial\Phi_\ell/\partial r|$ uniformly bounded in r.

Next, we pick a positive integer k and k positive numbers $N_1, \ldots, N_k$, with $N_j > N_*(\Phi_s)$, for all j. For $\varepsilon = 0$ and $N > N_*(\Phi_s)$, we define

$$\delta_N := \sqrt{N^{-1} \int d^3x \, Q_N(x)^2 x^2} \,. \tag{42}$$

We consider a configuration of k solitons $Q_{N_j}(x - q_j)$, $q_j \in \mathbb{R}^3$, $j = 1, \ldots, k$, with the following properties: Each $Q_{N_j}(x)$ is a solution of equations (34), with $N = N_j$, for $\Phi_c = \Phi_s + \Phi_\ell(0)$, i.e., $\varepsilon = 0$; and

$$\big(\max_{j=1,\ldots,k} \delta_{N_j}\big) / \big(\min_{1 \le i < j \le k} |q_i - q_j|\big) \le \varepsilon \,. \tag{43}$$

Our goal is to construct a solution ψ_ε of the non-linear Hartree equation (16) of the form

$$\psi_\varepsilon(x,t) = \sum_{j=1}^{k} Q_{N_j}(x - q_j(t)) \, e^{i\theta_j(x,t)} + h_\varepsilon(x,t) \,, \tag{44}$$

where $q_j(0) = q_j$, $\dot{q}_j(0) = v_j$, $j = 1, \ldots, k$, with the following properties: There is a positive constant T such that, for all times t with $|t| < \frac{T}{\varepsilon}$,

$$\text{(i)} \quad \| h_\varepsilon(\cdot, t) \|_{H^1} = o(\varepsilon) \,,$$
$$\text{(ii)} \quad \theta_j(x,t) = \dot{q}_j(t) \cdot [x - q_j(t)] + \vartheta_j(t) \,,$$

where $\vartheta_j(t)$ is independent of x; the trajectories $q_1(t), \ldots, q_k(t)$ and the phases $\vartheta_1(t), \ldots, \vartheta_k(t)$ will turn out to be determined by the non-linear Hartree equations (16) and properties (i) and (ii).

In order to convince ourselves that the ansatz (44) with properties (i) and (ii) is plausible, we evaluate the action functional $S(\bar\psi, \psi)$ introduced in Equation (30) on the ansatz functions $(\bar\psi, \psi) = (\bar\psi_\varepsilon, \psi_\varepsilon)$, with ψ_ε as in (44). We shall regard the soliton trajectories $q_1(t), \ldots, q_k(t)$, the phases $\vartheta_1(t), \ldots, \vartheta_k(t)$, and the small amplitude wave h_ε as variational parameters of the ansatz $(\bar{\psi}_\varepsilon, \psi_\varepsilon)$. After a moderately lengthy calculation one finds that, for a certain choice of $\dot{\vartheta}_j(t)$,

$$S(\bar{\psi}_\varepsilon, \psi_\varepsilon) = \int_{-\tau}^{\tau} dt \sum_{j=1}^{k} \Big[\frac{N_j}{2}\, \dot{q}_j(t)^2 - N_j V(\varepsilon q_j(t)) + \sum_{i:i<j} N_i N_j \Phi_\ell(\varepsilon(q_i(t) - q_j(t))) + o(\varepsilon)\Big] \tag{45}$$

for $\tau < \frac{T}{\varepsilon}$. Up to error terms $\sim o(\varepsilon)$, which are the source of a variety of mathematical subtleties, this is the action for a system of k Newtonian point particles with masses $N_1, \ldots, N_k$ moving under the influence of an external acceleration field with potential $W^{(\varepsilon)}$ and of two-body forces with potential $-N_i N_j \Phi_\ell(\varepsilon(q_i - q_j))$, $1 \le i < j \le k$; $q_j(t)$ is the position of the jth particle at time t.

Expression (45) for $S(\bar{\psi}_\varepsilon, \psi_\varepsilon)$ suggests that, for $S(\bar{\psi}, \psi)$ to have a critical point at $(\bar{\psi}, \psi) = (\bar{\psi}_\varepsilon, \psi_\varepsilon)$, with ψ_ε given by (41) and satisfying properties (i) and (ii), the particle trajectories $\{q_j(t)\}_{j=1}^{k}$ must be solutions of Newton's equations of motion

$$\ddot{q}_j(t) = -\varepsilon\left[\nabla V(\varepsilon q_j(t)) - \sum_{i \neq j} N_i \nabla \Phi_\ell(\varepsilon(q_i(t) - q_j(t)))\right] + a_j(t)\,, \tag{46}$$

where $j = 1, \ldots, k$, and the *"friction force"* $a_j(t)$ satisfies a bound $|a_j(t)| = o(\varepsilon)$, for $|t| < \frac{T}{\varepsilon}$.

These ideas are made precise (in the form of mathematical theorems) in [2], using methods partly based on the work of Weinstein [8], for just one soliton in an external potential $W^{(\varepsilon)}$. But the methods in [2] can be extended to general k.

Apparently, we have gained insight into how Newtonian point-particle mechanics with small friction forces can emerge from the non-linear Hartree equation with *attractive* two-body forces of short range (Φ_s) and weak attractive *or* repulsive two-body forces of very long range $(\Phi_\ell(\varepsilon\,\cdot))$ by considering the evolution of configurations of far-separated solitons over large, finite intervals of time. The friction forces $(a_j(t))$ appearing in Newton's equations of motion originate in the emission of small-amplitude, dispersive radiation from the solitons. This is an example of a general and quite fundamental phenomenon: *friction* (or dissipation) through *emission of radiation* in *conservative* (classical and quantum mechanical) *systems* of infinitely many degrees of freedom.

For similar considerations in the context of vortex dynamics in a Ginzburg–Landau superfluid see [9].

The results discussed above appear to be a special case of more general results on the Newtonian limit of the non-linear Hartree equation. One might imagine that it is possible to construct *"rotating-body"* solutions of the non-linear Hartree equations with non-zero angular momentum relative to their center of mass. Such solutions are, in general, unstable. Nevertheless, one could attempt to construct solutions of the non-linear Hartree equation which, over a large finite interval of time, describe a configuration of finitely many rotating bodies, whose motion is described by a system of Euler-type equations with small friction terms, emitting small-amplitude dispersive radiation.

The results reviewed in Equations (39) through (46) and the speculations just described raise the issue of *asymptotic properties* of the dynamics determined by the non-linear Hartree equation in the limit, as time tends to $\pm\infty$. Such problems belong to the realm of non-linear *scattering theory*. In [2], a result on the scattering of small-amplitude waves of a single soliton has been established. It requires $W(x)$ and $\Phi_c(x)$ to be smooth and of short range. We consider an "asymptotic profile" given by

$$\psi_{as}(x,t) = Q_N(x-vt)\, e^{i(x\cdot v - [\frac{v^2}{2}+E]t)} + h(x,t)\,, \tag{47}$$

where h is a solution of the free-particle Schrödinger equation

$$i\partial_t h(x,t) = -\frac{1}{2}\Delta h(x,t)\,,$$

with $h(x,0) \in H^3(\mathbb{R}^3) \cap W^{3,1}\left(\mathbb{R}^3, (1+|x|^2)\, d^3x\right)$, and $\hat{h}(v,0) = 0$. The claim is that there are solutions, $\psi_\pm(x,t)$, of the non-linear Hartree equation (16) such that

$$\psi_\pm(x,t) \underset{t\to+\infty}{\longrightarrow} \psi_{as}(x,t) \tag{48}$$

in $H^1(\mathbb{R}^3)$.

Thus, (non-linear) *Møller wave maps* $\Omega_\pm$ exist (as symplectic maps) on asymptotic profiles of the form (47); (see also [7] for background material on non-linear scattering theory). Of course, the results described here do not bring us anywhere close to an understanding of asymptotic completeness of the scattering map.

5. More about open problems, conclusions

In this last section, we draw the reader's attention to some interesting open problems.

(1) Clearly, a deeper understanding of the non-linear scattering theory of solitons and waves for the non-linear Hartree equation would be highly desirable. In particular, it would be important to gain more detailed insight into the domain of definition and the *image* of the Møller wave maps $\Omega_\pm$.

A comparatively simple situation has been studied by Soffer and Weinstein [10]. They consider a smooth external potential W of short range with exactly one quantum mechanical bound state, i.e., the spectrum of the operator $-\frac{1}{2}\Delta + W$ consists of one simple eigenvalue $\varepsilon < 0$ and of purely absolutely continuous spectrum covering $[0,\infty)$. One must also assume that there is no zero-energy resonance. They consider a local non-linearity (NL Schrödinger equation); but their techniques can be extended to the NL Hartree equation with Φ_c of short range. They construct small-amplitude non-linear bound states (stationary solutions of (16)) by standard bifurcation theory techniques. They then exhibit a class of initial conditions with the property that the asymptotic behavior of solutions of (16) with these initial conditions is that of a non-linear bound state plus a purely dispersive wave which tends to zero (pointwise) at the free dispersion rate, as $t \to \pm\infty$. They also show that initial conditions close to a (small-amplitude) non-linear bound state have this property.

Obviously, one would like to extend these results to external potentials W with more than one bound state and understand the radiative decay of initial conditions close to an *excited* boundstate of $-\frac{1}{2}\,\Delta + W$ to a non-linear ground state plus dispersive radiation. One would also like to go *beyond* results for small-amplitude initial conditions. In this case, it can happen that the restriction of the Hamilton functional $\mathscr{H}(\bar{\psi},\psi)$ (see Equation (21)) of the non-linear Hartree equation (16) to a sphere $\mathscr{S}_N$ in phase space Γ (see (22)) has several distinct local quadratic minima, provided N is large enough; (choose W to have several minima separated by large barriers and $-\Phi_c$ attractive). One would then like to understand the shape of the *"basins of attraction"* in phase space of these local minima. [The forward (backward) basin of attraction of a one-parameter family of local minima of $\mathscr{H}|_{\mathscr{S}_N}$ parametrized by N consists of all initial conditions in phase space Γ which approach an element of this family plus dispersive radiation decaying to zero, as $t \to +\infty$ ($t \to -\infty$, respectively). More generally, one may introduce a notion of a *center manifold* of (asymptotically) attracting configurations of solitons (freely moving, for $W = 0$) to which solutions of the non-linear Hartree equation with initial conditions sufficiently close to the center manifold converge, as $t \to +\infty$, or $t \to -\infty$. In this connection, see e.g. [11].]

(2) Consider a configuration ψ_ε of, say, just two far-separated solitons with initial positions q_1, q_2 and initial velocities $v_1 = \dot{q}_1(0)$, $v_2 = \dot{q}_2(0)$, as described in Equation (44), with $h_\varepsilon(x,0) = 0$. Let us assume that the two-body force with potential $-\Phi_\ell$ is purely attractive and of short range. We choose the initial conditions, (N_1, q_1, v_1) and (N_2, q_2, v_2) in such a way that the two solitons form a bound system, in the sense that

$$\frac{N_1}{2}\,v_1^2 + \frac{N_2}{2}\,v_2^2 - N_1 N_2 \Phi_\ell\,(\varepsilon\,(q_1 - q_2)) < 0\,. \tag{49}$$

We would like to understand the fate of this configuration, as $t \to \infty$. To begin with, we would like to estimate the power $P_R(t)$ of dispersive radiation through a sphere of radius $R \gg \max\,(|q_1|, |q_2|)$; see [9] for some results on a related problem. Furthermore, we would like to show that a configuration of two bound solitons collapses to a single soliton moving through space with a constant velocity, for $W = 0$ (!), plus a dispersive outgoing wave tending to O (pointwise) at the free dispersion rate; ("radiative collapse of a binary system").

(3) It appears reasonable to imagine that if the potential Φ_ℓ is chosen to be given by

$$\Phi_\ell(x) \simeq |x|^{-1},$$

for $|x|$ large, (Φ_s as in Section 4 and $W = 0$), the non-linear Hartree equation (16) can be used to describe Newtonian gravity of a giant cloud of bosonic atoms, with built-in dispersive friction. It would be interesting to understand how, on intermediate time scales, inhomogeneities in the initial conditions grow to form a structure consisting of rotating bodies plus dispersive radiation ("structure formation"), before the system collapses into a number of far separated solitons escaping from each other. Perhaps, one may gain some insight into such problems with the help of numerical simulations; see e.g. [12].

(4) It is of interest to study transport phenomena described by the non-linear Hartree equation with a *random* external potential W whose distribution is homogeneous and has correlations of short range. For some results in this direction see [13].

Perhaps most importantly, we would like to understand much more about how non-linear equations of classical physics emerge as descriptions of quantum systems in certain asymptotic regimes.

Progress in theoretical physics is achieved in many different ways. If one is very lucky (and ingenious) one may discover new Laws of Nature or *new* physical theories or new mathematical structure useful to formulate such theories. More modestly, one may attempt to improve the mathematical understanding of *known* physical theories or derive new consequences of such theories. It appears safe to predict that we will be sailing well into the 21st century with plenty of interesting questions about physical theories discovered during the past three hundred years to worry about, and that "analysis and physics" will be a worthy partner of the currently more popular "geometry and physics".

Acknowledgements. We thank J. Bourgain and A. Soffer and, especially, I.M. Sigal and T. Spencer for very helpful discussions. The work of H.-T. Yau has been partially supported by the US NSF undergrant DMS-9703752.

References

1. Dimock, J.: The non-relativistic limit of $P(\phi)_2$ quantum field theories: two particle phenomena, *Comm. Math. Phys.* **57** (1977), 51–66; QED_2 in the Coulomb gauge, *Ann. Inst. H. Poincaré* **43** (1985), 167–179.
2. Fröhlich, J., Tsai, T.-P., and Yau, H.-T.: The point-particle limit of the non-linear Hartree equation, preprint to appear.
3. Lieb, E. H.: The Stability of Matter: From Atoms to Stars, (Selecta of Elliott H. Lieb), W. Thirring, (ed.), Springer-Verlag, Berlin-Heidelberg-New York, 1991.

4. Feldman, J., Salmhofer, M., and Trubowitz, E.: Renormalization of the Fermi surface, in: D. De Wit et al., (eds.), *XIIth International Congress of Mathematical Physics (ICMP '97) (Brisbane)*, Internat. Press, Cambridge, MA, 1999, pp. 24–34; and references to previous work given there.
5. Chen, T., Fröhlich, J., and Seifert, M.: Renormalization group methods: Landau-Fermi liquid and BCS superconductor, in: F. David, P. Ginsparg, and J. Zinn-Justin, (eds.), *Fluctuating geometries in statistical mechanics and field theory (Les Houches, 1994)*, North-Holland, Amsterdam, 1996, pp. 913–970.
6. Hepp, K.: The classical limit for quantum mechanical correlation functions, *Comm. Math. Phys.* **35** (1974), 265–277. Some of the ideas in this paper can be traced back to: Schrödinger, E.: Der stetige Übergang von der Mikro- zur Makromechanik, *Die Naturwissenschaften* **28** (1926), 664–666.
7. Ginibre, J. and Velo, G.: On a class of nonlinear Schrödinger equations with nonlocal interaction, *Math. Z.* **170** (1980), 109–136; see also: Ginibre, J. and Velo, G.: Scattering theory in the energy space for a class of nonlinear Schrödinger equations, *J. Math. Pure Appl.* **64** (1985), 363–401; Scattering theory in the energy space for a class of Hartree equations, preprint 1998.
8. Weinstein, M. I.: Modulation stability of ground states of nonlinear Schrödinger equations, *SIAM J. Math. Annal.* **16** (1985), 472–490.
9. Ovchinnikov, Yu. N. and Sigal, I. M.: Dynamics of localized structures, *Physica A* **261** (1998), 143–158; and references given there, e.g., Ovchinnikov, Yu. N. and Sigal, I. M.: Ginzburg-Landau equation I. Static vortices, in: P. C. Greiner et al., (eds.), *Partial differential equations and their applications (Toronto 1995)*, CRM Proc. Lecture Notes, 12, Amer. Math. Soc., Providence, RI, 1997, pp. 199–220.
10. Soffer, A. and Weinstein, M. I.: Multichannel nonlinear scattering for nonintegrable equations, *Comm. Math. Phys.* **133** (1990), 119–146; Multichannel nonlinear scattering for nonintegrable equations II. the case of anisotropic potentials and data, *J. Differential Equations* **98** (1992), 376–390.
11. Pillet, C.-A. and Wayne, C. E.: Invariant manifolds for a class of dispersive, Hamiltonian partial differential equations, *J. Differential Equations* **141** (1997), 310–326.
12. Mikhail, A.V., Kuznetsov, E.A., Newell, A.C., and Zakharov, V.E., (eds.): *The nonlinear Schrödinger equation*, Proceedings of the conference held in Chernogolovka (1994), Physics D **87**, no. 1–4, North Holland, Amsterdam, 1995.
13. Albanese C. and Fröhlich, J.: Periodic solutions of some infinite-dimensional Hamiltonian systems associated with non-linear partial difference equations I, *Comm. Math. Phys.* **116** (1988), 475–502; Albanese, C., Fröhlich, J., and Spencer, T.: Periodic solutions of some infinite-dimensional Hamiltonian systems associated with non-linear partial difference equations II, *Comm. Math. Phys.* **119** (1988), 677–699.

Singletons and neutrinos

Christian Frønsdal
Physics Department, University of California, Los Angeles, CA 90095-1547, USA
e-mail: `fronsdal@physics.ucla.edu`

Abstract. The first half is a rapid review of 30 years of work on physics in anti-de Sitter space, with heavy emphasis on singletons. Principal topics are the kinematical basis for regarding singletons as the constituents of massless particles, and the effect of (negative) curvature in the infrared domain. Ideas that lead to an alternative to Big Bang cosmology are merely sketched. The second half presents new ideas inspired by experimental results on neutrino oscillations. Since leptons are massless before symmetry breaking it is natural to view them as composite states consisting of one Bose singleton (the Rac) and one Fermi singleton (the Di). This gives rise to a particular formulation of the phenomenology of electroweak physics, and strong suggestions for an expansion of the Standard Model. An expansion of the Higgs sector seems inevitable, and flavor changing symmetry, complete with a new set of heavy vector mesons, is a very attractive possibility.

Keywords: singleton, neutrino, flavor symmetry

Mathematics Subject Classifications (2000): 81T10, 81T20

Dedicated to the memory of my friend Moshé Flato

1. Introduction

This talk is intended as a modest offering to the memory of Moshé Flato, the most original physicist that I have known, and my best friend for more than 25 years.

Perhaps you will say that Moshé was a mathematician as much as he was a physicist. Indeed, he was a professor of mathematics and he created, from nothing, one of the finest departments of mathematical physics anywhere; nevertheless, his great passion was physics. Others will tell you about some of Moshé's greatest accomplishments in mathematics and in mathematical physics; it falls on me to summarize his work on the **physics** of spaces of constant curvature, an endeavor that is entirely in the realm of physics.

As it happens, this work, on which I shall report, was done in collaboration with others, including Daniel Sternheimer and myself, and the early part of it was initiated before we started to publish together, but Moshé's influence dates back to the beginning.

2. Physics in anti-de Sitter space

The initial problematic was very simple: The cosmological constant is small, but it can never be shown, experimentally, to be exactly zero. Any value other than zero is incompatible with the idea that the Poincaré group is the basic symmetry

G. Dito and D. Sternheimer (eds.), Conférence Moshé Flato 1999, Vol. 1, 209 – 216.

group of space time, but it is consistent with a symmetry group that has the same dimension as the Poincaré group; all of the most important concepts of flat space physics can be extended to a space of constant curvature that admits the global action of one of the two de Sitter groups, $SO(3,2)$ or $SO(4,1)$.

This idea was not new in the mid-sixties (see e.g. [34, 21, 25]). But up to that time, of the two possibilities, $SO(4,1)$ had received the most attention although it was known, already, that symmetry under this group leads unavoidably to spontaneous creation of matter (see e.g. [26]).

Both of us had a great admiration for Wigner's work [33] on the unitary representations of the Poincaré group that led to the first classification of elementary particles, in terms of mass and spin (or helicity). It was very natural to apply the same philosophy to the classification of elementary particles in (anti-) de Sitter space. The unitary representations of both groups were essentially known at the time (cf. [32, 8] for $SO(4,1)$ and [9] for the discrete series of $SO(3,2)$); the full unitary dual was calculated only at the end of the seventies [1]), but the interpretation in terms of elementary particles was not yet well developed.

When the problem is approached in this way, then immediately one is led to favor $SO(3,2)$, the symmetry group of anti-de Sitter space (so called later on, and now abbreviated as AdS), over the other alternative. This is because $SO(3,2)$ has representations that can be associated with elementary particles, while $SO(4,1)$ does not. The energy spectrum of every unitary representation of $SO(4,1)$, except the trivial one, is the real line, unbounded in both directions [30], and if there is one physical principle that has stood the test of time, it is the requirement that energy must be bounded below. Consequently, all our work was concerned exclusively with $SO(3,2)$ and anti-de Sitter space time.

The choice of $SO(3,2)$ implies a negative curvature. This curvature, in the cosmological context, is very small, and it was never expected to be measurable. Our project was not concerned with the *magnitude* of the cosmological constant, but merely with the fact that, as a matter of principle, it *need not be zero*. It was expected that physics in a space of constant, negative curvature was possible and above all that its elaboration would be very educational. That it would actually turn out to suggest new types of physical phenomena was a great surprise.

The first indication of new physics appeared in connection with "massless" particles. It turns out that "masslessness" is a term that can be applied to certain irreducible representations of $SO(3,2)$ with as much justification as in the context of the Poincaré group [3]. One instance of masslessness appears in anti-de Sitter electrodynamics [15].

For the sake of simplicity, I shall not deal directly with realistic Maxwell theory, but instead with the theory of a spinless, massless field. All the interesting features of a gauge theory are then lost, but the features that I want to discuss are not. (The realistic picture involves Gupta-Bleuler triplets and indecomposable representations [15, 12].) In flat space free (spinless) photons are associated with an irreducible representation of the Poincaré group; this particular representation has the interesting property of having a unique extension to the conformal group [2]. The conformal group in anti-de Sitter space time is the same as the conformal group in flat space, locally isomorphic to $SO(4,2)$, and the same irreducible representation of the conformal group appears in (scalar) electrodynamics, in both cases. But in the case of anti-de Sitter space this irreducible representation breaks up into two inequivalent representations of $SO(3,2)$. This means that there are two kinds of photons, with different propagators. In order to incorporate conformal invariance into the theory of free quantum fields, one must quantize the field in such a way that both types propagate. The big surprise is that this is incompatible with the self-adjointness of the Hamiltonian. In other words, energy is not conserved. To conserve energy, one must use only one of the two types of photons; this amounts to spontaneous breakdown of conformal symmetry.

Now I should like to present here for the first time an idea that would have been elaborated by Moshé and myself if Moshé had been given more time. It is a radical idea, for it suggests an alternative to the Big Bang, and thus it is likely to irritate some people; but I will risk it. The idea is to accept the lack of energy conservation that is implied by conformal invariance of QED. This will lead to spontaneous creation of energy, locally throughout the universe. Conceivably, the amount of energy (and matter) created could balance the loss occasioned by the divergence of matter in the form of visible galaxies, and lead to a kind of "steady state" model of cosmology, characterized by a mean mass density that is constant in time. Note that thermal equilibrium would not be reached, and some of the ideas [29] that attempt to explain the preponderance of baryons would become more viable. Some people (including Fred Hoyle [22]) are more comfortable with this scenario.

Before we turn to the more spectacular aspects of anti-de Sitter physics, it is worth while to stress, one more time, the theoretical benefits of negative curvature, however small. The energy spectra of elementary particles in anti-de Sitter space are positive definite; thus there are no infrared singularities. Given the pivotal role of the infrared catastrophe in QCD, I marvel at the fact that no attempt has yet been made to investigate the effect of negative curvature on the confinement problem.

3. Singletons

By far, the most dramatic consequence of allowing for a small negative curvature is the existence of singletons. These are highly degenerate representations of $SO(3,2)$, with positive energy and thus at first sight associated with elementary particles. They were discovered by Dirac in 1963 [7]. As elementary particles they were at first dismissed by us on the grounds that these representations have too few states to allow for the formation of localized wave packets. (One manifestation of this is the singleton black body spectrum: it turns out to be that of ordinary particles in 3-dimensional space-time [15].) But this property is precisely what gives singletons their fascinating properties [11]. A free singleton with fixed energy has a well defined angular momentum. If the energy is large enough to be measurable then the angular momentum is enormous and the state may be associated with sloshing modes of the universe; if the angular momentum is small then the energy is of the order of the curvature and thus too small to be observed.

Our first important observation was that all 2-singleton states are massless; in fact the action of $SO(3,2)$ on the space of 2-singleton states breaks up into an infinite direct sum of massless representations. For example, if $\varphi(x)$ creates a singleton, then $\varphi(x)\varphi(x)$ creates a massless particle with spin zero, and $\varphi(x)\partial_\mu\varphi(x)$ creates a photon. This property of singletons, for which there is no analog in flat space, suggests a model of massless particles as 2-singleton composites. No interaction and no binding energy is associated with this type of compositeness; it is just a kinematical fact. Composite electrodynamics was presented in [12] and the linear approximation to composite gravity in [13].

4. Singletons and electroweak interactions

Rather than continuing with this review of work that has been published (references may be found in our last paper with Moshé [14]) I prefer to present some new ideas. Moshé had a very strong belief in the physical role of singletons. As a tribute to Moshé, Daniel Sternheimer and I have done our best to vindicate this idea – feeling that it is something we owe our friend.

The Standard Model is based on "the weak group", $S_W = SU(2)\otimes U(1)$, and more precisely on the Glashow representation of this group, that is carried by the triplet $(\nu_e, e_L; e_R)$ and by each of the other generations of leptons. Let us now suppose that

(a) There are three bosonic singletons $(R^N R^L; R^R) = (R^A)_{A=N,L,R}$ (three "Rac"s) that carry the Glashow representation of S_W;

(b) There are three spinorial singletons $(D_\varepsilon, D_\mu; D_\tau) = (D_\alpha)_{\alpha=\varepsilon,\mu,\tau}$ (three "Di"s). They are insensitive to S_W but transform as a Glashow triplet with respect to another group S_F (the "flavor group"), isomorphic to S_W; (c) The vector mesons of the standard model are Rac-Rac composites, the leptons are Di-Rac composites, and there is a set of vector mesons that are Di-Di composites and that play exactly the same role for S_F as the weak vector bosons do for S_W: $W_A^B = \bar{R}^B R_A$, $L_\beta^A = R^A D_\beta$, $F_\beta^\alpha = \bar{D}_\beta D^\alpha$.

The vector mesons are associated with conserved currents and fall into a category of composite particles that was described in Moshé's last paper [13]. There is not any strong evidence, at this time, that S_F is isomorphic to S_W, only that S_F has a representation of dimension 3. The assignments of transformation properties of Di's and Rac's can be interchanged.

We propose a slightly more economical model; namely we shall identify the two $U(1)$s with each other. There is only one $U(1)$; the symmetry group is

$SU(2)_W \otimes U(1) \otimes SU(2)_F$. The subgroup $SU(2)_W$ acts on the Racs, $SU(2)_F$ acts on the Dis, and the hypercharge generator of $U(1)$ acts on both.

Let us concentrate on the leptons ($A = N, L, R$; $\beta = \varepsilon, \mu, \tau$)

$$(L^A_\beta) = \begin{pmatrix} \nu_e & e_L & e_R \\ \nu_\mu & \mu_L & \mu_R \\ \nu_\tau & \tau_L & \tau_R \end{pmatrix} . \tag{1}$$

It is a remarkable fact that there are very good reasons to believe that this collection of leptons, precisely three complete generations, is complete. If leptons are composite, and if lepton fields are related to bilinears, then the constituents must include both bosons and fermions and the factorization $L^A_\beta = R^A D_\beta$ is strongly urged upon us by the nature of the phenomenological summary in Equation 1.

Fields in the first two columns couple horizontally to make the standard electroweak current, those in the last two pair off to make Dirac mass-terms. Particles in the first two rows combine to make the (neutral) flavor current and couple to the flavor vector mesons.

The Higgs fields have a Yukawa coupling to lepton currents,

$$\mathscr{L}_{\text{Yu}} = -g_{\text{Yu}} \bar{L}^\beta_A L^B_\alpha H^{\alpha A}_{\beta B}. \tag{2}$$

The Standard Model was constructed with a single generation in mind, hence it assumes a single Higgs doublet, and must therefore introduce three independent Yukawa coupling constants, $g_{\text{Yu}} H^{\alpha A}_{\beta B} \to \delta^\alpha_\beta g_\alpha H^A_B$.

However, an early and remarkable property of weak interaction phenomenology was electron-muon universality. In a theory based on intermediary vector mesons, this is expressed as an equality of the coupling constants of the basic interaction lepton-lepton-meson. In Weinberg-Salam theory, with its spontaneous breakdown of Yang-Mills symmetry, this equality of coupling constants is natural and of geometric origin. If, as has been proposed, the Higgs field also has a geometrical meaning [6, 5], then it is natural to suppose that the Yukawa couplings lepton-lepton-Higgs are also characterized by a universal coupling constant; the same for electron, muon and (?) tau. This symmetry between leptons is broken only spontaneously, by the spread of the vacuum expectation values (VeVs) of the Higgs field.

Symmetry with respect to the group $SU(2)_W \otimes U(1) \otimes SU(2)_F$ does not justify Equation 2 with a single Yukawa coupling constant. We shall return to this point below. It is assumed that all the neutrinos are lefthanded.

If the great number of components of Higgs fields is unwelcome, it should be kept in mind that the usual Higgs field is not widely believed to correspond to an elementary particle. If it is composite, then there is no reason to expect it to have only two components. In our model, where leptons and vector mesons are composed of singletons, it would be natural to suppose that the Higgs field is likewise composed of singletons, and the topological structure of singleton field theory even suggests that there may be no elementary particles associated with the Higgs field in the full dynamical theory of the future.

Nonvanishing vacuum expectation values of the neutral components are directly related to charged lepton masses,

$$g_{\mathrm{Yu}}\langle H_{\alpha R}^{\alpha L}\rangle = m_\alpha, \ \alpha = \varepsilon, \mu, \tau. \tag{3}$$

The coupling constants in m_ε and m_μ must be equal, but the third one, associated with the tau, can be different. If these are the only components with nonvanishing VeV, then the following masses are induced for the weak vector mesons:

$$m^2(W^\pm) = g^2 \sum_\alpha \langle H_{\alpha R}^{\alpha L}\rangle^2, \quad m^2(Z) = (g^2 + g'^2) \sum_\alpha \langle H_{\alpha R}^{\alpha L}\rangle^2, \tag{4}$$

where g and g' are the two fundamental coupling constants of the Standard Model ($\tan\theta_W = g'/g$), and

$$m^2(C^\pm) = h^2(\langle H_{\mu R}^{\mu L}\rangle^2 - \langle H_{\varepsilon R}^{\varepsilon L}\rangle^2), \tag{5}$$

where h is the gauge coupling constant associated with the flavor group $SU(2)_F$. The theory is still invariant under the three abelian groups associated with the three lepton numbers, but only $L_\mu - L_\varepsilon$ is gauged, by the flavor gauge boson C_3 that remains massless, so far. The hypercharges of the leptons are the same as in the Standard Model and the hypercharges of the Higgs field are determined by the postulated invariance of the interaction.

5. The new developments

Neutrino oscillations, especially between the two neutrinos associated to the muon and to the tau, appear to have been firmly established [31, 4, 27, 10, 28]. This suggests non-vanishing neutrino masses but does not imply it, especially if new flavor changing interactions are not ruled out [20, 19, 23]. Nevertheless, there are several attractive mechanisms that account for most of the data and give masses to at least some of the neutrinos, without introducing any additional leptons. The most economical assumption is that the masses of ν_μ and ν_τ are of the order of 0.1 eV and that of ν_e even smaller, possibly zero. It was pointed out, almost 20 years ago [17, 16, 24], that there is room for an additional Higgs field, an $SU(2)$ triplet, and that this implies non-vanishing neutrino masses. Our proposal is slightly different.

We postulate additional Higgs fields, coupled to leptons in the following way,

$$\mathscr{L}'_{\mathrm{Yu}} = h_{\mathrm{Yu}} L_\alpha^A L_\beta^B K_{AB}^{\alpha\beta} + \text{h.c.}. \tag{6}$$

As in the case of Equation 3, this formula should not be taken quite literally, for symmetry does not require that all the couplings have the same strength.

We now give a nonvanishing vacuum expectation values to the neutral components (generalizing [GR])

$$\langle K_{NN}^{\alpha\beta}\rangle, \ \alpha, \beta = \varepsilon, \mu, \tau. \tag{7}$$

This implies a general, symmetric neutrino mass matrix. In addition, one of the two remaining massless vector mesons becomes massive. The masses are

$$\begin{aligned} m^2(W^\pm) &= g^2\Big(\sum_\alpha \langle H^{\alpha L}_{\alpha R}\rangle^2 + 2\sum_\alpha \langle K^{\alpha\tau}_{NN}\rangle^2\Big),\\ m^2(C^\pm) &= 2h^2(\langle H^{\mu L}_{\mu R}\rangle - H^{\varepsilon L}_{\varepsilon R}\rangle)^2,\\ m^2(C^3) &= h^2\sum_\alpha \langle K^{\alpha\tau}_{NN}\rangle^2,\\ m^2(Z) &= (g^2+g'^2)\Big(\sum \langle H^{\alpha L}_{\alpha R}\rangle^2 + 4\sum_\alpha \langle K^{\alpha\tau}_{NN}\rangle^2\Big). \end{aligned}$$

6. Other predictions of the model

The interpretation of the atmospheric neutrino experiments must take into account, not only neutrino masses and mixing, but also the effect of the flavor-changing interactions induced by the new vector mesons. To account for the observed smallness of flavor changing interactions it is necessary that the new VeV be at least of the order of magnitude of 100 GeV, so that h_{Yu} must be very small. Strong constraints on h_{Yu}, and h are imposed by past and future experiments involving neutral currents. As these become more accurate the model may fail.

References

1. Angelopoulos, E.: $\overline{\mathrm{SO}}_0(3,2)$: linear and unitary irreducible representations, in: *Quantum theory, groups, fields and particles*, Math. Phys. Stud. **4**, D. Reidel, Dordrecht-Boston, 1983, pp. 101–148; Sur les représentations unitaires irréductibles de $\overline{\mathrm{SO}}_0(p,2)$, *C. R. Acad. Sci. Paris Sér. I Math.* **292** (1981), 469–471.
2. Angelopoulos, E. and Flato, M.: On unitary implementability of conformal transformations, *Lett. Math. Phys.* **2** (1978), 405–412.
3. Angelopoulos, E., Flato, M., Fronsdal, C., and Sternheimer, D.: Massless particles, conformal group and de Sitter universe, *Phys. Rev. D* **23** (1981), 1278–1289.
4. Bilenky, S. M., Giunti, C., and Grimus, W.: Phenomenology of Neutrino Oscillations, *Prog. Part. Nucl. Phys.* **43** (1999), 1–86.
5. Connes, A.: Interprétation géométrique du modèle standard de la physique des particules et structure fine de l'espace-temps, *C. R. Acad. Sci. Paris Sér. Gén. Vie Sci.* **10** (1993), 223–234.
6. Connes, A. and Lott, J.: Particle models and noncommutative geometry, in: *Recent advances in field theory (Annecy-le-Vieux, 1990), Nuclear Phys. B Proc. Suppl.* **18B** (1990), 29–47.
7. Dirac, P. A. M.: A remarkable representation of the $3+2$ de Sitter group, *J. Math. Phys.* **4** (1963), 901–909.
8. Dixmier, J.: Représentations intégrables du groupe de de Sitter, *Bull. Soc. Math. France* **89** (1961), 9–41.
9. Evans, N.T.: Discrete series for the universal covering group of the $3+2$ de Sitter group, *J. Math. Phys.* **8** (1967), 170–184.
10. Fisher, P., Kayser, B., and McFarland, K.S.: Neutrino Mass and Oscillation, in: *Annual Review of Nuclear and Particle Science*, **49** (1999). Preprint hep-ph/9906244.
11. Flato, M. and Fronsdal, C.: On Dis and Racs, *Phys. Lett. B* **97** (1980), 236–240.
12. Flato, M. and Fronsdal, C.: Composite Electrodynamics, *J. Geom. Phys.* **5** (1988), 37–61.

13. Flato, M. and Frønsdal, C.: Interacting Singletons, *Lett. Math. Phys.* **44** (1998), 249–259.
14. Flato, M., Frønsdal, C., and Sternheimer, D. : Singletons, physics in AdS universe and oscillations of composite neutrinos, *Lett. Math. Phys.* **48** (1999), 109–119.
15. Fronsdal, C.: Elementary particles in a curved space. IV. Massless particles, *Phys. Rev. D* **12** (1975), 3819–3830.
16. Georgi, H., Glashow, S. L., and Nussinov, S.: Unconventional model of neutrino masses, *Nucl. Phys. B* **193** (1983) 297.
17. Gelmini, G.B. and Roncadelli, M.: Left-handed neutrino mass scale and spontaneously broken lepton number, *Phys. Lett. B* **98** (1981), 411–415.
18. Grimus, W.: Phenomenology of Neutrino Masses and Mixing, preprint hep-ph/9910340.
19. Grimus, W., Pfeiffer, R., and Schwetz, T.: A 4-neutrino model with a Higgs triplet, preprint hep-ph/9905320.
20. Guzzo, M. M., Nunokawa, H., Peres, O. L. G., Pleitez, V., and Zukanovich Funchal R.: Flavor changing models with strictly massless neutrinos, preprint hep-th/9908038.
21. Gürsey, F. and Lee, T.D.: Spin $1/2$ wave equation in de Sitter space, Proc. Nat. Acad. Sci. U.S.A. **49** (1963), 179–186.
22. Hoyle, F.: *The Nature of the Universe*, New Am. Lib., New York, 1955.
23. Johnson, L. M. and McKay, D. W.: Revising Neutrino Oscillation Parameter Space With Direct Flavor-Changing Interactions, preprint hep-ph/9909355.
24. Kim, J.E.: Light pseudo-Particles, particle physics and cosmology, *Phys. Rep.* **150** (1987), 1.
25. Nachtmann, O.: Quantum theory in de Sitter space, *Comm. Math. Phys.* **6** (1967), 1–16.
26. Nachtmann, O.: Continuous creation in a closed world order, *Z. Phys.* **208** (1968), 113.
27. Ohlsson, T. and Snellman, H.: Neutrino oscillations and mixings with three flavors, *Phys. Rev. D* **60** (1999), 3007; Three flavor neutrino oscillations in matter, preprint hep-ph/9910546.
28. Peccei, R.D.: Neutrino Physics, in: *Proceedings of the VIII Escuela Mexicana de Particulas y Campos*, hep-ph/9906509.
29. Sakharov, A. D.: Violation of CP invariance, C asymmetry, and baryon asymetry of the universe, *Pisma Zh. Eksp. Teor. Fiz.* **5** (1967), 32–35.
30. Sternheimer, D.: Propriétés spectrales dans les représentations de groupes de Lie, *J. Math. Pures Appl.* **47** (1968), 289–319.
31. The SuperKamiokande Collaboration: Evidence for oscillation of atmospheric neutrinos, *Phys. Rev. Lett.* **81** (1998), 1562–1567; Measurement of the flux and zenith-angle distribution of upward through-going muons by Super-Kamiokande, *Phys. Rev. Lett.* **82** (1999), 2644–2648; Neutrino-induced upward stopping muons in Super-Kamiokande, preprint hep-ex/9908049.
32. Thomas, L.H.: On unitary representations of the group of de Sitter space, Ann. of Math. (2) **42** (1941), 113–126.
33. Wigner, E. P.: On unitary representations of the inhomogeneous Lorentz group, *Ann. of Math.* **40** (1939), 149–204.
34. Wigner, E. P.: Some remarks on the infinite de Sitter space, *Proc. Nat. Acad. Sci. U.S.A.* **36** (1950), 184–188.

Variations on deformation quantization

Simone Gutt*
Université Libre de Bruxelles, Campus Plaine, CP 218, 1050 Brussels, Belgium
and
Université de Metz, Ile du Saulcy, 57045 Metz Cedex 01, France
e-mail: `sgutt@ulb.ac.be`

Abstract. I was asked by the organizers to present some aspects of deformation quantization. Moshé has pursued, for more than 25 years, a research program based on the idea that physics progresses in stages, and one goes from one level of the theory to the next one by a deformation, in the mathematical sense of the word, to be defined in an appropriate category. His study of deformation theory applied to mechanics started in 1974 and led to spectacular developments with the deformation quantization program.

I first met Moshé at a conference in Liège in 1977. A few months later he became my thesis "codirecteur". Since then he has been one of my closest friends, present at all stages of my personal and mathematical life. I miss him....

I have chosen, in this presentation of deformation quantization, to focus on 3 points: the uniqueness – up to equivalence – of a universal star-product (universal in the sense of Kontsevich) on the dual of a Lie algebra, the cohomology classes introduced by Deligne for equivalence classes of differential star-products on a symplectic manifold, and the construction of some convergent star-products on Hermitian symmetric spaces. Those subjects will appear in a promenade through the history of existence and equivalence in deformation quantization.

Keywords: deformation, star-product

Mathematics Subject Classifications (2000): 53D55

1. Introduction

Quantization of a classical system is a way to pass from classical to quantum results.

Classical mechanics, in its Hamiltonian formulation on the motion space, has for framework a symplectic manifold (or more generally a Poisson manifold). Observables are families of smooth functions on that manifold M. The dynamics is defined in terms of a Hamiltonian $H \in C^\infty(M)$ and the time evolution of an observable $f_t \in C^\infty(M \times \mathbb{R})$ is governed by the equation:

$$\frac{d}{dt} f_t = -\{H, f_t\}.$$

Quantum mechanics, in its usual Heisenberg's formulation, has for framework a Hilbert space (states are rays in that space). Observables are families of self-adjoint operators on the Hilbert space. The dynamics is defined in terms of a Hamiltonian

* Research supported by the Communauté française de Belgique, through an Action de Recherche Concertée de la Direction de la Recherche Scientifique.

G. Dito and D. Sternheimer (eds.), Conférence Moshé Flato 1999, Vol. 1, 217 – 254.

H, which is a self-adjoint operator, and the time evolution of an observable A_t is governed by the equation:

$$\frac{dA_t}{dt} = \frac{i}{\hbar}[H, A_t].$$

A natural suggestion for quantization is a correspondence $Q\colon f \mapsto Q(f)$ mapping a function f to a self-adjoint operator $Q(f)$ on a Hilbert space $\mathscr{H}$ in such a way that $Q(1) = \mathrm{Id}$ and

$$[Q(f), Q(g)] = i\hbar Q(\{f, g\}).$$

There is no such correspondence defined on all smooth functions on M when one puts an irreducibility requirement which is necessary not to violate Heisenberg's principle.

Different mathematical treatments of quantization appeared to deal with this problem:

- Geometric Quantization of Kostant and Souriau. This proceeds in two steps; first prequantization of a symplectic manifold (M, ω) where one builds a Hilbert space and a correspondence Q as above defined on all smooth functions on M but with no irreducibility, then polarization to "cut down the number of variables". One succeeds to quantize only a small class of functions.

- Berezin's quantization where one builds on a particular class of Kähler manifolds a family of associative algebras using a symbolic calculus, i.e., a dequantization procedure.

- Deformation Quantization introduced by Flato, Lichnerowicz, and Sternheimer in [47] and developed in [10] where they "suggest that quantization be understood as a deformation of the structure of the algebra of classical observables rather than a radical change in the nature of the observables."

This deformation approach to quantization is part of a general deformation approach to physics. This was one of the seminal ideas stressed by Moshé [45]: One looks at some level of a theory in physics as a deformation of another level.

Deformation quantization is defined in terms of a star-product which is a formal deformation of the algebraic structure of the space of smooth functions on a Poisson manifold. The associative structure given by the usual product of functions and the Lie structure given by the Poisson bracket are simultaneously deformed.

The plan of this presentation is the following:

- Examples and existence of star-products. After some definitions, I recall the history of the proofs of existence and the history of some constructions of star-products, starting from the Moyal star-product on a vector space endowed with a constant Poisson structure. I give the standard star-product on the dual of a Lie algebra, and recall Kontsevich star-product for any Poisson structure on a vector space. I prove by elementary methods that all universal star-products on the dual of a Lie algebra are essentially equivalent.

- Equivalence of star-products. I first wander through the history of the parametrization of equivalence classes of star-products. I define Deligne's cohomology classes associated to differential star-products on symplectic manifolds and show by Čech methods how this yields an intrinsic parametrization of the equivalence classes of differential star-products and how this allows to study automorphisms of a star-product (in the symplectic framework). Finally, I define a generalized moment map for star-products.

- Convergence of star-products. I study mostly the convergence of a Berezin type star-product on Hermitian symmetric spaces. The construction of such a star-product involves a correspondence between operators (on the Hilbert space given by geometric quantization) and functions (their Berezin symbols). A parameter is introduced in the construction (generalizing the power of the line bundle in geometric quantization). Asymptotic expansion in this parameter on a large algebra of functions yields a deformed product. One proves associativity and convergence of this product.

I would like to thank my friends Georges Pinczon and Daniel Sternheimer who brought many improvements to this presentation.

2. Examples and existence of star-products

Definition 1 *A **Poisson bracket** defined on the space of smooth functions on a manifold M, is an $\mathbb{R}$-bilinear map on $C^\infty(M)$, $(u,v) \mapsto \{u,v\}$ such that for any $u,v,w \in C^\infty(M)$:*
- $\{u,v\} = -\{v,u\}$;
- $\{\{u,v\},w\} + \{\{v,w\},u\} + \{\{w,u\},v\} = 0$;
- $\{u,vw\} = \{u,v\}w + \{u,w\}v$.

*A Poisson bracket is given in terms of a contravariant skew-symmetric 2-tensor P on M, called the **Poisson tensor**, by*

$$\{u,v\} = P(du \wedge dv).$$

The Jacobi identity for the Poisson bracket Lie algebra is equivalent to the vanishing of the Schouten bracket:

$$[P,P] = 0.$$

*(The Schouten bracket is the extension – as a graded derivation for the exterior product – of the bracket of vector fields to skew-symmetric contravariant tensor fields.) A **Poisson manifold**, denoted (M,P), is a manifold M with a Poisson bracket defined by the Poisson tensor P.*

A particular class of Poisson manifolds, essential in classical mechanics, is the class of **symplectic manifolds.** If (M,ω) is a symplectic manifold (i.e., ω is a closed nondegenerate 2-form on M) and if $u,v \in C^\infty(M)$, the Poisson bracket of u and v is defined by

$$\{u,v\} := X_u(v) = \omega(X_v,X_u),$$

where X_u denotes the Hamiltonian vector field corresponding to the function u, i.e., such that $i(X_u)\omega = du$. In coordinates the components of the corresponding Poisson tensor P^{ij} form the inverse matrix of the components ω_{ij} of ω.

Duals of Lie algebras form the class of linear Poisson manifolds. If $\mathfrak{g}$ is a Lie algebra then its dual $\mathfrak{g}^*$ is endowed with the Poisson tensor P defined by

$$P_\xi(X,Y) := \xi([X,Y])$$

for $X,Y \in \mathfrak{g} = (\mathfrak{g}^*)^* \sim (T_\xi \mathfrak{g}^*)^*$.

Definition 2. (Bayen et al. [10]) *A **star-product** on (M,P) is a bilinear map*

$$N \times N \to N[\![\mu]\!], \qquad (u,v) \mapsto u * v = u *_\mu v := \sum_{r \geq 0} \mu^r C_r(u,v)$$

where $N = C^\infty(M)$, such that

1) When the map is extended μ-linearly (and continuously in the μ-adic topology) to $N[\![\mu]\!] \times N[\![\mu]\!]$, it is formally associative:

$$(u * v) * w = u * (v * w);$$

2) (a) $C_0(u,v) = uv$, (b) $C_1(u,v) - C_1(v,u) = \{u,v\}$;

3) $1 * u = u * 1 = u$.

When the C_r's are bidifferential operators on M, one speaks of a **differential star-product.**

Remark 3 A star-product can also be defined not on the whole of $C^\infty(M)$ but on any subspace N of it which is stable under pointwise multiplication and Poisson bracket.

Requiring differentiability of the cochains is essentially the same as requiring them to be local [20].

In (b) we follow Deligne's normalization for C_1: Its skew-symmetric part is $\frac{1}{2}\{\,,\,\}$. In the original definition it was equal to the Poisson bracket. One finds in the literature other normalizations such as $\frac{i}{2}\{\,,\,\}$. All these amount to a rescaling of the parameter μ.

One assumed also the parity condition $C_r(u,v) = (-1)^r C_r(v,u)$ in the earliest definition.

Property (b) above implies that the center of $C^\infty(M)[[\mu]]$, when the latter is viewed as an algebra with multiplication $*$, is a series whose terms Poisson commute with all functions, so is an element of $\mathbb{R}[[\mu]]$ when M is symplectic and connected.

Properties (a) and (b) of Definition 2 imply that the **star-commutator** defined by $[u,v]_* = u*v - v*u$, which obviously makes $C^\infty(M)[[\mu]]$ into a Lie algebra, has the form $[u,v]_* = \mu\{u,v\} + \cdots$ so that repeated bracketing leads to higher and higher order terms. This makes $C^\infty(M)[[\mu]]$ an example of a **pronilpotent Lie algebra**. We denote the **star adjoint representation** by $ad_*u\,(v) = [u,v]_*$.

2.1. The Moyal star-product on $\mathbb{R}^n$

The simplest example of a deformation quantization is the Moyal product for the Poisson structure P on a vector space $V = \mathbb{R}^m$ with constant coefficients:

$$P = \sum_{i,j} P^{ij}\partial_i \wedge \partial_j, \quad P^{ij} = -P^{ji} \in \mathbb{R},$$

where $\partial_i = \partial/\partial x^i$ is the partial derivative in the direction of the coordinate x^i, $i = 1,\ldots,n$. The formula for the **Moyal product** is

$$(u *_M v)(z) = \exp\left(\frac{\mu}{2}P^{rs}\partial_{x^r}\partial_{y^s}\right)(u(x)v(y))\Big|_{x=y=z}. \tag{2.1}$$

When P is nondegenerate (so $V = \mathbb{R}^{2n}$), the space of formal power series of polynomials on V with Moyal product is called the Weyl algebra $W = (S(V)[[\mu]], *_M)$.

This example comes from the composition of operators via Weyl's quantization. Weyl's correspondence associates to certain functions f on $\mathbb{R}^{2n}$ an operator $W(f)$ on $L^2(\mathbb{R}^n)$

$$W(f) = \int \tilde{f}(\xi,\eta)\exp\left(\frac{i(\xi P + \eta Q)}{\hbar}\right) d\xi\, d\eta$$

where $\tilde{f}$ is the inverse Fourier transform of f. Then

$$W(f) \circ W(g) = W(f *_M g) \quad (\mu = i\hbar).$$

In fact, Moyal had used in 1949 the deformed bracket which corresponds to the commutator of operators to study quantum statistical mechanics. The Moyal product first appeared in Groenewold.

In 1974, Flato, Lichnerowicz, and Sternheimer [46] studied deformations of the Lie algebra structure defined by the Poisson bracket on the algebra N of smooth functions on a symplectic manifold; they studied 1-differential deformations because the relevant cohomology, i.e., the 1-differential Chevalley cohomology of the Lie algebra N with values in N for the adjoint representation, was known [61].

In 1975, Vey [86] pursued their work in the differential context: He constructed a differential deformation on $M = \mathbb{R}^{2n}$ which turns out to be the Moyal bracket

$\{u,v\}_M := \frac{1}{\mu}(u *_M v - v *_M u)$ and he proved that there exists a differential deformation on a symplectic manifold when its third de Rham cohomology space is trivial ($b_3(M) := \dim H^3(M;\mathbb{R}) = 0$). He also reconstructed the Moyal product on $\mathbb{R}^{2n}$.

This opened the path to deformation quantization presented in 1976 by Flato, Lichnerowicz, and Sternheimer [47].

In 1978, in their seminal paper about deformation quantization [10], Bayen, Flato, Fronsdal, Lichnerowicz, and Sternheimer proved that the Moyal star-product can be defined on any symplectic manifold (M,ω) which admits a symplectic connection ∇ (i.e., a linear connection such that $\nabla\omega = 0$ and the torsion of ∇ vanishes) with no curvature. They also built star-products on some quotient of $\mathbb{R}^{2n}$.

In 1979, Neroslavsky and Vlassov [68] proved with Lichnerowicz that on any symplectic manifold with $b_3(M) = 0$, there exists a differential star-product.

2.2. Hochschild cohomology

The study of star-products on a manifold M used Gerstenhaber theory of deformations [49] of associative algebras. This uses the Hochschild cohomology of the algebra, here $C^\infty(M)$ with values in $C^\infty(M)$, where p-cochains are p-linear maps from $(C^\infty(M))^p$ to $C^\infty(M)$ and where the **Hochschild coboundary operator** maps the p-cochain C to the $p+1$-cochain

$$(\partial C)(u_0,\dots,u_p) = u_0 C(u_1,\dots,u_p) + \sum_{r=1}^{p}(-1)^r C(u_0,\dots,u_{r-1}u_r,\dots,u_p) + (-1)^{p+1}C(u_0,\dots,u_{p-1})u_p.$$

For differential star-products, we consider differential cochains, i.e., given by differential operators in each argument. The associativity condition for a star-product at order k in the parameter μ reads

$$(\partial C_k)(u,v,w) = \sum_{\substack{r+s=k\\ r,s>0}} \left(C_r(C_s(u,v),w) - C_r(u,C_s(v,w)) \right).$$

If one has cochains C_j, $j < k$, such that the star-product they define is associative up to order $k-1$, then the right-hand side above is a cocycle (∂(RHS)$= 0$) and one can extend the star-product to order k if it is a coboundary (RHS$= \partial(C_k)$).

Denoting by m the usual multiplication of functions, and writing $* = m + C$ where C is a formal series of multidifferential operators ($C \in D_{poly}(M)[[\mu]]$) the associativity also reads $\partial C = [C,C]$ where the bracket on the right-hand side is the graded Lie algebra bracket on $D_{poly}(M)[[\mu]]$.

Theorem 4. (Vey [86]) *Every differential p-cocycle C on a manifold M is the sum of the coboundary of a differential (p-1)-cochain and a 1-differential skew-symmetric p-cocycle A:*

$$C = \partial B + A.$$

In particular, a cocycle is a coboundary if and only if its total skew-symmetrization, which is automatically 1*-differential in each argument, vanishes. Also*

$$H^p_{\text{diff}}(C^\infty(M), C^\infty(M)) = \Gamma(\Lambda^p TM).$$

Furthermore ([24]), given a connection ∇ *on* M, B *can be defined from* C *by universal formulas.*

By universal, we mean the following: Any p-differential operator D of order maximum k in each argument can be written

$$D(u_1, \dots, u_p) = \sum_{|\alpha_1|<k, \dots, |\alpha_p|<k} D^{\alpha_1 \cdots \alpha_p}_{|\alpha_1|, \dots, |\alpha_p|} \nabla_{\alpha_1} u_1 \cdots \nabla_{\alpha_p} u_p$$

where α's are multi-indices, $D_{|\alpha_1|, \dots, |\alpha_p|}$ are tensors (symmetric in each of the p groups of indices) and $\nabla_\alpha u = (\nabla \cdots (\nabla u))(\frac{\partial}{\partial x^{i_1}}, \dots, \frac{\partial}{\partial x^{i_q}})$ when $\alpha = (i_1, \dots, i_q)$. We claim that there is a B such that the tensors defining B are universally defined as linear combinations of the tensors defining C, universally meaning in a way which is independent of the form of C. Note that requiring differentiability of the cochains is essentially the same as requiring them to be local [20]. (An elementary proof of the above theorem can be found in [54].)

Remark 5 Behind Theorem 4 above, there are the following stronger results about Hochschild cohomology:

Theorem 6 *Let* $\mathscr{A} = C^\infty(M)$, *let* $\mathscr{C}(\mathscr{A})$ *be the space of continuous cochains and* $\mathscr{C}_{diff}(\mathscr{A})$ *the space of differential cochains. Then*

1) $\Gamma(\Lambda^p TM) \subset H^p(C^\infty(M), C^\infty(M))$;

2) the inclusions $\Gamma(\Lambda^p TM) \subset \mathscr{C}_{diff}(\mathscr{A}) \subset \mathscr{C}(\mathscr{A})$ *induce isomorphisms in cohomology.*

Point 1 follows from the fact that any cochain which is 1-differential in each argument is a cocycle and that the skew-symmetric part of a coboundary always vanishes. The fact that the inclusion $\Gamma(\Lambda TM) \subset \mathscr{C}_{diff}(\mathscr{A})$ induces an isomorphism in cohomology is proven by Vey [86]; it gives theorem 4. The general result about continuous cochains is due to Connes [29]. Another proof of Connes result was given by Nadaud in [67]. In the somewhat pathological case of completely general cochains the full cohomology does not seem to be known.

Examples of star-products were built using inductively the explicit formulas for coboundaries on locally symmetric symplectic manifolds in [24]; the cochains are given by universal expressions in the symplectic 2-form, its inverse (i.e., the Poisson tensor) and the curvature tensor of the symmetric symplectic connection.

We also showed that assuming a homogeneity condition on the cochains proves the existence of a star-product on the cotangent bundle to any parallelizable manifold [22] and gave explicit formulas for the cotangent bundle to a Lie group. The vertical part of this gives a deformation quantization of any linear Poisson manifold, i.e., any dual of a Lie algebra [52].

2.3. The standard *-product on $\mathfrak{g}^*$

Let $\mathfrak{g}^*$ be the dual of a Lie algebra $\mathfrak{g}$. The algebra of polynomials on $\mathfrak{g}^*$ is identified with the symmetric algebra $S(\mathfrak{g})$. One defines a new associative law on this algebra by a transfer of the product $\circ$ in the universal enveloping algebra $U(\mathfrak{g})$, via the bijection between $S(\mathfrak{g})$ and $U(\mathfrak{g})$ given by the total symmetrization σ:

$$\sigma : S(\mathfrak{g}) \to U(\mathfrak{g}), \quad X_1 \cdots X_k \mapsto \frac{1}{k!} \sum_{\rho \in S_k} X_{\rho(1)} \circ \cdots \circ X_{\rho(k)}.$$

Then $U(\mathfrak{g}) = \oplus_{n \geq 0} U_n$ where $U_n := \sigma(S^n(\mathfrak{g}))$ and we decompose an element $u \in U(\mathfrak{g})$ accordingly $u = \sum u_n$. We define for $P \in S^p(\mathfrak{g})$ and $Q \in S^q(\mathfrak{g})$

$$P * Q = \sum_{n \geq 0} (\mu)^n \sigma^{-1}((\sigma(P) \circ \sigma(Q))_{p+q-n}). \tag{2.2}$$

This yields a differential star-product on $\mathfrak{g}^*$ [52]. Using Vergne's result on the multiplication in $U(\mathfrak{g})$, this star-product is characterized by

$$X * X_1 \ldots X_k = X X_1 \cdots X_k + \sum_{j=1}^{k} \frac{(-1)^j}{j!} \mu^j B j[[[X, X_{r_1}], \cdots], X_{r_j}] X_1 \cdots \widehat{X_{r_1}} \cdots \widehat{X_{r_j}} \cdots X_k,$$

where B_j are the Bernouilli numbers. This star-product can be written with an integral formula (for $\mu = 2\pi i$) [37]:

$$u * v(\xi) = \int_{\mathfrak{g} \times \mathfrak{g}} \hat{u}(X) \hat{v}(Y) e^{2i\pi \langle \xi, CBH(X,Y) \rangle} dX dY,$$

where $\hat{u}(X) = \int_{\mathfrak{g}^*} u(\eta) e^{-2i\pi \langle \eta, X \rangle}$ and CBH denotes Campbell-Baker-Hausdorff formula for the product of elements in the group in a logarithmic chart: $\exp X \exp Y = \exp CBH(X,Y), \forall X, Y \in \mathfrak{g}$.

We call this the **standard (or CBH) star-product** on the dual of a Lie algebra.

An important property of this star-product is its covariance, i.e., the fact that

$$X * Y - Y * X = \mu [X, Y], \quad \forall X, Y \in \mathfrak{g}.$$

In general, when there is a Hamiltonian action of a connected Lie group G on a manifold M, denoting by λ_X the function on M corresponding to $X \in \mathfrak{g}$, a star-product on M is **covariant** under G if $\lambda_X * \lambda_Y - \lambda_Y * \lambda_X = \mu \lambda_{\{X,Y\}}, \forall X, Y \in \mathfrak{g}$.

Arnal, Cortet, Molin, and Pinczon have proven in [7] that for any covariant star-product there is a representation of G into the automorphisms of the star-product. We shall come back to this notion of covariance in §3.2.

A program to study representations through deformation quantization methods was introduced in 1978 by Bayen, Flato, Fronsdal, Lichnerowicz, and Sternheimer in [10, 82] and in Fronsdal [48]. The notion of covariant star-products is essential for this. The program involves a beautiful notion of adapted Fourier transform (or $*$ exponential). I shall not speak about this aspect of deformation quantization; let me just mention that the program to study representations through deformations was carried first in the case of nilpotent Lie groups in the mid-eighties by Arnal and Cortet [5] and that the adapted Fourier transform they found [6] gives a Kirillov type formula for the character of the representation and a localization of the (packets of) coefficients of representations on the corresponding orbits in the dual of the Lie algebra. This theory of star-representations was developed for all representations of compact groups and exponential groups, and for some series of representations of semi-simple groups.

The standard star-product on $\mathfrak{g}^*$ does not restrict to orbits (except for the Heisenberg group) so other algebraic constructions of star-products on $S(\mathfrak{g})$ were considered (with Michel Cahen in [23], with Cahen and Arnal in [4], by Arnal, Ludwig, and Masmoudi in [8] and more recently by Fioresi and Lledo in [44]). For instance, when $\mathfrak{g}$ is semisimple, if $\mathscr{H}$ is the space of harmonic polynomials and if $I_1,\ldots,I_r$ are generators of the space of invariant polynomials, then any polynomial $P \in S(\mathfrak{g})$ can be uniquely written as a sum $P = \sum_{a_1\ldots a_r} I_1^{a_1}\ldots I_r^{a_r} h_{a_1\ldots a_r}$ where $h_{a_1\ldots a_r} \in \mathscr{H}$. One considers the isomorphism σ' between $S(\mathfrak{g})$ and $U(\mathfrak{g})$ induced by this decomposition

$$\sigma'(P) = \sum_{a_1\ldots a_r} (\sigma(I_1)\circ)^{a_1}\cdots(\sigma(I_r)\circ)^{a_r}\circ\sigma(h_{a_1\ldots a_r}).$$

This gives a star-product on $S(\mathfrak{g})$ which is not defined by differential operators. In fact, with Cahen and Rawnsley, we proved [27] that if $\mathfrak{g}$ is semisimple, there is no differential star-product on any neighbourhood of 0 in $\mathfrak{g}^*$ such that $C * u = Cu$ for the quadratic invariant polynomial $C \in S(\mathfrak{g})$ and $\forall u \in S(\mathfrak{g})$ (thus no differential star-product which is tangential to the orbits).

In 1983, De Wilde and Lecomte proved [33] that on any symplectic manifold there exists a differential star-product. This was obtained by imagining a very clever generalization of a homogeneity condition consisting here in building at the same time the star-product and a set of local special derivations of it. A very nice presentation of this proof appears in [32]. Their technique works to prove the existence of a differential star-product on a regular Poisson manifold [64].

In 1985, but appearing only in the West in the nineties [40], Fedosov gave a recursive construction of a star-product on a symplectic manifold (M,ω) by:

– Considering the Weyl bundle $\mathscr{W}$ which is the bundle on M in associative Weyl algebras W associated to the principal bundle of symplectic frames;

– Building, from a symplectic connection ∇ on M, a covariant derivative on $\Gamma(\mathscr{W})$, $\partial = \partial^\nabla + [r,.]$ $(r \in \Gamma(T^*M \otimes \mathscr{W}))$, such that $\partial \circ \partial = 0$;

– Identifying $C^\infty(M)[[\mu]]$ with $\{s \in \Gamma(\mathscr{W}) \mid \partial s = 0\}$ and transferring the associative pointwise product of sections to an associative product on $C^\infty(M)[[\mu]]$.

In 1994, he extended this result to give a recursive construction in the context of regular Poisson manifold [41].

Independently, also using the framework of Weyl bundles, Omori, Maeda and Yoshioka [72] gave an alternative proof of existence of a differential star-product on a symplectic manifold by gluing local Moyal star-products.

In 1986, Drinfeld [37] proposed a program of quantization for Poisson-Lie groups. One way to consider the problem is to study deformations of the Hopf algebra $U(\mathfrak{g})$ of a finite-dimensional Lie bialgebra $(\mathfrak{g}, p)$, where one deforms the product and the coproduct, the deformation of the coproduct being driven by the cocycle p. By duality this is the quantization of formal Poisson-Lie groups. For the standard structures on semisimple Lie algebras, this was solved by Drinfeld; it is the construction of classical **quantum groups**. Drinfeld showed that the quantization is preferred, i.e., the product in $U(\mathfrak{g})$ is unchanged (or, by duality, the coproduct on formal series – which are the functions on the formal Poisson-Lie group – is unchanged). Etingof and Kazhdan [38] proved that one can deform $U(\mathfrak{g})$ for any Lie bialgebra $(\mathfrak{g}, p)$, so one can always quantize formal Poisson-Lie groups. They proved that the deformation is differential. Pinczon [77] proved that any preferred quantization of a formal Poisson-Lie group is differential.

Quantum groups became very popular and had fundamental applications. It is not my goal to describe those developments but they brought new interest in the question of deformation quantization of general Poisson manifolds.

Indeed, another way to consider the problem of quantization of a Poisson-Lie group (G,P) is to study deformations of the Hopf algebra $C^\infty(G)$, where the deformation of the product of functions is driven by P.

In the papers [17] by Bonneau, Flato, Gerstenhaber, and Pinczon and [16] by Bidegain and Pinczon, the duality between those two approaches is proven in the framework of topological deformations. In particular, any classical quantum group gives, by topological duality, a differential deformation of $C^\infty(G)$. In [15], Bidegain and Pinczon have built a differential preferred deformation of $C^\infty(G)$ for any Poisson-Lie group (G,P) with G semisimple. They also show that any preferred deformation of $C^\infty(G)$ is automatically differential. Then Etingof and Kazhdan [39] proved that there exists a differential deformation of $C^\infty(G)$ for any Poisson-Lie group (G,P), deformation which is preferred in the quasi-triangular case.

The existence of a star-product on other classes of Poisson manifolds was studied by various authors (Omori-Maeda-Yoshioka [75], Tamarkin [84], Asin [9].)

In 1997, Kontsevich [59] gave a proof of the existence of a star-product on any Poisson manifold and gave an explicit formula for a star-product for any Poisson structure on $V = \mathbb{R}^m$. This appeared as a consequence of the proof of his formality theorem. Tamarkin [85] gave a version of the proof in the framework of the theory of operads.

2.4. Kontsevich star-product on V

Let M be a domain in $V = \mathbb{R}^m$ and P be any Poisson structure on M. Kontsevich builds a star-product on (M,P), where the star-product of two functions u and v is given in terms of some universal polydifferential operators applied to the coefficients of the bivector field P and to the functions u, v; the formula is invariant under affine transformations of $\mathbb{R}^m$ and the description of the kth cochain uses a special class G_k of oriented labelled graphs.

An (oriented) graph Γ is a pair (V_Γ, E_Γ) of two finite sets such that E_Γ is a subset of $V_\Gamma \times V_\Gamma$; elements of V_Γ are vertices of Γ, elements of E_Γ are edges of Γ. If $e = (v_1, v_2) \in E_\Gamma \subseteq V_\Gamma \times V_\Gamma$ is an edge it is said that e starts at v_1 and ends at v_2.

A graph Γ belongs to G_k if Γ has $k+2$ vertices $V_\Gamma = \{1,\ldots,k\} \sqcup \{L,R\}$ and $2k$ labelled edges (with no multiple edges and no edge of the form (v,v) for $v \in V_\Gamma$), $E_\Gamma = \{e_1^1, e_1^2, e_2^1, e_2^2, \ldots, e_k^1, e_k^2\}$ where e_j^1 and e_j^2 start at j.

To each labelled graph $\Gamma \in G_k$ and each skew-symmetric 2-tensor P is associated a bidifferential operator $C_\Gamma(P)$ on $M \subset \mathbb{R}^m$:

$$(C_\Gamma(P))(u,v) : = \sum_{I:E_\Gamma \longrightarrow \{1,\ldots,m\}} [\prod_{j=1}^{k} (\prod_{e \in E_\Gamma | e=(*,j)} \partial_{I(e)}) \; P^{I(e_j^1)I(e_j^2)}] \times (\prod_{e \in E_\Gamma | e=(*,L)} \partial_{I(e)}) \; u \times (\prod_{e \in E_\Gamma | e=(*,R)} \partial_{I(e)}) \; v.$$

A weight $w_\Gamma \in \mathbb{R}$ is associated with each graph $\Gamma \in G_k$. Denote by $\mathcal{H}_k$ the space of configurations of k numbered distinct points in the upper half plane $\mathcal{H} = \{z \in \mathbb{C} | \; Im(z) > 0\}$:

$$\mathcal{H}_k = \{(p_1, \ldots, p_k) \mid p_j \in \mathcal{H} \quad p_i \neq p_j \quad \text{for} \quad i \neq j\}.$$

For $p \in \mathcal{H}$, $q \in \mathcal{H} \sqcup \mathbb{R}$ define

$$\phi^h(p,q) = \frac{1}{2i} Log \left(\frac{(q-p)(\bar{q}-p)}{(q-\bar{p})(\bar{q}-\bar{p})} \right).$$

If $\Gamma \in G_k$ is a graph as above, assign a point $p_j \in \mathcal{H}$ to the vertex j for $1 \leq j \leq k$, point $0 \in \mathbb{R} \subset \mathbb{C}$ to the vertex L, and point $1 \in \mathbb{R} \subset \mathbb{C}$ to the vertex R. Every edge $e \in E_\Gamma$ defines an ordered pair (p,q) of points on $\mathcal{H} \sqcup \mathbb{R}$, thus a function $\phi_e^h = \phi^h(p,q)$ on $\mathcal{H}_k$ with values in $\mathbb{R}/2\pi\mathbb{Z}$.

The weight of Γ is defined by

$$w_\Gamma = \frac{1}{k!(2\pi)^{2k}} \int_{\mathcal{H}_k} \wedge_{i=1}^{k} (d\phi_{e_i^1}^h \wedge d\phi_{e_i^2}^h).$$

Theorem 7. (Kontsevich [59]) *Let P be a Poisson tensor on a domain of $\mathbb{R}^m$. Then*

$$u *_\mu v := \sum_{k=0}^{\infty} \mu^k \sum_{\Gamma \in G_k} w_\Gamma(C_\Gamma(P))(u,v)$$

defines a differential star-product.

Definition 8 *A star-product on $\mathbb{R}^m$ is* **universal** *if it is given, for any Poisson tensor P by*

$$u *_\mu v = uv + \sum_{n>0} \mu^n (C_n(P))(u,v) \quad \text{with} \quad C_n(P) = \sum_{\Gamma \in G} w'(\Gamma) C_\Gamma(P),$$

where the $w'(\Gamma)$ are scalars which are independent of P.

Proposition 9 *Two universal star-products $*_\mu$ and $*'_\mu$ on $\mathfrak{g}^*$ are always equivalent modulo a change of parameter. This means that there exists a series $T(\mu) = \mathrm{Id} + \sum_{r=1}^{\infty} \mu^r T_r$ of linear universal differential operators and a series $\mu' = \mu + \sum_{n>1} f_n \mu^n$ such that*

$$u *'_{\mu'} v = T^{-1}(\mu)(T(\mu)u *_\mu T(\mu)v).$$

This was first proven by Arnal et al. [3]; we give here an elementary proof of this fact.

Proof. Assume by induction that $*$ and $*'$ coincide at order $n-1$. Associativity relation at order n implies $\partial C_n(P) = \partial C'_n(P)$; thus $C_n(P) = C'_n(P) + \partial E_n(P) + A_n(P)$ where E_n is a universal differential operator and $A_n(P)$ is universal, skew-symmetric and of order $(1,1)$. Clearly, if A_n corresponds to a graph in G_k, it has $2k$ arrows and $k+2$ vertices, but since P is linear on $\mathfrak{g}^*$, at most one arrow can end at any of the first k vertices and since A_n is 1-differential in each argument, exactly one arrow ends at each of the last 2 vertices. Hence $2k \leq k+2$ so $k \leq 2$; but if $k=1$ the only graph yields the Poisson bracket of functions so this can be cancelled by a change of parameter and if $k=2$ the graph corresponds to a symmetric bidifferential operator, hence does not yield an $A_n(P)$. So we can assume that A_n vanishes but then the equivalence through $T(\mu) = \mathrm{Id} + \mu^n E_n(P)$ builds a star-product which is universal and coincide with $*$ at order n.

Remark 10 When a covariance condition or a homogeneity condition ($C_n(sP) = s^n C_n(P)$) is added, clearly there is no need for a change of parameter since the graph in G_1 can only arise in C_1 so two such universal star-products are always equivalent.

In particular, the star-product of Kontsevich and the standard CBH star-product on $\mathfrak{g}^*$, which in general are not the same, are equivalent. The equivalence is given by universal differential operators, hence combinations of wheels which are graphs consisting of $k+1$ points $1,\ldots,k$ and L, and $2k$ arrows: $(1,2),(2,3),\ldots,(k-1,k)$, $(k,1)$ and (j,L), $1 \leq j \leq k$. Such a wheel clearly vanishes when the Poisson structure corresponds to a nilpotent Lie algebra so there is only one covariant universal star-product on the dual of a nilpotent Lie algebra (this appears in Arnal [2] and Kathotia [58]).

The equivalence between Kontsevich and CBH star-product has been explicitly constructed (see Arnal et al. [3] and Dito [36]) and gives an integral formula for Kontsevich star-product:

$$u * v(\xi) = \int \tilde{u}(X)\tilde{v}(Y)\frac{F(X)F(Y)}{F(CBH(X,Y))}\exp(2\pi i\langle \xi, CBH(X,Y)\rangle)dX\,dY,$$

where F is some formal function on $\mathfrak{g}$ written as a sum of products of traces of powers of $\mathrm{ad}(X)$.

This star-product has been used recently by Andler, Dvorsky and Sahi [1] to establish a conjecture of Kashiwara and Vergne which, in turn, gives a new proof of Duflo's result on the local solvability of bi-invariant differential operators on a Lie group.

3. Equivalence of star-products

Definition 11 *Two star-products $*$ and $*'$ on (M,P) are said to be* ***equivalent*** *if there is a series*

$$T = \mathrm{Id} + \sum_{r=1}^{\infty} \mu^r T_r,$$

where the T_r are linear operators on $C^\infty(M)$, such that

$$T(f*g) = Tf *' Tg. \tag{3.1}$$

Remark that the T_r automatically vanish on constants since 1 is a unit for $*$ and for $*'$. Using in a similar way linear operators which do not necessarily vanish on constants, one can pass from any associative deformation of the product of functions on a Poisson manifold (M,P) to another such deformation with 1 being a unit.

Definition 12 *A* ***Poisson deformation*** *of the Poisson bracket on a Poisson manifold (M,P) is a Lie algebra deformation of $(C^\infty(M),\{\ ,\ \})$ which is a derivation in each argument, i.e., of the form $\{u,v\}_\mu = P_\mu(du,dv)$ where $P_\mu = P + \sum \mu^k P_k$ is a series of skew-symmetric contravariant 2-tensors on M (such that $[P_\mu,P_\mu]=0$).*

Two Poisson deformations P_μ and P'_μ of the Poisson bracket P on a Poisson manifold (M,P) are ***equivalent*** *if there exists a formal path in the diffeomorphism group of M, starting at the identity, i.e., a series $T = \exp D = \mathrm{Id} + \sum_j \frac{1}{j!}D^j$ for $D = \sum_{r\geq 1} \mu^r D_r$ where the D_r are vector fields on M, such that*

$$T\{u,v\}_\mu = \{Tu,Tv\}'_\mu$$

where $\{u,v\}_\mu = P_\mu(du,dv)$ and $\{u,v\}'_\mu = P'_\mu(du,dv)$.

In the general theory of deformations, Gerstenhaber [49] showed how equivalence is linked to some second cohomology space.

For symplectic manifolds, Flato, Lichnerowicz, and Sternheimer in 1974 studied 1-differential deformations of the Poisson bracket [47]; it follows from their work, and appears in Lecomte [60], that:

Proposition 13 *The equivalence classes of Poisson deformations of the Poisson bracket P on a symplectic manifold (M,ω) are parametrized by $H^2(M;\mathbb{R})[[\mu]]$.*

Indeed, one first shows that any Poisson deformation P_μ of the Poisson bracket P on a symplectic manifold (M,ω) is of the form P^Ω for a series $\Omega = \omega + \sum_{k\geq 1}\mu^k\omega_k$ where the ω_k are closed 2-forms, and $P^\Omega(du,dv) = -\Omega(X_u^\Omega, X_v^\Omega)$ where $X_u^\Omega = X_u + \mu(\dots) \in \Gamma(TM)[[\mu]]$ is the element defined by $i(X_u^\Omega)\Omega = du$.

> Observe that for any series Ω as above the series of 1-differential 2-cochains P^Ω satisfies $[P^\Omega, P^\Omega] = 0$ because Ω is a closed 2-form, so defines indeed a Poisson deformation of P.
>
> Conversely, given a 1-differential deformation $P_\mu = P + \sum_j \mu^j P_j$, assume it coincides up to order k ($k \geq 0$) with P^{Ω_k} for some $\Omega_k = \omega + \sum_{j\geq 1}^k \mu^j\omega_j$ then $[P_\mu, P_\mu] - [P^{\Omega_k}, P^{\Omega_k}] = 0$ imply at order $k+1$ that $[P, P_{k+1} - (P^{\Omega_k})_{k+1} = 0]$ so that there exists a closed 2-form ω_{k+1} such that $(P_{k+1} - (P^{\Omega_k})_{k+1})(du,dv) = \omega_{k+1}(X_u, X_v)$. Hence P_μ coincides up to order $k+1$ with $P^{\Omega_{k+1}}$ where $\Omega_{k+1} = \omega + \sum_{j\geq 1}^{k+1}\mu^j\omega_j$. By induction, any Poisson deformation of P is of the form P^Ω for a series $\Omega = \omega + \sum_{k\geq 1}\mu^k\omega_k$ where the ω_k are closed 2-forms.

One then shows that two Poisson deformations P^Ω and $P^{\Omega'}$ are equivalent if and only if ω_k and ω_k' are cohomologous for all $k \geq 1$.

> If Ω and Ω' coincide to order $k-1$ and $\omega_k - \omega_k' = dF_k$, then $\{u,v\}_\mu$ and $T^{-1}\{Tu, Tv\}'_\mu$, with $T = \exp D$, $Du = \mu^{k-1}F_k(X_u)$, correspond to forms which coincide to order k so one has equivalence if all forms are cohomologous.
>
> Conversely, if P^Ω and $P^{\Omega'}$ are equivalent and Ω and Ω' coincide to order $k-1$, then the equivalence can be written $T = \exp D$, $Du = \sum_{j\geq k}\mu^j F_j(X_u)$, and $\omega_k - \omega_k' = dF_k$. Indeed if $Tu = u + \mu^r F_r(X_u) + \cdots$, then the relation $P^\Omega(du,dv) = T^{-1}P^{\Omega'}(dTu, dTv)$ at order $r \leq k$ yields $\omega_r - \omega_r' = dF_r$. When $r < k$ this means $dF_r = 0$; locally, on a contractible set U, $F_{r|U} = df_U$. The map $D_r : u \mapsto P^{\Omega'}(df_U, u)$ is globally defined and is a derivation of $\{\ ,\ \}'$ so $T' = \exp -\mu^r D_r \circ T = \mathrm{Id} + \mu^{r+1}F_{r+1} + \cdots$ is still an equivalence. By induction, one gets the result.

In 1978, Bayen, Flato, Fronsdal, Lichnerowicz, and Sternheimer [11] stressed that different orderings in physics lead to equivalent star-products on $\mathbb{R}^{2n}$. This shows that the notion of mathematical equivalence is not the same as the notion of physical equivalence (i.e., two star-products leading to the same spectrum for each observable); we studied this difference with Cahen, Flato, and Sternheimer in [21]. Bayen et al. also proved in [10] that the Moyal star-product is the only one whose cochains are given by polynomials in the Poisson structure P; this was the first consideration of some universality property to build and classify star-products.

Recall that a star-product $*$ on (M,ω) is called differential if the 2-cochains $C_r(u,v)$ giving it are bidifferential operators. As was observed by Lichnerowicz [63] and Deligne [31]:

Proposition 14 *If $*$ and $*'$ are differential star-products and $T(u)=\sum_{r\geq 0}\mu^r T_r(u)$ is an equivalence so that $T(u*v)=T(u)*'T(v)$ then the T_r are differential operators.*

Proof. Indeed if $T=\mathrm{Id}+\mu^k T_k+\ldots$ then $\partial T_k=C'_k-C_k$ is differential so C'_k-C_k is a differential 2-cocycle with vanishing skew-symmetric part but then, using Vey's formula, it is the coboundary of a differential 1-cochain E and T_k-E, being a 1-cocycle, is a vector field so T_k is differential. One then proceeds by induction, considering $T'=(\mathrm{Id}+\mu^k T_k)^{-1}\circ T=\mathrm{Id}+\mu^{k+1}T'_{k+1}+\cdots$ and the two differential star-products $*$ and $*''$, where $u*''v=(\mathrm{Id}+\mu^k T_k)^{-1}((\mathrm{Id}+\mu^k T_k)u*'(\mathrm{Id}+\mu^k T_k)v)$, which are equivalent through T' (i.e., $T'(u*v)=T'(u)*''T'(v)$).

A differential star-product is equivalent to one with linear term in μ given by $\frac{1}{2}\{u,v\}$. Indeed $C_1(u,v)$ is a Hochschild cocycle with antisymmetric part given by $\frac{1}{2}\{u,v\}$ so $C_1=\frac{1}{2}P+\partial B$ for a differential 1-cochain B. Setting $T(u)=u+\mu B(u)$ and $u*'v=T(T^{-1}(u)*T^{-1}(v))$, this equivalent star-product $*'$ has the required form.

In 1979, we proved [50] that all differential deformed brackets on $\mathbb{R}^{2n}$ (or on any symplectic manifold such that $b_2=0$) are equivalent modulo a change of the parameter, and this implies a similar result for star-products; this was proven by direct methods by Lichnerowicz [62]:

Proposition 15 *Let $*$ and $*'$ be two differential star-products on (M,ω) and suppose that $H^2(M;\mathbb{R})=0$. Then there exists a local equivalence $T=\mathrm{Id}+\sum_{k\geq 1}\mu^k T_k$ on $C^\infty(M)[\![\mu]\!]$ such that $u*'v=T(T^{-1}u*T^{-1}v)$ for all $u,v\in C^\infty(M)[\![\mu]\!]$.*

Proof. Let us suppose that, modulo some equivalence, the two star-products $*$ and $*'$ coincide up to order k. Then associativity at order k shows that $C_k-C'_k$ is a Hochschild 2-cocycle and so by (4) can be written as $(C_k-C'_k)(u,v)=(\partial B)(u,v)+A(X_u,X_v)$ for a 2-form A. The total skew-symmetrization of the associativity relation at order $k+1$ shows that A is a closed 2-form. Since the second cohomology vanishes, A is exact, $A=dF$. Transforming by the equivalence defined by $Tu=$

$u+\mu^{k-1}2F(X_u)$, we can assume that the skew-symmetric part of $C_k - C'_k$ vanishes. Then $C_k - C'_k = \partial B$ where B is a differential operator. Using the equivalence defined by $T = I + \mu^k B$ we can assume that the star-products coincide, modulo an equivalence, up to order $k+1$ and the result follows from induction since two star-products always agree in their leading term.

It followed from the above proof and results similar to [50] (i.e., two star-products which are equivalent and coincide at order k differ at order $k+1$ by a Hochschild 2-cocycle whose skew-symmetric part corresponds to an exact 2-form) that at each step in μ, equivalence classes of differential star-products on a symplectic manifold (M,ω) are parametrized by $H^2(M;\mathbb{R})$, if all such deformations exist. The general existence was proven by De Wilde and Lecomte. At that time, one assumed the parity condition $C_n(u,v) = (-1)^n C_n(v,u)$, so equivalence classes of such differential star-products were parametrized by series $H^2(M;\mathbb{R})[[\mu^2]]$. The parametrization was not canonical.

In 1994, Fedosov proved that his recursive construction works in a more general setting: Given any series of closed 2-forms on a symplectic manifold (M,ω), he could build a connection on the Weyl bundle whose curvature is linked to that series and a star-product whose equivalence class only depends on the element in $H^2(M;\mathbb{R})[[\mu]]$ corresponding to that series of forms.

In 1995, Nest and Tsygan [69], then Deligne [31] and Bertelson [13] proved that any differential star-product on a symplectic manifold (M,ω) is equivalent to a Fedosov star-product and that its equivalence class is parametrized by the corresponding element in $H^2(M;\mathbb{R})[[\mu]]$.

In 1997, Kontsevich [59] proved that the coincidence of the set of equivalence classes of star and Poisson deformations is true for general Poisson manifolds:

Theorem 16 *The set of equivalence classes of differential star-products on a Poisson manifold (M,P) can be naturally identified with the set of equivalence classes of Poisson deformations of P:*

$$P_\mu = P\mu + P_2\mu^2 + \cdots \in \Gamma(X,\wedge^2 T_X)[[\mu]], \quad [P_\mu, P_\mu] = 0.$$

Remark that all results concerning parametrization of equivalence classes of differential star-products are still valid for star-products defined by local cochains or for star-products defined by continuous cochains ([53], Pinczon [77]). Parametrization of equivalence classes of special star-products have been obtained: star-products with separation of variables (by Karabegov [56]), invariant star-products on a symplectic manifold when there exists an invariant symplectic connection (with Bertelson and Bieliavsky [14]), algebraic star-products (Chloup [28], Kontsevich [59]).

The association of an element in $H^2(M;\mathbb{R})[[\mu]]$ to the equivalence class of a star-product on a symplectic manifold is one way to associate an invariant to a star-product; other such associations are obtained by star version of index theorems and trace functionals on the algebra $(C^\infty(M)[[\mu]],*)$. I shall not develop that aspect here. It was first considered by Connes, Flato, and Sternheimer in [30] where they introduce the notion of closed star-product, i.e., such that

$$\int_M a*b\,\omega^n = \int_M b*a\,\omega^n \pmod{\mu^n}\ \forall a,b\in C^\infty(M)[[\mu]],$$

and show how their classification is linked to cyclic cohomology.

They obtain, for the cotangent bundle to a compact Riemannian manifold, its Todd class as the "character" associated to the star-product corresponding to normal ordering.

The notion of a trace for star-products and the star version of index theorems have been studied by Fedosov [41, 42, 43], and by Nest and Tsygan [69, 70].

3.1. Deligne's cohomology classes associated to differential star-products on symplectic manifolds

Deligne defines two cohomological classes associated to differential star-products on a symplectic manifold. This leads to an intrinsic way to parametrize the equivalence class of such a differential star-product. Although the question makes sense more generally for Poisson manifolds, Deligne's method depends crucially on the Darboux theorem and the uniqueness of the Moyal star-product on $\mathbb{R}^{2n}$ so the methods do not extend to general Poisson manifolds.

The first class is a relative class; fixing a star-product on the manifold, it intrinsically associates to any equivalence class of star-products an element in $H^2(M;\mathbb{R})[\![\mu]\!]$. This is done in Čech cohomology by looking at the obstruction to gluing local equivalences.

Deligne's second class is built from special local derivations of a star-product. The same derivations played a special role in the first general existence theorem [33] for a star-product on a symplectic manifold. Deligne used some properties of Fedosov's construction and central curvature class to relate his two classes and to see how to characterize an equivalence class of star-products by the derivation related class and some extra data obtained from the second term in the deformation. With John Rawnsley [54], we did this by direct Čech methods which I shall present here.

3.1.1. *The relative class*

Let $*$ and $*'$ be two differential star-products on (M,ω). Let U be a contractible open subset of M and $N_U = C^\infty(U)$. Remark that any differential star-product on M restricts to U and $H^2(M;\mathbb{R})(U)=0$, hence, by Proposition 15, there exists a local equivalence $T=\mathrm{Id}+\sum_{k\geq 1}\mu^k T_k$ on $N_U[\![\mu]\!]$ so that $u*'v = T(T^{-1}u * T^{-1}v)$ for all $u,v\in N_U[\![\mu]\!]$.

Proposition 17 *Let $*$ be a differential star-product on (M,ω) and suppose that $H^1(M;\mathbb{R})$ vanishes.*

– *Any self-equivalence* $A = \mathrm{Id} + \sum_{k\geq 1} \mu^k A_k$ *of* $*$ *is inner:* $A = \exp \mathrm{ad}_* a$ *for some* $a \in C^\infty(M)[[\mu]]$.

– *Any* μ*-linear derivation of* $*$ *is of the form* $D = \sum_{i\geq 0} \mu^i D_i$ *where each* D_i *corresponds to a symplectic vector field* X_i *and is given on a contractible open set* U *by*

$$D_i u|_U = \frac{1}{\mu}\left(f_i^U * u - u * f_i^U\right)$$

if $X_i u|_U = \{f_i^U, u\}|_U$.

Indeed, one builds a recursively; assuming $A = \mathrm{Id} + \sum_{r\geq k} \mu^r A_r$ and $k \geq 1$, the condition $A(u*v) = Au * Av$ implies at order k in μ that $A_k(uv) + C_k(u,v) = A_k(u)v + uA_k(v) + C_k(u,v)$ so that A_k is a vector field. Taking the skew part of the terms in μ^{k+1} we have that A_k is a derivation of the Poisson bracket. Since $H^1(M;\mathbb{R}) = 0$, one can write $A_k(u) = \{a_{k-1}, u\}$ for some function a_{k-1}. Then $(\exp - \mathrm{ad}_* \mu^{k-1} a_{k-1}) \circ A = \mathrm{Id} + O(\mu^{k+1})$ and the induction proceeds. The proof for μ-linear derivation is similar.

The above results can be applied to the restriction of a differential star-product on (M,ω) to a contractible open set U. Set, as above, $N_U = C^\infty(U)$. If $A = \mathrm{Id} + \sum_{k\geq 1} \mu^k A_k$ is a formal linear operator on $N_U[[\mu]]$ which preserves the differential star-product $*$, then there is $a \in N_U[[\mu]]$ with $A = \exp \mathrm{ad}_* a$. Similarly, any local μ-linear derivation D_U of $*$ on $N_U[[\mu]]$ is essentially inner: $D_U = \frac{1}{\mu} \mathrm{ad}_* d_U$ for some $d_U \in N_U[[\mu]]$.

It is convenient to write the composition of automorphisms of the form $\exp \mathrm{ad}_* a$ in terms of a. In a pronilpotent situation this is done with the **Campbell–Baker–Hausdorff composition** which is denoted by $a \circ_* b$:

$$a \circ_* b = a + \int_0^1 \psi(\exp \mathrm{ad}_* a \circ \exp t\, \mathrm{ad}_* b) b\, dt$$

where

$$\psi(z) = \frac{z \log(z)}{z-1} = \sum_{n\geq 1} \left(\frac{(-1)^n}{n+1} + \frac{(-1)^{n+1}}{n}\right)(z-1)^n.$$

Notice that the formula is well-defined (at any given order in μ, only a finite number of terms arise) and it is given by the usual series

$$a \circ_* b = a + b + \frac{1}{2}[a,b]_* + \frac{1}{12}([a,[a,b]_*]_* + [b,[b,a]_*]_*) \cdots .$$

The following results are standard (N. Bourbaki, Groupes et algèbres de Lie, *Éléments de Mathématique*, Livre 9, Chapitre 2, §6):

– $\circ_*$ is an associative composition law;

– $\exp \mathrm{ad}_*(a \circ_* b) = \exp \mathrm{ad}_* a \circ \exp \mathrm{ad}_* b$;

- $a \circ_* b \circ_* (-a) = \exp(\mathrm{ad}_* a)\, b$;
- $-(a \circ_* b) = (-b) \circ_* (-a)$;
- $\left.\frac{d}{dt}\right|_0 (-a) \circ_* (a+tb) = \frac{1-\exp(-\mathrm{ad}_* a)}{\mathrm{ad}_* a}(b)$.

Let (M,ω) be a symplectic manifold. We fix a locally finite open cover $\mathscr{U} = \{U_\alpha\}_{\alpha\in I}$ by Darboux coordinate charts such that the U_α and all their non-empty intersections are contractible, and we fix a partition of unity $\{\theta_\alpha\}_{\alpha\in I}$ subordinate to $\mathscr{U}$. Set $N_\alpha = C^\infty(U_\alpha)$, $N_{\alpha\beta} = C^\infty(U_\alpha \cap U_\beta)$, and so on.

Now suppose that $*$ and $*'$ are two differential star-products on (M,ω). We have seen that their restrictions to $N_\alpha[\![\mu]\!]$ are equivalent so there exist formal differential operators $T_\alpha \colon N_\alpha[\![\mu]\!] \to N_\alpha[\![\mu]\!]$ such that

$$T_\alpha(u * v) = T_\alpha(u) *' T_\alpha(v), \qquad u,v \in N_\alpha[\![\mu]\!].$$

On $U_\alpha \cap U_\beta$, $T_\beta^{-1} \circ T_\alpha$ will be a self-equivalence of $*$ on $N_{\alpha\beta}[\![\mu]\!]$ and so there will be elements $t_{\beta\alpha} = -t_{\alpha\beta}$ in $N_{\alpha\beta}[\![\mu]\!]$ with

$$T_\beta^{-1} \circ T_\alpha = \exp \mathrm{ad}_* t_{\beta\alpha}.$$

On $U_\alpha \cap U_\beta \cap U_\gamma$ the element

$$t_{\gamma\beta\alpha} = t_{\alpha\gamma} \circ_* t_{\gamma\beta} \circ_* t_{\beta\alpha}$$

induces the identity automorphism and hence is in the center $\mathbb{R}[\![\mu]\!]$ of $N_{\alpha\beta\gamma}[\![\mu]\!]$. The family of $t_{\gamma\beta\alpha}$ is thus a Čech 2-cocycle for the covering $\mathscr{U}$ with values in $\mathbb{R}[\![\mu]\!]$. The standard arguments show that its class does not depend on the choices made, and is compatible with refinements. Since every open cover has a refinement of the kind considered it follows that $t_{\gamma\beta\alpha}$ determines a unique Čech cohomology class $[t_{\gamma\beta\alpha}] \in H^2(M;\mathbb{R})[\![\mu]\!]$.

Definition 18 $t(*',*) = [t_{\gamma\beta\alpha}] \in H^2(M;\mathbb{R})[\![\mu]\!]$ *is* ***Deligne's relative class.***

It is easy to see, using the fact that the cohomology of the sheaf of smooth functions is trivial:

Theorem 19. (Deligne) *Fixing a differential star-product $*$, the class $t(*',*)$ in $H^2(M;\mathbb{R})[\![\mu]\!]$ depends only on the equivalence class of the differential star-product $*'$, and sets up a bijection between the set of equivalence classes of differential star-products and $H^2(M;\mathbb{R})[\![\mu]\!]$.*

If $*$, $*'$, $*''$ are three differential star-products on (M,ω) then

$$t(*'',*) = t(*'',*') + t(*',*). \tag{3.2}$$

3.1.2. *The derivation related class*

The addition formula above suggests that $t(*',*)$ should be a difference of classes $c(*'),c(*)\in H^2(M;\mathbb{R})[\![\mu]\!]$. Moreover, the class $c(*)$ should determine the star-product $*$ up to equivalence.

Definition 20 *Let U be an open set of M. Say that a derivation D of $(C^\infty(U)[\![\mu]\!],*)$ is **μ-Euler** if it has the form*

$$D=\mu\frac{\partial}{\partial\mu}+X+D', \tag{3.3}$$

where X is conformally symplectic on U ($\mathscr{L}_X\omega|_U=\omega|_U$) and $D'=\sum_{r\geq 1}\mu^r D'_r$ with the D'_r differential operators on U.

Proposition 21 *Let $*$ be a differential star-product on (M,ω). For each $U_\alpha\in\mathscr{U}$ there exists a μ-Euler derivation $D_\alpha=\mu\frac{\partial}{\partial\mu}+X_\alpha+D'_\alpha$ of the algebra $(N_\alpha[\![\mu]\!],*)$.*

Proof. On an open set in $\mathbb{R}^{2n}$ with the standard symplectic structure Ω, denote the Poisson bracket by P. Let X be a conformal vector field so $\mathscr{L}_X\Omega=\Omega$. The Moyal star-product $*_M$ is given by $u*_M v=uv+\sum_{r\geq 1}(\frac{\mu}{2})^r/r!P^r(u,v)$ and $D=\mu\frac{\partial}{\partial\mu}+X$ is a derivation of $*_M$.

Now (U_α,ω) is symplectomorphic to an open set in $\mathbb{R}^{2n}$ and any differential star-product on this open set is equivalent to $*_M$ so we can pull back D and $*_M$ to U_α by a symplectomorphism to give a star-product $*'$ with a derivation of the form $\mu\frac{\partial}{\partial\mu}+X_\alpha$. If T is an equivalence of $*$ with $*'$ on U_α then $D_\alpha=T^{-1}\circ(\mu\frac{\partial}{\partial\mu}+X_\alpha)\circ T$ is a derivation of the required form.

We take such a collection of derivations D_α given by Proposition 21 and on $U_\alpha\cap U_\beta$ we consider the differences $D_\beta-D_\alpha$. They are derivations of $*$ and the μ derivatives cancel out, so $D_\beta-D_\alpha$ is a μ-linear derivation of $N_{\alpha\beta}[\![\mu]\!]$. Any μ-linear derivation is of the form $\frac{1}{\mu}\,\mathrm{ad}_*d$, so there are $d_{\beta\alpha}\in N_{\alpha\beta}[\![\mu]\!]$ with

$$D_\beta-D_\alpha=\frac{1}{\mu}\,\mathrm{ad}_*\,d_{\beta\alpha}, \tag{3.4}$$

with $d_{\beta\alpha}$ unique up to a central element. On $U_\alpha\cap U_\beta\cap U_\gamma$ the combination $d_{\alpha\gamma}+d_{\gamma\beta}+d_{\beta\alpha}$ must be central and hence defines $d_{\gamma\beta\alpha}\in\mathbb{R}[\![\mu]\!]$. It is easy to see that $d_{\gamma\beta\alpha}$ is a 2-cocycle whose Čech class $[d_{\gamma\beta\alpha}]\in H^2(M;\mathbb{R})[\![\mu]\!]$ does not depend on any of the choices made.

Definition 22 $d(*)=[d_{\gamma\beta\alpha}]\in H^2(M;\mathbb{R})[\![\mu]\!]$ *is **Deligne's intrinsic derivation related class.***

- In fact the class considered by Deligne is actually $\frac{1}{\mu}d(*)$. A purely Čech-theoretic accounts of this class is given in Karabegov [56].
- If $*$ and $*'$ are equivalent then $d(*') = d(*)$;
- If $d(*) = \sum_{r\geq 0}\mu^r d^r(*)$ then $d^0(*) = [\omega]$ under the de Rham isomorphism, and $d^1(*) = 0$;

Consider two differential star-products $*$ and $*'$ on (M,ω) with local equivalences T_α and local μ-Euler derivations D_α for $*$. Then $D'_\alpha = T_\alpha \circ D_\alpha \circ T_\alpha^{-1}$ are local μ-Euler derivations for $*'$. Let $D_\beta - D_\alpha = \frac{1}{\mu}\,\mathrm{ad}_* d_{\beta\alpha}$ and $T_\beta^{-1}\circ T_\alpha = \exp \mathrm{ad}_* t_{\beta\alpha}$ on $U_\alpha \cap U_\beta$. Then $D'_\beta - D'_\alpha = \frac{1}{\mu}\mathrm{ad}_{*'} d'_{\beta\alpha}$ where

$$d'_{\beta\alpha} = T_\beta d_{\beta\alpha} - \mu T_\beta \circ \left(\frac{1-\exp(-\,\mathrm{ad}_* t_{\alpha\beta})}{\mathrm{ad}_* t_{\alpha\beta}}\right) \circ D_\alpha t_{\alpha\beta}.$$

In this situation

$$d'_{\gamma\beta\alpha} = T_\alpha(d_{\gamma\beta\alpha} + \mu^2\frac{\partial}{\partial\mu}t_{\gamma\beta\alpha}).$$

This gives a direct proof of:

Theorem 23. (Deligne) *The relative class and the intrinsic derivation related classes of two differential star-products $*$ and $*'$ are related by*

$$\mu^2\frac{\partial}{\partial\mu}t(*',*) = d(*') - d(*). \tag{3.5}$$

3.1.3. *The characteristic class*

The formula above shows that the information which is "lost" in $d(*') - d(*)$ corresponds to the zeroth order term in μ of $t(*',*)$.

Remark 24 In [51, 35] it was shown that any bidifferential operator C, vanishing on constants, which is a 2-cocycle for the Chevalley cohomology of $(C^\infty(M),\{\ ,\ \})$ with values in $C^\infty(M)$ associated to the adjoint representation (i.e., such that $S_{u,v,w}[\{u,C(v,w)\} - C(\{u,v\},w)] = 0$ where $S_{u,v,w}$ denotes the sum over cyclic permutations of u,v and w) can be written as

$$C(u,v) = aS^3_\Gamma(u,v) + A(X_u,X_v) + [\{u,Ev\} + \{Eu,v\} - E(\{u,v\})]$$

where $a\in\mathbb{R}$, where S^3_Γ is a bidifferential 2-cocycle introduced in [10] (which vanishes on constants and is never a coboundary and whose symbol is of order 3 in each argument), where A is a closed 2-form on M and where E is a differential operator vanishing on constants. Hence

$$H^2_{\mathrm{Chev,nc}}(C^\infty(M),C^\infty(M)) = \mathbb{R}\oplus H^2(M;\mathbb{R})$$

and we define the # operator as the projection on the second factor relative to this decomposition.

Proposition 25 *Given two differential star-products $*$ and $*'$, the term of order zero in Deligne's relative class $t(*',*) = \sum_{r\geq 0}\mu^r t^r(*',*)$ is given by*

$$t^0(*',*) = -2(C'^-_2)^\# + 2(C^-_2)^\#.$$

If $C_1 = \frac{1}{2}\{\,,\,\}$, then $C^-_2(u,v) = A(X_u,X_v)$ where A is a closed 2-form and $(C^-_2)^\# = [A]$ so it "is" the skew-symmetric part of C_2.

It follows from what we did before that the association to a differential star-product of $(C^-_2)^\#$ and $d(*)$ completely determines its equivalence class.

Definition 26 *The **characteristic class** of a differential star-product $*$ on (M,ω) is the element $c(*)$ of the affine space $\frac{-[\omega]}{\mu} + H^2(M;\mathbb{R})[[\mu]]$ defined by*

$$c(*)^0 = -2(C^-_2)^\#,$$
$$\frac{\partial}{\partial\mu}c(*)(\mu) = \frac{1}{\mu^2}d(*).$$

Theorem 27 *The characteristic class has the following properties:*

– *The relative class is given by*

$$t(*',*) = c(*') - c(*); \tag{3.6}$$

– *The map C from equivalence classes of star-products on (M,ω) to the affine space $\frac{-[\omega]}{\mu} + H^2(M;\mathbb{R})[[\mu]]$ mapping $[*]$ to $c(*)$ is a bijection.*

– *If $\psi\colon M \to M'$ is a diffeomorphism and if $*$ is a star-product on (M,ω) then $u *' v = (\psi^{-1})^*(\psi^* u * \psi^* v)$ defines a star-product denoted $*' = (\psi^{-1})^* *$ on (M',ω') where $\omega' = (\psi^{-1})^*\omega$. The characteristic class is natural relative to diffeomorphisms:*

$$c((\psi^{-1})^* *) = (\psi^{-1})^* c(*). \tag{3.7}$$

– *Consider a change of parameter $f(\mu) = \sum_{r\geq 1}\mu^r f_r$ where $f_r \in \mathbb{R}$ and $f_1 \neq 0$ and let $*'$ be the star-product obtained from $*$ by this change of parameter, i.e., $u *' v = u.v + \sum_{r\geq 1}(f(\mu))^r C_r(u,v) = u.v + f_1\mu C_1(u,v) + \mu^2((f_1)^2 C_2(u,v) + f_2 C_1(u,v)) + \cdots$. Then $*'$ is a differential star-product on (M,ω') where $\omega' = \frac{1}{f_1}\omega$ and we have equivariance under a change of parameter:*

$$c(*')(\mu) = c(*)(f(\mu)). \tag{3.8}$$

The characteristic class $c(*)$ coincides (cf. Deligne [31] and Neumaier [71]) for Fedosov-type star-products with their characteristic class introduced by Fedosov as the de Rham class of the curvature of the generalized connection used to build them (up to a sign and factors of 2). That characteristic class is also studied by

Weinstein and Xu in [88]. The fact that $d(*)$ and $(C_2^-)^\#$ completely characterize the equivalence class of a star-product is also proven by Čech methods in De Wilde [32].

3.2. Automorphisms of a star-product and generalized moment map

The above proposition allows to study automorphisms of star-products on a symplectic manifold ([78], [54]).

Definition 28 *An* ***isomorphism*** *from a differential star-product $*$ on (M,ω) to a differential star-product $*'$ on (M',ω') is an $\mathbb{R}$-linear bijective map $A\colon C^\infty(M)[[\mu]] \to C^\infty(M')[[\mu]]$, continuous in the μ-adic topology (i.e., $A(\sum_r \mu^r u_r)$ is the limit of $\sum_{r\leq N} A(\mu^r u_r)$), such that*

$$A(u*v) = Au*'Av.$$

Notice that if A is such an isomorphism, then $A(\mu)$ is central for $*'$ so that $A(\mu) = f(\mu)$ where $f(\mu) \in \mathbb{R}[[\mu]]$ is without constant term to get the μ-adic continuity. Let us denote by $*''$ the differential star-product on $(M, \omega_1 = \frac{1}{f_1}\omega)$ obtained by a change of parameter

$$u \overset{''}{*_\mu} v = u *_{f(\mu)} v = F(F^{-1}u * F^{-1}v),$$

where $F\colon C^\infty(M)[[\mu]] \to C^\infty(M)[[\mu]]$ is defined by $\sum_r \mu^r u_r \mapsto \sum_r f(\mu)^r u_r$. Define $A' : C^\infty(M)[[\mu]] \to C^\infty(M')[[\mu]]$ by $A = A' \circ F$. Then A' is a μ-linear isomorphism between $*''$ and $*'$:

$$A'(u*''v) = A'u*'A'v.$$

At order zero in μ this yields $A'_0(u.v) = A'_0u.A'_0v$ so that there exists a diffeomorphism $\psi : M' \to M$ with $A'_0u = \psi^*u$. The skew-symmetric part of the isomorphism relation at order 1 in μ implies that $\psi^*\omega_1 = \omega'$. Let us denote by $*'''$ the differential star-product on (M, ω_1) obtained by pullback via ψ of $*'$:

$$u*'''v = (\psi^{-1})^*(\psi^*u*'\psi^*v)$$

and define $B : C^\infty(M)[[\mu]] \to C^\infty(M)[[\mu]]$ so that $A' = \psi^* \circ B$. Then B is μ-linear, starts with the identity and

$$B(u*''v) = Bu*'''Bv$$

so that B is an equivalence – in the usual sense – between $*''$ and $*'''$. Hence [54]

Proposition 29 *Any isomorphism between two differential star-products on symplectic manifolds is the combination of a change of parameter and a μ-linear isomorphism. Any μ-linear isomorphism between two star-products $*$ on (M,ω) and $*'$ on (M',ω') is the combination of the action on functions of a symplectomorphism $\psi : M' \to M$ and an equivalence between $*$ and the pullback via ψ of $*'$. In particular, it exists if and only if those two star-products are equivalent, i.e., if and only if $(\psi^{-1})^*c(*') = c(*)$, where here $(\psi^{-1})^*$ denotes the action on the second de Rham cohomology space.*

In particular, two differential star-products $*$ on (M,ω) and $*'$ on (M',ω') are isomorphic if and only if there exist $f(\mu) = \sum_{r\geq 1}\mu^r f_r \in \mathbb{R}[\![\mu]\!]$ with $f_1 \neq 0$ and $\psi\colon M' \to M$, a symplectomorphism, such that $(\psi^{-1})^* c(*')(f(\mu)) = c(*)(\mu)$. In particular [50]: if $H^2(M;\mathbb{R}) = \mathbb{R}[\omega]$ then there is only one star-product up to equivalence and change of parameter.

Omori et al. [73] also show that when reparametrizations are allowed then there is only one star-product on $\mathbb{C}P^n$.

A special case of Proposition 29 gives:

Proposition 30 *A symplectomorphism ψ of a symplectic manifold can be extended to a μ-linear automorphism of a given differential star-product on (M,ω) if and only if $(\psi)^* c(*) = c(*)$.*

Notice that this is always the case if ψ can be connected to the identity by a path of symplectomorphisms (and this result was in Fedosov [41]).

If G is a connected Lie group acting on the symplectic manifold (M,ω) by symplectomorphisms, each element of G can be lifted to an automorphism of a star-product on M. The group G acts on the quantum level if there is a homomorphism ρ from G into the automorphism group of $*$ such that $\rho(g) = g^* + \mu\ldots\ \forall g \in G$. At the Lie algebra level, one considers a homomorphism σ from the Lie algebra $\mathfrak{g}$ of G to the algebra of derivations of $*$ such that $\sigma(X) = X^* + \mu\ldots\ \forall X \in \mathfrak{g}$ where X^* is the fundamental vector field on M associated to the action of G on M (i.e., $X^*_x = \frac{d}{dt}\exp -tX\cdot x_{|_0}$). Now any local μ-linear derivation of $*$ can be locally written on a contractible set U as $\mathrm{ad}_*(\frac{1}{\mu}f)$ for some $f \in C^\infty(U)[[\mu]]$.

Definition 31 *Given a star-product $*$ on a Poisson manifold (M,P) and given a connected Lie group G acting on M, the star-product gives* **a quantization of the action of the Lie algebra $\mathfrak{g}$** *if there exists a* **generalized moment map** *i.e., a map*

$$\tau : \mathfrak{g} \to \frac{1}{\mu}C^\infty(M)[[\mu]] : X \mapsto \tau(X)$$

such that

$$\tau(X) * \tau(Y) - \tau(Y) * \tau(X) = \tau([X,Y]), \quad \forall X,Y \in \mathfrak{g} \tag{3.9}$$

and

$$\tau(X) * u - u * \tau(X) = X^* u + \mu\cdots, \quad \forall X \in \mathfrak{g}. \tag{3.10}$$

Remark that the two conditions imply for the term of τ of order -1 in μ that the action of G on M admits a moment map, i.e., there is a map $\lambda : \mathfrak{g} \to C^\infty(M)$ such that $\{\lambda(Y),u\} = Y^* u$ and $\{\lambda(Y),\lambda(Y)\} = \lambda(\{X,Y\})$, and

$$\tau(X) = \frac{1}{\mu}\lambda(X) + \cdots.$$

When one can choose $\tau(X) = \frac{1}{\mu}\lambda(X)$, $\forall X \in \mathfrak{g}$, the notion of a star-product which is a quantization of the action of $\mathfrak{g}$ coincides with the notion of a covariant star-product.

When one can choose τ so that $\tau(X) * u - u * \tau(X) = X^*u \ \ \forall X \in \mathfrak{g}$, it implies that $X^*(u*v) = (X^*u)*v + u*(X^*v)$, so that

$$g^*(u*v) = g^*u * g^*v \ \forall g \in G$$

which are the conditions in Bayen et al [10] for the star-product to be **geometrically invariant**; in that situation, the notion of generalized moment map coincides with the notion of quantum moment map introduced by Xu in [89].

In the general case, when one has found a map τ at order k in μ satisfying condition (3.9) at order k, then one can extend things one order further if a Chevalley 2-cocycle from $\mathfrak{g}$ with values in $C^\infty(M)$ (for the representation of $Y \in \mathfrak{g}$ given by Y^*) is a 2-coboundary. When one deals with geometrically invariant star-products, one can always find a τ such that $\tau(X)*u - u*\tau(X) = X^*u$, $\forall X \in \mathfrak{g}$, if $H^2(\mathfrak{g},\mathbb{R}) = 0$ and $H^1(M;\mathbb{R}) = 0$ as was found in [89].

4. Convergence of star-products

Remark 32 Let (M,P) be a Poisson manifold and let $*$ be a differential star-product on it with 1 acting as the identity. Observe that if there exists a value k of μ such that

$$u*v = \sum_{r=0}^{\infty} \mu^r C_r(u,v)$$

converges (for the pointwise convergence of functions), for all $u,v \in C^\infty(M)$, to $F_k(u,v)$ in such a way that F_k is associative, then $F_k(u,v) = uv$.

So assuming "too much" convergence kills all deformations. On the other hand, in any physical situation, one needs some convergence properties to be able to compute the spectrum of quantum observables in terms of a star-product as was done for some observables already in Bayen, Flato, Fronsdal, Lichnerowicz, and Sternheimer [11].

Now consider the example of the Moyal star-product on the symplectic vector space $(\mathbb{R}^{2n},\omega)$. The formal formula for the Moyal star-product

$$(u *_M v)(z) = \exp\left(\frac{\mu}{2} P^{rs} \partial_{x^r} \partial_{y^s}\right) (u(x)v(y))\Big|_{x=y=z}$$

obviously converges when u and v are polynomials.

On the other hand, there is an integral formula for Moyal star-product given by

$$(u*v)(\xi) = (\pi\hbar)^{-2n} \int u(\xi')v(\xi'')e^{\frac{2i}{\hbar}(\omega(\xi,\xi'')+\omega(\xi'',\xi')+\omega(\xi',\xi))} d\xi' d\xi'',$$

and this product $*$ gives a structure of associative algebra on the space of rapidly decreasing functions $\mathscr{S}(\mathbb{R}^{2n})$.

The formal formula converges (for $\mu = i\hbar$) in the topology of $\mathscr{S}'$ for u and v with compact Fourier transform.

With Michel Cahen and John Rawnsley, we used the method of quantization of Kähler manifolds due to Berezin, [12], as the inverse of taking symbols of operators, to construct on Hermitian symmetric spaces star-products which are convergent on a large class of functions on the manifold. I shall develop this construction in this last part of my talk.

Let me mention, before closing this introduction about convergence, the work of Rieffel [81] where he introduces the notion of strict deformation quantization. An example of strict Fréchet quantization has been recently given by Omori, Maeda, Miyazaki, and Yoshioka in [76].

Also very important are the constructions of operator representations of star-products, in particular the works of Fedosov [41] and of Bordemann, Neumaier, and Waldmann [18, 19].

4.1. Convergence of Berezin type star-products on Hermitian symmetric spaces

The method to construct a star-product involves making a correspondence between operators and functions (their Berezin symbols), transferring the operator composition to the symbols, introducing a suitable parameter into the Berezin composition of symbols, taking the asymptotic expansion in this parameter on a large algebra of functions and then showing that the coefficients of this expansion satisfy the cocycle conditions to define a star-product on the smooth functions. The idea of an asymptotic expansion appeared already in Berezin [12] and in Moreno and Ortega-Navarro [65, 66].

In [25] we show that this asymptotic expansion exists for compact M, and defines an associative multiplication on formal power series in k^{-1} with coefficients in $C^\infty(M)$ for compact coadjoint orbits. We also show that this formal power series converges on the space of symbols for M a Hermitian symmetric space of compact type.

In [55], Karabegov proves convergence for general compact coadjoint orbits (i.e., flag manifolds).

In [26] we study general Hermitian symmetric spaces of non-compact type, and use their realization as bounded domains to define an analogous algebra of symbols of polynomial differential operators.

Recently Reshetikhin and Takhtajan have announced [80] an associative formal star-product given by an asymptotic expansion on any Kähler manifold. This they do in two steps, first by building an associative product for which 1 is not a unit element and then by passing to a star-product.

4.1.1. *Berezin symbols*

We denote by (L, ∇, h) a quantization bundle for the Kähler manifold (M, ω, J) (i.e., a holomorphic line bundle L with connection ∇ admitting an invariant hermitian structure h, such that the curvature is $\operatorname{curv}(\nabla) = -2i\pi\omega$). We denote by $\mathcal{H}$ the Hilbert space of square-integrable holomorphic sections of L which we assume to be non-trivial. The coherent states are vectors $e_q \in \mathcal{H}$ such that

$$s(x) = \langle s, e_q \rangle q, \qquad \forall q \in \mathcal{L}_x, \quad x \in M, \quad s \in \mathcal{H},$$

where $\mathcal{L}$ denotes the complement of the zero-section in L. The function

$$\varepsilon(x) = |q|^2 \|e_q\|^2, \qquad q \in \mathcal{L}_x,$$

is well-defined and real analytic.

We introduce also the 2-point function

$$\psi(x,y) = \frac{|\langle e_{q'}, e_q \rangle|^2}{\|e_{q'}\|^2 \|e_q\|^2}, \qquad q \in \mathcal{L}_x, \quad q' \in \mathcal{L}_y,$$

which is a globally defined real analytic function on $M \times M$ provided ε has no zeros. It is a consequence of the Cauchy–Schwartz inequality that $\psi(x,y) \leq 1$ everywhere, with equality where the lines spanned by e_q and $e_{q'}$ coincide ($q \in \mathcal{L}_x$, $q' \in \mathcal{L}_y$).

Let $A : \mathcal{H} \to \mathcal{H}$ be a bounded linear operator and let

$$\widehat{A}(x) = \frac{\langle Ae_q, e_q \rangle}{\langle e_q, e_q \rangle}, \qquad q \in \mathcal{L}_x, \quad x \in M,$$

be its symbol. The function $\widehat{A}$ has an analytic continuation to an open neighbourhood of the diagonal in $M \times \overline{M}$ given by

$$\widehat{A}(x,y) = \frac{\langle Ae_{q'}, e_q \rangle}{\langle e_{q'}, e_q \rangle}, \qquad q \in \mathcal{L}_x, \quad q' \in \mathcal{L}_y,$$

which is holomorphic in x and antiholomorphic in y. We denote by $\hat{E}(L)$ the space of symbols of bounded operators on $\mathcal{H}$. We can extend this definition of symbols to some unbounded operators provided everything is well-defined.

4.1.2. *Composition of operators - Parameter*

The composition of operators on $\mathcal{H}$ gives rise to a product for the corresponding symbols, which is associative and which we shall denote by $*$ following Berezin, [12]. The product $*$ of symbols is given in terms of the symbols by the integral formula

$$(\widehat{A} * \widehat{B})(x) = \int_M \widehat{A}(x,y) \widehat{B}(y,x) \psi(x,y) \varepsilon(y) \frac{\omega^n(y)}{n!}.$$

This formula is derived by use of the adjoint A^* of A so to apply it to the case where the operators are unbounded we need to be able to use the adjoint of A on coherent states. To be able to take the symbol of the composition, the result of applying B to a coherent state must be in the domain of A.

Example. The identity map has symbol $\widehat{I}=1$ and so $\widehat{A}*1=1*\widehat{A}=\widehat{A}$ for any operator A. In particular $1*1=1$.

Let k be a positive integer. The bundle $(L^k=\otimes^k L,\nabla^k,h^k)$ is a quantization bundle for $(M,k\omega,J)$ and we denote by $\mathscr{H}^k$ the corresponding space of holomorphic sections and by $\hat{E}(L^k)$ the space of symbols of linear operators on $\mathscr{H}^k$. We let $\varepsilon^{(k)}$ be the corresponding function. We say that the **quantization is regular** if $\varepsilon^{(k)}$ is a non-zero constant for all nonnegative k and if $\psi(x,y)=1$ implies $x=y$. The significance of these conditions has been explained in [25].

Let G be a Lie group of isometries of the Kähler manifold (M,ω,J) which lifts to a group of automorphisms of the quantization bundle (L,∇,h). This automorphism group acts naturally on $(L^k,\nabla^{(k)},h^k)$; if $g\in G$ and if $e_q^{(l)}$ is a coherent state of L^l, then $g.e_q^{(l)}=e_{gq}^{(l)}$ so the function $\varepsilon^{(k)}$ is invariant under G. In particular, if the quantization is homogeneous, all $\varepsilon^{(k)}$ are constants.

In the regular case, the function ψ in the integral defining the composition of symbols for powers L^k gets replaced by powers ψ^k.

When the manifold is compact, we have proven the following facts:

– When $\varepsilon^{(k)}$ is constant for all k one has the nesting property $\hat{E}(L^k)\subset\hat{E}(L^{k+1})$;

– With the same assumption $\cup_k\hat{E}(L^k)$ is dense in $\mathscr{C}^\circ(M)$.

From the nesting property, one sees that if $\hat{A},\hat{B}$ belong to $\hat{E}(L^l)$ and if $k\geq l$ one may define

$$(\widehat{A}*_k\widehat{B})(x)=\int_M\hat{A}(x,y)\hat{B}(y,x)\psi^k(x,y)\varepsilon^{(k)}\frac{k^n\omega^n}{n!}. \tag{4.1}$$

More generally, if $\widehat{A}$ is a symbol of an operator then its analytic continuation $\widehat{A}(x,y)$ may have singularities where $\psi(x,y)=0$ but $\widehat{A}\psi$ is always globally defined on $M\times M$. If M is not compact $\widehat{A}\psi$ may not be bounded, so we introduce the class $\mathscr{B}\subset C^\infty(M)$ of functions f which have an analytic continuation off the diagonal in $M\times\overline{M}$ so that $f(x,y)\psi(x,y)^l$ is globally defined, smooth and bounded on $K\times M$ and on $M\times K$ for each compact subset K of M for some positive power l and denote by $\mathscr{B}_l$ those for which the power l suffices. Since ψ is smooth and bounded it is clear that $\mathscr{B}$ is a subalgebra of $C^\infty(M)$. In the case M is compact we obviously have $\hat{E}(L^l)\subset\mathscr{B}_l$. If $\hat{A},\hat{B}$ belong to $\mathscr{B}$, formula (4.1) is well defined for k large enough.

We study the behaviour of the integral (4.1) in terms of k.

4.1.3. *An asymptotic formula*

In order to localize the integral (4.1) we use a version of the Morse Lemma adapted from Combet, as in Moreno and Ortega-Navarro.

Let (M,ω,J) be a Kähler manifold with metric g. We denote by $\exp_x X$ the exponential at x of $X \in T_xM$. If g is not complete the exponential map may not be defined for all x and X, but in any case there is an open subset $V \subset TM$ where it is defined and which contains the zero-section. The differential of the exponential map at 0 is the identity so the map $\alpha : V \to M \times M$ given by $\alpha(X) = (p(X), \exp_{p(X)} X)$ where p is the projection in the tangent bundle $p\colon TM \to M$ is a diffeomorphism near the zero-section. At any point of the zero-section the differential of α is the identity.

Proposition 33 *Let (M,ω,J) be a Kähler manifold with metric g and $\alpha\colon V \to M \times M$ be the map defined above. Let (L,∇,h) be a regular quantization bundle over M and let ψ be the corresponding 2-point function on $M \times M$. Then there exists an open neighbourhood $W \subset V$ of the zero-section in TM and a smooth open embedding $\mu : W \to TM$ such that*

$$(-\log \psi\circ\alpha\circ\mu)(X) = \pi g_{p(X)}(X,X), \quad X \in W,$$

and the differential of μ at any point of the zero-section is the identity.

Denote by $\widetilde{\mathscr{B}}$ the set of functions f on $M \times M \setminus \psi^{-1}(0)$ such that $f(x,y)\psi(x,y)^l$ has a smooth extension to all of $M \times M$ which is bounded on $K \times M$ for each compact subset $K \subset M$ for some l and denote by $\widetilde{\mathscr{B}}_l$ those for which the power l suffices. If f, g are in $\mathscr{B}$ then $f(x,y)g(y,x)$ is in $\widetilde{\mathscr{B}}$. Note also that if $f \in \widetilde{\mathscr{B}}$ then its restriction to the diagonal $\widehat{f}(x) = f(x,x)$ is smooth.

For any f belonging to $\widetilde{\mathscr{B}}_l$, the integral

$$F_k(x) = \int_M f(x,y)\psi(x,y)^k k^n \frac{\omega^n(y)}{n!}, \quad \text{for k} \geq 1+1,$$

admits an asymptotic expansion

$$F_k(x) \sim \sum_{r\geq 0} k^{-r} C_r(\widehat{f})(x),$$

where C_r is a smooth differential operator of order $2r$ depending only on the geometry of M. The leading term is given by $C_0(\widehat{f})(x) = \widehat{f}$.

We are not claiming that for this very general class of functions $f(x,y)$ the integral depends smoothly on x, only that the coefficients of the asymptotic expansion do.

In the regular case $\varepsilon^{(k)}$ has an asymptotic expansion $\sum_{r\geq 0} \varepsilon_r/k^r$ as k tends to infinity with $\varepsilon_0 = 1$. Indeed,

$$1 = 1 *_k 1 = \varepsilon^{(k)} \int_M \psi(x,y)^k k^n \frac{\omega^n(y)}{n!},$$

has an asymptotic expansion in k^{-1} with leading term 1 by the previous proposition and we can then invert the asymptotic expansion to obtain one for $\varepsilon^{(k)}$.

Theorem 34 *Let (M,ω,J) be a Kähler manifold and (L,∇,h) be a regular quantization bundle over M. Let $\widehat{A},\widehat{B}$ be in $\mathscr{B}$. Then*

$$(\widehat{A} *_k \widehat{B})(x) = \int_M \widehat{A}(x,y)\widehat{B}(y,x)\psi^k(x,y)\varepsilon^{(k)}k^n\frac{\omega^n(y)}{n!}(y),$$

defined for k sufficiently large, admits an asymptotic expansion in k^{-1} as $k \to \infty$

$$(\widehat{A} *_k \widehat{B})(x) \sim \sum_{r\geq 0} k^{-r}C_r(\widehat{A},\widehat{B})(x),$$

and the cochains C_r are smooth bidifferential operators, invariant under the automorphisms of the quantization and determined by the geometry alone. Furthermore

$$C_0(\widehat{A},\widehat{B}) = \widehat{A}\widehat{B},$$

and

$$C_1(\widehat{A},\widehat{B}) - C_1(\widehat{B},\widehat{A}) = \frac{i}{\pi}\{\widehat{A},\widehat{B}\}.$$

4.1.4. *A convergent star-product for flag manifolds*
We first would like to show that the asymptotic expansion obtained above defines an associative formal star-product. For this we assume that (M,ω,J) is a flag manifold. Reshetikhin and Takhtajan [80] have announced an analogous result for general Kähler manifolds. Observe that if G is a Lie group of isometries of the Kähler manifold (M,ω,J) which lifts to a group of automorphisms of the quantization bundle (L,∇,h) it acts naturally on $(L^k,\nabla^{(k)},h^k)$:

$$g^*(\widehat{A} *_k \widehat{B})(x) = (g^*\hat{A} * g^*\hat{B})(x),$$

for any $\hat{A},\hat{B}$ in $\hat{E}(L^l)$ and any $k \geq l$. Observe also that the bidifferential operators C_r depend on the geometry alone thus are invariant under G.

Lemma 35 *Let (M,ω,J) be a flag manifold with $M = G/K$ where G is a compact simply-connected Lie group and K the centralizer of a torus. Assume the geometric quantization conditions are satisfied and let (L,∇,h) be a quantization bundle over M. Let $\mathscr{C}_L = \cup_k \hat{E}(L^k)$ be the union of the symbol spaces. Then $\mathscr{C}_L$ coincides with the space E of vectors in $C^\infty(M)$ whose G-orbit is contained in a finite-dimensional subspace.*

In the case of a flag manifold as above, the group G lifts to a group of automorphisms of the quantization bundle $(L^k,\nabla^{(k)},h^k)$ hence the map $\hat{E}(L^l)\otimes\hat{E}(L^l) \to C^\infty(M)$ given by $\hat{A}\otimes\hat{B} \mapsto \widehat{A} *_k \widehat{B}$ intertwines the action of G and the bidifferential operators C_r are invariant under G. Thus, if $\hat{A}$, $\hat{B}$ belong to $\hat{E}(L^l)$, there exists an integer $a(l)$ such that $\widehat{A} *_k \widehat{B}$ belongs to $\hat{E}(L^{a(l)})$ for all $k \geq l$, and such that $C_r(\hat{A}, \hat{B})$ belongs to $\hat{E}(L^{a(l)})$ for every integer r.

Consider now the asymptotic development:

$$\widehat{A} *_k \widehat{B} = \sum_{r=0}^{N} k^{-r} C_r(\hat{A}, \hat{B}) + R_N(\hat{A}, \hat{B}, k),$$

where

$$\lim_{k \to \infty} k^N R_N(\hat{A}, \hat{B}, k) = 0.$$

The above tells us that $R_N(\hat{A}, \hat{B}, k)$ belongs to $\hat{E}(L^{a(l)})$ where $a(l)$ is independent of k. So, we can write

$$\begin{aligned}(\widehat{A} *_k \widehat{B}) *_k \hat{C} &= \sum_{r=0}^{N} k^{-r} C_r(\hat{A}, \hat{B}) *_k \hat{C} + R_N(\hat{A}, \hat{B}) *_k \hat{C} \\ &= \sum_{r,s=0}^{N} k^{-r-s} C_s(C_r(\hat{A}, \hat{B}), \hat{C}) \\ &\quad + \sum_{r=0}^{N} k^{-r} R_N(C_r(\hat{A}, \hat{B}), \hat{C}, k) + R_N(\hat{A}, \hat{B}, k) *_k \hat{C}.\end{aligned} \tag{4.2}$$

The last two terms multiplied by k^N tend to zero when k tends to infinity.

Theorem 36 *The asymptotic expansion $\sum_{r \geq 0} k^{-r} C_r(u, v)$ yields a formal associative deformation of the usual product of functions in $\mathscr{C}_L$. It is a formal star-product which extends to all of $C^\infty(M)$, using uniform convergence.*

We prove that $\widehat{A} *_k \widehat{B}$ is a rational function of k with no pole at infinity, when the flag manifold is a hermitian symmetric space, by using structure theory of these spaces.

Theorem 37 *Let M be a compact hermitian symmetric space and let (L, ∇, h) be a quantization bundle over M. Let $L^k = \otimes^k L$ and let $\mathscr{H}^k$ be the space of holomorphic sections of L^k. Let $\hat{E}(L^k)$ be the space of symbols of operators on $\mathscr{H}^k$. If $\hat{A}$, $\hat{B}$ belong to $\hat{E}(L^l)$ and $k \geq l$, the product $\hat{A} *_k \hat{B}$ depends rationally on k and has no pole at infinity, hence the asymptotic expansion of $\widehat{A} *_k \widehat{B}$ is convergent.*

This result is generalized by Karabegov [55]:

Theorem 38 *For any generalized flag manifold the $*_k$ product of two symbols is a rational function of k without pole at infinity.*

4.1.5. *Star-product on bounded symmetric domains*

Bounded symmetric domains. Let $\mathscr{D}$ denote a bounded symmetric domain. We shall use the Harish-Chandra embedding to realise $\mathscr{D}$ as a bounded subset of its Lie algebra of automorphisms. More precisely, if G_0 is the connected component of the group of holomorphic isometries then $\mathscr{D}$ is the homogeneous space G_0/K_0 where G_0 is a non-compact semi-simple Lie group and K_0 is a maximal compact subgroup. Let $\mathfrak{g}$ be the Lie algebra of G_0, $\mathfrak{k}$ the subalgebra corresponding with K_0, $\mathfrak{g}$ and $\mathfrak{k}$ the complexifications, and G, K the corresponding complex Lie groups containing G_0 and K_0. The complex structure on $\mathscr{D}$ is determined by K_0-invariant abelian subalgebras $\mathfrak{m}_+$ and $\mathfrak{m}_-$ with

$$\mathfrak{g} = \mathfrak{m}_+ + \mathfrak{k} + \mathfrak{m}_-, \qquad [\mathfrak{m}_+, \mathfrak{m}_-] \subset \mathfrak{k}, \qquad \overline{\mathfrak{m}_+} = \mathfrak{m}_-,$$

where $^-$ denotes conjugation over the real form $\mathfrak{g}$ of $\mathfrak{g}$. The exponential map sends $\mathfrak{m}_\pm$ diffeomorphically onto subgroups $M_\pm$ of G such that M_+KM_- is an open set in G containing G_0 and the multiplication map

$$M_+ \times K \times M_- \to M_+KM_-$$

is a diffeomorphism. KM_- is a parabolic subgroup of G and the quotient G/KM_- a generalized flag manifold. The G_0-orbit of the identity coset can be identified with G_0/K_0 and lies inside $M_+KM_-/KM_- \cong M_+$. Composing this identification with the inverse of the exponential map gives the desired Harish-Chandra embedding of $\mathscr{D}$ as a bounded open subset of $\mathfrak{m}_+$. We shall assume from now on that $\mathscr{D} \subset \mathfrak{m}_+$ via this embedding. In this realization it is clear that the action of K_0 on $\mathscr{D}$ coincides with the adjoint action of K_0 on $\mathfrak{m}_+$.

Following Satake, we define maps

$$k\colon \mathscr{D} \times \mathscr{D} \to K, \qquad m_\pm\colon \mathscr{D} \times \mathscr{D} \to \mathfrak{m}_\pm,$$

by

$$\exp -\overline{Z'} \exp Z = \exp m_+(Z,Z')\, k(Z,Z')^{-1} \exp m_-(Z,Z'),$$

for $Z, Z' \in \mathscr{D}$. They satisfy

$$\overline{k(Z,Z')} = k(Z',Z)^{-1}, \qquad \overline{m_+(Z,Z')} = -m_-(Z',Z).$$

The holomorphic quantization. For any unitary character χ of K_0 there is a Hermitian holomorphic line bundle L over $\mathscr{D}$ whose curvature is the Kähler form of an invariant Hermitian metric on $\mathscr{D}$. If χ also denotes the holomorphic extension to K then the Hermitian metric has Kähler potential $\log \chi(k(Z,Z))$ and L has a zero-free holomorphic section s_0 with

$$|s_0(Z)|^2 = \chi(k(Z,Z))^{-1}.$$

For χ sufficiently positive s_0 is square-integrable and the representation U of G_0 on the space $\mathscr{H}$ of square-integrable sections of L is one of Harish-Chandra's holomorphic discrete series. s_0 is a highest weight vector for the extremal K-type so is a smooth vector for the representation. We form the coherent states e_q and see that $e_{s_0(0)}$ transforms the same way as s_0 and so they must be equal up to a multiple. This means that the coherent states are also smooth vectors of the representation. Further, since the quantization is homogeneous, ε will be constant. Thus

$$\langle e_{s_0(Z')}, e_{s_0(Z)} \rangle = \varepsilon \chi(k(Z,Z')).$$

Lemma 39 *Up to a constant (determined by the normalization of Haar measure on G_0) ε is the formal degree d_U of the discrete series representation U, hence by Harish-Chandra's formula, it is a polynomial function of the differential $d\chi$ of the character χ.*

The two-point function ψ is given by

$$\psi(Z,Z') = \frac{|\chi(k(Z,Z'))|^2}{\chi(k(Z,Z))\chi(k(Z',Z'))};$$

it takes the value 1 only on the diagonal.
Polynomial differential operators and symbols. We let $\mathscr{A}$ denote the algebra of holomorphic differential operators on functions on $\mathscr{D}$ with polynomial coefficients. We filter $\mathscr{A}$ by both the orders of the differentiation and the degrees of the coefficients: $\mathscr{A}_{p,q}$ denotes the subspace of operators of order at most p with coefficients of degree at most q. Obviously, the composition of operators gives a map

$$\mathscr{A}_{p,q} \times \mathscr{A}_{p',q'} \to \mathscr{A}_{p+p',q+q'}.$$

The global trivialization by s_0 of the holomorphic line bundle L corresponding with the character χ allows us to transport the above operators to act on sections of L by sending $D \in \mathscr{A}$ to D^χ where

$$D^\chi(fs_0) = (Df)s_0.$$

Let $\mathscr{A}(\chi)$ denote the resulting algebra of operators on sections of L and $\mathscr{A}_{p,q}(\chi)$ the corresponding subspaces.

In this non-compact situation elements of $\mathscr{A}(\chi)$ do not define bounded operators on the Hilbert space $\mathscr{H}$, but the fact that the coherent states are smooth vectors of the holomorphic discrete series representation and that polynomials are bounded on $\mathscr{D}$ means that each operator in $\mathscr{A}(\chi)$ maps the coherent states into $\mathscr{H}$ so that it makes sense to speak of the symbols of these operators.

Lemma 40 *The analytically continued symbol $\widehat{D^\chi}(Z,Z')$ of an operator D^χ in $\mathscr{A}_{p,q}(\chi)$ is a polynomial in Z and $m_-(Z,Z')$ of bidegree p,q.*

The space of symbols of the operators in $\mathscr{A}_{k,l}(\chi)$ is the space of polynomials in Z and $m_-(Z,Z')$ of bidegree k,l so, in particular, is independent of χ.

Denote by $\mathscr{E}_{p,q}$ the space of polynomials in Z and $m_-(Z,Z')$ of bidegree p,q and by $\mathscr{E}$ the union of these spaces. $\mathscr{E}$ is an algebra under pointwise multiplication, the algebra of symbols . If we take a symbol in $\mathscr{E}_{p,q}$ then it is the symbol of an operator in $\mathscr{A}_{p,q}(\chi)$. Taking two such operators and composing them corresponds with the composition of two polynomial operators in $\mathscr{A}_{p,q}$ and so can be expressed in terms of a basis for $\mathscr{A}_{2p,2q}$ as a rational function of $d\chi$. In other words the Berezin product $f*g$ of two symbols f, g in $\mathscr{E}_{p,q}$ is a symbol in $\mathscr{E}_{2p,2q}$ depending rationally on $d\chi$. *The star-product.* We construct a formal deformation of the algebra $C^\infty(\mathscr{D})$ by first constructing it on the subalgebra $\mathscr{E}$.

We consider the powers L^k of the line bundle L which correspond with the powers χ^k of χ. These powers have differentials $kd\chi$, so the Berezin product $f *_k g$ of two symbols f, g in $\mathscr{E}_{p,q}$ is a rational function of k by the results of the previous section.

The symbols in $\mathscr{E}_{p,q}$ are in $\mathscr{B}_0$ (it is enough to show that $\mathrm{ad}\, m_-(Z,Z')$ is bounded on $X \times \mathscr{D}$ and $\mathscr{D} \times X$ for any compact subset X of $\mathscr{D}$); thus the asymptotic expansion exists.

Since the Berezin product is associative for each k, the same argument as in the compact situation shows that its asymptotic expansion in k^{-1} is an associative formal deformation on $\mathscr{E}$ with bidifferential operators as coefficients. To see that it extends to all of $C^\infty(\mathscr{D})$ we show that $\mathscr{E}$ contains enough functions to determine these operators, hence the asymptotic expansion of $f *_k g$ has bidifferential operator coefficients which satisfy the cocycle conditions to define a formal product on $C^\infty(\mathscr{D})$ which is associative.

Theorem 41 *Let $\mathscr{D}$ be a bounded symmetric domain and $\mathscr{E}$ the algebra of symbols of polynomial differential operators on a homogeneous holomorphic line bundle L over $\mathscr{D}$ which gives a realization of a holomorphic discrete series representation of G_0 (i.e., $\mathscr{E}$ is the algebra of functions on $\mathscr{D}$ which are polynomials in Z and $m_-(Z,Z)$), then for f and g in $\mathscr{E}$ the Berezin product $f *_k g$ has an asymptotic expansion in powers of k^{-1} which converges to a rational function of k. The coefficients of the asymptotic expansion are bidifferential operators which define an invariant and covariant star-product on $C^\infty(\mathscr{D})$.*

References

1. Andler, M., Dvorsky, A., and Sahi, S.: Kontsevich quantization and invariant distributions on Lie groups, preprints math/9910104 and math/9905065.
2. Arnal, D.: Le produit star de Kontsevich sur le dual d'une algèbre de Lie nilpotente, *C. R. Acad. Sci. Paris Sér. I Math.* **327** (1998), 823–826.
3. Arnal, D., Ben Amar N., and Masmoudi, M.: Cohomology of good graphs and Kontsevich linear star products, *Lett. Math. Phys.* **48** (1999), 291–306.
4. Arnal, D., Cahen, M., and Gutt, S.: Deformations on coadjoint orbits, *J. Geom. Phys.* **3** (1986), 327–351.

5. Arnal, D.: $*$ products and representations of nilpotent groups, *Pacific J. Math.* **114** (1984), 285–308; Arnal, D. and Cortet, J.-C.: $*$-products in the method of orbits for nilpotent groups, *J. Geom. Phys.* **2** (1985), 83–116
6. Arnal, D. and Cortet, J.-C.: Nilpotent Fourier transform and applications, *Lett. Math. Phys.* **9** (1985), 25–34; Arnal, D. and Gutt, S.: Décomposition de $L^2(G)$ et transformation de Fourier adaptée pour un groupe G nilpotent, *C. R. Acad. Sci. Paris Sér. I Math.* **306** (1988), 25–28.
7. Arnal, D., Cortet, J.-C., Molin, P., and Pinczon, G.: Covariance and geometrical invariance in star-quantization, *J. Math. Phys.* **24** (1983), 276–283.
8. Arnal, D., Ludwig, J., and Masmoudi, M.: Déformations covariantes sur les orbites polarisées d'un groupe de Lie, *J. Geom. Phys.* **14** (1994), 309–331.
9. Asin, S.: Ph.D. thesis, Warwick University, 1998.
10. Bayen, F., Flato, M., Fronsdal, C., Lichnerowicz, A., and Sternheimer, D.: Quantum mechanics as a deformation of classical mechanics, *Lett. Math. Phys.* **1** (1975/77), 521–530; Deformation theory and quantization. I. Deformations of symplectic structures, *Ann. Physics* **111** (1978), 61–110.
11. Bayen, F., Flato, M., Fronsdal, C., Lichnerowicz, A., and Sternheimer, D.: Deformation theory and quantization. II. Physical applications, *Ann. Physics* **111** (1978), 111–151
12. Berezin, F.A.: General concept of quantization, *Comm. Math. Phys.* **40** (1975), 153–174.
13. Bertelson, M.: Equivalence de produits star, *Mémoire de Licence* U.L.B. (1995); Bertelson, M., Cahen, M., and Gutt, S.: Equivalence of star products, *Classical Quantum Gravity* **14** (1997), A93–A107.
14. Bertelson, M., Bieliavsky, P., and Gutt, S.: Parametrizing equivalence classes of invariant star products, *Lett. Math. Phys.* **46** (1998), 339–345.
15. Bidegain, F. and Pinczon, G.: Quantization of Poisson-Lie groups and applications, *Comm. Math. Phys.* **179** (1996), 295–332.
16. Bidegain, F. and Pinczon, G.: A star-product approach to non-compact quantum groups, *Lett. Math. Phys.* **33** (1995), 231–240.
17. Bonneau, P., Flato, M., Gerstenhaber, M., and Pinczon, G.: The hidden group structure of quantum groups: strong duality, rigidity and preferred deformations, *Comm. Math. Phys.* **161** (1994), 125–156.
18. Bordemann, M., Neumaier, N., and Waldmann, S.: Homogeneous Fedosov star products on cotangent bundles. I, *Comm. Math. Phys.* **198** (1998), 363–396.
19. Bordemann, M., Neumaier, N., and Waldmann, S.: Homogeneous Fedosov star products on cotangent bundles. II, *J. Geom. Phys.* **29** (1999), 199–234.
20. Cahen, M., De Wilde, M., and Gutt, S.: Local cohomology of the algebra of C^∞ functions on a connected manifold, *Lett. Math. Phys.* **4** (1980), 157–167.
21. Cahen, M., Flato, M., Gutt, S., and Sternheimer, D.: Do different deformations lead to the same spectrum?, *J. Geom. Phys.* **2** (1985), 35–48.
22. Cahen, M. and Gutt, S.: Regular $*$ representations of Lie Algebras, *Lett. Math. Phys.* **6** (1982), 395–404.
23. Cahen, M. and Gutt, S.: Produits $*$ sur les orbites des groupes semi-simples de rang 1, *C. R. Acad. Sci. Paris Sér. I Math.* **296** (1983), 821–823; An algebraic construction of $*$ product on the regular orbits of semisimple Lie groups, in: *Gravitation and geometry*, Bibliopolis, Naples, 1987, pp. 71–82.
24. Cahen, M. and Gutt, S.: Produits $*$ sur les espaces affins symplectiques localement symétriques, *C. R. Acad. Sci. Paris Sér. I Math.* **297** (1983), 417–420.
25. Cahen, M., Gutt, S., and Rawnsley, J.: Quantization of Kähler manifolds II, *Trans. Amer. Math. Soc.* **337** (1993), 73–98.
26. Cahen, M., Gutt, S., and Rawnsley, J.: Quantization of Kähler manifolds III, *Lett. Math. Phys.* **30** (1994), 291–305; Quantization of Kähler manifolds IV, Lett. Math. Phys. **34** (1995), 159–168.

27. Cahen, M., Gutt, S., and Rawnsley, J.: On tangential star products for the coadjoint Poisson structure, *Comm. Math. Phys.* **180** (1996), 99–108.
28. Chloup, V.: Star products on the algebra of polynomials on the dual of a semi-simple Lie algebra, *Acad. Roy. Belg. Bull. Cl. Sci.* **8** (1997), 263–269.
29. Connes, A.: Noncommutative differential geometry, *Inst. Hautes Études Sci. Publ. Math.* 62 (1985), 257–360.
30. Connes, A., Flato, M., and Sternheimer, D.: Closed star products and cyclic cohomology, *Lett. Math. Phys.* **24** (1992), 1–12.
31. Deligne, P.: Déformations de l'algèbre des fonctions d'une variété symplectique: comparaison entre Fedosov et De Wilde, Lecomte, *Selecta Math. (N.S.)* **1** (1995), 667–697.
32. De Wilde, M.: Deformations of the algebra of functions on a symplectic manifold: a simple cohomological approach, Publication no. 96.005, Institut de Mathématiques, Université de Liège, 1996.
33. De Wilde, M. and Lecomte, P.: Existence of star-products and of formal deformations of the Poisson Lie algebra of arbitrary symplectic manifolds, *Lett. Math. Phys.* **7** (1983), 487–496.
34. De Wilde, M. and Lecomte, P.: Formal deformations of the Poisson Lie algebra of a symplectic manifold and star-products: existence, equivalence, derivations, in: M. Hazewinkel and M. Gerstenhaber, (eds.), *Deformation theory of algebras and structures and applications*, Kluwer Acad. Publ., Dordrecht, 1988, pp. 897–960.
35. De Wilde, M., Gutt, S., and Lecomte, P. B. A.: À propos des deuxième et troisième espaces de cohomologie de l'algèbre de Lie de Poisson d'une variété symplectique, *Ann. Inst. H. Poincaré Sect. A (N.S.)* **40** (1984), 77–83.
36. Dito, G.: Kontsevich star product on the dual of a Lie algebra, *Lett. Math. Phys.* **48** (1999), 307–322.
37. Drinfel'd, V. G.: Quantum Groups, in: *Proceedings of the International Congress of Mathematicians, (Berkeley, Calif., 1986), Proc. ICM86, Berkeley, Amer. Math. Soc.*, Amer. Math. Soc., Providence, RI, 1987, pp. 798–820.
38. Etingof, P. and Kazhdan, D.: Quantization of Lie Bialgebras I, *Selecta Math. (N.S.)* **2** (1996), 1–41.
39. Etingof, P. and Kazhdan, D.: Quantization of Poisson algebraic groups and Poisson homogeneous spaces, in: A. Connes, K. Gawedzki, and J. Zinn-Justin, (eds.), *Symétries quantiques (Les Houches, 1995)*, North-Holland, Amsterdam, 1998, pp. 935–946.
40. Fedosov, B. V.: A simple geometrical construction of deformation quantization, *J. Diff. Geom.* **40** (1994), 213–238.
41. Fedosov, B. V.: *Deformation quantization and index theory*, Mathematical Topics **9**, Akademie Verlag, Berlin, 1996.
42. Fedosov, B. V.: The index theorem for deformation quantization, in: M. Demuth, E. Schrohe, and B.-W. Schulze, (eds.), *Boundary value problems, Schrödinger operators, deformation quantization*, Mathematical Topics **8**, Akademie Verlag, Berlin, 1996, pp. 206–318.
43. Fedosov, B. V.: On G-Trace and G-Index in deformation quantization, in: G. Dito and D. Sternheimer, (eds.), *Conférence Moshé Flato 1999*, Kluwer Acad. Publ., Dordrecht, 2000, pp. 157–176.
44. Fioresi, R. and Lledó, M. A.: On the deformation quantization of coadjoint orbits of semisimple groups, preprint math/9906104.
45. Flato, M.: Deformation view of physical theories, *Czech. J. Phys. B* **32** (1982), 472–475.
46. Flato, M., Lichnerowicz, A., and Sternheimer, D.: Déformations 1-différentiables d'algèbres de Lie attachées à une variété symplectique ou de contact, *C. R. Acad. Sci. Paris Sér. A* **279** (1974), 877–881; *Compositio Math.* **31** (1975), 47–82.
47. Flato, M., Lichnerowicz, A., and Sternheimer, D.: Crochet de Moyal–Vey et quantification, *C. R. Acad. Sci. Paris Sér. A-B* **283** (1976), 19–24.
48. Fronsdal, C.: Some ideas about quantization, *Rep. Math. Phys.* **15** (1978), 111–145.

49. Gerstenhaber, M.: On the deformation of rings and algebras, *Ann. of Math.* **79** (1964), 59–103.
50. Gutt, S.: Equivalence of deformations and associated $*$-products, *Lett. Math. Phys.* **3** (1979), 297–309.
51. Gutt, S.: Second et troisième espaces de cohomologie différentiable de l'algèbre de Lie de Poisson d'une variété symplectique, *Ann. Inst. H. Poincaré Sect. A (N.S.)* **33** (1980), 1–31.
52. Gutt, S.: An explicit $*$-product on the cotangent bundle of a Lie group, *Lett. Math . Phys.* **7** (1983), 249–258.
53. Gutt, S.: On some second Hochschild cohomology spaces for algebras of functions on a manifold, *Lett. Math. Phys.* **39** (1997), 157–162.
54. Gutt, S. and Rawnsley, J.: Equivalence of star products on a symplectic manifold; an introduction to Deligne's Čech cohomology classes, *J. Geom. Phys.* **29** (1999), 347–392.
55. Karabegov, A.: Berezin's quantization on flag manifolds and spherical modules, *Trans. Amer. Math. Soc.* **350** (1998), 1467–1479.
56. Karabegov, A.: Cohomological classification of deformation quantizations with separation of variables, *Lett. Math. Phys.* **43** (1998), 347–357.
57. Karabegov, A.: On the canonical normalization of a trace density of deformation quantization, *Lett. Math. Phys.* **45** (1999), 217–228.
58. Kathotia, V.: Kontsevich universal formula for deformation quantization and the CBH formula, preprint math/9811174.
59. Kontsevich, M.: Deformation quantization of Poisson manifolds I, preprint q-alg/9709040.
60. Lecomte, P. B. A.: Application of the cohomology of graded Lie algebras to formal deformations of Lie algebras, *Lett. Math. Phys.* **13** (1987), 157–166.
61. Lichnerowicz, A.: Cohomologie 1-différentiable des algèbres de Lie attachées à une variété symplectique ou de contact, *J. Math. Pures Appl.* **53** (1974), 459–484.
62. Lichnerowicz, A.: Existence and equivalence of twisted products on a symplectic manifold, *Lett. Math. Phys.* **3** (1979), 495–502.
63. Lichnerowicz, A.: Déformations d'algèbres associées à une variété symplectique (les $*_\nu$-produits), *Ann. Inst. Fourier (Grenoble)* **32** (1982), 157–209.
64. Masmoudi, M.: Tangential formal deformations of the Poisson bracket and tangential star products on a regular Poisson manifold, *J. Geom. Phys.* **9** (1992), 155–171.
65. Moreno, C. and Ortega-Navarro, P.: $*$-products on $D^1(C)$, S^2 and related spectral analysis, *Lett. Math. Phys.* **7** (1983), 181–193.
66. Moreno, C.: *-products on some Kähler-manifolds, *Lett. Math. Phys.* **11** (1986), 361–372.
67. Nadaud, F.: On continuous and differential Hochschild cohomology, *Lett. Math. Phys.* **47** (1999), 85–95.
68. Neroslavsky, O. M. and Vlassov, A. T.: Sur les déformations de l'algèbre des fonctions d'une variété symplectique, *C. R. Acad. Sci. Paris Sér. I Math.* **292** (1981), 71–73.
69. Nest, R. and Tsygan, B.: Algebraic index theorem for families, *Adv. in Math.* **113** (1995), 151–205.
70. Nest, R. and Tsygan, B.: Algebraic index theorem, *Comm. Math. Phys.* **172** (1995), 223–262.
71. Neumaier, N.: Local ν-Euler derivations and Deligne's characteristic class of Fedosov star products and star products of special type, preprint math/9905176.
72. Omori, H., Maeda, Y., and Yoshioka, A.: Weyl manifolds and deformation quantization, *Adv. Math.* **85** (1991), 224–255.
73. Omori, H., Maeda, Y., and Yoshioka, A.: The uniqueness of star-products on $P_n(\mathbf{C})$, in: C. H. Gu, H. S. Hu, and Y. L. Xin, (eds.), *Differential geometry (Shanghai, 1991)*, World Sci. Publishing, River Edge, NJ, 1993, pp 170–176.
74. Omori, H., Maeda, Y., and Yoshioka, A.: Existence of a closed star product, *Lett. Math. Phys.* **26** (1992), 285–294.
75. Omori, H., Maeda, Y., and Yoshioka, A.: Deformation quantizations of Poisson algebras, in: Y. Maeda, H. Omori, and A. Weinstein, (eds.), *Symplectic geometry and quantization (Sanda*

and Yokohama, 1993) Contemp. Math. **179**, Amer. Math. Soc., Providence, RI, 1994, pp. 213–240.

76. Omori, H., Maeda, Y., Miyazaki, N., and Yoshioka, A.: An example of strict Fréchet deformation quantization, preprint 1999.
77. Pinczon, G.: On the equivalence between continuous and differential deformation theories, *Lett. Math. Phys.* **39** (1997), 143–156.
78. Rauch, D.: Equivalence de produits star et classes de Deligne, *Mémoire de Licence* U.L.B. (1998).
79. Rawnsley, J., Cahen, M., and Gutt, S.: Quantization of Kähler manifolds I, *J. Geom. Phys.* **7** (1990), 45–62.
80. Reshetikhin, N. and Takhtajan, L.: Deformation quantization of Kähler manifolds, preprint math/9907171.
81. Rieffel, M.: Questions on quantization, in: L. Ge, H. Lin, Z.-J. Ruan, D. Zhang, and S. Zhang, (eds.), *Operator algebras and operator theory (Shanghai, 1997)*, Contemp. Math. **228**, Amer. Math. Soc., Providence, RI, 1998, pp. 315–326.
82. Sternheimer, D.: Phase-space representations, in: M. Flato, P. Sally, and G. Zuckerman, (eds.), *Applications of group theory in physics and mathematical physics (Chicago, 1982)*, Lectures in Appl. Math. **21**, Amer. Math. Soc., Providence, RI, 1985, pp. 255–267.
83. Sternheimer, D.: Deformation quantization: Twenty years after, in: J. Rembieliński, (ed.), *Particles, Fields, and Gravitation (Lodz 1998)* AIP Press, New York, 1998, pp. 107–145.
84. Tamarkin, D.: Quantization of Poisson structures on $\mathbb{R}^2$, *Electron. Res. Announc. Amer. Math. Soc.* **3** (1997), 119–120.
85. Tamarkin, D.: Another proof of M. Kontsevich formality theorem, preprint math/9803025; Formality of chain operad of small squares, preprint math/9809164.
86. Vey, J.: Déformation du crochet de Poisson sur une variété symplectique, *Comment. Math. Helvet.* **50** (1975), 421–454.
87. Weinstein, A.: Deformation quantization, Séminaire Bourbaki, 1993/94, *Astérisque* **227** (1995), 389–409.
88. Weinstein, A. and Xu, P.: Hochschild cohomology and characteristic classes for star-products, preprint q-alg/9709043.
89. Xu, P.: Fedosov $*$-products and quantum moment maps, *Comm. Math. Phys.* **197** (1998), 167–197.

Deformations of algebras over operads and the Deligne conjecture

Maxim Kontsevich
I.H.E.S., 35 route de Chartres, Bures-sur-Yvette 91440, France
e-mail: maxim@ihes.fr

Yan Soibelman
I.H.E.S., 35 route de Chartres, Bures-sur-Yvette 91440, France
and
Department of Mathematics, Kansas State University, Manhattan, KS 66506, USA
e-mail: soibel@ihes.fr

Abstract. In the present paper we develop the deformation theory of operads and algebras over operads. Free resolutions (constructed via Boardman-Vogt approach) are used in order to describe formal moduli spaces of deformations. We apply the general theory to the proof of the Deligne conjecture. The latter says that the Hochschild complex of an associative algebra carries a canonical structure of a dg-algebra over the chain operad of the little discs operad. In the course of the proof we construct an operad of geometric nature which acts on the Hochschild complex. It seems to be different from the brace operad (the latter was used in the previous approaches to the Deligne conjecture). It follows from our results that the Grothendieck-Teichmüller group acts (homotopically) on the moduli space of structures of 2-algebras on the Hochschild complex. In the Appendix we develop a theory of piecewise algebraic chains and forms. It is suitable for real semialgebraic manifolds with corners (like Fulton-MacPherson compactifications of configuration spaces of points).

Keywords: quantum algebra, rings and algebras, category theory, algebraic topology, algebraic geometry.

Mathematics Subject Classifications (2000): 18D50, 55P48, 16S80

Introduction

The deformation theory of associative algebras is a guide for developing the deformation theory of many algebraic structures. Conversely, all the concepts of what should be the "deformation theory of everything" must be tested in the case of associative algebras.

An associative algebra is an algebra over an operad. This fact, along with the observation that in many examples we are dealing with algebras over operads, "explains" the universality of associative algebras. It also suggests how to develop the deformation theory of algebras over operads. This theory is one of the topics of the present paper.

Another remarkable fact is the relationship of the deformation theory of associative algebras to the geometry of configuration spaces of points on surfaces. One

G. Dito and D. Sternheimer (eds.), Conférence Moshé Flato 1999, Vol. 1, 255 – 307.

of its incarnations is the Deligne conjecture (formulated in [4]). A brief history of the conjecture as well as its generalizations can be found in [14]. The Deligne conjecture is the second topic of our paper.

The theme which motivated the whole project is the Grothendieck-Teichmüller group (*GT* for short) and its role in deformation theory. The Grothendieck-Teichmüller group can be defined as the automorphism group of the tower of the pro-nilpotent completions of pure braid groups (see [5]). The pure braid group of n strings is the fundamental group of the configuration space of n points in the plane. The Deligne conjecture shows its relation to the Hochschild complex. More evidence for that relation has been found in deformation quantization. The coefficients in the explicit formulas for the deformed multiplication given in [12] are periods of algebraic varieties defined over the field of rational numbers. Through the hypothetical relation to mixed Tate motives, this fact leads to some conjectures about the action of the Grothendieck-Teichmüller group on the moduli space of quantized algebras of functions on a manifold (see [14]). The Deligne conjecture says that the Hochschild complex $C^\bullet(A,A)$ of an associative (or more generally, A_∞) algebra A carries a structure of 2-algebra (i.e., an algebra over the operad of chains of the little discs operad). It follows from our results that the Grothendieck-Teichmüller group acts (homotopically) on the moduli space of structures of 2-algebras on $C^\bullet(A,A)$. This action is closely related to the action of the motivic Galois group described in [14]. We hope to discuss this topic in detail elsewhere.

The paper is organized in the following way.

Section 1 is devoted to the review of operads. It includes a brief introduction to operads, language of trees, polynomial functors, etc.

The deformation theory of algebras over free operads is discussed in Section 2. The "general" deformation theory is based on the notion of a formal pointed dg-manifold (a formal $\mathbf{Z}$-graded manifold with a vector field d of degree $+1$ such that $[d,d]=0$). We do not use any deep results concerning dg-manifolds. All necessary facts and definitions can be found in [12] (note that in [12], [13] formal pointed dg-manifolds were called formal pointed Q-manifolds). We explain how to construct a formal pointed dg-manifold which controls the deformation theory of an algebra over an operad. In traditional language this corresponds to a construction of the deformation functor as a functor of the category of local Artinian rings. We explain how this approach leads to the (homotopy) action of the Grothendieck-Teichmüller group on the dg-manifold controlling the deformation theory of the Hochschild complex of an A_∞-algebra.

In Section 3 we outline the strategy and state the theorem (proved in Section 6) which explain why the *GT* group appears in the deformation theory of associative (and more generally A_∞) algebras.

Our approach is based on the notion of a free resolution of an operad. This is the subject of Section 4. We construct canonical free resolutions via an approach similar to the one developed by Boardman and Vogt in [3]. We use free resolutions of operads instead of the conventional approach which uses free resolutions of algebras. This approach to the deformation theory of algebras over operads is not used very often (see however [15]).

In Section 5 we construct an operad M which acts on the Hochschild complex of an A_∞-algebra. It turns out that M is closely related to the compactifications of configuration spaces of point introduced by Fulton and MacPherson in [6].

Section 6 is devoted to the proof of the theorem from Section 3.

Section 7 is devoted to the Deligne conjecture and its proof. The proof presented in this paper is based on the general deformation theory developed in the previous sections. The strategy is explained in detail in Section 7. We also suggest certain generalizations of the original Deligne conjecture as well as some conjectures about the cell structure of the spaces which appear in the course of its study. It seems that the proof admits a generalization to the higher-dimensional version of the Deligne conjecture proposed in [14].

Other approaches to the original Deligne conjecture were proposed in [18], [21], [22].

The Appendix is devoted to a theory of piecewise algebraic chains. It is suitable for real semialgebraic manifolds with corners. A typical example is the compactification of the configuration space of points in the plane (Fulton-MacPherson compactification). This theory is useful for the proof of formality of the operad of chains of the little discs operad (in the approach of [14]). It is also useful (but not necessary) in the proof of the Deligne conjecture. The usual techniques of simplicial homology will do the job. At the same time we feel that the theory of piecewise algebraic chains is appropriate to the nature of the topic. For this reason we have decided to include it in the paper.

Acknowledgements. The second author acknowledges the support from the Clay Mathematical Institute and IHES. He thanks IHES for hospitality and a stimulating research atmosphere.

1. Generalities on operads

1.1. Polynomial functors, operads, algebras

The material of this subsection is very well-known (see e.g. [9], [7], [8], [17]). We present it here for completeness and in order to fix the notation.

Let k be a field of characteristic zero. All vector spaces below will be k-vector spaces unless we say otherwise.

We fix a category $\mathscr{C}$ which is assumed to be k-linear abelian symmetric monoidal and closed under infinite sums and products. We will also assume that it has internal $Hom's$. Our main examples will be the category of k-vector spaces, the category $Vect_{\mathbf{Z}}$ of $\mathbf{Z}$-graded vector spaces (with Koszul rule of signs), and the category of complexes of k-vector spaces.

Suppose we have a collection of representations $F = (F_n)_{n\geq 0}$ of the symmetric groups S_n, $n \geq 0$, in $\mathscr{C}$ (i.e., we have a sequence of objects F_n together with an action of the group S_n on F_n for each n).

Definition 1 *A polynomial functor $F:\mathscr{C}\to\mathscr{C}$ is defined on objects by the formula*

$$F(V)=\oplus_{n\geq 0}(F_n\otimes V^{\otimes n})_{S_n}$$

where for a group H and an H-module W we denote by W_H the space of coinvariants. The functor F is defined on morphisms in an obvious way.

Notice that having a sequence F_n as above we can define F_I for any finite set I using isomorphisms of I with the standard set $\{1,\dots,|I|\}$, where $|I|$ is the cardinality of I. Thus $F_{\{1,\dots,n\}}=F_n$. Technically speaking, we consider a functor Φ from the groupoid of finite sets (morphisms are bijections) to the symmetric monoidal category $\mathscr{C}$. Then we set $F_I=\Phi(I)$.

Polynomial functors on $\mathscr{C}$ form a category $\mathscr{PF}$ if we define morphisms between two such functors F and G as a vector space of S_n-intertwiners

$$Hom(F,G)=\prod_{n=0}^{\infty}Hom_{S_n}(F_n,G_n).$$

There is a composition operation $\circ$ on polynomial functors such that $(F\circ G)(V)$ is naturally isomorphic to $F(G(V))$ for any $V\in\mathscr{C}$. We also have a polynomial functor $\mathbf{1}$ such that $\mathbf{1}_1=1_{\mathscr{C}}$ and $\mathbf{1}_n=0$ for all $n\neq 1$. Here $1_{\mathscr{C}}$ is the unit object in the monoidal category $\mathscr{C}$. It is easy to see that in this way we get a monoidal structure on $\mathscr{PF}$.

Definition 2 *An operad in $\mathscr{C}$ is a monoid in the monoidal category $\mathscr{PF}$. In other words it is a polynomial functor $R\in\mathscr{PF}$ together with morphisms $m:R\circ R\to R$ and $u:\mathbf{1}\to R$ satisfying the associativity and the unit axioms.*

To shorten the notation we will denote the operad (R,m,u) simply by R. An operad R gives rise to a triple in the category $\mathscr{C}$. There is the notion of an algebra over a triple in a category. Hence we can use it in order to give a definition of an algebra in $\mathscr{C}$ over the operad R. It is given by an object $V\in\mathscr{C}$ and a morphism $R(V)\to V$ satisfying natural properties of compatibility with the structure of a triple. Equivalently, V is an R-algebra iff there is a morphism of operads $R\to\mathscr{E}nd(V)$, where $\mathscr{E}nd(V)$ is the endomorphism operad of V defined by $(\mathscr{E}nd(V))_n=\underline{Hom}(V^{\otimes n},V)$, and $\underline{Hom}$ denotes the internal Hom in $\mathscr{C}$. The category of R-algebras will be denoted by $R-alg$. There are two adjoint functors $Forget_R: R-alg\to\mathscr{C}$ and $Free_R:\mathscr{C}\to R-alg$ such that $Forget_R\circ Free_R=R$.

Definition 3 *For $X\in Ob(\mathscr{C})$, we call $Free_R(X)$ the free R-algebra generated by X.*

We remind the reader that there are operads *As*, *Lie*, *Comm* such that the algebras over them in the category of vector spaces are associative, Lie and commutative algebras correspondingly.

1.2. Colored operads

There is a generalization of the notion of operad. It is useful in order to describe in operadic terms pairs *(associative algebra A, A-module)*, homomorphisms of algebras over operads, etc.

Let I be set. We consider the category $\mathscr{C}^I$ consisting of families $(V_i)_{i\in I}$ of objects of $\mathscr{C}$.

A polynomial functor $F:\mathscr{C}^I\to\mathscr{C}^I$ is defined by the following formula:

$$(F((V_i)_{i\in I}))_j=\oplus_{a:I\to\mathbf{Z}_{\geq 0}}F_{a,j}\otimes_{\prod_i S_{a(i)}}\otimes_{i\in I}(V_i^{\otimes a(i)}),$$

where $a:I\to\mathbf{Z}_{\geq 0}$ is a map with the finite support, and $F_{a,j}$ is a representation in $\mathscr{C}$ of the group $\prod_{i\in I}S_{a(i)}$.

Polynomial functors in $\mathscr{C}^I$ form a monoidal category with the tensor product given by the composition of functors.

Definition 4 *A colored operad is a monoid in this category.*

Similarly, as in the case of usual operads, it defines a triple in the category $\mathscr{C}^I$. Therefore we have the notion of an algebra over a colored operad.

There exists a colored operad $\mathscr{OP}$ such that the category of $\mathscr{OP}$-algebras is equivalent to the category of operads.

Namely, let us consider the forgetful functor $Operads\to\mathscr{PF}$. It has a left adjoint functor. Thus we have a triple in $\mathscr{PF}$. As we have noticed before, the category $\mathscr{PF}$ can be described as a category of sequences $(P_n)_{n\geq 0}$ of S_n-modules. Then using the representation theory of symmetric groups, we conclude that the category $\mathscr{PF}$ is equivalent to the category $\mathscr{C}^{I_0}$, where I_0 is the set of all Young diagrams (partitions). Hence a polynomial functor $F:\mathscr{PF}\to\mathscr{PF}$ can be described as a collection $F_{(m_i),n}$ of the representations of the groups $S_{n,(m_k)}:=S_n\times\prod_{k\geq 0}(S_{m_k}\rhd S_k^{m_k})$, where $\rhd$ denotes the semidirect product of groups.

Having these data we can express any polynomial functor F on $\mathscr{PF}$ by the formula:

$$(F((U_k)_{k\geq 0}))_n=\bigoplus_{(m_k)}F_{(m_k),n}\otimes_{S_{1,(m_k)}}\bigotimes_{k\geq 0}(U_k^{\otimes m_k}).$$

In particular, one has a functor $\mathscr{OP}:\mathscr{PF}\to\mathscr{PF}$, which is the composition of the forgetful functor $Operads\to\mathscr{PF}$ with its adjoint. It gives rise to an I_0-colored operad $\mathscr{OP}=(\mathscr{OP}_{(m_i),n})$. We will describe it explicitly in the subsection devoted to trees.

1.3. Non-linear operads

We remark that operads and algebras over operads can be defined for any symmetric monoidal category $\mathscr{C}$, not necessarily k-linear. In particular, we are going to use operads in the categories of sets, topological spaces, etc.

Namely, an operad in $\mathscr{C}$ is a collection $(F_n)_{n\geq 0}$ of objects in $\mathscr{C}$, each equipped with an S_n-action, as well as composition maps:

$$F_n \otimes F_{k_1} \otimes \cdots \otimes F_{k_n} \to F_{k_1+\cdots+k_n},$$

for any $n \geq 0$ and $k_1,\ldots,k_n \geq 0$. Another datum is the unit, which is a morphism $\mathbf{1}_{\mathscr{C}} \to F_1$. All the data are required to satisfy certain axioms (see [17]). Analogously one describes colored operads and algebras over operads. Notice that in this framework one cannot speak about polynomial functors and free algebras.

This approach has some advantages and drawbacks (like the description of analytic functions in terms of Taylor series vs their description in terms of Taylor coefficients).

1.4. Pseudo-tensor categories

The notion of colored operad has been rediscovered many times. In [1] the notion of *pseudo-tensor category* was introduced as a generalization of the notion of symmetric monoidal (=tensor) category. The terminology stresses the similarity of operads with tensor categories. Similar notion was introduced in [2] under the name *multi-linear category.*

This notion is essentially equivalent to the notion of colored operad. We recall it below, using the name suggested in [1].

Definition 5 *A pseudo-tensor category is given by the following data:*

1) A class $\mathscr{A}$ called the class of objects, and a symmetric monoidal category $\mathscr{V}$ called the category of operations.

2) For every finite set I, a family $(X_i)_{i\in I}$ of objects, and an object Y, an object $P_I((X_i),Y) \in \mathscr{V}$ called the space of operations from $(X_i)_{i\in I}$ to Y.

3) For any map of finite sets $\pi : J \to I$, two families of objects $(Y_i)_{i\in I}, (X_j)_{j\in J}$ and an object Z, a morphism in $\mathscr{V}$

$$P_I((Y_i),Z) \otimes (\otimes_i P_{\pi^{-1}(i)}((X_{j_i}),Y_i)) \to P_J((X_j),Z),$$

called composition of operations. Here we denote by $\otimes$ the tensor product in $\mathscr{M}$.

4) For an 1-element set $\{pt\}$ and an object X, a unit morphism $\mathbf{1}_{\mathscr{V}} \to P_{\{pt\}}((X),X)$.

These data are required to satisfy natural conditions. In particular, compositions of operations are associative with respect to morphisms of finite sets, and the unit morphisms satisfy properties analogous to those of the identity morphisms (see [1] or [19] for details).

If $\mathscr{A}$ is a set, then a pseudo-tensor category is exactly the same as an $\mathscr{A}$-colored operad in the tensor category $\mathscr{V}$.

If we take $\mathscr{V}$ to be the category of sets, and take I above to be 1-element sets only, we obtain a category with the class of objects equal to $\mathscr{A}$.

Colored operad with one color gives rise to an ordinary operad. A symmetric monoidal category $\mathscr{A}$ produces the colored operad with

$$P_I((X_i),Y) = Hom_{\mathscr{A}}(\otimes_i X_i, Y).$$

The notion of pseudo-tensor category admits a generalization to the case when no action of the symmetric group is assumed. This means that we consider *sequences* of objects instead of *families* (see [19]). The new notion generalizes monoidal categories. In the terms of the next subsection, this would mean that one uses planar trees instead of all trees. One can go one step further by generalizing braided categories. This leads to colored braided operads (or pseudo-braided categories). In this case, trees in $\mathbf{R}^3$ should be used. The deformation theory of such structures can be developed along the lines of the present paper. It will be explained in detail elsewhere.

1.5. Trees

Definition 6 *A tree T is defined by the following data:*

1) A finite set $V(T)$ whose elements are called vertices;

2) A distinguished element $root_T \in V(T)$ called root vertex;

3) Subsets $V_i(T)$ and $V_t(T)$ of $V(T)\setminus\{root_T\}$ called the set of internal vertices and the set of tails respectively. Their elements are called internal and tail vertices respectively;

4) A map $N = N_T : V(T) \to V(T)$.

These data are required to satisfy the following properties:

a) $V(T) = \{root_T\} \sqcup V_i(T) \sqcup V_t(T)$;

b) $N_T(root_T) = root_T$, and $N_T^k(v) = root_T$ for all $v \in V(T)$ and $k \gg 1$;

c) $N_T(V(T)) \cap V_t(T) = \emptyset$;

d) There exists a unique vertex $v \in V(T), v \neq root_T$ such that $N_T(v) = root_T$.

We denote by $|v|$ the *valency* of a vertex v, which we understand as the cardinality of the set $N_T^{-1}(v)$.

We call the pairs $(v, N(v))$ *edges* in the case where $v \neq root_T$. If both elements of the pair belong to $V_i(T)$ we call the corresponding edge *internal*. The only edge e_r defined by the condition d) above is called the *root edge*. All edges of the type $(v,N(v)), v \in V_t(T)$ are called *tail edges*. We use the notation $E_i(T)$ and $E_t(T)$ for the sets of internal and tail edges respectively. We have a decomposition of the set of all edges $E(T) = E_i(T) \sqcup (E_t(T) \cup \{e_r\})$. There is a unique tree T_e such that $|V_t(T_e)| = 1$ and $|V_i(T_e)| = 0$. It has the only tail edge which is also the root edge.

A *numbered* tree with n tails is by definition a tree T together with a bijection of sets $\{1,\dots,n\} \to V_t(T)$. We can picture trees as in Figure 1.

Let R be an operad. Any tree T gives a natural way to compose elements of R, $comp_T : \otimes_{i \in V_i(T)} R_{N^{-1}(v)} \to R_{V_t(T)}$.

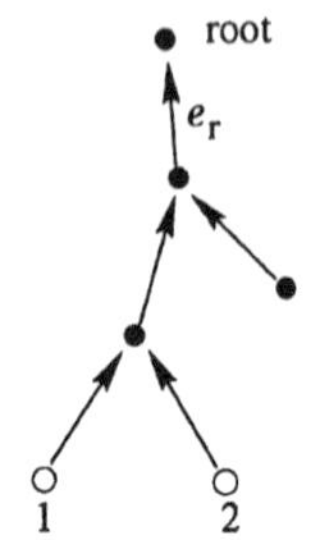

Figure 1. Numbered tree, non-numbered vertices are black.

Let us return to the colored operad $\mathcal{OP}$ and give its description using the language of trees.

Namely, $\mathcal{OP}_{(m_i),n}$ is a k-vector space generated by the isomorphism classes of trees T such that:

a) T has n tails numbered from 1 to n;

b) T has $\sum_i m_i$ internal vertices all numbered in such a way that first m_0 vertices have valency 0, and they are numbered from 1 to m_0, next m_1 internal vertices have valency 1, and they are numbered from 1 to m_1, and so on;

c) for every internal vertex $v \in V_i(T)$ the set of incoming edges $N_T^{-1}(v)$ is also numbered.

An action of the group $S_{(m_k),n}$ is defined naturally: the factor S_n permutes numbered tails, the factor S_{m_k} permutes numbered internal vertices and the factor $S_k^{m_k}$ permutes their incoming edges numbered from 1 to k.

The composition is given by the procedure of inserting of a tree into an internal vertex of another one. The new numeration is clear. We leave these details as well as the proof of the following proposition to the reader.

Proposition 1 *The category of $\mathcal{OP}$-algebras is equivalent to the category of k-linear operads.*

Let F be a polynomial functor on $\mathcal{C}$ (see Section 1.1). Let us consider a category $\mathcal{C}_F$, objects of which are pairs $(V, \phi : F(V) \to V)$ where V is an object of $\mathcal{C}$ and ϕ is a morphism in $\mathcal{C}$. Morphisms of pairs are defined in the natural way. The following lemma is easy to prove.

Lemma 1 *The category $\mathcal{C}_F$ is equivalent to the category of $Free_{\mathcal{OP}}(F)$-algebras.*

We call $P = Free_{\mathcal{OP}}(F)$ the *free operad* generated by F.

Components P_n of the functor P can be defined explicitly as follows.

Let $Tree(n)$ denotes the groupoid of numbered trees with n tails, $|Tree(n)|$ denotes the set of classes of isomorphisms of these trees. We denote the class of isomorphism of T by $[T]$. Then we have

$$P_n = Free_{\mathscr{O}\mathscr{P}}(F)_n = \oplus_{[T]\in|Tree(n)|}(\otimes_{v\in V_i(T)} F_{N^{-1}(v)})_{AutT}.$$

2. Deformations and differentials in free operads

Let F be a polynomial functor, $P = Free_{\mathscr{O}\mathscr{P}}(F)$ be the corresponding free operad. Let g_P be the Lie algebra (in the symmetric monoidal category $\mathscr{C}$) of derivations of the operad P. Then, as an object of $\mathscr{C}$:

$$g_P = \prod_{n\geq 0} \underline{Hom}(F_n, P_n)^{S_n}$$

where W^H denotes the space of H-invariants of an H-module W and $\underline{Hom}$ denotes the internal Hom in $\mathscr{C}$. This observation follows from the fact that

$$Hom_{\mathscr{P}\mathscr{F}}(F, Forget_{\mathscr{O}\mathscr{P}}(G)) = Hom_{\mathscr{O}\mathscr{P}-alg}(Free_{\mathscr{O}\mathscr{P}}(F), G).$$

From now on we suppose that $\mathscr{C}$ is the category $Vect_{\mathbf{Z}}$ of $\mathbf{Z}$-graded vector spaces. Then g_P is a graded Lie algebra with the graded components g_P^n.

Definition 7 *A differential on P is an element $d_P \in g_P^1$ such that $[d_P, d_P] = 0$.*

If d_P is a differential then $\hat{P} := (P, d_P)$ is an operad in the symmetric monoidal category of complexes. We should distinguish it from P, which is a free operad in the category $Vect_{\mathbf{Z}}$.

For the purposes of the deformation theory it is very convenient to work with operads of the type $\hat{P}$. A general dg-operad could be replaced by its free resolution.

Definition 8 *A dg-algebra over (P, d_P) is an algebra over $\hat{P}$ in the category of complexes.*

Later we will sometimes abuse notations speaking about dg-algebras over P meaning that the differential d_P is obvious from the context.

We recall that usually a deformation theory of a mathematical structure is controlled by a formal pointed dg-manifold (see [12]). In what follows we shall define deformation theories of operads and algebras by declaring the dg-manifolds which control them. Thus:

Definition 9 *The deformation theory of $\hat{P}$ is controlled by the formal dg-manifold associated in the usual way with the differential graded Lie algebra $(g_P, [d_P, \cdot])$.*

The motivation for this definition is that in order to deform the pair (P, d_P) it is enough to deform d_P (since P itself is free and therefore rigid).

Now we are going to describe the deformation theory of dg-algebras over P ($\widehat{P}$-algebras). In what follows we will often speak about points of $\mathbf{Z}$-graded manifolds.

It will always mean Λ-points, where Λ is an auxiliary $\mathbf{Z}$-graded commutative associative algebra (in general without the unit). In the case of formal manifolds we take Λ to be nilpotent.

Let us describe the formal pointed dg-manifold controlling deformations of a P-algebra V. We have the following graded vector space

$$\mathcal{M} = \mathcal{M}(P,V) = (\underline{Hom}(V,V))[1] \oplus \underline{Hom}(F(V),V).$$

We denote by $\mathcal{M}^n, n \in \mathbf{Z}$, the graded components of $\mathcal{M}$.

The structures of a complex on a graded vector space V and an action of P on V define a point $(d_V,\rho) \in \mathcal{M}^0 = Hom_{Vect_{\mathbf{Z}}}(k,\mathcal{M})$. We consider here d_V and ρ as morphisms of graded vector spaces. The equation $d_V^2 = 0$ and the condition of compatibility of d_V and ρ can be written in the form $d_{\mathcal{M}}(d_V,\rho) = 0$, where $d_{\mathcal{M}}(d_V,\rho) = (d_V^2, \xi(d_V,\rho)) \in \mathcal{M}^1$, for some $\xi(d_V,\rho) \in \underline{Hom}(F(V),V)$. It is easy to see that the assignment $(d_V,\rho) \mapsto d_{\mathcal{M}}(d_V,\rho)$ defines an odd vector field $d_{\mathcal{M}}$ on the "infinite-dimensional graded manifold" $\mathcal{M}$. A zero of this vector field corresponds to a structure of a complex on V together with a compatible structure of a dg-algebra over P. This gives a bijection between the set of zeros and the set of such structures.

It is easy to check that $[d_{\mathcal{M}}, d_{\mathcal{M}}] = 0$. Therefore a formal neighborhood of a fixed point (d_V,ρ) of $d_{\mathcal{M}}$ becomes a formal pointed dg-manifold.

Definition 10 *The deformation theory of a dg-algebra V is controlled by the formal pointed dg-manifold $M(P,V)$.*

Remark 1 Operads are algebras over the colored operad $\mathcal{O}\mathcal{P}$. One can show that deformation theories for an $\mathcal{O}\mathcal{P}$-algebra $\widehat{P}$ described in the last two definitions are in fact equivalent.

Let $P^{(0)}$ be an operad in $Vect_k$. In order to define the deformation theory of $P^{(0)}$-algebras, one needs to choose a resolution $\phi : P \to P^{(0)}$, where $P = Free_{\mathcal{O}\mathcal{P}}(F)$ as a $\mathbf{Z}$-graded operad, and ϕ is a quasi-isomorphism. One wants to be sure that the deformation theory does not depend on the choice of the resolution. This can be achieved by assuming that (see e.g. [16]):

a) F admits a filtration (as a polynomial functor) $F = \cup_{j\geq 1} F^{(j)}, F^{(j)} \subset F^{(j+1)}$ such that $d_P(F^{(0)}) = 0$ and $d_P(F^{(j)}) \subset Free_{\mathcal{O}\mathcal{P}}(F^{(j-1)}), j \geq 1$;

b) $\phi : P \to P^{(0)}$ is a an epimorphism.

Property a) means that F is a cofibrant object in the category of operads (see [10]). All resolutions used in the paper will satisfy these properties.

2.1. Example: A_∞-operad and A_∞-algebras

Let $V \in Vect_{\mathbf{Z}}$ and $m_n : V^{\otimes n} \to V[n-2]$, $n \geq 2$, be a sequence of morphisms. It gives rise to an action on V of the free operad $P = Free_{\mathcal{O}\mathcal{P}}(F)$ where

$$F(V) = \oplus_{n\geq 2} V^{\otimes n}[n-2].$$

Then $F_n = k[S_n]m_n \otimes k[1]^{\otimes(n-2)}$. This notation means that we consider F_n as a space (with the grading shifted by $n-2$) of the regular representation of the group algebra of the symmetric group S_n. This space is generated by an element which we denote by m_n.

The differential $d_P \in g_P$ (equivalently, a structure of a dg-operad on P) is defined by the standard formulas:

$$d_P(m_2) = 0,$$
$$d_P(m_n)(v_1 \otimes \cdots \otimes v_n) =$$
$$\sum_{k+l=n} \pm m_k(v_1 \otimes \cdots \otimes v_i \otimes m_l(v_{i+1} \otimes \cdots \otimes v_{i+l}) \otimes \cdots \otimes v_n), \quad n > 2.$$

We do not specify signs in these well-known formulas. In Section 5 we propose a general framework allowing to fix signs in the formulas like the one above.

Definition 11 *The dg-operad $\mathcal{A}_\infty = (P, d_P)$ is called the A_∞-operad. Algebras over this dg-operad are called A_∞-algebras.*

Deformations of an A_∞-algebra A are controlled by the truncated Hochschild complex

$$C_+^\bullet(A,A) = \prod_{n\geq 1} Hom_{Vect_{\mathbf{Z}}}(A^{\otimes n}, A)[-n].$$

More precisely, let A be a graded vector space. We define a graded vector space of all Hochschild cochains of A as

$$C^\bullet(A,A) = \prod_{n\geq 0} Hom_{Vect_{\mathbf{Z}}}(A^{\otimes n}, A)[-n].$$

Then $C^\bullet(A,A)[1]$ can be equipped with the structure of a graded Lie algebra with the Gerstenhaber bracket (the latter appears naturally if we interpret Hochschild cochains as coderivations of the free coalgebra $\oplus_{n>0}(A[1])^{\otimes n}$).

Let us consider an element $m = (m_1, m_2 \ldots) \in C_+^\bullet(A,A)[1]$ of degree $+1$ such that $[m,m] = 0$. Such an element defines a differential $d = m_1$ on A, and the sequence $(m_2, m_3, \ldots)$ gives rise to a structure of an A_∞-algebra on (A, m_1).

Then we can make $C^\bullet(A,A)$ into a complex (Hochschild complex) with the differential $d_m = [m, \cdot]$. It is easy to see that in this way we get a differential graded Lie algebra (DGLA for short) $(C^\bullet(A,A)[1], d_m)$. The truncated Hochschild complex $C_+^\bullet(A,A)[1]$ is a DGLA subalgebra. According to the general theory (see [12]) both DGLAs define formal pointed dg-manifolds, and therefore give rise to the deformation functors. This is a straightforward generalization of the well-known deformation theory of associative algebras.

In some sense, the full Hochschild complex controls deformations of the A_∞-category with one object, such that its endomorphism space is equal to A.

The deformation theory of A_∞-categories is not in the scope of present paper. It will be explained elsewhere. Nevertheless we will refer to the formal dg-manifold associated with $C^\bullet(A,A)[1]$ as to the *moduli space of A_∞-categories*. Similarly, the formal dg-manifold associated with $C^\bullet_+(A,A)[1]$ will be called the *moduli space of A_∞-algebras*. (All the terminology assumes that we deform a given A_∞-algebra A).

The moduli space of A_∞-algebras is the same as $\mathcal{M}(\mathcal{A}_\infty, A)$ in the previous notation. Similarly we will denote the moduli space of A_∞-categories by $\mathcal{M}_{cat}(\mathcal{A}_\infty, A)$. The natural inclusion of DGLAs $C^\bullet_+(A,A)[1] \to C^\bullet(A,A)[1]$ induces a dg-map (dg-map is a morphism of dg-manifolds):

$$\mathcal{M}(\mathcal{A}_\infty, A) \to \mathcal{M}_{cat}(\mathcal{A}_\infty, A).$$

Let us remark that the operad $\mathcal{A}_\infty$ gives rise to a free resolution of the operad As. Algebras over the latter are associative algebras without the unit.

Remark 2 It is interesting to describe deformation theories of free resolutions of the classical operads As, *Lie*, *Comm*. It seems that for an arbitrary free resolution P of either of these operads the following is true: $H^i(g_P) = 0$ for $i \neq 0$, $H^0(g_P) = k$. This one-dimensional vector space gives rise to the rescaling of operations, like $m_n \mapsto \lambda^n m_n$ in the case of A_∞-algebras.

2.2. Homotopical actions of Lie algebras

Let g be a Lie algebra acting on a formal dg-manifold (Y, d_Y). This means that we have a homomorphism of Lie algebras $g \to Der(Y), \gamma \mapsto \hat{\gamma}$ where $Der(Y)$ is the Lie algebra of vector fields on Y preserving $\mathbf{Z}$-grading an d_Y.

We can make $Z = Y \times g[1]$ into a formal dg-manifold by introducing an odd vector field by the following formula

$$d_Z(y, \gamma) = (d_Y(y) + \hat{\gamma}, [\gamma, \gamma]/2).$$

Then $[d_Z, d_Z] = 0$. We can make $g[1]$ into a formal dg-manifold by using the odd vector field $d_{g[1]}$ arising from the Lie bracket. Then the natural projection $(Z, d_Z) \to (g[1], d_{g[1]})$ becomes a dg-bundle (cf. [13]).

This construction of a dg-bundle out of a group or Lie algebra acting on a dg-manifold motivates the following definition.

Definition 12 *Let g be a Lie algebra. Homotopical g-action on a formal dg-manifold (Y, d_Y) is a dg-bundle $\pi : (Z, d_Z) \to (g[1], d_{g[1]})$ together with an isomorphism of dg-manifolds $(\pi^{-1}(0), d_Z) \simeq (Y, d_Y)$.*

Remark 3 It was pointed out in [13] that in this case g acts on the cohomology of all complexes naturally associated with (Y, d_Y) (like the tangent space at a zero point of d_Y, the space of formal functions on Y, etc.).

We remark also that if g is a DGLA then the same definition can be given. It can be also generalized to the case when the base of a equivariant dg-bundle is an arbitrary dg-manifold with a marked d-stable point.

Suppose that F is a polynomial functor in the category of $\mathbf{Z}$-graded vector spaces, $P = Free(F), V \in Vect_{\mathbf{Z}}$. We apply the general scheme outlined above to the case $Y = \mathcal{M}(P, V)$, $g = g_P$. Obviously g acts on the dg-manifold $\underline{Hom}(F(V), V)$ equipped with the trivial odd vector field.

Let us consider the graded vector space

$$\mathcal{N} = \underline{Hom}(V,V)[1] \oplus \underline{Hom}(F(V),V) \oplus g_P[1].$$

Let $d_V \in \underline{Hom}(V,V)[1]$, it makes V into a complex, $\gamma = d_P \in g_P[1]$ satisfies the equation $[d_P, d_P] = 0$ and $\rho \in \underline{Hom}(F(V), V)$ makes V into a dg-algebra over (P, d_P).

We consider the formal neighborhood of the point (d_V, ρ, d_P) in $\mathcal{N}$, and define an odd vector field by the formula

$$d_{\mathcal{N}}(d_V, \rho, d_P) = (d_V^2, \xi(d_V, \rho) + \hat{d}_P, [d_P, d_P]/2).$$

The notation here is compatible with the one for $\mathcal{M}$.

One can check that $[d_{\mathcal{N}}, d_{\mathcal{N}}] = 0$. Thus the formal neighborhood becomes a formal dg-manifold. It controls deformations of pairs *(an operad, an algebra over this operad)*.

The natural projection $\pi : \mathcal{N} \to g_P[1]$ is a morphism of formal dg-manifolds. Here on $g_P[1]$ we use the odd vector field $d_{g_P[1]}$ defined by the Lie bracket. Then the formal scheme of zeros of $d_{g_P[1]}$ corresponds to the structures of a dg-operad on P. The fiber over a fixed point $x \in g_P[1]$ is a dg-manifold with the differential induced from $\mathcal{N}$. Then the formal neighborhood of a fixed point in $\pi^{-1}(x)$ controls deformations of $\widehat{P}$-algebras.

We conclude that the Lie algebra of derivations of an operad acts homotopically on the moduli space of algebras over this operad.

3. Hochschild complex and operads

This section serves as a sort of a "second introduction", outlining objectives and the strategy of the rest of the paper.

One of our aims will be to construct a dg-operad of the type $\widehat{P}$ (i.e. free as a graded operad) acting naturally on the Hochschild complex of an arbitrary A_∞-algebra.

In Section 5 we are going to construct an operad M which acts naturally on the full Hochschild complex $C^\bullet(A, A)$ of an A_∞-algebra A as well as on $C_+^\bullet(A, A)$.

There is a natural free resolution P of the operad M, so that $C := C^\bullet(A,A)$ becomes a $\widehat{P}$-algebra. Then we can say that there is a dg-map of the moduli space of A_∞-categories to the moduli space $\mathcal{M}(P,C)$ of structures of $\widehat{P}$-algebras on the graded vector space C.

From the point of view of deformation theory it is not very natural to make constructions of the type *algebraic structure* $\to$ *another algebraic structure* (like our construction $A_\infty - algebras \to M - algebras$). It is more natural to extend them to morphisms between the formal pointed dg-manifolds controlling the deformation theories of structures.

We will construct an explicit dg-map $\mathcal{M}(\mathcal{A}_\infty, A) \to \mathcal{M}(P,C)$ as well as a dg-map $\mathcal{M}_{cat}(\mathcal{A}_\infty, A) \to \mathcal{M}(P,C)$, such that the one is obtained from another by the restriction from the moduli space of algebras to the moduli space of categories.

The operad $\mathcal{A}_\infty$ is augmented, i.e., equipped with a morphism of dg-operads $\eta: \mathcal{A}_\infty \to Free(0)$. Here $Free(0)$ is the trivial operad:

$$Free(0)_1 = k \quad \text{and} \quad Free(0)_{(n\neq 1)} = 0.$$

Since A (as any graded vector space) is an algebra over $Free(0)$, it becomes also an algebra over $\mathcal{A}_\infty$. Any structure of an $\mathcal{A}_\infty$-algebra on A can be considered as a deformation of this trivial structure. Notice also that in the previous notation the augmentation morphism defines a point in the dg-manifold $\mathcal{M}(\mathcal{A}_\infty, C)$, where $C = C^\bullet(A,A)$. Therefore it is sufficient to work in the formal neighborhood of this point.

Notice that we can also consider the moduli space of structures of a complex on the graded vector space $C^\bullet(A,A)$ where A is an arbitrary graded vector space. It gives rise to a formal dg-manifold. There are natural morphisms to it from the formal dg-manifold of the moduli space of A_∞-categories and from the formal dg-manifold of the moduli space of structures of $\widehat{P}$-algebras on $C^\bullet(A,A)$. Theorem 1 below combines all three morphisms discussed above into a commutative diagram. Let us make it more precise.

First we formulate a simple general lemma, which will be applied in the case $V = C[2]$.

Let V be an arbitrary graded vector space, d_V an odd vector field on V (considered as a graded manifold) such that $[d_V, d_V] = 0$. Thus we get a dg-manifold. The graded vector space $H = H(V) = \underline{Hom}(V,V)[1]$ is a dg-manifold with $d_H(\gamma) = \gamma^2$. To every point $v \in V$ we assign a point in H by taking the first Taylor coefficient $d_V^{(1)}(v)$ of d_V at v. In this way we obtain a map $\nu : V \to H$.

Lemma 2 *The map ν is a morphism of dg-manifolds.*

Proof. Let us write in local coordinates $x = (x_1, \ldots, x_n)$ the vector field $d_V = \sum_i \phi_i \partial_i$ where ∂_i denotes the partial derivative with respect to x_i, and ϕ_i are functions on V. Then the map ν assigns to a point x the matrix $M = (M_{ij}(x))$ with $M_{ij} = \partial_j \phi_i$. Then direct computation shows that the condition $[d_V, d_V] = 0$ implies that the vector field $\dot{x} = d_V(x)$ is mapped to the vector field $\dot{M} = d_H(M) = M^2$. ■

Let A be a graded vector space endowed with the trivial A_∞-structure, and $C = C^\bullet(A,A) = \prod_{n\geq 0} Hom_{Vect_{\mathbf{Z}}}(A^{\otimes n}, A)$ be the graded space of Hochschild cochains. Since $C[1]$ carries a structure of a graded Lie algebra when endowed with the Gerstenhaber bracket, it gives rise to the structure of a dg-manifold on $C[2]$, which is the same as $\mathcal{M}_{cat}(\mathcal{A}_\infty, A)$. We will denote it by (X, d_X) (or simply by X for short).

Even for the trivial A_∞-algebra structure on A, we get a non-trivial P-algebra structure on C. The corresponding moduli space $\mathcal{M}(P,C)$ is denoted by (Y, d_Y) (or Y for short).

There is a natural morphism of dg-manifolds $p : Y \to \underline{Hom}(C,C)[1] = H$ (projection of $Y = \mathcal{M}(P,C)$ to the first summand).

The following theorem will be proved later in the paper.

Theorem 1 *There exists a $GL(A)$-equivariant morphism of dg-manifolds $f : X \to Y$ such that $pf = \nu$.*

Moreover we will present an explicit construction of the morphism.

Suppose that A is an A_∞-algebra. Geometrically the structure of an A_∞-algebra on the graded vector space A gives rise to a point $\gamma \in X = C[2], C = C^\bullet(A,A)$ such that $d_X(\gamma) = 0$. Indeed, the definition can be written as $[\gamma, \gamma] = 0$. Thus we get a differential in C (commutator with γ) making it into a complex. The structure of a complex on the graded vector space C gives rise to a zero of the field d_H in the dg-manifold $H = \underline{Hom}(C,C)[1]$. Theorem 1 implies that $f(\gamma)$ is a zero of the vector field $d_{\mathcal{M}(P,C)}$. Therefore the Hochschild complex $(C, [\gamma, \cdot])$ carries a structure of a dg-algebra over P.

Our next aim is to uncover the geometric origin of the operad P. It will be related to the configuration space of discs inside of the unit discs in the plane. More precisely, the operad $Chains(E_2)$ of chains on the little discs operad (see [14] and Section 7 below) is quasi-isomorphic to $\widehat{P}$. Here we use either usual singular chains or piecewise algebraic chains (see the Appendix). In fact we are going to construct explicitly a morphism $\widehat{P} \to Chains(E_2)$ which gives the homotopy equivalence (to be more precise we will do that for the operad $Chains(FM_2)$ which is quasi-isomorphic to E_2). Then using the fact that both dg-operads are free as graded operads, we invert this quasi-isomorphism. This gives a structure of an $Chains(E_2)$-algebra on the Hochschild complex of an A_∞-algebra. This result is known as the Deligne conjecture.

Let us recall that there is a notion of d-algebra, $d \in \mathbf{Z}_+$ (see e.g. [14]). Namely, a graded vector space V is called a d-algebra if it is an algebra over the operad $Chains(E_d)$ (chains of the topological operad of little d-dimensional discs). Thus the moduli space $\mathcal{M}(P,C)$ can be thought as a moduli space of structures of a 2-algebra on a graded vector space C. Then our Theorem 1 says that there is a $GL(A)$-equivariant morphism of the moduli space of A_∞-categories with one object to the moduli space of 2-algebras.

Remark 4 In the unpublished paper [7] the name d-algebras was reserved for algebras over the operad $H_{\bullet}(Chains(E_d))$. It was proved by Tamarkin in [21] and by the first author in [14] that there exists a (non-canonical) quasi-isomorphism between the operad $Chains(E_d)$ and its homology operad $H_{\bullet}(Chains(E_d))$ (in other words the operad $Chains(E_d)$ is formal).

Let $g_P = \underline{DerP}$ denote as before the DGLA of derivations of $\widehat{P}$. Then g_P acts on the moduli space of $\widehat{P}$-algebras. On the other hand, from the point of view of deformation theory, we can replace the operad of little discs E_2 by the operad of configurations of points in the plane (properly compactified). This is the operad FM_2 mentioned above. There is a natural action of the Grothendieck-Teichmüller groups on the rational homotopy type of the latter. It gives rise to a morphism of L_∞-algebras $Lie(GT)[1] \to (g_P, [d_P, \cdot])$ where GT is the Grothendieck-Teichmüller group.

Therefore one has a homotopical action of the Lie algebra $Lie(GT)$ on the moduli space of $\widehat{P}$-algebras.

4. Free resolution of a dg-operad

In this section we recall well-known constructions of free resolutions of operads (see e.g. [7]).

Let k be a field as before, R be a dg-operad over k. The aim of this subsection is to construct canonically a dg-operad P_R over k, which is free as a graded operad, and a quasi-isomorphism $P_R \to R$. Then P_R will be a free resolution of R. In this subsection we will assume that R is non-trivial, which means that the unit operation from R_1 is not equal to zero.

4.1. Topological construction

We will mainly follow [3].

Let $O = (O_n)_{n\geq 0}$ be a topological operad (i.e., all O_n are S_n-topological spaces and all operadic morphisms are continuous). We describe (following [3]) a construction of a topological operad $B(O) = (B(O)_n)_{n\geq 0}$ together with a morphism of topological operads $B(O) \to O$ which is homotopy equivalence.

To simplify the exposition we assume that S_n acts freely on O_n for all n.

Each space $B(O)_n$ will be the quotient of

$$\overline{B(O)_n} = \bigsqcup_{[T], T\in Tree(n)} ([0,+\infty]^{E_i(T)} \times \prod_{v\in V_i(T)} O_{N^{-1}(v)})/AutT$$

under the relations described in the following way.

Let us consider the elements of $\overline{B(O)_n}$ as numbered trees with elements of O attached to the internal vertices, and the *length* $l(e) \in [0,+\infty]$ attached to every edge e. We require that all external edges (i.e., root edge and the tail edges) have lengths $+\infty$.

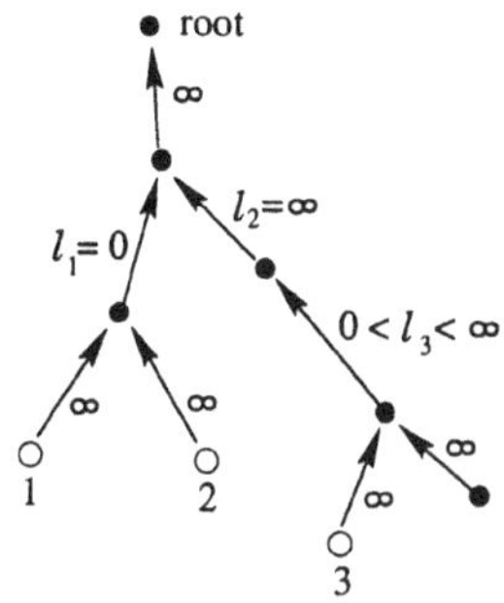

Figure 2. Tree with lengths.

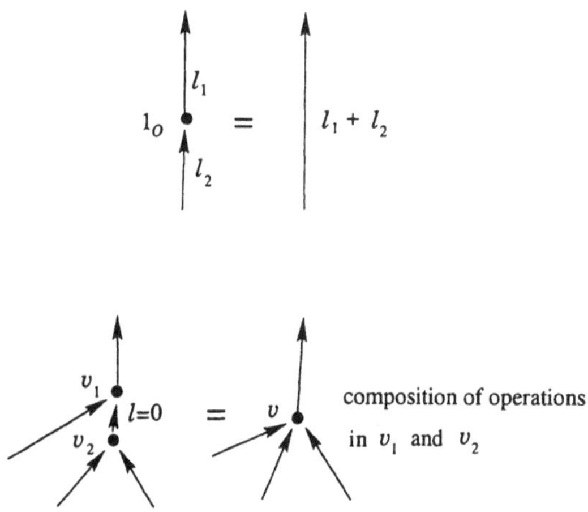

Figure 3. Two basic identifications between trees with lengths.

We impose two type of relations:

1) We can delete every vertex v of valency 1 if it contains the unit of the operad, replacing it and the attached two edges of lengths $l_i, i = 1, 2$ by the edge with the length $l_1 + l_2$. We use here the usual assumption: $l + \infty = \infty + l = \infty$.

2) We can contract every internal edge $e = (v_1, v_2), v_2 = N(v_1)$ of the length 0 and compose in O the operations attached to $v_i, i = 1, 2$.

We depict the trees and relations in Figures 2 and 3.

Let us describe how $B(O)$ can act naturally on a topological space.

Let X be a topological space, and $g^t : X \to X, t \in [0, +\infty)$ a 1-parameter semigroup of continuous maps acting on X.

We assume that the map $[0, +\infty) \times X \to X, (t, x) \mapsto g^t x$ extends continuously to $[0, +\infty] \times X$. We denote by Y the image of $g^\infty := lim_{t \to \infty} g^t$. Clearly the subspace Y is a homotopy retract of X.

Suppose that a topological operad O acts on X, i.e., that we are given continuous maps $O_n \times X^n \to X, n \geq 0$, satisfying the usual properties. We can construct an action of $B(O)$ on Y as follows. Let $\gamma \in O_n$, $t, t_i \in \mathbf{R}_+ \cup \{+\infty\}$, $x_i \in X$, $1 \leq i \leq n$. Then we assign to these data the point $x = g^t\gamma(g^{t_1}x_1, ..., g^{t_n}x_n)$ of X. We define the composition of such operations in the natural way.

We can interpret the parameters t, t_i above as lengths of edges of trees. Putting $t = +\infty$ we obtain an action of $B(O)$ on the homotopy retract Y.

4.2. Free resolutions of linear operads

Let R be a dg-operad over a field k. To describe its free resolution P_R we need a special class of trees described below.

For every $n \geq 0$ we introduce a groupoid $\mathcal{T}(n)$ of *marked trees* with n tails. An object of $\mathcal{T}(n)$ is a numbered tree $T \in Tree(n)$ and a map to a 3-element set $l = l_T : E(T) \to \{0, finite, +\infty\}$ such that $l_T(\{e_r\} \cup E_t(T)) = \{+\infty\}$. Notice that in the case of topological operads the component $\overline{B(O)_n}$ is stratified naturally with the strata labeled by equivalence classes $|\mathcal{T}(n)|$ of marked trees. The label of an edge e of the corresponding marked tree is 0 if $l(e) = 0$, is *finite* if $l(e) \in (0, +\infty)$ and is $+\infty$ if $l(e) = +\infty$. According to this description, we call them *zero, finite, infinite* edges respectively. We denote these sets of edges by E_{zero}, E_{finite} and $E_{infinite}$ correspondingly.

We will give three different but equivalent descriptions of the operad $P = P_R$ as a graded operad. Then we define a differential on P.

Description 1.

Let $\bar{P}_n = \bigoplus_{[T], T \in \mathcal{T}(n)} (\otimes_{v \in V(T)} R_{N^{-1}(v)}[J_T])_{AutT}$ where $AutT$ is the group of automorphisms of the tree T, $J_T = l_T^{-1}(finite)$, and for any graded vector space W and a finite set J we use the notation $W[J] = W \otimes k[1]^{\otimes J}$ (shift of the grading by J).

Notice that the dimension of the corresponding stratum of $\mathcal{T}(n)$ is the cardinality of the set J_T, i.e., the number of finite edges.

Then $(\bar{P}_n)_{n \geq 0}$ evidently form a graded operad $\bar{P}$. It is a k-linear analog of the operad $\overline{BO}$.

The operad $\bar{P}$ contains a subspace I_P generated by the following relations:

1) If the length of an edge (w, v) is 0 we contract it and make the composition in R of the operations attached to w and v (cf. description for $B(O)$);

2) If v is a vertex of $T \in \overline{P}$ such that $|v| = 1$ and $1_R \in R_1$ is attached to v, then T belongs to I_P if the following holds: Lengths of both edges attached to v are non-zero edges, and at least one of them is finite. If both edges are infinite, we remove the vertex and two attached edges, replacing them by an infinite edge.

One can check easily that I_P is a graded ideal in $\bar{P}$. We denote by P the quotient operad $\bar{P}/I_P$.

Description 2.

We define P_n in the same way, but making the summation over all trees without edges of zero length. We also drop the relation 1) from the list of imposed relations (there are no edges with $l = 0$).

Description 3.
We define an operad R' such as follows: $R'_n = R_n$ for $n \neq 1$, R'_1 is a complement to the subspace $k \cdot 1_R$ in R_1. Then we define P_n as in Description 2, but using R' instead of R and dropping both relations 1) and 2).

It is clear that this description defines a free graded operad. Equivalently, it can be described as a free graded operad P such that

$$P = Free(Cofree'(R'[1]))[-1].$$

Here $Cofree(L)$ means a dg-cooperad generated by L which is cofree as a graded co-operad, and $'$ denotes the procedure of taking a (non-canonical) complement to the subspace generated by the unit (or counit in the case of a co-operad) as described above in the case of R.

In this description the generators of P correspond to such trees T in $\mathcal{T} = (\mathcal{T}(n))_{n\geq 0}$ that every T has at least one internal vertex, all internal edges are finite and there are no zero-edges in T.

Proposition 2 *All three descriptions give rise to isomorphic graded free operads over k.*

Proof. Straightforward. ∎

We can make $\bar{P}$ into a dg-operad introducing a differential $d_{\bar{P}}$. We use the Description 1 for this purpose.

The differential $d_{\overline{P}}$ is naturally decomposed into the sum of two differentials: $d_{\overline{P}} = \tilde{d}_R + d_{\mathcal{T}}$ where:

a) The differential $\tilde{d}_R$ arises from the differential d_R in R;

b) The differential $d_{\mathcal{T}}$ arising from the stratification of $\mathcal{T}(n)$: It either contracts a finite edge or makes it into an infinite edge.

To be more precise, let us consider the following object Δ in $\mathcal{C} = Vect_{\mathbf{Z}}$: $\Delta^{-1} = 1_{\mathcal{C}}, \Delta^0 = 1_{\mathcal{C}} \oplus 1_{\mathcal{C}}$ where $1_{\mathcal{C}}$ is the unit object in the monoidal category $\mathcal{C}$. Then Δ can be made into a chain complex of the CW complex $[0, +\infty] = \{0\} \cup (0, +\infty) \cup \{+\infty\}$.

We see that as a graded space $\bar{P}_n$ is given by the formula

$$\bar{P}_n = \oplus_{[T], T\in Tree(n)} (\otimes_{v\in V_i(T)} R_{N^{-1}(v)} \otimes \Delta^{\otimes E_i(T)})_{AutT}.$$

Since we have here a tensor product of complexes, we get the corresponding differential $d_{\bar{P}}$ in $\bar{P}$.

Proposition 3 *The ideal I_P is preserved by $d_{\bar{P}}$.*

Proof. Straightforward computation. ■

Therefore $P = P_R$ is a dg-operad which is free as a graded operad.

There is a natural morphism of dg-operads $\phi : P \to R$. In terms of the Description 2 it can be defined as follows: ϕ sends to zero all generators of P corresponding to trees with at least one finite edge. Let $T \in P$ be a tree with all infinite edges. Then T gives rise to a natural rule of composing in R elements of $R_{N^{-1}(v)}$ assigned to the vertices of T. We define $\phi(T) \in R$ as the result of this composition.

It is easy to check that ϕ is a well-defined morphism of dg-operad.

Proposition 4 *The morphism ϕ is a quasi-isomorphism of dg-operads.*

Proof. Follows from the spectral sequence arising from the natural stratification of $\mathcal{T}$. To say it differently, let us consider the tautological embedding ψ of R into P. Then ψ is a right inverse to ϕ. It gives a splitting of P into the sum $P = \psi(R) \oplus P^{(0)}$. Here $P^{(0)}$ is spanned by the operations corresponding to trees with finite edges only. Such a tree can be contracted to a point which means that $P^{(0)}$ is contractible as a complex. Hence ϕ defines a quasi-isomorphism of complexes and dg-operads. ■

4.3. Example

Let us discuss an example when R is the operad of associative algebras without the unit. We denote it by As. Then for any $n \geq 1$ we have: As_n is isomorphic to the regular representation of the symmetric group S_n.

In this case the complex P_n from the previous subsection can be identified with the chain complex of the CW-complex K_n, $n \geq 2$, described below.

The cells of K_n are parametrized by *planar trees* with an additional structure on edges. By a planar tree here we understand a numbered tree T such that for any $v \in V_i(T)$ the cardinality of $N^{-1}(v)$ is at least 2 and this set is completely ordered. The additional structure is a map $E_i(T) \to \{finite, infinite\}$. We call an edge finite or infinite according to its image under this map. Dimension of the cell is equal to the number of finite edges of the corresponding planar tree.

We can either contract a finite edge or make it infinite. This defines an incidence relation on the set of cells.

An example of planar tree is given in Figure 4. Here the dashed lines show the complete orders on the set of incoming edges. We will not show them on other figures in the text. Instead, we will tacitly assume that for a given vertex the incoming edges are completely ordered from the left to the right.

In this way we obtain cubical subdivisions of the Stasheff polyhedra. The case $n = 4$ is represented in Figure 5.

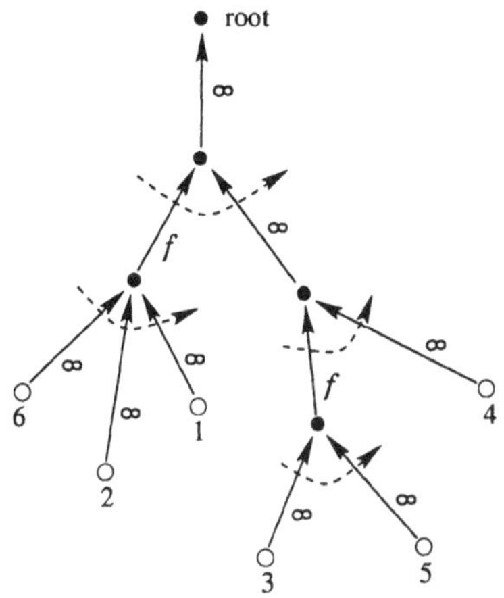

Figure 4. A planar tree which gives a 2-dimensional cell in K_6

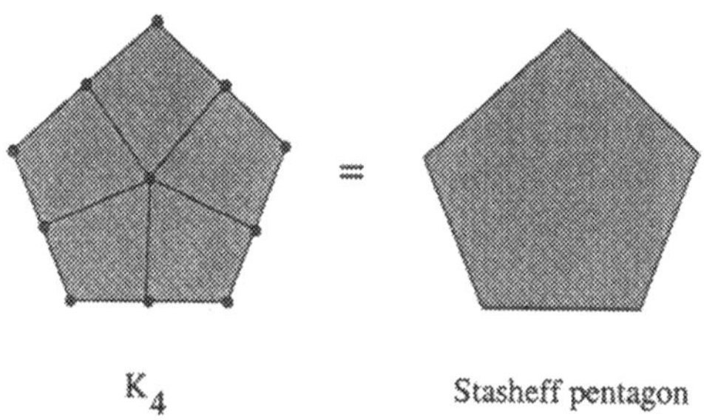

Figure 5. Cell complex K_4 has 11 vertices, 15 edges and 5 two-dimensional faces. It is homeomorphic to the Stasheff pentagon.

5. Minimal operad

In this section we describe a dg-operad $M = (M_n)_{n\geq 1}$ which acts naturally on the Hochschild complex of an A_∞-algebra. We call it *minimal* operad. Let us describe this operad informally. We treat elements $\gamma \in C^\bullet(A,A)$ as polylinear operations on A. For given operations $\gamma_1, \dots, \gamma_n \in C^\bullet(A,A)$, and an A_∞-structure $m \in C^\bullet(A,A)$ we can make compositions in all possible ways, reading the arguments from left to right. For example we can make an expression like this:

$$\gamma(a_1 \otimes \cdots \otimes a_5) = \gamma_2(a_1 \otimes a_2 \otimes m_2(a_3 \otimes \gamma_1(a_4)) \otimes \gamma_3(a_5)), \quad \text{etc.}$$

Such compositions can be depicted by planar trees. The operad M is spanned by operations corresponding to certain trees, which we call admissible (see below).

It seems that the operad M is close to what is described in [18] as "natural transformations $(C^\bullet(A,A))^{\otimes n} \to C^\bullet(A,A)$" (this terminology is confusing because the assignment $A \mapsto C^\bullet(A,A)$ is not a functor).

In all previous works dealing with the Deligne conjecture the authors used operads which act on $C^\bullet(A,A)$ and generated by operations called braces. It seems that the braces generate the operad M, but it is not clear what is the complete list of relations. The advantage of our operad M is that it is described directly, not as a quotient of a free operad.

The defining properties of a dg-operad (i.e., associativity of the composition, Leibniz rule for the differential) will become clear later. We will give two definitions of the operad M. The first one is suitable for the pure algebraic descriptions of the operadic composition and the differential. But the signs in the formulas are not very transparent. Second description takes care about "parity of edges", so the correct signs come out automatically. In some formulas related to the first description we will write $\pm$ having in mind that the correct sign follows from the second description.

5.1. Basis of M

Definition 13 *For a finite set I we define an I-labeled planar tree as a triple (T, lab, ord) where T is a tree in the sense of Section 1.2, $lab : I \hookrightarrow V_i(T)$ is an embedding, and ord is a complete order on the sets $N^{-1}(v), v \in V_i(T)$.*

We call *labeled* a vertex from the image of the map lab. All other internal vertices are called *non-labeled.*

Definition 14 *We will call an I-labeled tree admissible if it has no tail vertices, and for every non-labeled internal vertex v we have $|v| \geq 2$.*

We denote the set of isomorphism classes of I-labeled planar trees by $Tree^{(p)}(I)$ where the upper script p stands for "planar". Notice that the automorphism group of an I-labeled planar tree is trivial. For $I = \{1,...,n\}$ we will use the notation $Tree^{(p)}(n)$.

We are going to use admissible trees unless we say otherwise. If it will not lead to a confusion, we will simply call them trees. Slightly abusing the notation we will denote an I-labeled tree by T, skipping *lab* and *ord*. Notice that the terminology here is different from the one in Section 1.5. In particular, we label here internal vertices, not tails. Since we do not consider here numbered trees (in the terminology of Section 1.5), this change of terminology should not lead any confusion.

We can depict trees from $Tree^{(p)} = (Tree^{(p)}(n))_{n\geq 1}$ as in Figure 6.

Labeled vertices are depicted as circles with numbers inscribed, non-labeled vertices are depicted as black vertices.

We define M_I to be a k-vector space spanned by all elements of $Tree^{(p)}(I)$. For $I = \{1,...,n\}$ we will use the notation M_n. The symmetric group S_n acts on M_n by permuting labeled vertices.

Abusing the notation further, we will denote the element of M_n corresponding to a tree T simply by T. Thus we have: $M_0 = 0$ and M_1 is a 1-dimensional vector

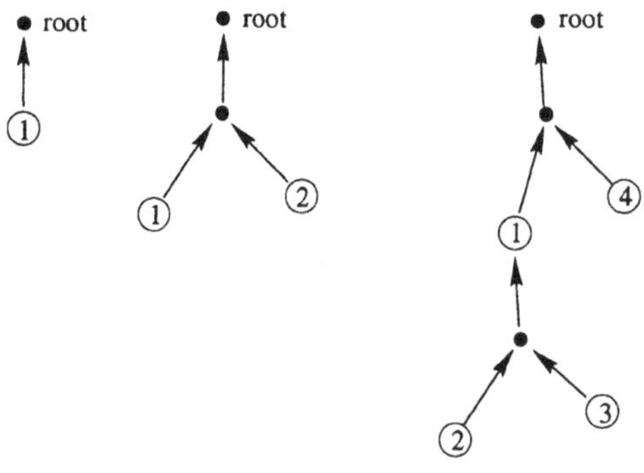

Figure 6. Three examples of admissible labeled planar trees.

space generated by T_e, the first tree on the figure above. In fact T_e corresponds to the unit $1_M \in M_1$ of the operad M.

The operadic composition in M and the differential will be described below.

The degree of the basis element corresponding to a tree T is equal to

$$deg(T) = \sum_{v \in V_{lab}(T)} (-|v|) + \sum_{v \in V_{nonl}(T)} (2-|v|),$$

where $V_{lab}(T)$ and $V_{nonl}(T)$ denote the sets of labeled and non-labeled vertices respectively , and $|v|$ is the cardinality of the set $N^{-1}(v)$.

5.2. Composition in M

We need to define an element of M which corresponds to a tree T_2 glued to a tree T_1 at a labeled vertex $v \in V_{lab}(T_1)$. The trees T_i, $i = 1,2$ correspond to some elements of M. The resulting element will be by definition their operadic composition. It is given by the sum

$$T_1 \circ_v T_2 = \sum_\beta \pm (T_1 \circ_v T_2)_\beta,$$

where the trees $(T_1 \circ_v T_2)_\beta$ are defined below.

First, with the tree T_2 we associate a set $A(T_2) = \bigsqcup_{v \in V_i(T_2)} \{0, ..., |v|\}$. We call it the set of *angles* of T_2. Obviously there is a natural map $\kappa : A(T_2) \to V_i(T_2)$. The path in $\mathbf{R}^2$ which goes from the left to the right and surrounds T_2 defines a complete order on $A(T_2)$. In Figure 7, angles are marked by asterisks.

The datum β above is a (non-strictly) monotonic map $\beta : N^{-1}(v) \to A(T_2)$. We will think of this map as about the way to insert a vertex $w \in N^{-1}(v)$ and the edge (w,v) inside of an angle formed by two edges incoming to $\beta(v)$. Schematically it is shown in Figure 8.

Let $T_\beta = (T_1 \circ_v T_2)_\beta$. Then we define the set of vertices $V(T_\beta)$ as $(V(T_1) \setminus \{v\}) \bigsqcup (V(T_2) \setminus \{root\})$. The map $N = N_{T_\beta}$ is defined as follows: For all $w \in V(T_1) \setminus$

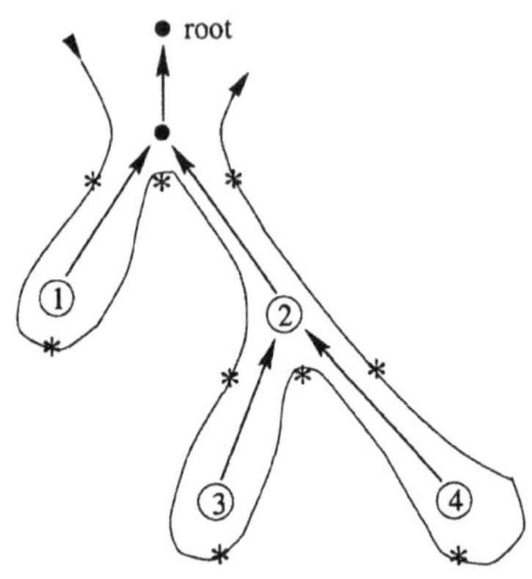

Figure 7. A path from the left to the right passing through all angles.

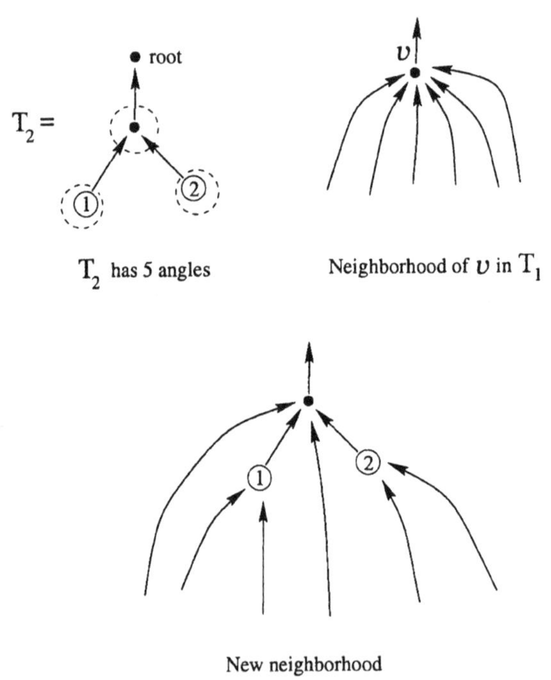

Figure 8. A way to insert tree T_2 in the place of vertex v in tree T_1.

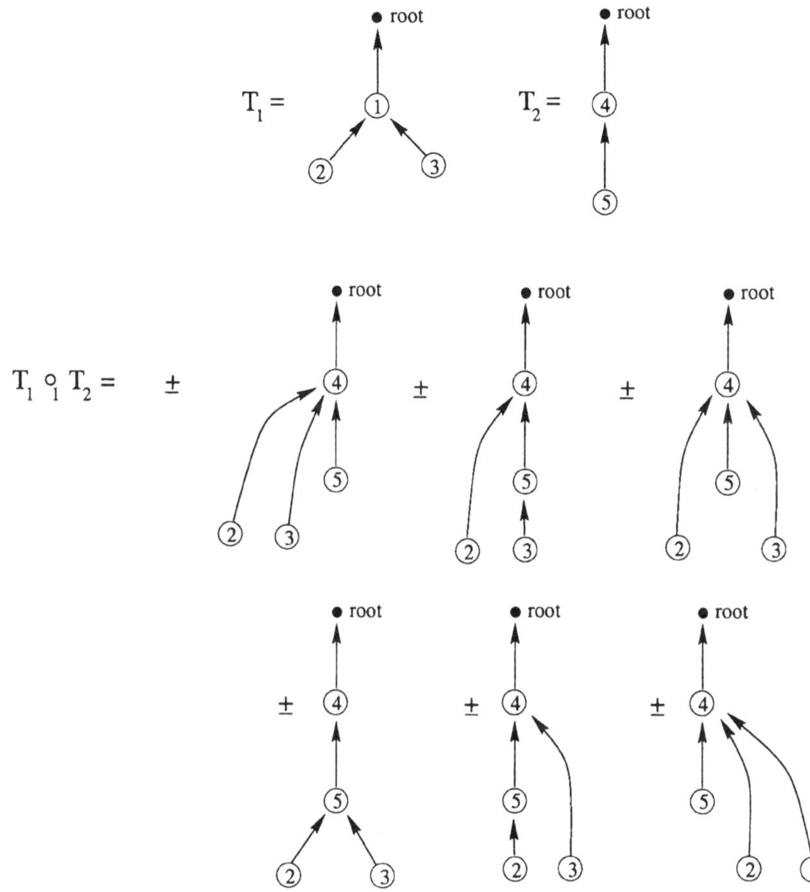

Figure 9. An example of the composition in M.

$\{v\}$ such that $N_{T_1}(w) \neq v$ we put $N(w) = N_{T_1}(w)$ in a self-explained notation. Similarly if $w \in V(T_2), N_{T_2}(w) = root$ we put $N(w) = N_{T_1}(v)$. If $w \in V(T_2), N_{T_2}(w) \neq root_{T_2}$ then we put $N(w) = N_{T_2}(w)$. Let us suppose that $w \in N_{T_1}^{-1}(v)$. Then we define $N(w)$ as $\kappa(\beta(w))$. The root vertex of T_β is the same as for T_1. The labeling and complete orders on the sets $N^{-1}(x)$ are defined in the natural way. Informally speaking, T_β is obtained by removing from T_1 the vertex v together with all incoming edges and vertices, and gluing T_2 to v. Then we use the map β in order to "insert" removed vertices. With such a composition we obtain the structure of a graded operad on M.

We depict an example of the composition in M in Figure 9.

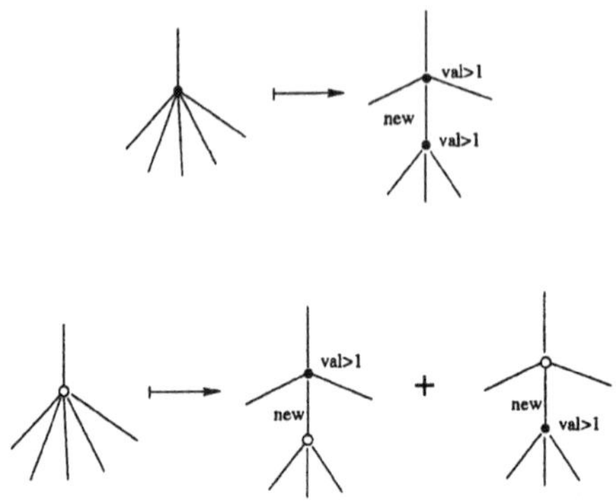

Figure 10. Two types of terms in the formula for differential in M.

5.3. Differential in M

For a generator T we define $d_M(T) = \sum_{v \in V_i(T)} d_v(T)$ where each $d_v(T)$ will be of the form

$$d_v(T) = \sum_{i,j} \pm d_{v,i,j}(T).$$

We need to explain the set of indices of summation and each summand. Let us recall that for every vertex $v \in V_i(T), |v| = k$ there is a bijection of sets $\{1, ..., k\} \to N^{-1}(v)$ which defines a complete order on the set $N^{-1}(v)$. We will identify an element of $N^{-1}(v)$ with the corresponding number. The indices i, j in the sum above will be half-integers: $1/2 \le i \le j \le k + 1/2$. For a pair i, j, we define $d_{v,i,j}(T)$ to be a tree T' such that:

a) $V(T') = (V(T) \setminus \{v\}) \cup \{v^{up}, v^{down}\}$ where $\{v^{up}, v^{down}\}$ are new vertices;

b) $root_{T'} = root_T$;

We put

$$N_{T'}(v^{down}) = v^{up};$$
$$N_{T'}(l) = v^{down} \quad \text{if } i < l < j;$$
$$N_{T'}(l) = v^{up}, \quad \text{if } l < i \text{ or } l > j.$$

For all other vertices u we put $N_{T'}(u) = N_T(u)$.

Complete orders on the sets $N_{T'}^{-1}(u)$ are defined in the natural way. The tree T' has one new edge (v^{down}, v^{up}) which we denote in pictures as "new".

We can depict these definitions as in Figure 10.

The range of summation is defined differently in the case of labeled and non-labeled vertices v. The idea is to keep admissible graphs only.

1) If v is non-labeled then we take sum over $1/2 \le i \le j \le k + 1/2$ such that $2 \le j - i$ and $1 \le k - (j - i)$, in $d_{v,i,j}(T)$ both vertices v^{up} and v^{down} are non-labeled.

2) If v is labeled, then $d_v(T) = d_v^{(1)}(T) + d_v^{(2)}(T)$ where the first summand $d_v^{(1)}(T)$ is the sum of $d_{v,i,j}(T)$ with $1 \leq k-(j-i)$ and with the vertex v^{down} appearing with the old label of v, and v^{up} being non-labeled. The second summand $d_v^{(2)}(T)$ is the sum of $d_{v,i,j}(T)$ with $2 \leq j-i$, the vertex v^{down} is non-labeled, and v^{up} has the same label as v in T. Labeling of all vertices different from v^{up} and v^{down} remains the same.

5.4. Action of *M* on the Hochschild complex

Let (A,m) be an A_∞-algebra. Our next step is to define an action of a dg-operad M on $C^\bullet(A,A)$ so the latter becomes a dg-algebra over M.

Let $T \in M_n, \gamma_i \in C^\bullet(A,A), 1 \leq i \leq n$. We need to define an element $T(\gamma_1,...,\gamma_n) \in C^\bullet(A,A)$. It can be expressed as a sequence of morphisms of graded vector spaces $T(\gamma_1,...,\gamma_n)_N : A^{\otimes k} \to A, k \geq 0$.

Let $T_{(k)}$ be a unique (non-admissible) planar tree with only one internal vertex v, which is labeled, and k tails, numbered from the left to the right.

Then we have the composition

$$T_{(k)} \circ_v T = \sum_\beta \pm (T_{(k)} \circ_v T)_\beta$$

defined in the same way as for admissible trees. Here as before $\beta : \{1,...,k\} \to A(T_{(k)} \circ_v T)$ is a monotonic map.

Every tree $(T_{(k)} \circ_v T)_\beta$ has k tails. For every $v \in V_i((T_{(k)} \circ_v T)_\beta)$ we define a polylinear map $\gamma_v : A^{\otimes |v|} \to A$ in the following way:

1) If v is labeled by j, $1 \leq j \leq n$, we define γ_v as the component of γ_j which belongs to $Hom_{Vect_{\mathbf{Z}}}(A^{\otimes |v|}, A)$;

2) If v is non-labeled we define $\gamma_v = m_{|v|}$.

The tree $(T_{(k)} \circ_v T)_\beta$ defines the way to compose operations γ_v into an operation $\gamma_\beta : A^{\otimes k} \to A$. We define $T(\gamma_1,...,\gamma_n)$ to be equal to the sum $\sum_\beta \pm \gamma_\beta$.

We claim that in this way we get on $C^\bullet(A,A)$ a structure of a dg-algebra over M.

This can be checked by a straightforward computation. It is more or less clear from the definitions that $C^\bullet(A,A)$ is an algebra over the graded operad M. Hence the question is about the compatibility with differentials. The latter follows from a more general result (the Theorem 1), and will be proved in Section 6.

5.5. Signs in the minimal operad

Now we would like to discuss the second description of the operad M. It is based on the following result from the theory of Strebel's differentials (see [20]).

Theorem 2 *Let $I = \{z_1,...,z_n\}$ be a finite non-empty subset of the complex line $\mathbf{C}$. Then there exists a unique quadratic differential $\alpha = f(z)(dz)^2$ with $f(z) = \sum_i \frac{a_i}{z-z_i}$, $a_i \in \mathbf{C}, i = 1,...,n$, $\sum_i a_i = 1$, satisfying the following property:*

There exists a unique tree $T \in Tree^{(p)}(I)$ and an embedding $j : T \to \mathbf{CP}^1$, $\{root\} \mapsto \{+i\infty\}$ such that the pair $(\mathbf{CP}^1 \setminus j(T), \alpha)$ is equivalent (as a complex curve with quadratic differential) to the lower half-plane $Im(z) < 0$ equipped with the quadratic differential $i(dz)^2$.

We will call such $\alpha = f(z)(dz)^2$ the Strebel differential associated with the finite set $I \subset \mathbf{C}$.

To every Strebel differential α we can assign canonically the $\{1,...,n\}$-labeled planar tree T. Conversely, having a tree $T \in Tree^{(p)}(n)$ we can look for sequences of pairwise distinct points $(z_1,...,z_n)$ in $\mathbf{C}$ which can appear as the sets of possible poles of the Strebel differential from the Theorem. We remark that the set of all possible $(z_1,...,z_n) \in \mathbf{C}^n \setminus \{diag\}$ form an open cell Str_T. This follows from the fact that this space is a vector bundle of rank 2 over a cell. The latter cell is defined by the lengths of internal edges (lengths are taken with respect to the metric $|\alpha|$ which is well-defined on $\mathbf{C} \setminus I$). Then the embedding of T is fixed up to parallel translations. The latter span the fiber of a vector bundle. The total space of this bundle is the cell $Str_T \subset \mathbf{C}^n \setminus \{diag\}$. It is easy to see that $codim(Str_T) = -deg(T) = -(|E_i(T)| + 2 - 2|V_{lab}(T)|)$.

One can show that there is a finite CW-complex $\Sigma_{(n)} \subset \mathbf{C}^n \setminus \{diag\}$ with the cells Σ_T labeled by $T \in Tree^{(p)}(n)$, and such that Σ_T intersects Str_T transversally and at exactly one point.

Now we can give the second definition of the operad M:

For $n \geq 2$ the complex M_n is defined as the chain complex of $\Sigma_{(n)}$.

Using our agreement about degrees of homological complexes, one can check again that every complex M_n is a finite-dimensional complex concentrated in degrees $\{-(n-1),...,0\}$.

We use the second definition in order to derive the following combinatorial description of the operad M.

One associates to a tree $T \in Tree^{(p)}(n)$ a one-dimensional vector space (in fact an abelian group, so everything can be done over integers). It is defined by the formula $U_T = ((\widetilde{H}_\bullet(\mathbf{R}))^*)^{E_i(T)} \otimes ((\widetilde{H}_\bullet(\mathbf{R}^2)^*) \otimes ((\widetilde{H}_\bullet(\mathbf{R}^2))^{V_{lab}(T)}$.

In the formula $\widetilde{H}_\bullet(X)$ denotes the reduced homology of the one-point compactification of X (Borel-Moore homology).

The graded vector space M_n coincides with the sum of U_T over all trees belonging to $Tree^{(p)}(n)$.

Next step is to interpret various shifts in complexes as tensor products with the reduced homology of vector spaces. For example the shift by 1 is the tensor product with the reduced homology of $\mathbf{R}$.

It is more convenient to use a different notation for the same spaces of reduced homology.

Let $L_1 := \widetilde{H}_\bullet(\mathbf{R})$ and $L_2 := \widetilde{H}_\bullet(\mathbf{R}^2)$. These are 1-dimensional graded vector spaces of pure degrees -1 and -2 correspondingly. For a finite set I we define a vector space

$$M_I := \oplus_{T \in Tree^{(P)}(I)} (L_1^*)^{E_i(T)} \otimes L_2^* \otimes L_2^{\otimes I} = \oplus_{T \in Tree^{(P)}(I)} U_T.$$

Geometrically, the new notation is related to the picture with the Strebel differentials. We think about $\mathbf{R}^2$ as about direct sum of two lines $\mathbf{R}^2 = \mathbf{R}_{hor} \oplus \mathbf{R}_{vert}$ (horizontal and vertical lines). The vertical line is the y-axis in $\mathbf{C} = \mathbf{R}^2$. The vertical line corresponds to L_1 in the notation above, and the horizontal line corresponds to $L_1^* \otimes L_2$ (we think of it as $\mathbf{R}^2/\mathbf{R}_{vert}$).

There is a natural structure of a complex on M_I. We define a linear map $d_I : M_I \to M_I \otimes L_1$ as a sum $d_I = \sum_{T_1 \to T} d_{T_1,T}$.

Here $T_1 \to T$ means that T_1 is obtained from T by adding an internal edge. The summand $d_{T_1,T}$ is a linear map from U_T to $U_{T_1} \otimes L_1 = (U_T \otimes L_1^*) \otimes L_1$. It is given by $id_{U_T} \otimes \varepsilon_{L_1}$ where $\varepsilon_{L_1} : \mathbf{1} \to L_1^* \otimes L_1$ is the canonical morphism in the symmetric monoidal category $Vect_{\mathbf{Z}}$.

Lemma 3 *The linear map d_I defines a differential in M_I.*

Proof. We need to prove that $(d_I \otimes id_{L_1}) \circ d_I = 0$. This can be checked directly using the fact that L_1 has degree -1 , so the commutativity constraint acts on $L_1 \otimes L_1$ as $-\sigma$ where σ is the permutation map. ∎

The composition maps for the operad M can be naturally described in the new notation. We remark that in order to glue a tree T_2 to a tree T_1 at a vertex v, first we need to define a sequence of trees $\{(T_1 \circ_v T_2)_\beta\}$ (they were also denoted by T_β in the previous description of the gluing). For every $(T_1 \circ_v T_2)_\beta$ the set of internal edges is the disjoint union of the corresponding sets for T_1 and T_2. The set of labeled vertices for $(T_1 \circ_v T_2)_\beta$ is $(V_{lab}(T_1) \cup V_{lab}(T_2)) \setminus \{v\}$. Then we need to define a morphism of vector spaces $U_{T_1} \otimes U_{T_2} \to U_{(T_1 \circ_v T_2)_\beta}$.

Equivalently we need to define a morphism of the graded vector space

$$(L_1^*)^{\otimes E_i(T_1)} \otimes L_2^* \otimes L_2^{\otimes I_1} \otimes (L_1^*)^{\otimes E_i(T_2)} \otimes L_2^* \otimes L_2^{\otimes I_2}$$

to the graded vector space

$$(L_1^*)^{\otimes (E_i(T_1) \sqcup E_i(T_2))} \otimes L_2^* \otimes L_2^{\otimes ((I_1 \setminus \{v\}) \sqcup I_2)}.$$

(The notation is self-explained).

We define the morphism to be the identity on the tensor factors marked by same edges or same elements of the sets $I_k, k = 1, 2$. For example $(L_1^*)^{\otimes E_i(T_1)} \otimes (L_1^*)^{\otimes E_i(T_2)}$ is identically mapped to $(L_1^*)^{\otimes (E_i(T_1) \sqcup E_i(T_2))}$.

Hence it is enough to define a morphism $L_2^* \otimes L_2^* \otimes L_2 \to L_2^*$, where first and third tensor factors in the LHS correspond to the tree T_1, middle tensor factor corresponds to the tree T_2 and the space L_2^* in the RHS corresponds to the tree

$(T_1 \circ_v T_2)_\beta$. Then one uses the canonical evaluation map $ev : L_2^* \otimes L_2 \to \mathbf{1}$ in the symmetric monoidal category $Vect_{\mathbf{Z}}$.

5.6. Signs in the Hochschild complex

In this subsection we are going to reformulate the definition of the Hochschild complex using the reduced homology spaces L_1 and L_2. Then one gets the sign agreement with the computations in the previous subsection.

For a graded vector space A we define

$$C = C^\bullet(A,A) = \oplus_I \underline{Hom}(A^{\otimes I}, A) \otimes (L_2^* \otimes L_1)^{\otimes I},$$

where the sum is taken over all non-empty completely ordered finite sets, and $\underline{Hom}$ is the internal Hom in the tensor category $Vect_{\mathbf{Z}}$.

Let us demonstrate how the structures in C can be reformulated by means of this language.

The Gerstenhaber bracket is a map $C \otimes C \to C \otimes (L_2^* \otimes L_1)$. Now the multiplication m can be described as a point of a dg-manifold: $m \in C \otimes L_2$. The bracket $[m,m]$ defines a point of the dg-manifold $C \otimes L_2^* \otimes L_1 \otimes L_2 \otimes L_2 = (C \otimes L_2) \otimes L_1$. If $[m,m] = 0$ then we have a differential $d_m = [m,\cdot] : C \to C \otimes L_1$.

The corresponding structure of DGLA on C can be reformulated as follows: Consider the class $\mathcal{F}$ of *simple forests*. A simple forest F is a finite collection of planar trees $F = \{T_\alpha\}_{\alpha \in \Omega}$ with no internal edges. For every $F \in \mathcal{F}$ we define the graded vector space

$$W_F := \otimes_{\alpha \in \Omega} Hom(A^{\otimes V_i(T_\alpha)}, A) \otimes (L_2^* \otimes L_1)^{\otimes V_i(T_\alpha)} \otimes L_2.$$

We have the natural groupoid structure on $\mathcal{F}$. The correspondence $F \mapsto W_F$ defines a functor from this groupoid to the category $Vect_{\mathbf{Z}}$. The free cocommutative coalgebra cogenerated by $C \otimes L_2$ can be described as a colimit of this functor: $Coalg(C \otimes L_2) = colim_{F \in \mathcal{F}} W_F = \oplus_{F/iso} (W_F)_{AutF}$.

The coalgebra structure on $Coalg(C \otimes L_2)$ can be described in these terms. Namely $\Delta \circ pr = \Sigma_{F_1 \subset F} (pr \otimes pr) \circ i(F_1, F_2)$.

Let us explain the notation.

Here Δ is the coproduct on the free cocommutative coalgebra cogenerated by $C \otimes L_2$, pr is the projection to the coinvariants. The sum is taken over all subforests F_1 (unions of some connected components of F), and we fix the splitting $i(F_1, F_2) : W_F \simeq W_{F_1} \otimes W_{F_2}$, where F_2 is the complementary forest. We leave to the reader straightforward reformulations of other structures on the Hochschild complex and checking the signs.

Remark 5 For an A_∞-algebra A one can define the opposite algebra A^{op}. In the geometric language of this section it corresponds to the antipodal involution on $\mathbf{R}_{hor}$.

6. Morphism of dg-manifolds

6.1. Generators of the free operad

The main purpose of this section is to prove the Theorem 1. It will be done in subsection 6.2. This subsection is devoted to some technical preparatory material.

Applying the general theory of Section 4 to the case of the operad M we obtain its free resolution P. It is a dg-operad which is free as a graded operad. Its graded components P_n can be described explicitly in terms of the operad M. We know that $P_0 = 0$ and $P_1 = k \cdot 1_P$. One can easily describe the space $G_n = (Free'(M'[1])[-1])_n$ of generators of $P_n, n \geq 2$.

Namely, G_n is a direct sum of 1-dimensional graded vector spaces $W_{\mathbf{T}}$. The sum is taken over the set of equivalence classes of collections $\mathbf{T} = (T, \{T_v\}_{v \in V_i(T)})$ where:

a) $T \in Tree(n), n \geq 2$ such that for every $v \in V_i(T)$ we have $|v| \geq 2$;

b) $T_v \in Tree^{(p)}(N_T^{-1}(v))$.

In order to be consistent with the notation of Section 7, we should also label all internal edges of $\mathbf{T}$ by an additional label *finite*. We omit the labeling since it will not be used in this section.

Clearly the automorphism group of any such $\mathbf{T}$ is trivial. The symmetric group S_n acts freely on G_n by permuting the tails of T. Any $\mathbf{T}$ gives rise to the 1-dimensional vector space $W_{\mathbf{T}} = L_1^{\otimes E_i(T)} \otimes U_{T_v}$, where the space $U_{T'}$ was defined in 5.5. In what follows we will identify $\mathbf{T}$ with the corresponding (up to a sign) generator of G_n.

Notice that if $E_i(T) = \emptyset$ then T has the only internal vertex v and $W_{\mathbf{T}} = U_{T_v}$. Moreover, the set $N^{-1}(v)$ is naturally identified with the set $\{1, ..., n\}$, so that we can write $T_v \in Tree^{(p)}(n)$.

There is a natural morphism of S_n-modules $pr_n : G_n \to M_n$, such that $pr_n(W_{\mathbf{T}}) = 0$ if $E_i(T) = \emptyset$, otherwise $pr_n(W_{\mathbf{T}}) = id_{T_v}$ where v is the only internal vertex of T.

A generator $\mathbf{T}$ is pictured by a tree T with n numbered tails, and with generators of M inscribed into all internal vertices of T. If $T_v \in M$ is inscribed into a vertex $v \in V_i(T)$ then the cardinality of $V_{lab}(T_v)$ is equal to the cardinality of $N^{-1}(v)$.

Now we can return to the Theorem 1. We have constructed the minimal operad M. The Hochschild complex $C^\bullet(A,A)$ of an A_∞-algebra A is an algebra over M (if we forget the differentials). Then the free resolution P of M acts on $C^\bullet(A,A)$, so the latter becomes an algebra over the graded operad P (again we forget the differentials).

On the other hand, let A be a graded vector space and $m \in C = C^\bullet(A,A)$, but not necessarily $[m,m] = 0$. Then the constructions of Section 5 give rise to a sequence of elements $\rho(m) = (\rho(m)_n)_{n \geq 1}$ of $\underline{Hom}(M_n \otimes_{S_n} C^{\otimes n}, C)$ (no conditions on m were imposed by the construction). The following lemma is easy to prove.

Lemma 4 *The sequence $\rho(m)$ defines a structure of a graded M-operad on C.*

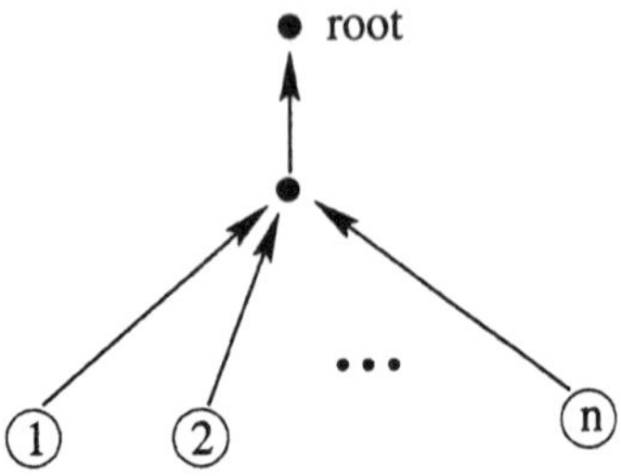

Figure 11. The unique abstract tree which is not killed by the projection from G_n to the space of generators.

Compositions $\gamma(m)_n = \rho(m)_n \circ pr_n$ for $n \geq 2$ give rise to a sequence of S_n-equivariant maps $G_n \otimes_{S_n} C \to C$. Since the graded operad P is freely generated by $G = (G_n)_{n\geq 2}$, these compositions define a point in $\mathcal{M}(P,C)$. We denote the sequence $(\gamma(m)_n)_{n\geq 2}$ by $\gamma(m)$. Thus we can write $\gamma(m) = \rho(m) \cdot pr$ where $pr = (pr_n)_{n\geq 2}$.

We can define an element $d_m \in \underline{Hom}(C,C)[1]$ in the natural way: $d_m = [m,\cdot]$. In general $d_m^2 \neq 0$. In any case it can be naturally extended to a map

$$f : \mathcal{M}_{cat}(\mathcal{A}_\infty, A) \to \mathcal{M}(P,C),$$

such that $f(m) = (d_m, \gamma(m))$.

We claim that this map is a morphism of dg-manifolds and it satisfies the condition $p \cdot f = \nu$ of the Theorem 1. It will be proved in the next subsection.

6.2. Proof of the theorem

In the course of the proof we will not pay much attention to the signs. The reason for that was explained in the previous section. Namely, our second description of the operad M (with reduced homology) gives automatically the agreement of signs.

Proof of the theorem will occupy the rest of this subsection. We start with some general considerations.

Let us recall that points of the moduli space $Y = \mathcal{M}(P,C)$ parametrize pairs (d_C, γ) where $d_C : C \to C \otimes L_1$ is a morphism in $Vect_{\mathbf{Z}}$, and γ is an action of the space of generators $G = (G_n)_{n\geq 2}$ on C. Having the action $\gamma(m) = \rho(m) \cdot pr$ of P on C we would like to compute the odd vector field d_Y at the point $(d_m, \gamma(m))$. Clearly $d_Y(d_C, \gamma) = (d_C^2, \bar{\gamma})$ where $\bar{\gamma} = (\bar{\gamma}_2, \bar{\gamma}_3, \ldots)$ and $\bar{\gamma}_n \in \underline{Hom}(G_n \otimes_{S_n} C^{\otimes n}, C \otimes L_1)$.

We need to compute $\bar{\gamma}$. One can extend d_C to $\underline{Hom}(C^{\otimes n}, C)$ using the Leibniz rule. We denote this extension by $d_C^{(n)}$. To write down $\bar{\gamma}$ we need to know the images of the generators of P under all γ_n. We know that the maps pr_n send to zero all of them except the trees with one non-labeled vertex and n numbered tail vertices. The latter is shown in Figure 11.

Then the direct computation together the previous lemma show that the following result holds.

Lemma 5 *The component $\bar{\gamma}_n$ of $d_Y(f(m))$ is equal to*

$$(d_m^{(n)} \cdot \rho(m)_n + \rho(m)_n \cdot d_{M_n}) \cdot pr_n,$$

where d_{M_n} is the differential d_M being restricted to M_n.

Proof. We sketch the proof. First we observe that all components of $\bar{\gamma}_n$, which corresponds to the generators of G_n with more than two internal edges, vanish. Using the previous lemma one can show that the generators with one internal edge give no input as well. After that one can make computations with elements of the operad M only. Then a direct computation proves the lemma. ■

We would like to prove that the map $f : X = \mathscr{M}_{cat}(\mathscr{A}_\infty, A) \to Y = \mathscr{M}(P, C)$ is a dg-map, i.e., it transforms the odd vector field d_X into the odd vector field d_Y. Vector field on X is given by $\dot{m} = d_X(m) = \frac{1}{2}[m,m]$.

The image of the map f belongs to the vector subspace

$$\underline{Hom}(C,C)[1] \oplus \underline{Hom}(M'(C),C)$$

of Y. Therefore the image $\dot{f}(m) = f_*(d_X(m))$ of the tangent vector $d_X(m)$ belongs to the same vector space. Thus, we have to show that the second component $\dot{\gamma}$ of $\dot{f}(m)$ is equal to $\overline{\gamma}$.

Its first component is equal to $\frac{1}{2}[\dot{m}, \cdot]$ which is the same as $\frac{1}{2}[[m,m], \cdot] = d_m^2$. Notice that $\overline{\gamma} - \dot{\gamma}$ can be considered as an action of an action of M' on C. We decompose it into the sum of terms corresponding to planar trees from $Tree^{(p)}$. For such a tree T the component of $\overline{\gamma} - \dot{\gamma}$ is a sum of four terms described below:

A. These terms correspond to the differential d_M of the operad M. We can schematically write them as $(d_M T)((c_i), m)$. We are inserting $c_i \in C$ in labeled vertices and m in non-labeled vertices.

B. These terms can be schematically written as $\sum_i T([m, c_i], (c_j)_{j \neq i}, m)$. We are inserting m in non-labeled vertices, elements c_j and $[m, c_i]$ in the corresponding labeled vertices.

C. These terms can be schematically written as $[m, T((c_i), m)]$ in the notation above. These terms appear when we apply the differential (= commutator with m) to the tree with c_i inserted in labeled vertices and m inserted in non-labeled vertices.

D. These terms correspond to $-\dot{\gamma}$. Each of them consists of the replacement of m in one non-labeled vertex v by $\dot{m}$. The latter can be in turn replaced by $\frac{1}{2}[m,m]$. We can schematically write the resulting sum as $\sum_{v \in V_{nonl}(T)} T((c_i), \frac{1}{2}[m,m])$. We are inserting the element m in all non-labeled vertices of T except v, the element $\frac{1}{2}[m,m]$ in the non-labeled vertex v and c_i to all labeled vertices.

Notice that although a tree T is always admissible, planar trees appearing in the decomposition of $\dot{\gamma}$ are not necessarily admissible. This means that valencies of some non-labeled vertices can be either 0 or 1.

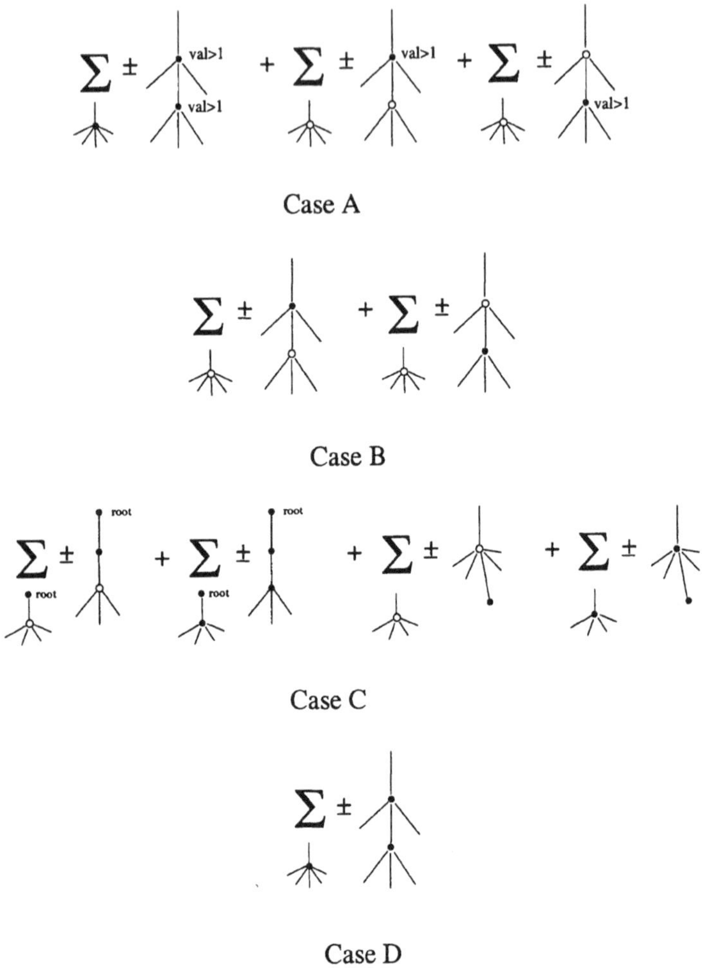

Figure 12. Four terms in the decomposition of $\overline{\gamma} - \dot{\gamma}$.

In order to depict all four cases we use the following notation: Composition of an operation α sitting in a vertex v with m produces a new tree with a new non-labeled vertex w, as well as a new edge (w,v) where $N(w) = v$. Similarly, a composition of m with α produces a new tree with a non-labeled vertex w, as well as a new edge (v,w) such that $N(v) = w$. The commutator $[\alpha, m]$ corresponds to the difference of the above-mentioned trees. This agreement will be used also in the case when $\alpha = m$ thus giving the way to depict trees with $\frac{1}{2}[m,m]$ inserted. In Figure 12, we show neighborhoods of vertices where the original tree changes.

Then we split these terms in the following way $A = A_1 + A_2 + A_3$, where:

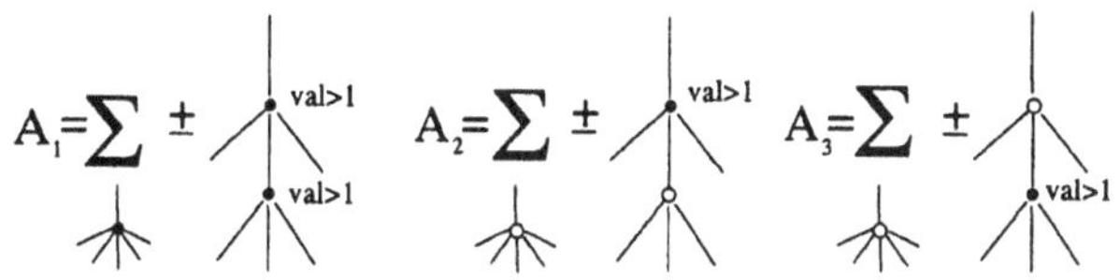

Figure 13. Decomposition of terms of type A.

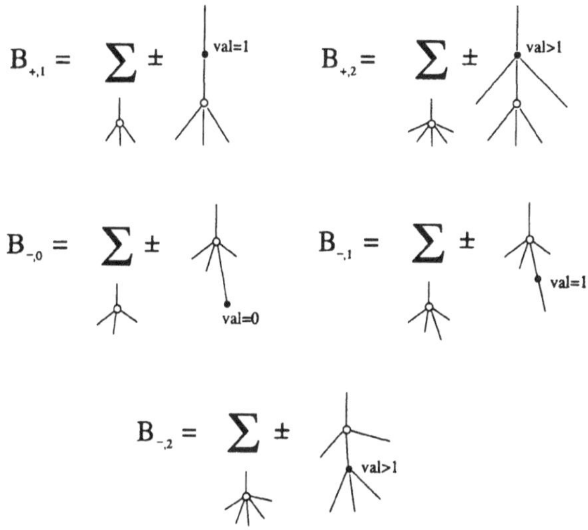

Figure 14. Decomposition of terms of type B.

a) The terms A_1 correspond to the two adjoint non-labeled vertices of valency ≥ 2 with m inserted in each;

b) The terms A_2 correspond to the labeled vertex adjoint to a non-labeled one of valency ≥ 2 with m inserted in the latter;

c) The terms A_3 correspond to the non-labeled vertex of valency ≥ 2 with m inserted in it adjoint to a labeled vertex.

These cases can be depicted as in Figure 13.

Similarly we split the terms B as follows: $B = B_{+,1} + B_{+,2} + B_{-,0} + B_{-,1} + B_{-,2}$, where individual summands are depicted in Figure 14.

We split the terms C such as follows: $C = C'_{root} + C''_{root} + C_\circ + C_\bullet$ where individual summands are depicted in Figure 15.

We split the terms D such as follows: $D = D_0 + D_{1,A} + D_{1,B} + D_2$ where individual summands are depicted in Figure 16. We see that $A_1 + D_2 = 0$, $A_2 + B_{+,2} = 0$, $A_3 + B_{-,2} = 0$, $B_{-,0} + C_\circ = 0$, $D_0 + C_\bullet = 0$.

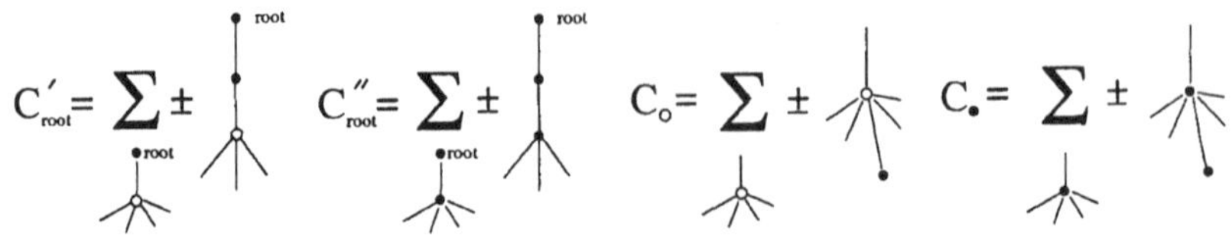

Figure 15. Decomposition of terms of type C.

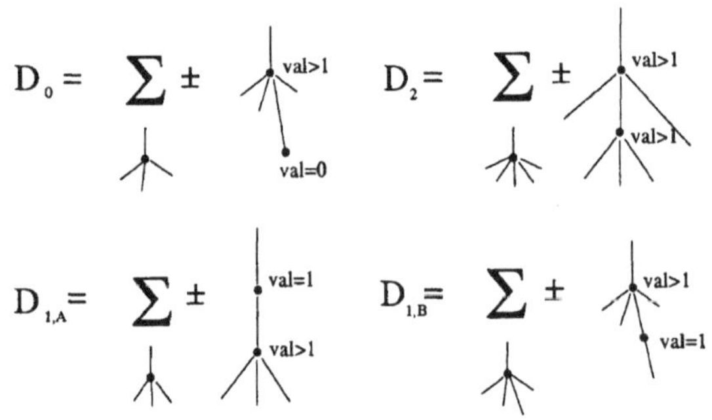

Figure 16. Decomposition of terms of type D.

Furthermore we can split each of the remaining terms into summands $B_{+,1} = B'_{+,1} + B''_{+,1} + B'''_{+,1}, B_{-,1} = B'_{-,1} + B''_{-,1}, D_{1,A} = D'_{1,A} + D''_{1,A} + D'''_{1,A}, D_{1,B} = D'_{1,B} + D''_{1,B}$. This splitting is depicted in Figure 17.

We combine these summands into six groups corresponding to the six types of edges depicted in Figure 18.

We see that $B'_{+,1} + C'_{root} = 0, C''_{root} + D'_{1,A} = 0, B''_{+,1} + B'_{-,1} = 0$, $B''_{-,1} + D''_{1,A} = 0, B'''_{+,1} + D'_{1,B} = 0, D'''_{1,A} + D''_{1,B} = 0$.

Then we conclude that $B_{+,1} + C'_{root} + C''_{root} + D_{1,A} + B_{-,1} + D_{1,B} = 0$. Thus $A + B + C + D = 0$. This means that the map $f : X \to Y$ is a dg-map. Obviously it is $GL(A)$-equivariant and satisfies the condition $pf = v$.

This concludes the proof of Theorem 1.

6.3. Remark about a generalization

Let A be an A_∞-algebra, and $C = C^\bullet(A,A)$ be its Hochschild complex. Admissible planar trees with n labeled vertices give rise to operations $C^{\otimes n} \to C$, $n \geq 1$. Analogously, planar trees with n labeled vertices and m tails, $n \geq 0$, $m \geq 1$ give rise to operations $C^{\otimes n} \otimes A^{\otimes m} \to A$.

Let us restrict ourselves to such trees that $|v| \geq 2$ for all non-labeled internal vertices v. Thus we obtain a colored operad $M^{(2)}$ with two colors *(Alg, Hoch)* acting on the set of pairs $(A, C^\bullet(A,A))$, where A is an A_∞-algebra. Clearly $M^{(2)}$ contains as suboperads both $\mathcal{A}_\infty$ and M. Presumably a result analogous to Theorem 1 holds for the colored operad $M^{(2)}$.

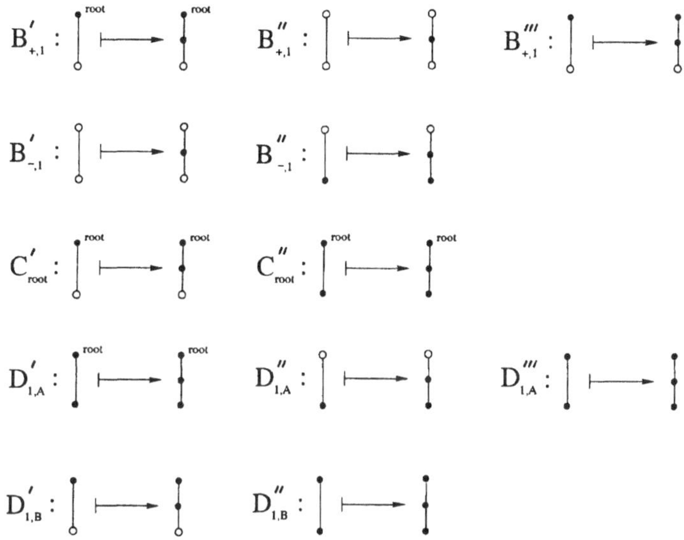

Figure 17. Further decompositions of terms $B_{+,1}, B_{-,1}, D_{1,A}, D_{1,B}$.

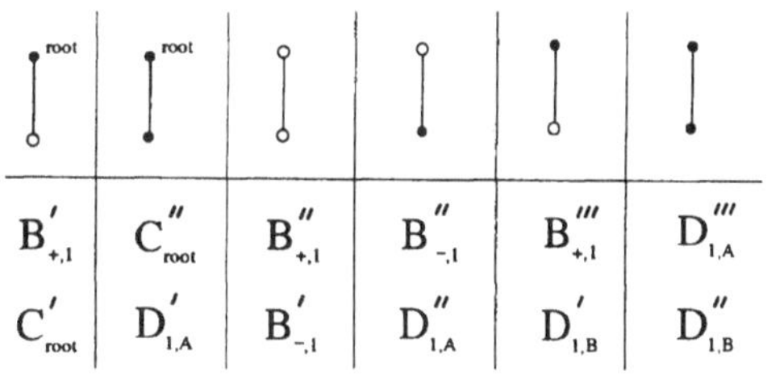

Figure 18. The rest, decomposed into six pairs of terms. In every pair terms cancel each other.

7. The Deligne conjecture

7.1. Preliminaries

We are going to prove the following result.

Theorem 3 *Let P be the free resolution of the minimal operad M constructed in Section 6. Then there is a homomorphism of dg-operads $P \to Chains(FM_2)$ which induces an isomorphism on cohomology (i.e., it is a quasi-isomorphism of dg-operads). Here $Chains(FM_2)$ is the chain operad for the Fulton-MacPherson operad of configurations of points in $\mathbf{R}^2$ (see Section 7.2 below).*

We will construct a homomorphism of the dg-operads which induces a quasi-isomorphism of the chain complexes. Such a homomorphism is not defined canonically. Different choices are naturally parametrized by a contractible topological space.

Using the fact that the operad $Chains(FM_2)$ is free as an operad (not as dg-operad), one can invert the quasi-isomorphism mentioned in the theorem. Since the Hochschild complex is a P-algebra, we obtain the following Corollary known as the Deligne conjecture (see for example [14], [22], [18]).

Corollary 1 *a) The Hochschild complex $C^\bullet(A,A)$ of an A_∞-algebra A can be equipped with a structure of an algebra over the operad $Chains(FM_2)$.*
b) The corresponding structure of a $H^\bullet(Chains(FM_2))$-algebra on the Hochschild cohomology $H^\bullet(A,A)$ coincides with the standard structure of a Gerstenhaber algebra on the Hochschild cohomology of an A_∞-algebra.

Same results remain true with $Chains(FM_2)$ being replaced by the operad $Chains(E_2)$ of chains on the little disc operad E_2 (see Section 7.2 for the definition).

Remark 6 It is easy to see that the operad P can be defined over $\mathbf{Z}$. It follows that such an operad acts on the Hochschild complex of an A_∞-algebra defined over a field of arbitrary characteristic. The quasi-isomorphism $P \to Chains(FM_2)$ can be also defined over the ring of integers (see also [18]).

The proofs of the Theorem and the Corollary will occupy the rest of the section. We are going to use the following strategy.

Let $\mathbf{T}$ be a "meta-tree" corresponding to a generator of P (see Section 6 and Definition 17 below).

1) To every $\mathbf{T}$ we are going to associate a contractible closed subspace $X_{\mathbf{T}}$ in the Fulton-MacPherson compactification of the configuration space of points in $\mathbf{R}^2$ modulo shifts and dilatations.

2) The collection of subspaces $(X)_{\mathbf{T}}$ will satisfy the following properties:
a) The correspondence $\mathbf{T} \mapsto X_{\mathbf{T}}$ is S_n-equivariant;
b) if a tree $\mathbf{T}'$ appears as a summand in the formula for $d_P(\mathbf{T})$ then $X_{\mathbf{T}'} \subset X_{\mathbf{T}}$;
c) if a composition of trees $\mathbf{T}_1 \circ \mathbf{T}_2 \in P$ appears as a summand in the formula for $d_P(\mathbf{T})$ then the operadic composition $X_{\mathbf{T}_1} \circ X_{\mathbf{T}_2}$ belongs to the stratum $X_{\mathbf{T}}$. The latter composition of the strata has meaning because the Fulton-MacPherson compactifications form a topological operad.

3) For any generator $\mathbf{T}$ of the operad P we will choose inductively chains $\gamma_{\mathbf{T}} \in Chains(FM_2)$ such that $Supp(\gamma_{\mathbf{T}}) \subset X_{\mathbf{T}}$, where $Supp$ means the support of a chain. In this way we obtain a homomorphism of dg-operads $P \to Chains(FM_2)$.

4) This homomorphism is a quasi-isomorphism. This will follow from the fact that P is quasi-isomorphic to the minimal operad M, and on the level of chain complexes (not operads) every M_n is quasi-isomorphic to $Chains(FM_2(n))$.

Remark 7 The theorem and its proof seem to admit a generalization to the case of higher dimensions.

7.2. Little discs operad and Fulton-MacPherson operad

We recall here the definitions of both operads following [14].

We fix the dimension $d \geq 1$. Let us denote by G_d the $(d+1)$-dimensional Lie group acting on $\mathbf{R}^d$ by affine transformations $u \mapsto \lambda u + v$, where $\lambda > 0$ is a real number and $v \in \mathbf{R}^d$ is a vector. This group acts simply transitively on the space of closed discs in $\mathbf{R}^d$ (in the usual Euclidean metric). The disc with center v and with radius λ is obtained from the standard disc

$$D_0 := \{(x_1, \ldots, x_d) \in \mathbf{R}^d | x_1^2 + \cdots + x_d^2 \leq 1\}$$

by a transformation from G_d with parameters (λ, v).

Definition 15 *The little discs operad $E_d = \{E_d(n)\}_{n\geq 0}$ is a topological operad defined as follows:*
1) $E_d(0) = \emptyset$,
2) $E_d(1) = \text{point} = \{\mathrm{id}_{E_d}\}$,
3) for $n \geq 2$ the space $E_d(n)$ is the space of configurations of n disjoint discs $(D_i)_{1\leq i\leq n}$ inside the standard disc D_0.

The composition $E_d(k) \times E_d(n_1) \times \cdots \times E_d(n_k) \to E_d(n_1 + \cdots + n_k)$ is obtained by applying elements from G_d associated with discs $(D_i)_{1\leq i\leq k}$ in the configuration in $E_d(k)$ to configurations in all $E_d(n_i)$, $i = 1, \ldots, k$ and putting the resulting configurations together. The action of the symmetric group S_n on $E_d(n)$ is given by renumeration of indices of discs $(D_i)_{1\leq i\leq n}$.

The space $E_d(n)$ is homotopy equivalent to the configuration space of n pairwise distinct points in $\mathbf{R}^d$.

There is an obvious continuous map $E_d(n) \to \mathsf{Conf}_n(Int(D_0))$ which associates to a collection of disjoint discs the collection of their centers. This map induces a homotopy equivalence because its fibers are contractible.

The little discs operad and homotopy equivalent little cubes operad were introduced in topology by J. P. May in order to describe homotopy types of iterated loop spaces.

The Fulton-MacPherson operad defined below is homotopy equivalent to the little discs operad.

For $n \geq 2$ we denote by $\tilde{E}_d(n)$ the quotient space of the configuration space of n points in $\mathbf{R}^d$

$$\mathsf{Conf}_n(\mathbf{R}^d) := \{(x_1, \ldots, x_n) \in (\mathbf{R}^d)^n | x_i \neq x_j \text{ for any } i \neq j\}$$

by the action of the group G_d. The space $\tilde{E}_d(n)$ is a smooth manifold of dimension $d(n-1) - 1$. For $n = 2$, the space $\tilde{E}_d(n)$ coincides with the $(d-1)$-dimensional sphere S^{d-1}. There is an obvious free action of S_n on $\tilde{E}_d(n)$. We define the spaces $\tilde{E}_d(0)$ and $\tilde{E}_d(1)$ to be empty. The collection of spaces $\tilde{E}_d(n)$ does not form an operad because there is no identity element, and compositions are not defined.

Now we are ready to define the operad $FM_d = \{FM_d(n)\}_{n\geq 0}$. The components of the operad FM_d are:

1) $FM_d(0) := \emptyset$,

2) $FM_d(1)$ =point,

3) $FM_d(2) = \tilde{E}_d(2) = S^{d-1}$,

4) for $n \geq 3$ the space $FM_d(n)$ is a manifold with corners, its interior is $\tilde{E}_d(n)$, and all boundary strata are certain products of copies of $\tilde{E}_d(n')$ for $n' < n$.

The spaces $FM_d(n)$, $n \geq 2$ can be defined explicitly.

Definition 16 *For $n \geq 2$, the manifold with corners $FM_d(n)$ is the closure of the image of $\tilde{E}_d(n)$ in the compact manifold $(S^{d-1})^{n(n-1)/2} \times [0,+\infty]^{n(n-1)(n-2)}$ under the map $G_d \cdot (x_1,\dots,x_n) \mapsto \left(\left(\frac{x_j-x_i}{|x_j-x_i|}\right)_{1\leq i<j\leq n}, \frac{|x_i-x_j|}{|x_i-x_k|}\right)$ where i,j,k are pairwise distinct indices.*

One can define the natural structure of operad on the collection of spaces $FM_d(n)$. We skip here the obvious definition.

It is easy to check that in this way we obtain a topological operad (in fact an operad in the category of real compact piecewise algebraic sets defined in the Appendix). We call it the *Fulton-MacPherson operad* and denote by FM_d.

Set-theoretically, the operad FM_d is the same as the free operad generated by the collection of sets $(\tilde{E}_d(n))_{n\geq 0}$ endowed with the S_n-actions described above.

Using piecewise algebraic chains from the Appendix we define dg-operads $Chains(E_d)$ and $Chains(FM_d)$ (they are operads in the symmetric monoidal category of complexes of abelian groups). Since we are working over the ground field k of characteristic zero, the dg-operads will be complexes of k-vector spaces.

Notice that when we write $Chains(Z)$ we mean the *cohomological* complex with graded components $Chains_i(Z)[i]$, $i \geq 0$ concentrated in *negative* degrees.

The complex P_n, $n \geq 2$ is concentrated in degrees $[-2n-3,\dots,0]$. The same is true for the complex $Chains(FM_2(n))$ (we use piecewise algebraic chains) because $dim_{\mathbf{R}} FM_2(n) = 2n-3$.

For every $n \geq 2$, the space P_n has a canonical (up to signs) basis, called the *standard basis*, with elements labeled by certain combinatorial objects, which we will call *meta-trees*.

Definition 17 *A meta-tree* **T** *with n tails is given by the following data:*

a) An abstract tree $T \in Tree(n)$ together with a marking $E_i(T) \to \{finite, infinite\}$;

b) For every internal vertex v of T we have: $|v| \geq 2$;

c) To every $v \in V_i(T)$ we assign an admissible labeled tree T_v together with a bijection $V_{lab}(T_v) \to N_{T_v}^{-1}(v)$.

We denote by $MT(n)$ the set of isomorphism classes of meta-trees with n tails.

Conjecture 1 *There is a piecewise algebraic (see Appendix) cell decomposition of the spaces $FM_2(n), n \geq 2$, with the cells $\sigma_{\mathbf{T}}$ labeled by $MT(n)$ such that:*

1) The correspondence $\mathbf{T} \to \sigma_{\mathbf{T}}$ is S_n-equivariant;

2) The operadic composition of any two cells is again a cell;

3) There is a morphism of dg-operads $Chains_{(\sigma_{\mathbf{T}})}(FM_2) \to P$ such that every cell $\sigma_{\mathbf{T}}$ is mapped (up to a sign) to the corresponding element of the standard basis of P. Here $Chains_{(\sigma_{\mathbf{T}})}(FM_2)$ denotes the chain subcomplex of $Chains(FM_2)$ formed by k-linear combinations of cells $\sigma_{\mathbf{T}}$;

4) For any $n \geq 2$ the cell decomposition of $FM_2(n)$ formed by cells $(\sigma_{\mathbf{T}}), \mathbf{T} \in MT(n)$ is regular, i.e. the closure of every cell is homeomorphic to a closed ball.

This conjecture implies the Deligne conjecture.

Let us now introduce a partial order on the set $MT(n)$, $n \geq 2$.

Definition 18 *Let $\mathbf{T}, \mathbf{T}'$ be meta-trees with n tails. We say that $\mathbf{T} < \mathbf{T}'$ if there exists a sequence of meta-trees $(\mathbf{T}_0, ..., \mathbf{T}_m)$ such that $\mathbf{T}_0 = \mathbf{T}, \mathbf{T}_m = \mathbf{T}'$ and for any i such that $0 \leq i \leq m-1$ we have: $\mathbf{T}_i$ appears as a summand in the decomposition of $d_P \mathbf{T}_{i+1}$ with respect to the standard basis of P_n.*

It follows from the condition 4) of Conjecture 1, that the nerve of the partially ordered set $(MT(n), <)$ is homeomorphic to $FM_2(n)$. The following conjecture also follows from Conjecture 1.

Conjecture 2 *The nerve of $(MT(n), <)$ is a PL-manifold with the boundary. The above-mentioned homeomorphism with $FM_2(n)$ is a homeomorphism of PL-manifolds with boundaries.*

We checked this conjecture for small n.

7.3. Partial orders induced by trees

We recall here the structure of the space G_n of generators of the components P_n, $n \geq 1$ of the free operad P (see Section 6).

There is a basis of G_n elements of which are (up to signs) parametrized by meta-trees $\mathbf{T} = (T, \{T_v\}_{v \in V_i(T)})$ such that all internal edges of T are finite. The degree of the generator $\mathbf{T} = (T, \{T_v\}_{v \in V_i(T)})$ is equal to $deg(\mathbf{T}) = \sum_{v \in V_i(T)} deg(T_v) - |E_i(T)|$.

Let $\mathbf{T}$ be a generator of P_n. We are going to introduce on the set $\{1, 2, ..., n\}$ two partial orders $<_{h,\mathbf{T}}$ and $<_{v,\mathbf{T}}$ (called *horizontal* and *vertical*). Although these orders will depend on $\mathbf{T}$, we will skip $\mathbf{T}$ from the notation if it does not lead to a confusion.

Let $i, j \in \{1, ..., n\}$, $i \neq j$. We have two tail vertices of T labeled by i and j respectively. Then there exists a unique internal vertex v of T satisfying the following properties:

a) $N_T^k(i) = N_T^l(j) = v$ where k, l are positive integers;

b) the vertex v is minimal among those satisfying a) (which means that k,l are both minimal in a).

Then there exist unique labeled vertices $x,y \in T_v$ such that $N_T^{k-1}(i) = x$ and $N_T^{l-1}(j) = y$. Since T_v is a planar tree, we can compare x and y with respect to exactly one of the following partial orders: "x is to the left of y in T_v" or "x is above y in T_v". We will call them the "horizontal" and "vertical" order respectively.

Let us describe the orders in T_v more precisely. We say that x is *above* y (or y is *below* x) in T_v if there exists a positive integer a such that $N_{T_v}^a(y) = x$.

We say that x is *to the left* of y (or y is *to the right* of x) if there exist positive integers a,b such that $N_{T_v}^a(x) = N_{T_v}^b(y) = w$, but $N_{T_v}^{a-1}(x) \neq N_{T_v}^{b-1}(y)$ and $N_{T_v}^{a-1}(x)$ precedes $N_{T_v}^{b-1}(y)$ in $N_{T_v}^{-1}(w)$ with respect to the order on the latter set given by the planar structure on T_v.

Thus, we have defined two partial orders on vertices of $(T_v)_{v \in V_i(T)}$. They induce partial orders $<_{h,\mathbf{T}}$ (horizontal) and $<_{v,\mathbf{T}}$ (vertical) on the set $\{1,...,n\}$. We have identified the latter with the set of tails of T. Namely, we say that $i <_{h,\mathbf{T}} j$ if (in the above notation) the vertex x is to the left of y. We say that $i <_{v,\mathbf{T}} j$ if y is above x (equivalently we say that x is below y).

The following lemma is easy to prove.

Lemma 6 *Both horizontal and vertical orders are indeed partial orders (i.e., they satisfy all the axioms of orders).*

Definition 19 *Let S be a set, $<_1$ and $<_2$ be partial orders on S. We will call them complementary if any two elements of S can be compared with respect to exactly one of them. This means that for any two elements $i,j \in S$, $i \neq j$ exactly one of the following properties holds: $i <_1 j$, $j <_1 i$, $i <_2 j$, $j <_2 i$.*

Suppose that we have two partial orders as in Definition 19. We define on S two new pre-orders $<_{1+2}$ and $<_{1-2}$ as follows:

(i) $x <_{1+2} y$ if $x <_1 y$ or $x <_2 y$;

(ii) $x <_{1-2} y$ if $x <_1 y$ or $y <_2 x$.

Notice that we can reconstruct $<_1$ and $<_2$ from these new orders. For example $x <_1 y$ is equivalent to the conjunction: $(x <_{1+2} y)$ and $(x <_{1-2} y)$.

Proposition 5 *Formulas (i) and (ii) define complete orders on the set S.*

Proof. Straightforward. ∎

Using this result one can easily prove the following one.

Proposition 6 *Let $<_1$ and $<_2$ be a pair of two complementary orders given on a finite set S. Then there exists a unique element $s_0 \in S$ such that for any $i \in S, i \neq s_0$ we have either $s_0 <_1 i$ or $s_0 <_2 i$.*

Proof. The element s_0 is minimal with respect to the complete order $<_{1+2}$. ■

Summarizing, we can say that on the set of tails of a generator of P_n we have defined two complementary partial orders (or, equivalently, two complete orders). These orders will be used below when we will construct the closed sets $X_{\mathbf{T}}$.

7.4. Closed sets $X_{\mathbf{T}}$

Let us recall that for any $n \geq 2$, $1 \leq i,\ j \leq n$, $i \neq j$ we have a natural projection $p_{i,j}\colon FM_2(n) \to FM_2(2)$ (forgetting all points in $(x_1,...,x_n)$ except x_i and x_j). As a topological space $FM_2(2)$ is identified to the unit circle $S^1 \subset \mathbf{R}^2$ via the map $G_2 \cdot (x_1,x_2) \mapsto \frac{x_2-x_1}{|x_2-x_1|}$.

We denote by $S^1_{+,v} \subset FM_2$ the closed upper-half circle, and by $S^1_{+,h} \subset S^1$ the one-element subset consisting of the point $\{(1,0)\}$. Then $S^1_{+,v}$ corresponds to the configurations $(x_1,x_2) \in Conf_2(\mathbf{R}^2)$ such that if we put $x_2 - x_1 = re^{i\alpha}, 0 \leq \alpha < 2\pi$ then $\alpha \in [0,\pi]$. Similarly the subset $S^1_{+,h}$ corresponds to the configurations (x_1,x_2) such that both x_j, $j = 1,2$ belong to the same horizontal line, and x_1 is positioned to the left of x_2.

Suppose that we are given two complementary orders $<_h$ and $<_v$ on the set $\{1,...,n\}, n \geq 2$. Then we define the following subset of $FM_2(n)$:

$$X_{<_h,<_v} = \{x \in FM_2(n) | i <_h j \Rightarrow p_{i,j}(x) \in S^1_{+,h}, i <_v j \Rightarrow p_{i,j}(x) \in S^1_{+,v}\}.$$

For a generator $\mathbf{T} \in P_n$ we define $X_{\mathbf{T}}$ as $X_{<_{h,\mathbf{T}},<_{v,\mathbf{T}}}$.

First of all we would like to prove that $X_{\mathbf{T}}$ is contractible. This is a special case of a more general statement.

Proposition 7 *For any pair of complementary orders $<_h$ and $<_v$ given on the set $\{1,...,n\}$, $n \geq 2$ the subspace $X_{<_h,<_v}$ is non-empty and contractible.*

Proof. It can be done by induction. For $n = 2$ the result is clear, since subsets $S^1_{+,v}$ and $S^1_{+,h}$ are contractible. Suppose that the Proposition is true for $n-1$ points. Let us take the element i_0 which is minimal in $I_n = \{1,...,n\}$ with respect to $<_{h+v}$. We have already proved that it exists. Let us consider the induced complementary partial orders $<_{h'}, <_{v'}$ on the set $I_n \setminus \{i_0\}$. They define the subset $X_{<_{h'},<_{v'}} \subset FM_2(n-1)$ which is non-empty and contractible by induction. Then the result is a corollary of the following observation: the fibers of the natural projection $\pi : X_{<_h,<_v} \to X_{<_{h'},<_{v'}}$ are contractible. Basically it follows from the fact that the point x_{i_0} is either left or below of all the points $x_i, 1 \leq i \leq n, i \neq i_0$. Let us prove that the fibers of π are non-empty. Indeed, for any $x' \in X_{<_{h'},<_{v'}}$ the operadic composition $(-1,0) \circ_j x'$ belongs to the fiber $\pi^{-1}(x')$. Here $(-1,0) \in S^1$ is considered as a point in $FM_2(\{i_0,j\})$, and j is an auxiliary index.

We leave to the reader the proof of contractibility of the fibers of π. ■

Remark 8 a) One can prove that all homotopies can be taken in the category of piecewise algebraic sets.
b) It follows from the construction that the map $\mathbf{T} \mapsto X_{\mathbf{T}}$ is S_n-equivariant.

Now we would like to explain Property 2 of $X_{\mathbf{T}}$ stated in Section 7.1. First of all, the S_n-equivariance is obvious.

Let $\mathbf{T} \in G$ be a generator of P. We recall that the differential in P can be written schematically (up to signs) as

$$d_P(\mathbf{T}) = \sum_{v \in V_i(T), l} \mathbf{T}_{v,l} + \sum_{\alpha \in E_{infinite}(T)} \mathbf{T}'_\alpha \circ \mathbf{T}''_\alpha + \sum_{\alpha \in E_{finite}(T), j} \mathbf{T}_{\alpha,j}.$$

Here meta-trees $\mathbf{T}_{v,l} \in G$ arise from the application of the differential of M to the tree T_v inscribed into the vertex $v \in V_i(T)$. This differential was described in Section 5. Index l runs over all possible insertions of a new edge.

The second summand corresponds to the tree, obtained from T by making a finite edge α into an infinite edge. The result is a composition of two generators of P which we denote by $\mathbf{T}'_\alpha$ and $\mathbf{T}''_\alpha$.

The last sum corresponds to the operation "contract a finite edge α" in T. Then the planar trees from M inscribed into the vertices which are endpoints of α must be composed and inscribed into the new vertex. The result is a sum of generators $\mathbf{T}_{\alpha,j}$ of P.

Proposition 8 *In the above notation we have* $X_{\mathbf{T}_{v,l}}, X_{\mathbf{T}'_\alpha} \circ X_{\mathbf{T}''_\alpha}, X_{\mathbf{T}_{\alpha,j}}$ *belong to* $X_{\mathbf{T}}$.

Proof. Straightforward check which uses the fact that $S^1_{+,h} \in S^1_{+,v}$. ∎

7.5. Morphism $P \to Chains(FM_2)$

We would like to construct the chains $\gamma_{\mathbf{T}} \in Chains(FM_2)$ where $\mathbf{T}$ runs through the set of generators of P. Let us explain the idea.

We will construct $\gamma_{\mathbf{T}}$ by induction in the degree of $\mathbf{T}$.

a) Let us assume that $deg(\mathbf{T}) = 0$. Then T has the only internal vertex v and the corresponding planar tree T_v is a binary tree. The corresponding element of the operad P is defined up to a sign. We make the following choice: It is just the composition of several copies of the operation $m_2 \in M_2$.

The corresponding chain $\gamma_{\mathbf{T}}$ should have the dimension zero. Therefore it must be a linear combination of points with integer coefficients. We choose $\gamma_{\mathbf{T}}$ to be just one point in $X_{\mathbf{T}}$ with the multiplicity equal to $+1$.

b) Let us assume that $deg(\mathbf{T}) = -1$. Then the formula for $d_P(\mathbf{T})$ contains two summands $\mathbf{T}', \mathbf{T}''$ corresponding to the two binary planar trees. It is not difficult to check that they appear with the opposite coefficients ± 1. According to a) these summands define 0-dimensional chains $\gamma_{\mathbf{T}'}, \gamma_{\mathbf{T}''}$. We define $\gamma_{\mathbf{T}}$ to be the only (up to a boundary) 1-chain having $\gamma_{\mathbf{T}'} - \gamma_{\mathbf{T}''}$ as the boundary. This 1-chain exists because of the condition imposed on multiplicities of 0-chains.

c) Suppose that we have a generator $\mathbf{T}$, $deg(\mathbf{T}) = -k, k \geq 2$.

Then we can find a chain $\gamma_{\mathbf{T}}$ of degree $-k$ such that

$$\partial(\gamma_{\mathbf{T}}) = \sum_{v,l} \gamma_{\mathbf{T}_{v,l}} + \sum_{\alpha} \gamma_{\mathbf{T}'_\alpha \circ \mathbf{T}''_\alpha} + \sum_{\alpha,j} \gamma_{\mathbf{T}_{\alpha,j}}$$

where ∂ is the boundary operator in the chain complex for FM_2, and rest of the notation is self-explained.

Indeed, the RHS of this formula is known by the induction assumption. We also know that it is a closed chain because $d_P^2 = 0$. Then one can always find a chain $\gamma_{\mathbf{T}}$ with the given boundary. Indeed, $X_{\mathbf{T}}$ is contractible, and we consider chains of negative degrees (zero degree case was considered in a)). The space parametrizing all different choices of $\gamma_{\mathbf{T}}$ is contractible, so our choice is unique in a given homology class.

The map $\mathbf{T} \to \gamma_{\mathbf{T}}$ extends to a homomorphism $\Phi : P \to Chains(FM_2)$ of graded operads in such a way that $\Phi(\mathbf{T}) = \gamma_{\mathbf{T}}$. We have checked that Φ is compatible with the differentials. Therefore it is a homomorphism of dg-operads.

Theorem 4 *The morphism Φ is a quasi-isomorphism of complexes.*

Proof. We will give only a sketch of the proof. The idea is to consider a subcomplex L of the complex P spanned by the trees without internal edges. This subcomplex is isomorphic to the complex of the minimal dg-operad M. Indeed, P can be decomposed into a direct sum of L and a contractible complex (see Proposition 4 in Section 4). Therefore it is enough to prove that the restriction $\Phi|_L : L \to Chains(FM_2)$ is a quasi-isomorphism of complexes.

In order to do that we need to describe chains from the subcomplex $\Phi(L)$. Our constructions were based on certain non-canonical choices of chains $\gamma_{\mathbf{T}}, \mathbf{T} \in G$. Now, for $\mathbf{T} \in L$ we will make specific choices of $\gamma_{\mathbf{T}}$.

Let $\mathbf{T} \in L$ corresponds to a labeled tree $T_v \in M$. We will denote it simply by T_v, where T_v is a labeled planar tree inscribed in the only internal vertex v of $\mathbf{T}$.

We associate with T_v an abstract tree $\widehat{T}_v$ such that $V_t(\widehat{T}_v) = V_i(T_v)$, and the valency of each internal vertex of $\widehat{T}_v$ is at least 2. Internal vertices of $\widehat{T}_v$ correspond to subsets $T_{\leq x_0}$ described below.

Let us choose an internal vertex $x_0 \in V_i(T_v)$ and consider the set of such $x \in V_i(T_v)$ that $N^j_{T_v}(x) = x_0$ for some $j \geq 0$. Then internal vertices of $\widehat{T}_v$ correspond to such sets $T_{\leq x_0}$ for which the cardinality $|T_{\leq x_0}| \geq 2$. Moreover, there is a path in $\widehat{T}_v$ from a tail vertex $v \in V_t(\widehat{T}_v)$ to an internal vertex $T_{\leq x_0}$ iff $v \in T_{\leq x_0}$.

It is well-known that any abstract tree $\widehat{T}$ with valencies of internal vertices at least 2 gives rise to a stratum $J(\widehat{T}) \in FM_2(|V_t(\widehat{T})|)$ (see [6]). It is the operadic composition of $FM_2(N^{-1}_{\widehat{T}_v}(u)), u \in V_i(\widehat{T}_v)$.

We are going to construct a subspace $X_{\widehat{T}} \subset J(\widehat{T}_v)$ The space $X_{\widehat{T}}$ will be constructed as the operadic composition of certain subspaces

$$X_u \subset FM_2(|N_{\widehat{T}_v}^{-1}(u)|), u \in V_i(\widehat{T}_v).$$

Let $u = T_{\leq x_0}$ be an internal vertex of $\widehat{T}$. The set of edges having u as an endpoint is in one-to-one correspondence with the set $\{x_0\} \sqcup N_{T_v}^{-1}(x_0)$. The subspace X_u consists of configurations of points $G_2 \cdot (p_{x_0}, (p_y)_{y \in N_T^{-1}(x_0)})$ such that $p_{x_0} = (0,1) \in \mathbf{R}^2$, all points p_y belong to the horizontal line $\{(x,0)|x \in \mathbf{R}\} \subset \mathbf{R}^2$, and their order on this line is the same as their order in $N_{T_v}^{-1}(x_0)$.

One can check that:

a) $X_{\widehat{T}_v}$ is an open cell.

b) The natural projection (forgetting map) $FM_2(|V_i(T_v)|) \to FM_2(|V_{lab}(T_v)|)$ maps $X_{\widehat{T}_v}$ onto an open cell $X_{T_v} \subset X_{\mathbf{T}}$.

c) The closure $\overline{X_{T_v}}$ is a manifold with corners (more precisely, a real piecewise algebraic manifold).

d) The boundary of $\overline{X_{T_v}}$ is the union of cells of the same type .

Using a)-d) one checks that the construction above gives rise to a homomorphism of complexes $\chi : L \to Chains(FM_2)$, and moreover, it coincides with the homomorphism $\Phi|_L$. To be more precise, one observes that the construction of Φ was not canonical. We made certain choices when constructed a chain with the prescribed boundary. The point is that one can choose inductively the chains in such a way that the restriction $\Phi|_L$ of the homomorphism $\Phi : P \to Chains(FM_2)$ coincides with χ.

We claim that χ induces a homotopy equivalence. This follows from the following lemma.

Lemma 7 *Let $MT_1(n)$ denotes the subset of $MT(n)$ consisting of meta-trees with only one internal vertex v. Then the natural embedding of the CW-complex $X_n = \cup_{\mathbf{T} \in MT_1(n)} X_{T_v}$ into $FM_2(n)$ is a homotopy equivalence for all $n \geq 2$.*

Proof. It is not difficult to show that X_n is isomorphic to the CW-complex $\Sigma_{(n)}$ constructed in Section 5.5 via Strebel differentials. It follows from the fact that both are regular complexes and posets of their cells are isomorphic. Moreover, both complexes are $K(\pi,1)$ spaces, classifying spaces for the pure braid group of n-strings. Moreover, the map Φ induces an equivalence of the fundamental groupoids. Hence it is a homotopy equivalence. ■

This Lemma concludes the proof of Theorem 2. ■

7.6. Proof of the Deligne conjecture

The results of the previous subsection establish Theorem 2. In this subsection we will prove the Corollary (Deligne conjecture). First of all, we remark that both P and $Chains(FM_2)$ are dg-operads which are free as graded operads. Therefore the quasi-isomorphism morphism of graded operads $\Phi : P \to Chains(FM_2)$ admits a homotopy inverse $\Psi : Chains(FM_2) \to P$ (see for ex. [16]).

We have already proved that the Hochschild complex of an A_∞-algebra is an algebra over the operad P. Indeed, P is a free resolution of M, and the latter dg-operad acts on the Hochschild complex. Using the morphism Ψ we make the Hochschild complex into an algebra over the dg-operad $Chains(FM_2)$. This concludes the proof of the Deligne conjecture.

8. Appendix: Singular chains and differential forms

In this appendix we are going to describe a theory of singular chains which is suitable for working with manifolds with corners. This formalism is helpful in the proof of formality of the operad of chains on the little disc operad (see [14]). We used it in the proof of the Deligne conjecture.

The idea of the theory of singular chains developed below is the following: One can construct chains which produce complexes quasi-isomorphic to the standard complexes of singular chains, and built out of some kind of "piecewise algebraic spaces".

8.1. Spaces

We recall that closed real semialgebraic sets in $\mathbf{R}^n$ are subsets defined by a finite number of polynomial equations and non-strict inequalities. Constructible sets are obtained from closed semialgebraic ones by boolean operations.

We define the category $\mathcal{P}$ of *compact piecewise algebraic spaces* (compact PA-spaces for short) in the following way.

Objects of $\mathcal{P}$ are pairs (X,n), $n = 1,2,\ldots$ such that $X \subset \mathbf{R}^n$ is a compact constructible set (it is the same as a compact real semialgebraic set).

For two objects (X,n) and (Y,m) the space of morphisms $Hom((X,n),(Y,m))$ is formed by continuous maps $f : X \to Y$ such that $graph(f) \subset \mathbf{R}^n \times \mathbf{R}^m$ is constructible.

In the future we will skip the index n in the notation (X,n) if it will not lead to a confusion.

Obviously we have a functor from $\mathcal{P}$ to the category of compact Hausdorff topological spaces. An isomorphism in $\mathcal{P}$ is a morphism which is a homeomorphism of topological spaces.

Let $X \in \mathcal{P}$. Then one can define a sheaf $\mathcal{O}_X$ of piecewise algebraic functions on X. To do this we note first that one can speak about constructible subsets of X. By definition they are constructible sets in the bigger space $\mathbf{R}^n$. Then for any open $U \subset X$ we define $\mathcal{O}_X$ to be an $\mathbf{R}$-algebra of continuous functions f on U such that for any compact constructible $V \subset U$ we have: $graph(f|_V)$ is constructible. It is easy to see that we get a sheaf $\mathcal{O}_X$ of algebras.

Lemma 8 *Morphisms from X to Y in the category $\mathcal{P}$ are in one-to-one correspondence with homomorphisms of algebras $\mathcal{O}(Y) \to \mathcal{O}(X)$.*

Proof. Exercise. ■

Similarly, we can consider the non compact case.

Definition 20 *Piecewise algebraic space (PA-space for short) is a locally compact Hausdorff topological space X, equipped with the sheaf $\mathcal{O}_X$ of $\mathbf{R}$-algebras which is locally isomorphic to $\mathcal{O}_{X'}$ for some compact $X' \subset \mathcal{P}$.*

Clearly, PA-spaces form a category. Compact PA-spaces are exactly objects of $\mathcal{P}$.

We define a d-dimensional PA-manifold with boundary as a PA-space X which is modeled locally on the closed half-space $\mathbf{R}^{d-1} \times \mathbf{R}_{\geq 0}$.

8.2. Singular chains

We would like to define an appropriate version of singular chains for PA-spaces. We will give two equivalent descriptions.

First description.

Let X be a PA-space. We define the group of n-chains $Chains_n(X) := C_n(X, \mathbf{Z})$ as an abelian group generated by equivalence classes of triples (M, or, f) such that:

i) M is a compact PA-manifold of dimension n;

ii) $f : M \to X$ is a morphism in $\mathcal{P}$;

iii) or is an orientation of M.

We need to define an equivalence relation. It is the same as to say when a finite linear combination $\sum_i n_i(M_i, or_i, f_i)$ is equal to zero in $C_n(X, \mathbf{Z})$.

Notice that $Y = \cup_i f_i(M_i)$ carries the structure of a compact PA-space, of dimension $\leq n$. Then there exists a constructible subset $Y_0 \subset Y$ such that $dim(Y \setminus Y_0) \leq n-1$, and for any point $y \in Y_0$, any i and $x_{i,\alpha} \in f_i^{-1}(y)$ there exists a neighborhood $U_{i,\alpha}$ of $x_{i,\alpha}$ such that $Y_0 \cap f_i(U_{i,\alpha})$ is a PA-manifold and the morphism $f_i|_{U_{i,\alpha}}$ is a homeomorphism of $U_{i,\alpha}$ onto its image.

We choose an orientation or_y of Y_0 near the point y. Then the above mentioned linear combination is declared to be zero iff for every point $y \in Y_0$ we have $\sum_{i,\alpha} n_i sgn(i, \alpha, y) = 0$ where $sgn(i, \alpha, y)$ is defined as $f_{i*}|_{U_{i,\alpha}}(or_i) = sgn(i, \alpha, y) or_y$, and sum is taken over all points $x_{i,\alpha}$ such that $f_i(x_{i,\alpha}) = y$. Notice that $sgn(i, \alpha, y)$ always takes values ± 1.

Second description.

We will define $C_n(X) = C_n(X, \mathbf{Z})$ as a quotient by certain equivalence relation of the set of quadruples $(Y, Y_0, or, mult)$ such that:

a) $Y \subset X$ is a compact PA-subspace, $dimY \leq n$, which contains an open dense constructible subset Y_0 without boundary which is a PA-manifold of dimension n;

b) or is an orientation of Y_0;

c) $mult : Y_0 \to \{1,2,...,\}$ is a locally constant map (multiplicity).

The equivalence relation is defined such as follows.

We say that $(Y, Y_0, or, mult)$ is equivalent to $(Y', Y_0', or', mult')$ iff
(i) $Y = Y'$ and there exists an open constructible $Y_0'' \subset Y_0' \cap Y_0''$ such that Y is equal to the closure of Y_0'';
(ii) Restrictions of orientations and multiplicity functions of Y_0 and Y_0' to Y_0'' coincide.

Using the pairing between chains and differential forms (see 8.3), one can show that the first and the second descriptions give canonically isomorphic sets of chains.

One can also check the following properties of singular chains:

1) There is a naturally defined differential $\partial : C_n(X, \mathbf{Z}) \to C_{n-1}(X, \mathbf{Z}), \partial^2 = 0$ (in Description 1 it is well-defined by the formula $\partial(M, or, f) = (\partial M, \partial(or), f|_{\partial M})$).

2) The correspondence $X \to C_\bullet(X, \mathbf{Z})$ is a functor from the category of PA-spaces to the category of abelian groups;

3) $C_i(X, \mathbf{Z})$ vanishes for $i < 0$ and $i > dimX$;

4) If (X, or) is a compact oriented PA-manifold, then there exists a canonically defined chain $[X] \in C_n(X, \mathbf{Z})$. In Description 1, it is defined by the formula $[X] = (X, or, id_X)$;

5) For any finite collection of PA-spaces $(X_i)_{i \in I}$ there is a natural homomorphism of complexes of abelian groups $\otimes_{i \in I} C_\bullet(X_i, \mathbf{Z}) \to C_\bullet(\prod_{i \in I} X_i, \mathbf{Z})$;

6) There is a naturally defined soft sheaf of complexes $\underline{C}_\bullet^{closed}$ on X such that for a compact X the abelian group $\Gamma(X, \underline{C}_\bullet^{closed})$ coincides with $C_\bullet(X, \mathbf{Z})$.

7) The homology of $C_\bullet(X, \mathbf{Z})$ is naturally isomorphic to the usual singular homology $H_\bullet(X, \mathbf{Z})$.

For non-compact PA-spaces we define locally finite chains as global sections of the sheaf $\underline{C}_\bullet^{closed}$. Then the homology of $C_\bullet^{closed}(X, \mathbf{Z})$ is isomorphic to the usual singular homology with locally compact support $H_\bullet^{closed}(X, \mathbf{Z})$.

Remark 9 a) Mayer-Vietoris sequence trivializes for PA chains: If $X = Y_1 \cup Y_2$ is a union of locally closed constructible subsets, then the corresponding short sequence of abelian groups

$$0 \to C_n(Y_1 \cap Y_2, \mathbf{Z}) \to C_n(Y_1, \mathbf{Z}) \oplus C_n(Y_2, \mathbf{Z}) \to C_n(X, \mathbf{Z}) \to 0$$

is exact.

b) The Property 5 is very convenient in order to formulate the Deligne conjecture. It allows to avoid Eilenberg-Zilber theorem.

c) Property 4 seems to be useful for the higher-dimensional generalization of the Deligne conjecture (see [14]). Namely, if A is a d-algebra (i.e., an algebra over the dg-operad $Chains(FM_d)$), then $A[d-1]$ carries a natural structure of an L_∞-algebra. More precisely, for any $n \geq 2$ the fundamental cycle $[FM_d(n)]$ gives the n-th higher bracket on $A[d-1]$.

8.3. Differential forms: First approximation

Let X be a PA-space. We would like to define an appropriate notion of differential form on X. In this subsection we construct a first approximation to the future algebra of differential forms.

Definition 21 *The sheaf $\Omega^k_{X,min}$ is locally defined as a vector subsubspace in $Hom(C_k(X,\mathbf{Z}),\mathbf{R})$ generated by functionals $l = (f_0, f_1, \dots, f_k), f_i \in \mathcal{O}_X$ such that*

$$l(M, or, \phi) = \int_{M_0} \phi^*(f_0 df_1 \wedge df_2 \dots \wedge df_k).$$

Here $M_0 = M$ is a dense open constructible subset. We can assume that M is a subspace of some $\mathbf{R}^N$, and all functions f_i are smooth on a smooth dense open submanifold $M_0 \subset M$.

Proposition 9 *The functional l from the definition above is well-defined (i.e., the integral absolutely converges).*

Proof. Proof. We can use a sequence of functions $(f_0, f_1, \dots, f_k)$ in order to map M to $\mathbf{R}^{k+1}$. Then we obtain a singular chain $\gamma \in C_k(\mathbf{R}^{k+1})$. The support of γ is a compact constructible subset of dimension $\leq k$. Moreover, the volume of γ with respect to the induced metric is finite. It follows that $\int_\gamma \omega$ absolutely converges for an arbitrary smooth differential k-form ω. In particular the integral in question converges. ■

It is easy to check that the differential $d : \Omega^k_{X,min} \to \Omega^{k+1}_{X,min}$ is well-defined (as the adjoint to the boundary operator on the chains). Moreover the following Stokes formula holds.

Lemma 9 *In the previous notation we have:*

$$(1, f_0, \dots, f_k)(M, or, \phi) = (f_0, \dots, f_k)(\partial M, or|_{\partial M}, \phi|_{\partial M}).$$

Hence we obtain a complex of sheaves $\Omega^k_{X,min}$ for every $k \geq 0$. Our (first approximation to) k-forms are sections of these sheaves. One can take pull-backs of forms.

Moreover, we can introduce wedge product of forms. Indeed the space of k-forms $\Omega^k_{min}(X)$ is naturally embedded into the direct limit of spaces of smooth

k-forms $\Omega^k_{C^\infty}(X_\alpha)$. Here $X_\alpha \subset X$ are open dense locally constructible subspaces, $X_\alpha = \sqcup_i X_{i,\alpha}$, where $X_{i,\alpha}$ is a PA-manifold. Moreover, it is assumed that we have fixed an embedding $X_{i,\alpha} \to \mathbf{R}^{N_{i,\alpha}}$ identifying $X_{i,\alpha}$ with a C^∞-submanifolds in $\mathbf{R}^{N_{i,\alpha}}$. Using the embeddings we define $\Omega^k_{C^\infty}(X_\alpha)$. Therefore the wedge product given by the formula $(f_0,\dots,f_k)\wedge(g_0,\dots,g_n) = (f_0g_0,f_1,\dots,f_k,g_1,\dots,g_n)$ is well-defined.

The above defined forms are sections of soft sheaves because one can use piecewise algebraic functions to produce a partition of unity, so the standard proofs work. One can show that Mayer-Vietoris sequence degenerates as in the case with chains above: If $X = Y_1 \cup Y_2$ is a union of locally closed piecewise algebraic subspaces then one has a short exact sequence

$$0 \to \Omega^k_{min}(X) \to \Omega^k_{min}(Y_1) \oplus \Omega^k_{min}(Y_2) \to \Omega^k_{min}(Y_1 \cap Y_2) \to 0.$$

In the next subsection we will extend algebras $\Omega^\bullet_{min}(X)$ adding push-forwards of such forms. The following example illustrates one of the reasons for that.

Example 1 *Let $Y = [0,1]\times[0,1]$, $X = [0,1]$ and $f : Y \to X$ be the map $(x,y) \mapsto t = xy$. Then, outside of $t = 1$, we have a bundle in the category $\mathcal{P}$ (i.e., base, fibers and total space are objects, projection is a morphism in $\mathcal{P}$). Take the form $\omega = xdy$. Then $f_*\omega = t(logt)$ is a continuous function, but does not belong to $\mathcal{O}(X)$.*

8.4. Full algebra of forms

We are going to define a dg-algebra $\Omega^\bullet_{PA}(X)$ which contains $\Omega^\bullet_{min}(X)$ and satisfies Poincaré lemma, etc. Elements of $\Omega^\bullet_{PA}(X)$ will be called PA-forms, although their coefficients are not PA-functions (see Example 1). It is interesting that only 0-forms will be forms with continuous coefficients. Higher order forms can have coefficients in $L_1^{loc}(X)$. But they are still closed under the wedge product.

We start with some preliminaries. Let $f : Y \to X$ be a proper PA-map of PA-spaces. For simplicity we will give the definition below in the case when X is compact.

Definition 22 *Continuous family of PA k-cycles (PA-family of cycles for short) is defined as an element of the abelian group $C_k(Y \to X)$ described below.*

The group $C_k(Y \to X)$ consists of maps $\Phi : X \to C_k(X)$ such that:

a) For any $x \in X$ the set $Supp\,\Phi(x)$ belongs to $f^{-1}(x)$.

b) The set $Z = \cup_{x\in X} Supp\,\Phi(x)$ is constructible with the compact closure $\overline{Z} \subset Y$.

c) There exists a dense constructible $Z_0 \subset \overline{Z}$ such that for any $x \in X$ the intersection $Z_0 \cap f^{-1}(x)$ is a PA-manifold without boundary, dense in $\overline{Z}\cap f^{-1}(x)$. Moreover, the chain $\Phi(x)$ is obtained from some orientation of $Z_0 \cap f^{-1}(x)$ and locally constant multiplicity map $Z_0 \cap f^{-1}(x) \to \mathbf{Z}$ as in the second description of PA-chains.

d) For any $x \in X$ and $z \in f^{-1}(x)\cap Z_0$, and any sequence $(x_i), x_i \to x$ the multiplicity at the point z is the "natural limit" of the multiplicities of $f^{-1}(x_i)\cap Z_0$.

In the example below, X is not compact. It easy to make necessary modifications of our definition, so it will be valid in the non compact case. In that case we get a sheaf on X.

Example 2 *Let $Y = \mathbf{R}^2, X = \mathbf{R}$ and $f : Y \to X$ is the projection $(x_1, x_2) \mapsto x_1$. Let us define $\Phi(x_1)$ to be equal $\{x_1\} + \{-x_1\}$ for $x_1 > 0$ and equal to $2\{0\}$ otherwise. Then this map gives an element of $C_0(Y \to X)$.*

We will briefly describe some properties of PA-families of cycles below.
1) PA-families $C_\bullet(Y \to X) = \{C_k(Y \to X)\}_{k\geq 0}$ form a complex of sheaves (abelian groups in the case when X is compact).
2) There are natural operations $C_k(Y \to X) \otimes C_l(X) \to C_{k+l}(Y)$. Informally we will denote them by $(\gamma, \alpha) \mapsto \gamma \times \alpha$.
3) For any Cartesian square:

$$\begin{array}{ccc} Y' & \to & Y \\ \downarrow & & \downarrow \\ X' & \to & X \end{array}$$

one has the natural morphisms

$$C_k(Y \to X) \to C_k(Y' \to X')$$

and

$$C_k(Y \to X) \otimes C_l(X' \to X) \to C_{k+l}(Y' \to X).$$

Now we are ready to define piecewise PA-forms.

Definition 23 *PA-forms of degree k are sections of soft sheaves which are locally given by functionals $l : C_k(X) \to \mathbf{R}$ such that*

$$l(\alpha) = \int_{\gamma\times\alpha} \omega$$

where $\gamma \in C_l(Y \to X), l \geq 0$ for some proper $Y \to X$, and $\omega \in \Omega^{k+l}(Y)_{min}$.

We will denote the space of PA-forms on X by $\Omega^\bullet_{PA}(X)$.

Similarly to the case of $\Omega^\bullet_{min}(X)$ one can see that at the "generic point" the space $\Omega^\bullet_{PA}(X)$ can be naturally embedded into the space of differential forms. In fact we obtain a soft sheaf of dg-algebras. It is closed under pushforwards f_* where $f : Y \to X$ is a locally trivial bundle in the category $\mathscr{P}$ with fibers which are compact oriented PA-manifolds.

The latter fact can be generalized further. One can consider a family $f : Y \to X$ where Y and X are oriented compact PA-manifolds, and all fibers of f have the same dimension $dim(Y) - dim(X)$. Then one get a continuous family of cycles ("fundamental cycle" of $f^{-1}(x), x \in X$) over X. This gives a pushforward of PA-forms for certain maps ("flat morphisms") which are not necessarily fibrations.

Notice that the Poincaré lemma holds for piecewise forms. One can imitate the usual proof based on the integration over rays $(x,t) \mapsto xt$, where $x \in U \subset \mathbf{R}^n$, U is a convex domain containing $0 \in \mathbf{R}^n$, and $t \in [0,1]$.

References

1. Beilinson, A. and Drinfel'd, V. G.: Chiral algebras, preprint (1995).
2. Borcherds, R. E.: Vertex algebras, in: M. Kashiwara, A. Matsuo, K. Saito, and I. Satake, (eds.), *Topological field theory, primitive forms and related topics*, Progr. Math. 160, Birkhäuser Boston Inc., Boston, 1998, pp. 35–77
3. Boardman, J. M. and Vogt, R. M.: *Homotopy invariant algebraic structures on topological spaces*, Lect. Notes Math. 347 Springer-Verlag, Berlin, 1973.
4. Deligne, P.: Letter to Stasheff, Gerstenhaber, May, Schechtman, Drinfel'd (1993).
5. Drinfel'd, V. G.: On quasitriangular quasi-Hopf algebras and a group closely connected with $\mathrm{Gal}(\overline{\mathbf{Q}}/\mathbf{Q})$, *Leningrad Math. J.* **2** (1991), 829–860.
6. Fulton, W. and MacPherson, R.: A compactification of configuration spaces, *Ann. of Math.* **139** (1994), 183–225.
7. Getzler, E. and Jones, J.: Operads, homotopy algebra, and iterated integrals for double loop spaces, preprint hep-th/9403055.
8. Getzler, E. and Kapranov, M. M.: Cyclic operads and cyclic homology, in: S.-T. Yau, (ed.), *Geometry, Topology & Physics (for Raul Bott)*, Conf. Proc. Lecture Notes Geom. Topology, VI, Internat. Press, Cambridge, MA, 1995, pp. 167–201.
9. Ginzburg, V. and Kapranov, M.: Koszul duality for operads, *Duke Math. J.* **76** (1994), 203–272.
10. Hinich, V.: Homological algebra of homotopy algebras, preprint q-alg/9702015.
11. Kontsevich, M. and Manin, Yu.: Gromov-Witten classes, quantum cohomology, and enumerative geometry, *Comm. Math. Phys.* **164** (1994), 525–562.
12. Kontsevich, M.: Deformation quantization of Poisson manifolds I, preprint q-alg/9709040.
13. Kontsevich, M.: Rozansky-Witten invariants via formal geometry, *Compositio Math.*, **115** (1999), 115–127.
14. Kontsevich, M.: Operads and motives in deformation quantization, *Lett. Math. Phys.*, **48** (1999), 35–72.
15. Markl, M.: Cotangent cohomology of a category and deformations. *J. Pure Appl. Algebra* **113** (1996), 195–218.
16. Markl, M.: Homotopy Algebras are Homotopy Algebras, preprint math.AT/9907138.
17. May, J. P.: Infinite loop space theory, *Bull. Amer. Math. Soc.* **83** (1977), 456–494.
18. McClure, J. and Smith, J.: A solution of Deligne's conjecture, preprint math.QA/9910126.
19. Soibelman, Y.: Meromorphic tensor categories, preprint q-alg/9709030. To appear in Contemporary Math.
20. Strebel, K.: *Quadratic differentials*, Springer-Verlag, Berlin, 1984.
21. Tamarkin, T.: Formality of chain operad of small squares, preprint math.QA/9809164.
22. Voronov, A.: Homotopy Gerstenhaber algebras, preprint math.QA/9908040.

A generalization of the Bott-Borel-Weil theorem and Euler number multiplets of representations

Bertram Kostant*
Department of Mathematics, MIT, Cambridge, MA 02139, USA
e-mail: `kostant@math.mit.edu`

Keywords: representation theory, Bott-Borel-Weil theorem, Euler number multiplets, nilpotent Lie algebra cohomology, spin representation, sheaf cohomology, Dirac operator

Mathematics Subject Classifications (2000): 22Exx, 22E46, 32Q15

Dedicated to the memory of a much missed friend, Moshé Flato

0. Introduction

0.1. In Remark 0.27 of [3] we have stated that results in [3] yield an extensive generalization of the Bott-Borel-Weil theorem (BBW). The purpose of the present paper is to explain and justify this statement in detail.

Let $G_{\mathbb{C}}$ be a simply-connected complex semisimple Lie group and let $\mathfrak{g} = Lie\, G_{\mathbb{C}}$. Let $P \subset G_{\mathbb{C}}$ be a parabolic subgroup and let $\mathfrak{p} = Lie\, P$. As one knows the complex $G_{\mathbb{C}}$-homogeneous space $X = G/P$ is a compact projective variety. Let E be a holomorphic homogeneous vector bundle over X defined by an irreducible representation of P. Then BBW, established as Theorem IV′ in [1], determines the multiplicity of any irreducible representation $\pi_\lambda : \mathfrak{g} \to End\, V_\lambda$ of $\mathfrak{g}$ in the cohomology $H^j(X, \mathcal{S}E)$ where $\mathcal{S}E$ is the sheaf of local holomorphic sections of E. The subscript λ is the highest weight of π_λ. Let

$$\mathfrak{p} = \mathfrak{g}_1 + \mathfrak{n}$$

be the Levi-decomposition of $\mathfrak{p}$ where $\mathfrak{g}_1$ is a Levi-factor of $\mathfrak{p}$ and $\mathfrak{n}$ is the nilradical of $\mathfrak{p}$. It is implicit in [1] and made more explicit in [2] that BBW is equivalent to knowing the $\mathfrak{g}_1$-module structure on the Lie algebra cohomology $H^j(\mathfrak{n}, V_\lambda)$. In [2] we obtained an algebraic proof of BBW by determining this $\mathfrak{g}_1$-structure. The technique was to introduce a certain Laplacian L_λ on the cochain complex $C(\mathfrak{n}, V_\lambda)$ which defines $H^*(\mathfrak{n}, V_\lambda)$ so that one has a graded $\mathfrak{g}_1$-module isomorphism

$$Ker\, L_\lambda \cong H^*(\mathfrak{n}, V_\lambda).$$

Now $C(\mathfrak{n}, V_\lambda) = V_\lambda \otimes \wedge \mathfrak{n}^*$. What was needed, and what was directly established in [2], was that the $\mathfrak{g}_1$-submodule $Ker\, L_\lambda$ of $V_\lambda \otimes \wedge \mathfrak{n}^*$ is multiplicity free and

* Research supported in part by NSF grant DMS-9625941 and in part by the KG&G Foundation.

G. Dito and D. Sternheimer (eds.), Conférence Moshé Flato 1999, Vol. 1, 309 – 325.

decomposes into a direct sum of $d = Eul(X)$ irreducibles

$$Ker L_\lambda = \sum_{\tau\in W^1} (Ker L_\lambda)^{\tau*\lambda}. \tag{0.1}$$

Here W^1 is a special set of coset representatives of the Weyl group of $\mathfrak{g}_1$ in the Weyl group of $\mathfrak{g}$. Also $\tau * \lambda = \tau(\lambda+\rho)-\rho$ where ρ has its usual meaning. The superscript denotes the primary $\sigma_{\tau*\lambda}$-component where, in general, σ_μ is an irreducible $\mathfrak{g}_1$-module with highest weight μ. In addition, what was needed and what was proved was the inclusion

$$(Ker L_\lambda)^{\tau*\lambda} \subset V_\lambda \otimes \wedge^{l(\tau)}\mathfrak{n}^*, \tag{0.2}$$

where $l(\tau)$ is the length of τ. Actually the result in [2] is stronger than (0.1) and (0.2). For one thing the result in [2] names the highest weight vector of $(Ker L_\lambda)^{\tau*\lambda}$. But more significantly, it established the equality $(Ker L_\lambda)^{\tau*\lambda} = C(\mathfrak{n},V_\lambda)^{\tau*\lambda}$ so that (0.1) can be replaced by the stronger statement

$$Ker L_\lambda = \sum_{\tau\in W^1} C(\mathfrak{n},V_\lambda)^{\tau*\lambda}. \tag{0.3}$$

Now $\mathfrak{g}_1$ is a reductive Lie subalgebra of $\mathfrak{g}$ with the same rank as $\mathfrak{g}$. Much more generally, let $\mathfrak{r}$ be an arbitrary reductive Lie subalgebra of $\mathfrak{g}$ with the same rank as $\mathfrak{g}$. We refer to the special case when $\mathfrak{r}=\mathfrak{g}_1$ as the Levi-factor case. Of course $\mathfrak{r}$ can be semisimple whereas $\mathfrak{g}_1$ is never semisimple — except in the trivial situation when $\mathfrak{g}_1=\mathfrak{g}$. Let $B_\mathfrak{g}$ be the Killing form on $\mathfrak{g}$ and let $B_\mathfrak{r} = B|\mathfrak{r}$. Let $\mathfrak{q}$ be the $B_\mathfrak{g}$-orthocomplement of $\mathfrak{r}$ in $\mathfrak{g}$ so that

$$\mathfrak{g} = \mathfrak{r} + \mathfrak{q}$$

and let $B_\mathfrak{q} = B_\mathfrak{g}|\mathfrak{q}$. Let $Cliff(\mathfrak{q})$ be the Clifford algebra over $\mathfrak{q}$ with respect to $B_\mathfrak{q}$. The dimension of $\mathfrak{q}$ is even so that $Cliff(\mathfrak{q})$ has a unique irreducible module S — the spin module. The natural action ν of $\mathfrak{r}$ on $\mathfrak{q}$ (since $[\mathfrak{r},\mathfrak{q}]\subset\mathfrak{q}$) defines a homomorphism $\nu_*:\mathfrak{r}\to Cliff(\mathfrak{q})$ and hence a representation $Spin\,\nu:\mathfrak{r}\to End\,S$. Taking the tensor product of $Spin\,\nu$ with $\pi_\lambda|\mathfrak{r}$ defines a representation

$$\zeta_\lambda : \mathfrak{r} \to End\,(V_\lambda\otimes S).$$

The point to be made here is that in the Levi-factor case $V_\lambda\otimes S$ differs from the cochain complex $C(\mathfrak{n},V_\lambda)$, as $\mathfrak{g}_1$-modules, only by tensoring with a 1-dimensional representation, $\sigma_{\rho(\mathfrak{n})}$, of $\mathfrak{g}_1$. That is, in the Levi-factor case one can readily go back and forth between $C(\mathfrak{n},V_\lambda)$ and $V_\lambda\otimes S$ whereas in the general case one has only $V_\lambda\otimes S$. (If $\mathfrak{r}$ is semisimple it has no nontrivial 1-dimensional representations.)

In the general case let G be a compact form of $G_\mathbb{C}$ chosen so that there exists a connected Lie subgroup $R\subset G$ which corresponds to a real form of $\mathfrak{r}$. The compact homogeneous space $X = G/R$ need no longer be complex but is, in fact, the most general simply-connected compact homogeneous space of positive Euler characteristic. To sense the generality consider the example of when X is a sphere. In the Levi-factor case one is looking only at the 2-sphere whereas in the general case all even dimensional spheres are examples.

In the general case one still has the set W^1 of Weyl group coset representatives and if $d = card\, W^1$ then $d = Eul(X)$. Also, as in the Levi-factor case, σ_μ will denote an irreducible representation of $\mathfrak{r}$ with highest weight μ. The paper [3] is much concerned with the d-multiplet of irreducible $\mathfrak{r}$-representations $\{\sigma_{\tau\bullet\lambda} \mid \tau \in W^1\}$ defined by any highest weight λ of $\mathfrak{g}$. Here $\tau \bullet \lambda = \tau(\lambda + \rho) - \rho_{\mathfrak{r}}$. Note that in the Levi-factor case

$$\tau \bullet \lambda = \rho(\mathfrak{n}) + \tau * \lambda.$$

Let $Cas_{\mathfrak{r}} \in Cent\, U(\mathfrak{r})$ be the Casimir element which corresponds to $B_{\mathfrak{r}}$. The generalization of (0.3) will be an easy consequence (Theorem B) of Theorem A below. Theorem A is part of the statement of Theorem 4.17 in [3].

Theorem A *Let λ be any highest weight for $\mathfrak{g}$ and let $\mathfrak{r}$ be any reductive subalgebra of $\mathfrak{g}$ having the same rank as $\mathfrak{g}$. Then the highest eigenvalue of the operator $\zeta_\lambda(Cas_{\mathfrak{r}})$ on $V_\lambda \otimes S$ is $|\lambda + \rho|^2 - |\rho_{\mathfrak{r}}|^2$. Furthermore if M is the corresponding eigenspace then M is a multiplicity-free $\mathfrak{r}$-module whose irreducible components are the elements of the d-multiplet $\{\sigma_{\tau\bullet\lambda} \mid \tau \in W^1\}$. In fact*

$$M = \sum_{\tau \in W^1} (V_\lambda \otimes S)^{\tau\bullet\lambda} \tag{0.4}$$

and the primary component $(V_\lambda \otimes S)^{\tau\bullet\lambda}$ is irreducible for any $\tau \in W^1$.

Assume the Levi-factor case and let $Z_{\rho(\mathfrak{n})}$ be the 1-dimensional $\mathfrak{g}_1$-module which affords the representation $\sigma_{\rho(\mathfrak{n})}$ so that as $\mathfrak{g}_1$-modules

$$V_\lambda \otimes S = C(\mathfrak{n}, V_\lambda) \otimes Z_{\rho(\mathfrak{n})}. \tag{0.5}$$

Recalling the Laplacian L_λ, let L^e_λ be the operator on $V_\lambda \otimes S$ defined by putting $L^e_\lambda = L_\lambda \otimes 1$ with respect to the right side of (0.5).

Theorem B *Let the notation be as in Theorem A except that now we assume the Levi-factor case ($\mathfrak{r} = \mathfrak{g}_1$). Then*

$$2L_\lambda = (|\lambda + \rho|^2 - |\rho_{\mathfrak{r}}|^2) - \zeta_{\mathfrak{r}}(Cas_{\mathfrak{r}}) \tag{0.6}$$

so that the $\mathfrak{g}_1$-module structure on $Ker\, L_\lambda$ (and hence on $H^*(\mathfrak{n}, V_\lambda)$) is given by the $\mathfrak{g}_1$-module structure on M. More precisely, as $\mathfrak{g}_1$-modules,

$$(Ker\, L_\lambda) \otimes Z_{\sigma(\mathfrak{n})} \cong M \tag{0.7}$$

and hence (0.3) follows from (0.4).

0.2. In the general case one does not have the Laplacian L_λ. Instead one has the cubic Dirac operator $\square_\lambda$, defined in [3]. The relation of $\square_\lambda$ to L_λ in the Levi-factor case is given in

Theorem C *With the assumptions of Theorem B one has*

$$2L_\lambda^e = \square_\lambda. \tag{0.8}$$

The equation (0.6) is Theorem 2.9 in this paper and Theorem C is Theorem 3.10.

1. The Bott-Borel-Weil theorem and nilpotent Lie algebra cohomology

1.1. Let $G_{\mathbb{C}}$ be a complex semisimple Lie group and let $\mathfrak{g} = Lie\,G_{\mathbb{C}}$. Let $l = rank\,\mathfrak{g}$. Let $\mathfrak{r}$ be a reductive Lie subalgebra of $\mathfrak{g}$ and let $\mathfrak{h}$ be a Cartan subalgebra of $\mathfrak{r}$. Assume $rank\,\mathfrak{r} = rank\,\mathfrak{g}$ so that $\mathfrak{h}$ is also a Cartan subalgebra of $\mathfrak{g}$ and $dim\,\mathfrak{h} = l$. Let $\mathfrak{h}^*$ be the dual space to $\mathfrak{h}$ and let $\Delta \subset \mathfrak{h}^*$ be the set of roots of $\mathfrak{g}$ with respect to $\mathfrak{h}$. Let $\Delta_\mathfrak{r} \subset \Delta$ be the set of roots of $\mathfrak{r}$ with respect to $\mathfrak{h}$. Let (z,z'), for $z,z' \in \mathfrak{g}$, be the Killing form, $B_\mathfrak{g}$, of $\mathfrak{g}$ and let $B_\mathfrak{r} = B_\mathfrak{g}|\mathfrak{r}$. For any $\varphi \in \Delta$ let $e_\varphi \in \mathfrak{g}$ be a corresponding root vector where the choice is normalized so that $(e_\varphi, e_{-\varphi}) = 1$ for any $\varphi \in \Delta$. Let $\mathfrak{h}_\mathbb{R}$ be the space of hyperbolic elements in $\mathfrak{h}$ and let $\mathfrak{h}_\mathbb{R}^*$ be the real span of Δ so that $\mathfrak{h}_\mathbb{R}$ (resp. $\mathfrak{h}_\mathbb{R}^*$) is a real form of $\mathfrak{h}$ (resp. $\mathfrak{h}^*$) and $\mathfrak{h}_\mathbb{R}^*$ can be taken as the real dual space to $\mathfrak{h}_\mathbb{R}$. The restriction $B_\mathfrak{g}|\mathfrak{h}$ is nonsingular and induces a nonsingular symmetric bilinear form (δ,γ) on $\mathfrak{h}^*$ that is positive definite on $\mathfrak{h}_\mathbb{R}^*$. We write $|\mu| = \sqrt{(\mu,\mu)}$ for $\mu \in \mathfrak{h}_\mathbb{R}^*$. Let $\Gamma \subset \mathfrak{h}_\mathbb{R}^*$ be the lattice of $\mathfrak{g}$-integral linear forms on $\mathfrak{h}$. That is, $\Gamma = \{\mu \in \mathfrak{h}^* \mid 2(\mu,\varphi)/(\varphi,\varphi) \in \mathbb{Z}, \forall \varphi \in \Delta\}$.

For any subspace $\mathfrak{m} \subset \mathfrak{g}$ which is stable under $ad\,\mathfrak{h}$ let $\Delta(\mathfrak{m}) = \{\varphi \in \Delta \mid e_\varphi \in \mathfrak{m}\}$. Let $\mathfrak{b}$ be a Borel subalgebra of $\mathfrak{g}$ which contains $\mathfrak{h}$. Then $\mathfrak{b}_\mathfrak{r} = \mathfrak{b} \cap \mathfrak{r}$ is a Borel subalgebra of $\mathfrak{r}$. Positive root systems for $\mathfrak{g}$ and $\mathfrak{r}$ can then be given respectively by letting $\Delta^+ = \Delta(\mathfrak{b})$ and $\Delta_\mathfrak{r}^+ = \Delta(\mathfrak{b}_\mathfrak{r})$. Let $\rho = \frac{1}{2}\sum_{\varphi\in\Delta^+}\varphi$ and $\rho_\mathfrak{r} = \frac{1}{2}\sum_{\varphi\in\Delta_\mathfrak{r}^+}\varphi$. One has $\rho \in \Gamma$ but $\rho_\mathfrak{r}$ is not necessarily in Γ. Of course $2\rho_\mathfrak{r} \in \Gamma$. Let $\Gamma_\mathfrak{r} \subset \mathfrak{h}_\mathbb{R}^*$ be the lattice generated by Γ and $\rho_\mathfrak{r}$. One of course has

$$2(\mu,\varphi)/(\varphi,\varphi) \in \mathbb{Z}, \quad \forall \varphi \in \Delta_\mathfrak{r} \text{ and } \forall \mu \in \Gamma_\mathfrak{r}.$$

Let D be the $\mathfrak{g}$-Weyl chamber in $\mathfrak{h}_\mathbb{R}^*$ defined by $D = \{\gamma \in \mathfrak{h}_\mathbb{R}^* \mid (\gamma,\varphi) \geq 0, \forall \varphi \in \Delta^+\}$ and let $D_\mathfrak{r}$ be the $\mathfrak{r}$-Weyl chamber in $\mathfrak{h}^*$ defined by $D_\mathfrak{r} = \{\gamma \in \mathfrak{h}_\mathbb{R}^* \mid (\gamma,\varphi) \geq 0, \forall \varphi \in \Delta_\mathfrak{r}^+\}$. Now let $\Lambda = D \cap \Gamma$ and let $\Lambda_\mathfrak{r} = D_\mathfrak{r} \cap \Gamma_\mathfrak{r}$. For each $\lambda \in \Lambda$ let

$$\pi_\lambda : \mathfrak{g} \to End\,V_\lambda$$

be an irreducible representation with highest weight λ and, for any $\mu \in \Lambda_\mathfrak{r}$, let

$$\sigma_\mu : \mathfrak{r} \to End\,Z_\mu$$

be an irreducible representation with highest weight μ. If Y is any completely reducible $\mathfrak{r}$-module we will let Y^μ be the σ_μ-primary component of Y.

Let $\mathfrak{q}$ be the $B_{\mathfrak{g}}$-orthocomplement of $\mathfrak{r}$ in $\mathfrak{g}$ and let $B_{\mathfrak{q}} = B_{\mathfrak{g}}|\mathfrak{q}$. Much of the notation of [2] will be adopted here. However we are writing $\mathfrak{q}$ here for $\mathfrak{p}$ in [2] to avoid confusion with the Lie algebra of a parabolic subalgebra to be considered in §1.2. Let $n = dim\,\mathfrak{q}$ and let $\nu : \mathfrak{r} \to Lie\,SO(\mathfrak{q})$ be the representation defined so that

$$\nu(x)y = [x,y] \tag{1.1}$$

for $x \in \mathfrak{r}$ and $y \in \mathfrak{q}$. Of course $\Delta(\mathfrak{q})$ is the complement of $\Delta(\mathfrak{r})$ in Δ and one of course has $card\,\Delta(\mathfrak{q}) = n$. Let $\Delta^+(\mathfrak{q}) = \Delta^+ \cap \Delta(\mathfrak{q})$ so that $\Delta^+(\mathfrak{q})$ is the complement of $\Delta_{\mathfrak{r}}^+$ in Δ^+. Put $m = card\,\Delta^+(\mathfrak{q})$ and since one readily has the partition

$$\Delta(\mathfrak{q}) = \Delta^+(\mathfrak{q}) \cup -\Delta^+(\mathfrak{q})$$

it follows that

$$n = 2m. \tag{1.2}$$

Let $\mathfrak{q}_+$ (resp. $\mathfrak{q}_-$) be the span of e_φ (resp. $e_{-\varphi}$) for $\varphi \in \Delta^+(\mathfrak{q})$ so that as linear spaces

$$\mathfrak{q} = \mathfrak{q}_+ \oplus \mathfrak{q}_-. \tag{1.3}$$

Let W (resp. $W_{\mathfrak{r}}$) be the Weyl group of $\mathfrak{g}$ (resp. $\mathfrak{r}$) operating as usual in both $\mathfrak{h}$ and $\mathfrak{h}^*$. Of course $W_{\mathfrak{r}}$ is a subgroup of W. Let d be the index of $W_{\mathfrak{r}}$ in W. A cross-section, W^1, of the set of right $W_{\mathfrak{r}}$ cosets is defined (using the Weyl chambers $D \subset D_{\mathfrak{r}}$, see §3.1) by putting

$$W^1 = \{\tau \in W \mid \tau(D) \subset D_{\mathfrak{r}}\}. \tag{1.4}$$

That is, the map

$$W_{\mathfrak{r}} \times W^1 \to W, \qquad (w,\tau) \mapsto w\tau \tag{1.5}$$

is clearly a bijection. In particular

$$card\,W^1 = d. \tag{1.6}$$

Let Γ^{reg} be the set of W-regular elements in the lattice $\Gamma \subset \mathfrak{h}_{\mathbb{R}}^*$. One has the partition

$$\Gamma^{reg} = \bigcup_{w \in W} w(\Lambda + \rho). \tag{1.7}$$

The partition (1.7) clearly defines a partition of the subset, $\Gamma^{reg} \cap D_{\mathfrak{r}}$, of elements in Γ^{reg} which are $\Delta_{\mathfrak{r}}^+$-dominant

$$\Gamma^{reg} \cap D_{\mathfrak{r}} = \bigcup_{\tau \in W^1} \tau(\Lambda + \rho). \tag{1.8}$$

Obviously $\Gamma^{reg} \cap D_{\mathfrak{r}} \subset \Lambda_{\mathfrak{r}}$. Furthermore it is also obvious that the elements in $\Gamma^{reg} \cap D_{\mathfrak{r}}$ are $W_{\mathfrak{r}}$-regular. For any $\tau \in W^1$ and $\lambda \in \Lambda$ let

$$\tau \bullet \lambda = \tau(\lambda + \rho) - \rho_{\mathfrak{r}}. \tag{1.9}$$

It then follows that $\tau \bullet \lambda \in \Lambda_{\mathfrak{r}}$. In fact this isolates a distinguished subset, $\Lambda_{\mathfrak{r}}^*$, of $\Lambda_{\mathfrak{r}}$. Let

$$\Lambda_{\mathfrak{r}}^* = \{\mu \in \Lambda_{\mathfrak{r}} \mid \mu + \rho_{\mathfrak{r}} \in \Gamma^{reg}\}. \tag{1.10}$$

The following proposition (stated as Proposition 3.31 in [2]) is an immediate consequence of the partition (1.8).

Proposition 1.1 *For any $\tau \in W^1$ and $\lambda \in \Lambda$ one has $\tau \bullet \lambda \in \Lambda_{\mathfrak{r}}^*$. Moreover the map*

$$W^1 \times \Lambda \to \Lambda_{\mathfrak{r}}^*, \qquad (\tau, \lambda) \mapsto \tau \bullet \lambda \tag{1.11}$$

is a bijection.

Now observe that the elements of Λ parameterize subsets of $\Lambda_{\mathfrak{r}}^*$. For each $\lambda \in \Lambda$ let

$$\Lambda_{\mathfrak{r}}^{\lambda} = \{\tau \bullet \lambda \mid \tau \in W^1\}. \tag{1.12}$$

From Proposition 3.3 one has

$$card\, \Lambda_{\mathfrak{r}}^{\lambda} = d \tag{1.13}$$

so that all the subsets have the same cardinality and

$$\Lambda_{\mathfrak{r}}^* = \bigcup_{\lambda \in \Lambda} \Lambda_{\mathfrak{r}}^{\lambda}. \tag{1.14}$$

is a partition. By abuse of notation we will use (as in [3]) the word multiplet (or d-multiplet) to refer, either to any subset of $\Lambda_{\mathfrak{r}}^*$ of the form $\Lambda_{\mathfrak{r}}^{\lambda}$, or to the corresponding d-set $\{\sigma_\mu\}$, $\mu \in \Lambda_{\mathfrak{r}}^{\lambda}$, of irreducible $\mathfrak{r}$-representations, or again, to the corresponding d-set $\{Z_\mu\}, \mu \in \Lambda_{\mathfrak{r}}^{\lambda}$, of irreducible $\mathfrak{r}$-modules.

1.2. The Bott-Borel-Weil theorem (BBW) has to do with a special choice of $\mathfrak{r}$ (and Δ_+). By abuse of notation we will say that $\mathfrak{r}$ is a Levi-factor if there exists $x \in \mathfrak{h}_{\mathbb{R}}$ such that $\mathfrak{r}$ is the centralizer of x in $\mathfrak{g}$. In this case, for $\varphi \in \Delta$, $\langle \varphi, x \rangle$ is zero or not zero according as $\varphi \in \Delta(\mathfrak{r})$ or $\varphi \in \Delta(\mathfrak{q})$. If $\mathfrak{r}$ is a Levi-factor we may (and shall) choose $\mathfrak{b}$ such that $\langle \varphi, x \rangle > 0$ for $\varphi \in \Delta^+(\mathfrak{q})$. When $\mathfrak{r}$ is a Levi-factor, to partially conform to the notation of [3], we will write $\mathfrak{g}_1$ for $\mathfrak{r}$, W_1 for $W_{\mathfrak{r}}$, and $\mathfrak{n}$ for $\mathfrak{q}_+$. In that case

$$\mathfrak{p} = \mathfrak{g}_1 \oplus \mathfrak{n} \tag{1.15}$$

is a parabolic subalgebra of $\mathfrak{g}$, $\mathfrak{g}_1$ is a Levi-factor of $\mathfrak{p}$ and $\mathfrak{n}$ is the nilradical of $\mathfrak{p}$. The most general parabolic Lie subalgebra of $\mathfrak{g}$ is of this form.

Assume we are in the Levi-factor case and $\mathfrak{p}$ is given by (1.15). Let P be the (parabolic) Lie subgroup of $G_{\mathbb{C}}$ corresponding to $\mathfrak{p}$. Thus $P = (G_1)_{\mathbb{C}} N$ where $(G_1)_{\mathbb{C}}$ (resp. N) is the subgroup of $G_{\mathbb{C}}$ corresponding to $\mathfrak{g}_1$ (resp. $\mathfrak{n}$). Let $X = G_{\mathbb{C}}/P$ so that X is a complex (in fact projective) compact simply-connected $G_{\mathbb{C}}$ homogeneous space, and for the Euler characteristic one has

$$Eul(X) = d.$$

For $\mathfrak{r} = \mathfrak{g}_1$ put $\Lambda_1 = \Lambda_{\mathfrak{r}}$. Also let $\Lambda_{(1)} = D_{\mathfrak{r}} \cap \Gamma$ so that

$$\Lambda_{(1)} \subset \Lambda_1.$$

Let $\xi \in \Lambda_{(1)}$. Note that the representation σ_ξ of $\mathfrak{g}_1$ integrates to a representation $\sigma_\xi : (G_1)_{\mathbb{C}} \to Aut\, Z_\xi$ and extends, by trivializing the action on N, to a representation $\sigma_\xi : P \to Aut\, Z_\xi$. Let $\sigma_\xi^* : P \to Aut\, Z_\xi^*$ be the corresponding contragredient representation of P on the dual space Z_ξ^* to Z_ξ. Let $E^{-\xi}$ be the homogeneous holomorphic vector bundle on X corresponding to σ_ξ^*. Let $\mathscr{S}E^{-\xi}$ be the sheaf, on X, of local holomorphic sections of $E^{-\xi}$ and let $H^*(X, \mathscr{S}E^{-\xi})$ be the cohomology of $\mathscr{S}E^{-\xi}$. See §6, p. 364, in [2]. Then $H^*(X, \mathscr{S}E^{-\xi})$ has the natural structure of a graded $G_{\mathbb{C}}$ (or equivalently $\mathfrak{g}$)-module. The Bott-Borel-Weil theorem explicitly describes this structure in representation theoretic terms.

In order to state the BBW Theorem (Theorem 1.3. below) we need to introduce a certain subset $\Lambda_{(1)}^*$ (written D_1^0 in [2]) of $\Lambda_{(1)}$. Let $\Lambda_{(1)}^* = \{\xi \in \Lambda_{(1)} \mid \xi + \rho \text{ is } W\text{-regular}\}$. (Note that ρ is written as g in [2] and in the definition of D_1^0 on p. 370 in [2], D_1 was misprinted as D). For any $\tau \in W^1$ and $\lambda \in \Lambda$, let $\tau * \lambda = \tau(\lambda + \rho) - \rho$. Let $\rho(\mathfrak{n}) = \frac{1}{2} \sum_{\varphi \in \Delta(\mathfrak{n})} \varphi$. Note that

$$\tau \bullet \lambda = \rho(\mathfrak{n}) + \tau * \lambda \tag{1.16}$$

for any $(\tau, \lambda) \in W^1 \times \Lambda$. The following result has been established as Lemma 6.4 in [2].

Proposition 1.2 *One has $\tau * \lambda \in \Lambda_{(1)}^*$ for any $(\tau, \lambda) \in W^1 \times \Lambda$ and the map*

$$W^1 \times \Lambda \to \Lambda_{(1)}^*, \qquad (\tau, \lambda) \mapsto \tau * \lambda \tag{1.17}$$

is a bijection.

For any $\sigma \in W$ let $l(\sigma)$ be the length of σ with respect to reflections defined by the simple roots in Δ^+. The Bott-Borel-Weil theorem is due to Bott and has been proved as Theorem IV′ in [1]. It was later reproved as Theorem 6.4, p. 371 in [2]. The notation in Theorem 1.3 below is essentially the same as in Theorem 6.4 in [2].

Theorem 1.3 *Let $\xi \in \Lambda_{(1)}$. Then $H^*(X, \mathscr{S}E^{-\xi}) = 0$ in case $\xi \notin \Lambda_{(1)}^*$. Assume $\xi \in \Lambda_{(1)}^*$ so that $\xi = \tau * \lambda$ uniquely for $(\tau, \lambda) \in W^1 \times \Lambda$. Then $H^j(X, \mathscr{S}E^{-\xi}) = 0$ if $j \neq l(\tau)$. On the other hand, as a $\mathfrak{g}$-module, $H^{l(\tau)}(X, \mathscr{S}E^{-\xi})$ is irreducible and affords the representation π_λ^* where π_λ^* is the representation contragredient to π_λ.*

The determination of the $\mathfrak{g}$-module structure of $H^*(X, \mathscr{S}E^{-\xi})$ may be given in terms of nilpotent Lie algebra cohomology. Let $\lambda \in \Lambda$. The restriction, $\pi_\lambda | \mathfrak{n}$ defines the structure of an $\mathfrak{n}$-module on V_λ and consequently one defines the cohomology $H^*(\mathfrak{n},V_\lambda)$. Since $\mathfrak{n}$ is normalized by $\mathfrak{g}_1$ the restriction $\pi_\lambda|\mathfrak{g}_1$ defines the structure of a graded $\mathfrak{g}_1$-module on $H^*(\mathfrak{n},V_\lambda)$. The following Frobenius-like multiplicity result, due to Bott, is implicit in (1.6) of [1]. Explicitly it is given as Proposition 6.3, p. 367 in [2].

Proposition 1.4 *Let $\xi \in \Lambda_{(1)}, \lambda \in \Lambda$ and $j \in \mathbb{Z}_+$. Then*

$$\text{mult. of } \pi_\lambda^* \text{ in } H^j(X,\mathscr{S}E^{-\xi}) = \text{mult. of } \sigma_\xi \text{ in } H^j(\mathfrak{n},V_\lambda). \tag{1.18}$$

As an immediate consequence of Proposition 1.4 one has

Proposition 1.5 *The Bott-Borel-Weil theorem is equivalent to the statement that, for any $\lambda \in \Lambda$,*
(1) $H(\mathfrak{n},V_\lambda)$ is a multiplicity free $\mathfrak{g}_1$-module such that, using the primary component notation of §1.1,
(2)

$$H(\mathfrak{n},V_\lambda) = \sum_{\tau\in W^1} H(\mathfrak{n},V_\lambda)^{\tau*\lambda}. \tag{1.19}$$

and
(3)

$$0 \neq H(\mathfrak{n},V_\lambda)^{\tau*\lambda} \subset H^{l(\tau)}(\mathfrak{n},V_\lambda). \tag{1.20}$$

2. The determination of $H^*(\mathfrak{n},V_\lambda)$ and an algebraic proof of BBW

2.1. Retain the notation of §1.2. Let $\lambda \in \Lambda$. We recall results in [2] and for suitability here we will make some changes in the notation of [2]. Let $C(\mathfrak{n},V_\lambda) = V_\lambda \otimes \wedge\mathfrak{n}^*$, where $\mathfrak{n}^*$ is the dual space to $\mathfrak{n}$, so that $C(\mathfrak{n},V_\lambda)$ has the structure of a cochain complex whose cohomology is $H^*(\mathfrak{n},V_\lambda)$. The coboundary operator is denoted here by d_λ. See §3.1 in [2] where d_λ is denoted by d_π and we have reversed the order of the factors V_λ and $\wedge\mathfrak{n}^*$ in $C(\mathfrak{n},V_\lambda)$. Let $d_\lambda^* \in End\, C(\mathfrak{n},V_\lambda)$ be the Hermitian adjoint of d_λ with respect to the positive definite inner product on $C(\mathfrak{n},V_\lambda)$ defined in §3.1 of [2]. Then the Laplacian $L_\lambda = d_\lambda d_\lambda^* + d_\lambda^* d_\lambda$ is a positive semidefinite operator of degree 0 on $C(\mathfrak{n},V_\lambda)$. Furthermore L_λ commutes with the natural $\mathfrak{g}_1$-module structure on $C(\mathfrak{n},V_\lambda)$. In particular one then has a graded $\mathfrak{g}_1$-module isomorphism

$$Ker\, L_\lambda \cong H^*(\mathfrak{n},V_\lambda). \tag{2.1}$$

Let $\Lambda_{(1)}(\lambda) = \{\xi \in \Lambda_{(1)} \mid C(\mathfrak{n},V_\lambda)^\xi \neq 0\}$. The following result is a restatement of Theorem 5.7, p. 253, in [2].

Theorem 2.1 *Let $\lambda \in \Lambda$. Then L_λ reduces to scalars on the $\mathfrak{g}_1$-primary components of $C(\mathfrak{n},V_\lambda)$. In fact for any $\xi \in \Lambda_{(1)}(\lambda)$,*

$$L_\lambda = \frac{1}{2}(|\lambda+\rho|^2 - |\xi+\rho|^2) \text{ on } C(\mathfrak{n},V_\lambda)^\xi. \tag{2.2}$$

From the positive semidefiniteness of L_λ, it follows from (2.2) that

$$|\lambda+\rho| \geq |\xi+\rho|, \qquad \forall\, \xi \in \Lambda_{(1)}(\lambda). \tag{2.3}$$

Let $\Lambda_{(1)}[\lambda] = \{\xi \in \Lambda_{(1)}(\lambda) \mid |\lambda+\rho| = |\xi+\rho|\}$. As a consequence of Theorem 2.1 one has

Theorem 2.2 *For any $\lambda \in \Lambda$,*

$$Ker L_\lambda = \sum_{\xi\in\Lambda_{(1)}[\lambda]} (Ker L_\lambda)^\xi \tag{2.4}$$

so that by (2.1)

$$H^*(\mathfrak{n}, V_\lambda) = \sum_{\xi\in\Lambda_{(1)}[\lambda]} H^*(\mathfrak{n}, V_\lambda)^\xi. \tag{2.5}$$

2.2. To introduce notation which will serve us now and also in Section 3 we return "momentarily" to the general case of §1.1. For any subset $\Phi \subset \Delta_+$ let $\langle\Phi\rangle \in \Gamma$ be defined by putting $\langle\Phi\rangle = \sum_{\varphi\in\Phi}\varphi$. Also for any $\sigma \in W$ let $\Phi_\sigma = \Delta_+ \cap \sigma(\Delta_-)$. One has

$$-\langle\Phi_\sigma\rangle = \sigma(\rho) - \rho \tag{2.6}$$

by (5.10.1) in [2]. Now order the roots in $\Delta(\mathfrak{q}_+)$ so that $\Delta(\mathfrak{q}_+) = \{\beta_1, \ldots, \beta_m\}$. Let $\Phi \subset \Delta(\mathfrak{q}_+)$ so that for a sequence $i_j \in \mathbb{N}$, $j = 1, \ldots, k$, in increasing order, where $k = card\,\Phi$, we can write $\Phi = \{\beta_{i_1}, \ldots, \beta_{i_k}\}$. Let $e_{-\Phi} \in \wedge^k \mathfrak{q}_-$ be defined by putting $e_{-\Phi} = e_{-\beta_{i_1}} \wedge \cdots \wedge e_{-\beta_{i_k}}$. It is clear that the set $\{e_{-\Phi}\}$, $\Phi \subset \Delta(\mathfrak{q}_+)$, is a basis of $\wedge\mathfrak{q}_-$.

Remark 2.3 If $\sigma \in W$ then one notes that $\Phi_\sigma \subset \Delta(\mathfrak{q}_+)$ if and only if $\sigma \in W^1$. This easy fact is established in the second from the bottom paragraph on p. 485 in [3]. In particular, for $\tau \in W^1$ one has

$$e_{-\Phi_\tau} \in \wedge^{l(\tau)}\mathfrak{q}_-. \tag{2.7}$$

Now return to the notation of §1.2 where $\mathfrak{r} = \mathfrak{g}_1$, $\mathfrak{q}_+ = \mathfrak{n}$. Put $\mathfrak{n}_- = \mathfrak{q}_-$. The Killing form defines a nonsingular pairing of $\mathfrak{n}$ and $\mathfrak{n}_-$, and using this pairing we will identify $\mathfrak{n}^*$ with $\mathfrak{n}_-$. Let $\lambda \in \Lambda$. Then $C(\mathfrak{n}, V_\lambda) = V_\lambda \otimes \wedge\mathfrak{n}_-$. If $\gamma \in \Gamma$ is an extremal weight of V_λ, let $v_\gamma \in V_\lambda$ be a nonzero γ-weight vector. For $\tau \in W^1$ let $c(\tau,\lambda) = v_{\tau(\lambda)} \otimes e_{-\Phi_\tau} \in C(\mathfrak{n}, V_\lambda)$.

Remark 2.4 Note that, by (2.6) and (2.7), $c(\tau,\lambda) \in C^{l(\tau)}(\mathfrak{n}, V_\lambda)$ is a weight vector of weight $\tau * \lambda$ with respect to the $\mathfrak{g}_1$-module structure on $C(\mathfrak{n}, V_\lambda)$.

The following result is a reformulation of Theorem 5.14 in [2]. Using Proposition 1.5 it establishes the Bott-Borel-Weil theorem.

Theorem 2.5 *Let the notation be as in Theorem 2.2. Then* $\Lambda_{(1)}[\lambda] = \{\tau * \lambda \mid \tau \in W^1\}$ *so that (2.4) becomes*

$$Ker L_\lambda = \sum_{\tau \in W^1} C(\mathfrak{n}, V_\lambda)^{\tau*\lambda}. \tag{2.8}$$

Furthermore $Ker L_\lambda$ *is a multiplicity-free* $\mathfrak{g}_1$*-module so that* $C(\mathfrak{n}, V_\lambda)^{\tau*\lambda}$ *is* $\mathfrak{g}_1$*-irreducible for any* $\tau \in W^1$*. In addition* $c(\tau, \lambda)$ *is a highest weight vector in* $C(\mathfrak{n}, V_\lambda)^{\tau*\lambda}$ *so that* $C(\mathfrak{n}, V_\lambda)^{\tau*\lambda} \subset C^{l(\tau)}(\mathfrak{n}, V_\lambda)$*. Finally, using (2.1), the conditions (1), (2), and (3) of Proposition 1.5 are consequently satisfied, thereby establishing BBW.*

Remark 2.6 Note that (2.8) trivially implies $(Ker L_\lambda)^{\tau*\lambda} = C(\mathfrak{n}, V_\lambda)^{\tau*\lambda}$ so that (2.8) implies the weaker equation

$$Ker L_\lambda = \sum_{\tau \in W^1} (Ker L_\lambda)^{\tau*\lambda}. \tag{2.9}$$

If one replaces (2.8) by (2.9) then note, using Proposition 1.5, that this weakened version of Theorem 2.5 is equivalent to BBW (except for the more refined statement in Theorem 2.5 explicitly naming the highest weight vector in $(Ker L_\lambda)^{\tau*\lambda}$, and not just the degree in which $(Ker L_\lambda)^{\tau*\lambda}$ occurs). For a further comment on the equivalence, in the case where X is the flag manifold (i.e. $\mathfrak{p} = \mathfrak{b}$), see the discussion of Schmid on p. 42 in [4]. We will reformulate Theorem 2.5 by introducing some slight changes. The altered form, Theorem 2.10, of Theorem 2.5 will be seen in Section 3 to be a special case of a more general theorem involving $\mathfrak{r}$ of §1.1 instead of the Levi-factor $\mathfrak{g}_1$.

2.3. In this paragraph adopt the notation of §1.1. Let $U(\mathfrak{r})$ be the universal enveloping algebra of $\mathfrak{r}$. Given a representation

$$\sigma : \mathfrak{r} \to End Y \tag{2.10}$$

we will also let $\sigma : U(\mathfrak{g}) \to End Y$ denote the homomorphism extending (2.10) to $U(\mathfrak{g})$. Let $Cas_\mathfrak{r} \in Cent\, U(\mathfrak{g})$ be the Casimir element with respect to $B_\mathfrak{r}$. Thus if $\{x_i\}$ is a basis of $\mathfrak{r}$ and $\{y_i\}$ is the dual basis with respect to $B_\mathfrak{r}$ then $Cas_\mathfrak{r} = \sum_i x_i y_i$. For any $\mu \in \Lambda_\mathfrak{r}$ let $\chi_\mu : Cent\, U(\mathfrak{r}) \to \mathbb{C}$ be the infinitesimal character of σ_μ. One knows that

$$\chi_\mu(Cas_\mathfrak{r}) = |\mu + \rho_\mathfrak{r}|^2 - |\rho_\mathfrak{r}|^2. \tag{2.11}$$

Return to the case of §1.2 so that $\mathfrak{r} = \mathfrak{g}_1$, $q_+ = \mathfrak{n}$, and $\mathfrak{q}_- = \mathfrak{n}_-$. We write Cas_1 for $Cas_\mathfrak{r}$ and ρ_1 for $\rho_\mathfrak{r}$. Recall the notation of (1.16). One clearly has

$$\rho = \rho_1 + \rho(\mathfrak{n}). \tag{2.12}$$

It is clear from (2.12) that $\rho(\mathfrak{n}) \in \Gamma_\mathfrak{r}$ for $\mathfrak{r} = \mathfrak{g}_1$ (recalling the definition of $\Gamma_\mathfrak{r}$ in §1.1). But more is true.

Proposition 2.7 *One has $\rho(\mathfrak{n}) \in \Lambda_1$ and the irreducible $\mathfrak{g}_1$-module $Z_{\rho(\mathfrak{n})}$ (defined by the representation $\sigma_{\rho(\mathfrak{n})}$) is 1-dimensional.*

Proof. Using notation in §1.1 clearly $\mathfrak{n}$ is stable under $\nu(\mathfrak{g}_1)$. Let $f : \mathfrak{g}_1 \to \mathbb{C}$ be defined by putting $f(x) = \frac{1}{2}tr(\nu(x)|\mathfrak{n})$. Since $f([\mathfrak{g}_1, \mathfrak{g}_1]) = 0$ it follows that f defines a representation of $\mathfrak{g}_1$ on $\mathbb{C}$. On the other hand note that $f(x) = \langle \rho(\mathfrak{n}), x\rangle$ for $x \in \mathfrak{h}$. Thus $\rho(\mathfrak{n})$ is the unique weight of this 1-dimensional representation. This clearly proves the proposition. ∎

If Y is a completely reducible $\mathfrak{g}_1$-module with respect to a representation σ we will let $Y^e = Y \otimes Z_{\rho(\mathfrak{n})}$ be the completely reducible $\mathfrak{g}_1$-module defined by the tensor product representation $\sigma \otimes \sigma_{\rho(\mathfrak{n})}$. We will write σ^e for $\sigma \otimes \sigma_{\rho(\mathfrak{n})}$. Also for $\xi \in \Lambda_1$ let $\xi^e = \xi + \rho(\mathfrak{n})$. Clearly $\xi^e \in \Lambda_1$.

Remark 2.8 Note that if $\Gamma_1(Y) \subset \mathfrak{h}^*$ is the set of set weights of Y then $\Gamma_1(Y) + \rho(n)$ is the set of weights of Y^e. Also if $\xi \in \Lambda_1$ then one has the equivalence

$$Z_\xi^e \cong Z_{\xi^e}$$

of irreducible $\mathfrak{g}_1$-modules.

Let $\lambda \in \Lambda$ and let $\eta_\lambda : \mathfrak{g}_1 \to End\, C(\mathfrak{n}, V_\lambda)$ be the representation which defines the given $\mathfrak{g}_1$-module structure on the cochain complex $C(\mathfrak{n}, V_\lambda)$. We can regard $\eta_0 : \mathfrak{g}_1 \to End \wedge \mathfrak{n}_-$ as the representation of $\mathfrak{g}_1$ on $\wedge\mathfrak{n}_-$ which defines the given $\mathfrak{g}_1$-module structure on $\wedge\mathfrak{n}_-$. To align notation here with that in [3], put $\zeta_{(\lambda)} = \eta_\lambda^e$ and $\zeta_{(0)} = \eta_0^e$ so that

$$\zeta_{(\lambda)} : \mathfrak{g}_1 \to End\,(C(\mathfrak{n}, V_\lambda) \otimes Z_{\rho(\mathfrak{n})}) \tag{2.13}$$

and

$$\zeta_{(0)} : \mathfrak{g}_1 \to End\,(\wedge\mathfrak{n}_- \otimes Z_{\rho(\mathfrak{n})}). \tag{2.14}$$

Recall the Laplacian L_λ defined in §2.1. Let L_λ^e be the operator on $C(\mathfrak{n}, V_\lambda)^e = C(\mathfrak{n}, V_\lambda) \otimes Z_{\rho(\mathfrak{n})}$ defined by putting $L_\lambda^e = L_\lambda \otimes 1$.

Theorem 2.9 *Let $\lambda \in \Lambda$. Then*

$$2L_\lambda^e = (|\lambda + \rho|^2 - |\rho_1|^2) - \zeta_{(\lambda)}(Cas_1). \tag{2.15}$$

Proof. Recalling the definition of $\Lambda_{(1)}(\lambda)$ in §2.1 one has the primary decomposition

$$C(\mathfrak{n}, V_\lambda)^e = \sum_{\xi \in \Lambda_{(1)}(\lambda)} (C(\mathfrak{n}, V_\lambda)^\xi)^e. \tag{2.16}$$

But $2L_\lambda^e$ reduces to the scalar $|\lambda + \rho|^2 - |\xi + \rho|^2$ on $C(\mathfrak{n}, V_\lambda)^e)^\xi$ by Theorem 2.2. But

$$\xi + \rho = \xi^e + \rho_1 \tag{2.17}$$

by (2.11). One the other hand

$$(C(\mathfrak{n},V_\lambda)^\xi)^e = (C(\mathfrak{n},V_\lambda)^e)^{\xi^e} \tag{2.18}$$

by (2.12) so that $\zeta_{(\lambda)}(Cas_1)$ reduces to the scalar $|\xi^e+\rho_1|^2-|\rho_1|^2$ on $(C(\mathfrak{n},V_\lambda)^\xi)^e$ by (2.11). But, by (2.17),

$$|\lambda+\rho|^2-|\xi+\rho|^2 = |\lambda+\rho|^2-|\rho_1|^2-(|\xi^e+\rho_1|^2-|\rho_1|^2)$$

so that the left and right sides of (2.15) take the same value on $(C(\mathfrak{n},V_\lambda)^\xi)^e$ for any $\xi \in \Lambda_{(1)}(\lambda)$. But this establishes (2.15). ■

Let $\lambda \in \Lambda$. By (2.18) $\{\chi_{\xi^e}(Cas_1)\}$, $\xi \in \Lambda_{(1)}(\lambda)$, is the set of eigenvalues of $\zeta_{(\lambda)}(Cas_1)$. Since L_λ is a positive semidefinite operator the equality (2.15) implies that for any $\xi \in \Lambda_{(1)}(\lambda)$,

$$|\lambda+\rho|^2-|\rho_1|^2 \geq \chi_{\xi^e}(Cas_1) \tag{2.19}$$

where, recalling Theorem 2.2, equality occurs if and only if $\xi \in \Lambda_{[1]}(\lambda)$.

Now recall that Theorem 2.5 yields BBW. In addition Theorem 2.5 determines $\Lambda_{[1]}(\lambda)$ for any $\lambda \in \Lambda$. In fact the theorem asserts $\Lambda_{[1]}(\lambda) = \{\tau * \lambda \mid \tau \in W^1\}$. On the other hand $(\tau * \lambda)^e = \tau \bullet \lambda$ by (1.16). In the notation of [3] the set $\{\tau \bullet \lambda \mid \tau \in W^1\}$ is called a multiplet. This multiplet plays a key role in Theorem 2.10 below. Using (2.15), Theorem 2.10 is just a reformulation of Theorem 2.5 and hence Theorem 2.10 also implies BBW. Let $0 \neq z_{\rho(\mathfrak{n})} \in Z_{\rho(\mathfrak{n})}$.

Theorem 2.10 *Let $\lambda \in \Lambda$. Then $|\lambda+\rho|^2-|\rho_1|^2$ is the maximal eigenvalue of $\zeta_{(\lambda)}(Cas_1)$ on $C(\mathfrak{n},V_\lambda)^e$. Let $M_1 \subset C(\mathfrak{n},V_\lambda)^e$ be the corresponding eigenspace. Then*

$$M_1 = \sum_{\tau\in W^1} (C(\mathfrak{n},V_\lambda)^e)^{\tau\bullet\lambda}. \tag{2.20}$$

Furthermore $C(\mathfrak{n},V_\lambda)^e)^{\tau\bullet\lambda}$ is irreducible for any $\tau \in W^1$ so that M_1 is a multiplicity free $\mathfrak{g}_1$-module whose irreducible components are a multiplet in the sense of [3]. Finally, in the notation of Theorem 2.5, $c(\tau,\lambda)\otimes z_{\tau,\lambda}$ is a highest weight vector in $C(\mathfrak{n},V_\lambda)^e)^{\tau\bullet\lambda}$.

It will be seen in Section 3 that Theorem 2.10 generalizes to the case where the Levi-factor $\mathfrak{g}_1$ is replaced by the much more general subalgebra $\mathfrak{r}$ of §1.1.

3. The spin module S and Euler number multiplets

3.1. We return to the notation and definitions of §1.1. We will also use results in [3]. We will use the notation of [3] with some exceptions. For example $\mathfrak{q}$ here replaces $\mathfrak{p}$ in [3] and $Cliff(\mathfrak{q})$ replaces $C(\mathfrak{p})$.

Let $Cliff(\mathfrak{q})$ be the Clifford algebra over $\mathfrak{q}$ with respect to $B_{\mathfrak{q}}$. We recall some of the relations arising from Chevalley's identification of the underlying vector spaces of the exterior algebra $\wedge\mathfrak{q}$ and the Clifford algebra $Cliff(\mathfrak{q})$. We think of $\wedge\mathfrak{q}$ as having two multiplicative structures. If $u,w \in \wedge\mathfrak{q}$ then $uw \in \wedge\mathfrak{q}$ denotes the Clifford product of u and w and $u \wedge w \in \wedge\mathfrak{q}$ is the exterior product of u and w. For more details see §2 in [3]. If $u,w \in \wedge\mathfrak{q}$ the natural extension, (see e.g. §2.1 in [3]) (u,w), of $B_{\mathfrak{q}}$ to $\wedge\mathfrak{q}$ is again denoted by $B_{\mathfrak{q}}$. If $A \in End \wedge \mathfrak{q}$ then $A^t \in End \wedge \mathfrak{q}$ denotes the transpose of A with respect to $B_{\mathfrak{q}}$. If $w \in \wedge\mathfrak{q}$ then $\varepsilon(w) \in End \wedge \mathfrak{q}$ denotes the operator of left exterior multiplication by w and $\iota(w) = \varepsilon(w)^t$. In particular $\iota(y)$, for $y \in \mathfrak{q}$, is the anti-derivation of degree -1 of the exterior algebra $\wedge\mathfrak{q}$ such that $\iota(y)y' = (y,y')$ if $y' \in \mathfrak{q}$. Clifford multiplication in $\wedge\mathfrak{q}$ is readily determined from the equality

$$yw = (\varepsilon(y) + \iota(y))w \tag{3.1}$$

for any $y \in \mathfrak{q}$ and $w \in \wedge\mathfrak{q}$.

For any $u \in \wedge^2\mathfrak{q}$ let $ad\,u \in End \wedge \mathfrak{q}$ be the operator defined so that $ad\,u(w) = uw - wu$ for any $w \in \wedge\mathfrak{q}$. Then $ad\,u$, clearly a derivation of the Clifford algebra structure of $\wedge^2\mathfrak{q}$, is also a derivation of degree 0 of the exterior algebra structure on $\wedge\mathfrak{g}$. In particular $\wedge^2\mathfrak{q}$ is a Lie algebra under Clifford product and

$$\wedge^2\mathfrak{q} \to End \wedge \mathfrak{q}, \quad u \mapsto ad\,u$$

is a Lie algebra representation. Next, one has that

$$t : \wedge^2\mathfrak{q} \to Lie\,SO(\mathfrak{q}) \tag{3.2}$$

is a Lie algebra isomorphism, where for $u \in \wedge^2\mathfrak{q}$ and $y \in \mathfrak{q}$, one has

$$\begin{aligned} t(u)(y) &= ad\,u(y) \\ &= -2\iota(y)u. \end{aligned} \tag{3.3}$$

For a proof of the above statements in this paragraph see Proposition 7 and Theorem 8 in [3].

Remark 3.1 The second equality in (3.3) is a special case of the general fact that if $w \in \wedge^k\mathfrak{q}$ and $y \in \mathfrak{q}$ then

$$yw - (-1)^k wy = 2\iota(y)w.$$

In particular $\iota(y)$ is an anti-derivation of the Clifford algebra structure of $\wedge\mathfrak{q}$ as well as the exterior algebra structure of $\wedge\mathfrak{q}$. See Lemma 5, p. 284 in [3].

It clearly follows from (3.3) that there exists a unique Lie algebra homomorphism

$$\nu_* : \mathfrak{r} \to \wedge^2 \mathfrak{q} \tag{3.4}$$

such that $t \circ \nu_* = \nu$. See Theorem 8 in [3]. Putting $u = \nu_*(x)$ in (3.3) one has

$$\nu(x)(y) = -2\iota(y)\nu_*(x).$$

For any $x \in \mathfrak{r}$ let $\theta_\nu(x)$ be the unique derivation of the exterior algebra $\wedge \mathfrak{q}$ which extends $\nu(x)$. Regard $\wedge \mathfrak{q}$ as an $\mathfrak{r}$-module where $x \in \mathfrak{r}$ maps to $\theta_\nu(x)$. One notes that the extended bilinear form $B_\mathfrak{q}$ is invariant under $\theta_\nu(x)$ and recalls (see (8) in [3] where $u = \nu_*(x)$) that

$$\theta_\nu(x) = ad\, \nu_*(x). \tag{3.5}$$

Remark 3.2 In the special case of §1.2 where $\mathfrak{g}_1 = \mathfrak{r}$ one obviously has $\mathfrak{n}_- \subset \mathfrak{q}$ and hence $\wedge \mathfrak{n}_- \subset \wedge \mathfrak{q}$. Note that the representation $\eta_0 : \mathfrak{g}_1 \to End \wedge \mathfrak{n}_-$ in §2.3 is just the restriction of θ_ν to the submodule $\wedge \mathfrak{n}_-$.

Since $n = 2m = dim\, \mathfrak{q}$ is even the Clifford algebra $Cliff(\mathfrak{q})$ is simple and hence has a unique (up to equivalence) faithful irreducible module S. One has $dim\, S = 2^m$. Let

$$\varepsilon : Cliff(\mathfrak{q}) \to End\, S \tag{3.6}$$

be the corresponding isomorphism. With this structure the space S is referred to as the spin module for $Cliff(\mathfrak{q})$. The Lie algebra homomorphism $\nu_* : \mathfrak{r} \to \wedge^2 \mathfrak{q}$, composed with ε, defines a representation

$$Spin\, \nu : \mathfrak{r} \to End\, S \tag{3.7}$$

which is referred to as the spin of ν.

For computational purposes it is convenient to take S to be a minimal left ideal in $Cliff(\mathfrak{q})$ and the map ε defining the action of $Cliff(\mathfrak{q})$ on S is left multiplication. Recall (1.3). Clearly $B_\mathfrak{q}|\mathfrak{q}_+ = 0$ and $B_\mathfrak{q}|\mathfrak{q}_- = 0$ so that Clifford multiplication and exterior multiplication are the same on both $\wedge \mathfrak{q}_+$ and $\wedge \mathfrak{q}_-$. On the other hand, if $w \in \wedge^k \mathfrak{q}$ and $w' \in \wedge^{k'} \mathfrak{q}$, it follows from (3.1) that

$$ww' - w \wedge w' \in \sum_{j=1}^{[\frac{k+k'}{2}]} \wedge^{k+k'-2j} \mathfrak{q}. \tag{3.8}$$

This readily implies that the map

$$\wedge \mathfrak{q}_- \otimes \wedge \mathfrak{q}_+ \to Cliff(\mathfrak{q}), \quad w \otimes w' \mapsto ww' \tag{3.9}$$

is a linear bijection. Recall (see §2.2) $\Delta(\mathfrak{q}_+) = \{\beta_1, \ldots, \beta_m\}$. Now put $e_{\mathfrak{q}_+} = e_{\beta_1} \cdots e_{\beta_m}$. Clearly since (Clifford multiplication) $\wedge \mathfrak{q}_+ e_{\mathfrak{q}_+} = \mathbb{C} e_{\mathfrak{q}_+}$, one has that

$$C(\mathfrak{q})\, e_{\mathfrak{q}_+} = \wedge \mathfrak{q}_-\, e_{\mathfrak{q}_+} \tag{3.10}$$

so that the left ideal $C(\mathfrak{q})\, e_{\mathfrak{q}_+}$ has dimension 2^m and consequently is a minimal ideal. We can therefore take

$$\begin{aligned} S &= Cliff\,(\mathfrak{q})\, e_{\mathfrak{q}_+} \\ &= \wedge \mathfrak{q}_-\; e_{\mathfrak{q}_+}. \end{aligned} \tag{3.11}$$

Let $\rho_{\mathfrak{q}} = \frac{1}{2}\sum_{\varphi \in \Delta(\mathfrak{q}_+)} \varphi$ so that

$$\rho = \rho_{\mathfrak{r}} + \rho_{\mathfrak{q}}. \tag{3.12}$$

Remark 3.3 In the special case of §1.2 where $\mathfrak{r} = \mathfrak{g}_1$ note that $\rho_{\mathfrak{q}} = \rho(\mathfrak{n})$ recalling the notation of (1.16).

Recall the notation of §2.2. For any $\Phi \subset \Delta(\mathfrak{q}_+)$ let $s_\Phi = e_{-\Phi}\, e_{\mathfrak{q}_+}$. The following result is part of Proposition 3.17, p. 477, in [3].

Proposition 3.4 *The set* $\{s_\Phi\}$, $\Phi \subset \Delta(\mathfrak{q}_+)$ *is a weight basis of S where the weight of* s_Φ *is* $\rho_{\mathfrak{q}} - \langle \Phi \rangle$.

Note that Proposition 3.4 implies that $e_{\mathfrak{q}_+}$ is a highest weight vector in S. This has special significance in the case of §1.2 where $\mathfrak{r} = \mathfrak{g}_1$.

Proposition 3.5 *Assume the case of §1.2 where* $\mathfrak{r} = \mathfrak{g}_1$. *Then* $\mathbb{C}e_{\mathfrak{q}_+}$ *is stable under (left Clifford multiplication) the action of* $Spin\, \nu(\mathfrak{g}_1)$ *and affords the 1-dimensional representation* $\sigma_{\rho(n)}$.

Proof. Since $e_{\mathfrak{q}_+}$ is a highest weight vector and the weight is $\rho(\mathfrak{n})$, the $\mathfrak{g}_1$-module generated, under $Spin\,\nu$, by $e_{\mathfrak{q}_+}$ is irreducible and equivalent to $Z_{\rho(\mathfrak{n})}$. But $Z_{\rho(\mathfrak{n})}$ is 1-dimensional by Proposition 2.6. ■

As a corollary one has

Proposition 3.6 *Retain the notation and assumptions of Proposition 3.5. Recall also the notation of §2.3 (especially Theorem 2.9). Let*

$$\gamma : \wedge \mathfrak{n}_-^e \to S \tag{3.13}$$

be the linear isomorphism defined so that for $u \in \wedge \mathfrak{n}_-$,

$$\gamma(u \otimes z_{\rho(\mathfrak{n})}) = u\, e_{\mathfrak{q}_+}. \tag{3.14}$$

Then γ *intertwines the representations (see (2.14))* $\zeta_{(0)}$ *and* $Spin\,\nu$ *of* $\mathfrak{g}_1$.

Proof. Let $x \in \mathfrak{g}_1$ and let $u \in \wedge\mathfrak{q}_-$. Let $f : \mathfrak{g}_1 \to \mathbb{C}$ be defined as in the proof of Proposition 2.6. Then by (3.5), Remark 3.2 and Proposition 3.5 one has

$$\begin{aligned} Spin\,\nu(\gamma(u\otimes z_{\rho(\mathfrak{n})})) &= Spin\,\nu(x)(u e_{\mathfrak{q}_+}) \\ &= \nu_*(x)u e_{\mathfrak{q}_+} \\ &= [\nu_*(x),u]e_{\mathfrak{q}_+} + u\,\nu_*(x)e_{\mathfrak{q}_+} \\ &= \eta_0(x)(u)\,e_{\mathfrak{q}_+} + f(x)u e_{\mathfrak{q}_+} \\ &= \gamma(\zeta_{(0)}(u\otimes z_{\rho(\mathfrak{n})})) \end{aligned}$$

■

Let $\lambda \in \Lambda$. Assume the general case of §1.1. Then in the notation of (2.20) in [3],

$$\zeta_\lambda : \mathfrak{r} \to End\,V_\lambda \otimes S$$

is the tensor product of the representations $\pi_\lambda|\mathfrak{r}$ and $Spin\,\nu$. The result in [3] which generalizes Theorem 2.10 (and hence generalizes BBW) will be a statement, Theorem 3.8 below, about ζ_λ. Proposition 3.7 below insures that Theorem 3.8 generalizes Theorem 2.10.

Retain the notation and assumptions of Proposition 3.6. Clearly we can write

$$C(\mathfrak{n},V_\lambda)^e = V_\lambda \otimes \wedge\mathfrak{n}_-^e. \tag{3.15}$$

Let

$$\gamma_\lambda : C(\mathfrak{n},V_\lambda)^e \to V_\lambda \otimes S \tag{3.16}$$

be the linear isomorphism defined so that if $v \in V_\lambda$ and $w \in \wedge\mathfrak{n}_-^e$ then $\gamma_\lambda(v\otimes w) = v\otimes\gamma(w)$. The following corollary of Proposition 3.6 is immediate.

Proposition 3.7 *Let $\lambda \in \Lambda$. Retain the notation and assumptions of Proposition 3.6. Recall (2.13). Then γ_λ intertwines the representations $\zeta_{(\lambda)}$ and ζ_λ of $\mathfrak{g}_1$.*

3.2. Return to the general case of §1.1. Let $\lambda \in \Lambda$. Then $V_\lambda \otimes S$ is an $\mathfrak{r}$-module with respect to ζ_λ. If $\delta \in \Gamma$ is an extremal weight of V_λ then, as in §2.2, let $v_\delta \in V_\lambda$ be a nonzero δ-weight vector. For any $\tau \in W^1$ recall that $\tau\bullet\lambda = \tau(\lambda+\rho)-\rho_{\mathfrak{r}}$. Recall (2.7). Let $w_{\tau\bullet\rho} = v_{\tau(\lambda)}\otimes s_{\Phi_\tau}$ so that, by (2.6) and Proposition 3.4, $w_{\tau\bullet\rho} \in V_\lambda \otimes S$ is a weight vector of weight $\tau\bullet\lambda$. The following result is part of Theorem 4.17 in [3]

Theorem 3.8 *Let $\lambda \in \Lambda$. Then for any $\tau \in W^1$ the multiplicity of $\sigma_{\tau\bullet\lambda}$ in $V_\lambda \otimes S$ is 1 so that there is an equivalence*

$$(V_\lambda \otimes S)^{\tau\bullet\lambda} \cong Z_{\tau\bullet\lambda} \tag{3.17}$$

of $\mathfrak{r}$-modules. In particular

$$\{(V_\lambda \otimes S)^{\tau\bullet\lambda} \mid \tau \in W^1\} \tag{3.18}$$

is a $\Lambda_{\mathfrak{r}}^{\lambda}$-multiplet (see (3.3) in [3]). Moreover $w_{\tau\bullet\lambda}$ is a highest weight vector of $(V_\lambda \otimes S)^{\tau\bullet\lambda}$.

Furthermore the highest eigenvalue of $\zeta_\lambda(Cas_r)$ is $|\lambda+\rho|^2 - |\rho_{\mathfrak{r}}|^2$ and the sum

$$\sum_{\tau\in W^1} (V_\lambda \otimes S)^{\tau\bullet\lambda} \tag{3.19}$$

of the components of the $\Lambda_{\mathfrak{r}}^{\lambda}$-multiplet (3.18) is the corresponding eigenspace.

Remark 3.9 Recall that Theorem 2.10 implies BBW. It is obvious from Proposition 3.7 that Theorem 3.8 reduces to Theorem 2.10 under the special Levi-factor assumptions of §1.2.

Again under the special Levi-factor assumptions of §1.2 one also recovers a familiar relation between the square of a Dirac operator and Laplacians. Let $\lambda \in \Lambda$. Identify $C(\mathfrak{n}, V_\lambda)^e$ with $V_\lambda \otimes S$ using the isomorphism (3.16). But then the Laplacian L_λ^e (see (2.15)) can be regarded as an operator on $V_\lambda \otimes S$. Let $\square_\lambda$ be the cubic Dirac operator on $V_\lambda \otimes S$ defined by (2.19) in [3]. (The following result uses Theorem 2.21 in [3]. The statement of the latter result misprinted the word Theorem 2.16 and wrote Theorem 2.2 instead.)

Theorem 3.10 *Let $\lambda \in \Lambda$. Assume the special Levi-factor case of §1.2 so that $\mathfrak{r} = \mathfrak{g}_1$. Then*

$$2L_\lambda^e = (\square_\lambda)^2. \tag{3.20}$$

Proof. The result follows immediately from (2.15), Proposition 3.7 and Theorem 2.21 in [3]. ∎

Remark 3.11 In the general case the maximal eigenspace, (3.19), of $\zeta_\lambda(Cas_r)$ is the kernel of the Dirac operator $\square_\lambda$. See Theorem 4.24 in [3].

References

1. Bott, R.: Homogeneous vector bundles, *Ann. of Math.* **66** (1957), 203–248.
2. Kostant, B.: Lie algebra cohomology and the generalized Borel-Weil theorem, *Ann. of Math.* **74** (1961), 329–387.
3. Kostant, B.: A cubic Dirac operator and the emergence of Euler number multiplets of representations for equal rank subgroups, *Duke Math. J.* **100** (1999), 447–501.
4. Schmid, W.: The mathematical legacy of the paper "Homogeneous Vector Bundles", in: *Raoul Bott: Collected Papers, Topology and Lie Groups, Vol. 1.* Birkhäuser, Boston, MA, 1994, pp 40–47.

Self-energy of electrons in non-perturbative QED[‡]

Elliott H. Lieb *
Departments of Mathematics and Physics, Princeton University, Princeton, New Jersey 08544-0708 USA
e-mail: `lieb@princeton.edu`

Michael Loss **
School of Mathematics, Georgia Institute of Technology, Atlanta, Georgia 30332-0160 USA
e-mail: `loss@math.gatech.edu`

Abstract. Various models of charged particles interacting with a quantized, ultraviolet cutoff radiation field (but not with each other) are investigated. Upper and lower bounds are found for the self- or ground state-energies without mass renormalization. For N fermions the bounds are proportional to N, but for bosons they are sublinear, which implies 'binding', and hence that 'free' bosons are never free. Both 'relativistic' and non-relativistic kinematics are considered. Our bounds are non-perturbative and differ significantly from the predictions of perturbation theory.

Keywords: QED, quantum electrodynamics, non-perturbative, ultraviolet cutoff, electron, ground state energy, self energy, electromagnetic field, radiation, vector potential, non-relativistic, relativistic

Mathematics Subject Classifications (1991): 81V10, 81V70, 81V45, 81T08, 81T16

1. Introduction

Quantum electrodynamics (QED), the theory of electrons interacting with photons (at least for small energies) is one of the great successes of physics. Among its major achievements is the explanation of the Lamb shift and the anomalous magnetic moment of the electron. Nevertheless, its computations, which are entirely based on perturbation theory, created some uneasiness among the practitioners. The occurrence of infinities was and is especially vexing. Moreover, a truly non-trivial, 3+1-dimensional example of a relativistically invariant field theory has not yet been achieved.

There are, however, unresolved issues at a much earlier stage of QED that hark back to black-body radiation, the simplest and historically first problem involving the interaction of matter with radiation. The conceptual problems stemming from black-body radiation were partly resolved by quantum mechanics, i.e., by

‡ This work was presented in a lecture by E. H. L. at the 'Conférence Moshé Flato' in Dijon in September, 1999. It was presented earlier at the University of Alabama, Birmingham-Georgia Tech International Conference on Differential Equations and Mathematical Physics, Birmingham, March 15-19, 1999 and first appears in the proceedings of that conference published by International Press. Some typos have been corrected.

* Supported in part by NSF Grant PHY-98 20650.

** Supported in part by NSF Grant DMS-95 00840.

G. Dito and D. Sternheimer (eds.), Conférence Moshé Flato 1999, Vol. 1, 327 – 344.

the non-relativistic Schrödinger equation, which is, undoubtedly, one of the most successful of theories, for it describes matter at low energies almost completely. It is mathematically consistent and there are techniques available to compute relevant quantities. Moreover, it allows us to explain certain facts about bulk matter such as its stability, it extensivity, and the existence of thermodynamic functions. What has not been as successful, so far, is the incorporation of radiation phenomena, the very problem quantum mechanics set out to explain.

It ought to be possible to find a mathematically consistent theory, free of infinities, that describes the interaction of non-relativistic matter with radiation at moderate energies, such as atomic binding energies. It should not be necessary, as some physicists believe, to embed QED as a low energy part of a consistent high energy theory.

From such a theory one could learn a number of things that have not been explained rigorously. i) The decay of excited states in atoms. This problem has been investigated in some ultraviolet cutoff models in [2] and in a massive photon model in [10]. See also the review of Hogreve [5]. ii) Non-relativistic QED could be a playground for truly non-perturbative calculations and it could shed light on renormalization procedures. In fact, this was the route historically taken by Kramers that led to the renormalization program of Dyson, Feynman, Schwinger and Tomonaga. iii) Last but not least, one could formulate and answer the problems of stability of bulk matter interacting with the radiation field.

It has been proved in [3], [7] that stability of non-relativistic matter (with the Pauli Hamiltonian) interacting with classical magnetic fields holds provided that the fine-structure constant, $\alpha = e^2/\hbar c$, is small enough. It is certain, that the intricacies and difficulties of this classical field model will persist and presumably magnify in QED.

The same may be expected from a relativistic QED since replacing the Pauli Hamiltonian by a Dirac operator leads to a similar requirement on α [8]. Indeed, stability of matter in this model (the Brown-Ravenhall model) requires that the electron (positron) be defined in terms of the positive (negative) spectral subspace of the Dirac operator *with* the magnetic vector potential $A(x)$, instead of the free Dirac operator without $A(x)$. This observation, that perturbation theory, if there is one, must start from the dressed electrons rather than the electrons unclothed by its magnetic field, might ultimately be important in a non-perturbative QED.

The first, humble step is to understand electrons that interact with the radiation field but which are free otherwise. In order for this model to make sense an ultraviolet cutoff has to be imposed that limits the energy of photon modes. The simplest question, which is the one we address in this paper, is the behavior of the self-energy of the electron as the cutoff tends to infinity (with the bare mass of the electron fixed). The self-energy of the electron diverges as the cutoff tends to infinity and it has to be subtracted for each electron in any interacting theory. The total energy will still depend strongly on the cutoff because of the interactions. This dependence will, hopefully, enter through an effective mass which will be set equal to the physical mass (mass renormalization). The resulting theory should be essentially Schrödinger's mechanics, but slightly modified by so-called radiative corrections.

Lest the reader think that the self-energy problem is just a mathematical exercise, consideration of the many-body problem will provide a counterexample. Imagine N charged bosons interacting with the radiation field, but neglect any interaction among them such as the Coulomb repulsion. We say that these particles bind if the energy of the combined particles is less than the energy of infinitely separated particles. As we shall show, charged bosons indeed bind and they do it in such a massive way that it will be very likely that this cannot be overcome by the Coulomb repulsion. In particular, the energy of a charged many-boson system is *not* extensive, and from this perspective it is fortunate that stable, charged bosons do not exist in nature.

The situation is very different for fermions. We are not able to show that they do not bind but we can show — and this is one of the main results of our paper — that the self-energy is extensive, i.e., bounded above and below by a constant times N.

We thus have strong evidence that there is *no* consistent description of a system of stable charged relativistic or non-relativistic bosons interacting with the radiation field, while the Pauli exclusion principle, on the other hand, is able to prevent the above mentioned pathology.

In the remainder of the section we explain our notation and state the results. In the subsequent sections we sketch the proof of some of them but for details we refer the reader to [6].

We measure the energy in units of mc^2 where m is the bare mass of the electron, the length in units of the Compton wave length $\ell_C = \hbar/mc$ of the bare electron. We further choose $\ell_C^{-1}\sqrt{\hbar c}$ as the unit for the vector potential A and $\ell_C^{-2}\sqrt{\hbar c}$ as the unit for the magnetic field B. The argument is the dimensionless quantity $\ell_C^{-1}x$. As usual, $\alpha = e^2/\hbar c \approx 1/137.04$ is the fine structure constant.

In the expression below, $A(x)$ denotes an ultraviolet cutoff radiation field localized in a box $L \times L \times L$ with volume $V = L^3$,

$$A(x) = \frac{1}{\sqrt{2V}} \sum_{|k|<\Lambda} \sum_{\lambda=1,2} \frac{1}{\sqrt{|k|}} \varepsilon_\lambda(k) \left[a_\lambda(k) e^{ix\cdot k} + a_\lambda^*(k) e^{-ix\cdot k} \right] . \tag{1.1}$$

The index $k = 2\pi n/L$ where $n \in \mathbb{Z}^3$, and the word cutoff refers to the restriction to all values of k with $|k| < \Lambda$.

The vectors $\varepsilon_\lambda(k)$ are the polarization vectors and are normalized in such a way that

$$\varepsilon_i(k) \cdot \varepsilon_j(k) = \delta_{i,j} \, , \; \varepsilon_i(k) \cdot k = 0 \, . \tag{1.2}$$

The operators $a_\lambda(k)$ and $a_\lambda^*(k)$ satisfy the commutation relation

$$\left[a_\lambda(k), a_{\lambda'}^*(k') \right] = \delta_{\lambda,\lambda'} \delta(k,k') \, , \tag{1.3}$$

while all others commute with each other.

The energy of the radiation field can now be conveniently written as

$$H_f = \sum_{|k|<\Lambda} \sum_{\lambda=1,2} |k| a_\lambda^*(k) a_\lambda(k) \, . \tag{1.4}$$

These operators act on the Hilbert space generated by the polynomials in $a_\lambda^*(k)$ acting on the vacuum $|0\rangle$.

The self energy of (one or more) particles is the **ground state energy** of the Hamiltonian

$$H = \text{kinetic energy} \; + \; H_f \, . \tag{1.5}$$

where, as usual, the ground state energy of H is defined to be

$$E_0 = \inf_\Psi \frac{\langle \Psi, H\,\Psi \rangle}{\langle \Psi, \Psi \rangle} \, . \tag{1.6}$$

Typically, in the inquiry about the self–energy problem, i.e., the problem of computing the self–energy for fixed, albeit small, α and for large Λ, one proceeds via perturbation theory. First order perturbation theory will predict an energy of the order of $\alpha\Lambda^2$, and a higher order power counting argument confirms the asymptotically large Λ dependence of that calculation. Our theorems below show that the predictions of perturbation theory for the self–energy problem are wrong, if one is interested in the large Λ asymptotics of the energy. If perturbation theory works at all, then it works only for a range of α that vanishes as Λ increases. In fact we deduce from the upper bound in Theorem 1.1 that the size of this range shrinks at least as $\Lambda^{-2/5}$.

All the theorems below are asymptotic statements for large Λ and for fixed α. For actual bounds we refer the reader to [6]. The first result concerns the self energy of a nonrelativistic electron interacting with the radiation field. The Hamiltonian is given by

$$H = \frac{1}{2}(p + \sqrt{\alpha} A(x))^2 + H_f \, , \tag{1.7}$$

where $p = -i\nabla$ and acts on $L^2(\mathbb{R}^3) \otimes \mathcal{F}$, where $\mathcal{F}$ denotes the photon Fock space.

Theorem 1.1 *The ground state energy, E_0, of the operator (1.7) satisfies the bounds*

$$C_1 \alpha^{1/2} \Lambda^{3/2} \; < \; E_0 \; < \; C_2 \alpha^{2/7} \Lambda^{12/7} \tag{1.8}$$

We do not know how to get upper and lower bounds that are of the same order in Λ, but we suspect that $\Lambda^{12/7}$ is the right exponent. This is supported by the following theorem in which the $p \cdot A$ term is omitted.

Theorem 1.2 *The ground state energy E_0 of the operator*

$$\frac{1}{2}\left[p^2 + \alpha A(x)^2\right] + H_f \tag{1.9}$$

satisfies the bounds

$$C_1 \alpha^{2/7} \Lambda^{12/7} \leq E_0 \leq C_2 \alpha^{2/7} \Lambda^{12/7} \tag{1.10}$$

While these results are not of direct physical relevance (since E_0 is not observable), the many-body problem is of importance since it reveals a dramatic difference between bosons and fermions.

Theorem 1.3 *The ground state energy of N bosons, $E_0^{\text{boson}}(N)$, with Hamiltonian*

$$H(N) = \sum_{j=1}^{N} \frac{1}{2}(p_j + \sqrt{\alpha} A(x_j))^2 + H_f \tag{1.11}$$

satisfies the bounds

$$C_1 \sqrt{N} \sqrt{\alpha} \Lambda^{3/2} \leq E_0^{\text{boson}}(N) \leq C_2 N^{5/7} \alpha^{2/7} \Lambda^{12/7} \tag{1.12}$$

Thus, the energy $E_0^{\text{boson}}(N)$ is not extensive, i.e., it costs a huge energy to separate bosons. This has to be contrasted with the next theorem about fermions. The Hamiltonian is the same as before but it acts on the Hilbert space

$$\mathscr{F} \otimes \wedge_{j=1}^{N} L^2(\mathbb{R}^3; \mathbb{C}^2)\,, \tag{1.13}$$

where the wedge product indicates that the antisymmetric tensor product is taken.

Theorem 1.4 *The ground state energy, $E_0^{\text{fermion}}(N)$, of N charged fermions interacting with the radiation field satisfies*

$$C_1 \alpha^{1/2} \Lambda^{3/2} N \;\leq\; E_0^{\text{fermion}}(N) \;\leq\; C_2 \alpha^{2/7} \Lambda^{12/7} N \tag{1.14}$$

The **"relativistic" kinetic energy** for an electron is

$$T^{\text{rel}} = |p + \sqrt{\alpha} A(x)| = \sqrt{[p + \sqrt{\alpha} A(x)]^2} \tag{1.15}$$

with $p = -i\nabla$. (Really, we should take $\sqrt{[p + \sqrt{\alpha} A(x)]^2 + 1}$, but since $x < \sqrt{x^2+1} < x+1$, the difference is bounded by N.)

Consider, first, the $N = 1$ body problem with the Hamiltonian

$$H = T^{\rm rel} + H_f \, . \tag{1.16}$$

By simple length scaling (with a simultaneous scaling of the volume V) we easily see that $E_0 = \inf \operatorname{spec}(H) = C\Lambda$. Our goal here is to show that the constant, C, is strictly positive and to give an effective lower bound for it. But we would like to do more, namely investigate the dependence of this constant on α. We also want to show, later on, that for N fermions the energy is bounded below by a positive constant times $N\Lambda$. Our proof will contain some novel – even bizarre – features.

Theorem 1.5 *For the Hamiltonian in 1.16 there are positive constants, C, C', C'' such that*

$$\begin{aligned} E_0 &\leq C\sqrt{\alpha}\Lambda \\ E_0 &\geq C'\sqrt{\alpha}\Lambda \text{ for small } \alpha \\ E_0 &\geq C''\Lambda \text{ for large } \alpha \, . \end{aligned}$$

The generalization of this to N fermions is similar to the nonrelativistic generalization, except that the power of Λ is the same on both sides of the inequalities.

Theorem 1.6 *For N fermions with Hamiltonian*

$$H_N = \sum_{i=1}^{N} T^{\rm rel}(x_i) + H_f$$

there are positive constants C, C', C'', independent of α and N, such that

$$\begin{aligned} E_0 &\leq CN\sqrt{\alpha}\Lambda \\ E_0 &\geq C'N\sqrt{\alpha}\Lambda \quad \text{for small } \alpha \\ E_0 &\geq C''N\Lambda \quad \text{for large } \alpha \end{aligned} \tag{1.17}$$

We close this introduction by mentioning one last result about the Pauli operator. The kinetic energy expression is given by

$$T^{\rm Pauli} = [\sigma \cdot (p + \sqrt{\alpha}A(x))]^2 = (p + \sqrt{\alpha}A(x))^2 + \sqrt{\alpha}\, \sigma \cdot B(x) \, . \tag{1.18}$$

where σ denotes the vector consisting of the Pauli matrices. Observe that this term automatically accounts for the spin–field interaction. Our result for the self energy of a Pauli electron is the following.

Theorem 1.7 *The ground state energy E_0 of the Hamiltonian with Pauli energy,*

$$\frac{1}{2}[\sigma\cdot(p+\sqrt{\alpha}A(x))]^2+H_f, \tag{1.19}$$

satisfies the bounds

$$\begin{aligned} E_0 &\leq C_3\sqrt{\alpha}\Lambda^{3/2} \\ E_0 &\geq C_1\alpha\Lambda \quad \text{for small } \alpha \\ E_0 &\geq C_2\alpha^{1/3}\Lambda \quad \text{for large } \alpha \end{aligned} \tag{1.20}$$

For N fermions, the bounds above are multiplied by N (and the constants are changed).

For the details of the proof, we refer the reader to [6]. We believe that the upper bound is closer to the truth since the main contributions to the self energy should come from the fluctuations of the A^2 term.

Theorem 1.7 has the following consequence for stability of matter interacting with quantized fields. It was shown in [7] that a system of electrons and nuclei interacting with Coulomb forces, with the Pauli kinetic energy for the electrons and with a *classical* magnetic field energy is stable (i.e., the ground state energy is bounded below by N) if and only if α is small enough. In [1], [4] this result was extended to *quantized*, ultraviolet cutoff magnetic fields (as here). Among other things, it was shown in [4] that the ground state energy, E_0, of the electrons and nuclei problem is bounded below by $-\alpha^2\Lambda N$ for small α. Theorem 1.7 implies, as a corollary, that for small α the *total energy* (including Coulomb energies) is bounded below by $+\alpha\Lambda N$. In other words, among the three components of energy (kinetic, field and Coulomb), the first two overwhelm the third – for small α, at least.

All of these statements are true without mass renormalization and the situation could conceivably be more dramatic when the mass is renormalized. In any case, the true physical questions concern energy differences, and this question remains to be addressed.

2. Non-relativistic energy bounds

Theorem 1.1: We sketch a proof of Theorem 1.1. It is clear by taking the state $V^{-1/2}/\otimes|0\rangle$ that the ground state energy is bounded above by $(\mathrm{const})\alpha\Lambda^2$, which is the same result one gets from perturbation theory. Since the field energy in this state vanishes, such a computation ignores the tradeoff between the kinetic energy of the electron and the field energy. Thus, it is important to quantify this tradeoff. The main idea is to estimate the field energy in terms of selected modes.

Consider the operators (field modes), parametrized by $y\in\mathbb{R}^3$,

$$L(y)=\frac{1}{\sqrt{2V}}\sum_{|k|<\Lambda,\lambda}\sqrt{|k|}a_\lambda(k)\bar{v}_\lambda(k)e^{ik\cdot y}, \tag{2.1}$$

with some arbitrary complex coefficients $v_\lambda(k)$. The following lemma is elementary

Lemma 2.1

$$H_f \geq \int w(y) L^*(y) L(y) \mathrm{d}y \tag{2.2}$$

provided that the functions $v_\lambda(k)$ and w are chosen such that, as matrices,

$$|k| \delta_{\lambda',\lambda} \delta(k',k) \geq \frac{1}{2V} \sqrt{|k|} \overline{v}_\lambda(k) \widehat{w}(k-k') v_{\lambda'}(k') \sqrt{|k'|}\,, \tag{2.3}$$

or equivalently, that

$$\sum_{|k|<\Lambda,\lambda} \frac{|f_\lambda(k)|^2}{|v_\lambda(k)|^2} \geq \sum_{|k|,|k'|<\Lambda,\lambda,\lambda'} \frac{1}{2V} \overline{f_\lambda(k)} f_{\lambda'}(k') \widehat{w}(k-k') \tag{2.4}$$

for all $f_\lambda(k)$, where $\widehat{w}(k) = \int e^{ik\cdot x} w(x) \mathrm{d}x$.

For the proof, one simply notes that both sides of (2.2) are quadratic forms in the creation and annihilation operators, and hence (2.3) and (2.4) are necessary and sufficient conditions for (2.2) to be true. ■

Corollary 2.2

$$H_f \geq -\frac{1}{2V} \sum_{|k|<\Lambda,\lambda} |k| |v_\lambda(k)|^2 \int |w(y)| \mathrm{d}y + \frac{1}{4} \begin{cases} \int w(y)(L(y)+L^*(y))^2 \mathrm{d}y \\ -\int w(y)(L(y)-L^*(y))^2 \mathrm{d}y \end{cases} \tag{2.5}$$

To prove this, note that

$$L^* L = L L^* - \frac{1}{2V} \sum_{|k|<\Lambda,\lambda} |k| |v_\lambda(k)|^2\,, \tag{2.6}$$

and, quite generally for operators,

$$LL + L^* L^* \leq L^* L + L L^*\,. \tag{2.7}$$

■

Returning to the proof of Theorem 1.1 we start with the lower bound. Denote by

$$\Pi(x) = \frac{-i}{\sqrt{2V}} \sum_{|k|<\Lambda,\lambda} \sqrt{|k|} \varepsilon_\lambda(k) \left(a_\lambda(k) e^{ik\cdot x} - a_\lambda^*(k) e^{-ik\cdot x} \right)\,. \tag{2.8}$$

This operator is canonically conjugate to $A(x)$ in the sense that we have the commutation relations

$$i[\Pi_i(x), A_j(x)] = \delta_{i,j} \frac{1}{(3\pi)^2} \Lambda^3 . \tag{2.9}$$

For our calculation, it is important to note that

$$\text{div}\, \Pi(x) = 0 . \tag{2.10}$$

Hence from (2.9) and (2.10) we get that

$$\sum_{j=1}^{3} [p_j + \sqrt{\alpha} A_j(x), \Pi_j(x)] = \sqrt{\alpha} \frac{i}{(3\pi)^2} \Lambda^3 . \tag{2.11}$$

The inequality

$$\frac{1}{2}(p + \sqrt{\alpha} A(x))^2 + 2a^2 \Pi(x)^2 \geq -ai \sum_{j=1}^{3} [p_j + \sqrt{\alpha} A_j(x), \Pi_j(x)] , \tag{2.12}$$

valid for all positive numbers a, yields

$$\frac{1}{2}(p + \sqrt{\alpha} A(x))^2 + H_f \geq a\sqrt{\alpha} \frac{1}{(3\pi)^2} \Lambda^3 + H_f - 2a^2 \Pi(x)^2 . \tag{2.13}$$

Now, with

$$v_\lambda(k) = (3\pi) \frac{\varepsilon_\lambda(k)}{\Lambda^{3/2}} \tag{2.14}$$

and

$$w(y) = \delta(x - y) , \tag{2.15}$$

it is elementary to see that (2.4) is satisfied. Hence Corollary 2.2 yields

$$H_f \geq \frac{9\pi^2}{4\Lambda^3} \Pi(x)^2 - \frac{9}{8} \Lambda \tag{2.16}$$

Choosing $a = (3\pi)/(\sqrt{2}\Lambda^{3/2})$ yields the *lower bound*

$$H \geq \frac{1}{3\pi} \sqrt{\frac{\alpha}{2}} \Lambda^{3/2} - \frac{9}{8} \Lambda \tag{2.17}$$

The idea of using a commutator, as in (2.12), (2.13) to estimate the ground state energy, goes back to the study of the polaron [9].

For the upper bound we take a simple trial function of the form

$$\phi(x) \otimes \Psi \tag{2.18}$$

where $\Psi \in \mathscr{F}$ is normalized and $\phi(x)$ is a *real* function normalized in $L^2(\mathbb{R}^3)$. An upper bound to the energy is thus given by

$$\frac{1}{2} \int |\nabla \phi(x)|^2 dx + \frac{\alpha}{2} \int \phi(x)^2 \, (\Psi, A(x)^2 \Psi) \, dx + (\Psi, H_f \Psi) . \tag{2.19}$$

It is not very difficult to see that the last two terms can be concatenated into the following expression.

$$\frac{1}{2}\int \left(\Psi, \left[\Pi(x)^2+\alpha A(x)(-\Delta+\phi(x)^2)A(x)\right]\Psi\right) \mathrm{d}x - \frac{1}{2}\mathrm{Tr}\sqrt{P-\Delta P}\,. \tag{2.20}$$

Here, P is the projection onto the divergence free vector fields with ultraviolet cutoff Λ. This can be deduced by writing the field energy in terms of $\Pi(x)$ and $A(x)$. The first term in (2.20) is a sum of harmonic oscillators whose zero point energy is given by

$$\frac{1}{2}\mathrm{Tr}\sqrt{P(-\Delta+\alpha\phi(x)^2)P} \tag{2.21}$$

and hence

$$\frac{1}{2}\mathrm{Tr}\sqrt{P(-\Delta+\alpha\phi(x)^2)P}-\frac{1}{2}\mathrm{Tr}\sqrt{P(-\Delta)P}\,, \tag{2.22}$$

is an exact expression for the ground state energy. Using the operator monotonicity of the square root we get as an upper bound on (2.20)

$$\frac{1}{2}\int|\nabla\phi(x)|^2\mathrm{d}x+\frac{1}{2}\sqrt{\alpha}\;\mathrm{Tr}\sqrt{P\phi(x)^2P}\,. \tag{2.23}$$

As a trial function we use the positive function (note that this really is positive)

$$\phi(x) = \mathrm{const.}K^{-3/2}\int\left(1-\frac{|k|}{K}\right)_+^3 e^{ik\cdot x}\mathrm{d}x\,. \tag{2.24}$$

Optimizing the resulting expression over K yields the stated result. For details we refer the reader to [6]. ■

It is natural to ask, how good this upper bound is. If we neglect the cross terms in $(p+A)^2$, i.e., we replace the kinetic energy by $p^2+\alpha A(x)^2$, then we have Theorem 1.2, which we prove next.

Theorem 1.2: The upper bound was already given in Theorem 1.1 because $< p\cdot A >= 0$ in the state (2.18). Loosely speaking equation (2.9) expresses the Heisenberg uncertainty principle for the field operators. An uncertainty principle that is quite a bit more useful is the following.

Lemma 2.3 *The following inequality holds in the sense of quadratic forms*

$$\Pi(x)^2 \geq \frac{1}{4}\frac{1}{(3\pi)^4}\Lambda^6\frac{1}{A(x)^2}\,. \tag{2.25}$$

For the proof note that $[A_j(x),A_k(y)]=0$ and compute

$$i[\Pi(x)_j,\frac{A_j(x)}{A(x)^2}] = \frac{1}{(3\pi)^2}\Lambda^3\left[\frac{1}{A(x)^2}-2\left(\frac{A_j(x)}{A(x)^2}\right)^2\right]\,, \tag{2.26}$$

and summing over j we obtain that

$$i\sum_{j=1}^{3}[\Pi(x)_j, \frac{A_j(x)}{A(x)^2}] = \frac{1}{(3\pi)^2}\Lambda^3\frac{1}{A(x)^2} \tag{2.27}$$

Our statement follows from the Schwarz inequality. ■

To prove Theorem 1.2 we return to Lemma 2.1 and choose $v_\lambda(k) = \varepsilon_\lambda(k)$ and $w(x)$ any function ≤ 1. Corollary 2.2 applied to each of the 3 components of $\Pi(x)$ then yields

$$H_f \geq \frac{1}{4}\int w(x-y)\Pi(y)^2 \mathrm{d}y - \Lambda^4\frac{3}{8\pi^2}\int w(y)\mathrm{d}y\,, \tag{2.28}$$

for every $x \in \mathbb{R}^3$. By Lemma 2.1 the right side is bounded below by

$$\Lambda^6\int w(x-y)\frac{1}{A(y)^2}\mathrm{d}y - \Lambda^4\int w(y)\mathrm{d}y\,, \tag{2.29}$$

and hence

$$\begin{aligned}\langle\Psi, H\Psi\rangle \;\geq\; & \frac{1}{2}\int\langle\nabla\Psi(x), \nabla\Psi(x)\rangle\mathrm{d}x + \frac{\alpha}{2}\int\langle\Psi(x), A(x)^2\Psi(x)\rangle\mathrm{d}x \\ &+ \Lambda^6\int w(x-y)\langle\Psi(y), \frac{1}{A(x)^2}\Psi(y)\rangle\mathrm{d}y\mathrm{d}x \\ &- \Lambda^4\int w(y)\mathrm{d}y\int\langle\Psi(x), \Psi(x)\rangle\mathrm{d}x\,.\end{aligned} \tag{2.30}$$

By Schwarz's inequality the second and third term together are bounded below by

$$\sqrt{\frac{\alpha}{2}}\Lambda^3\int\langle\Psi(x), \Psi(y)\rangle\frac{w(x-y)}{\sqrt{\int w(z)\mathrm{d}z}}\mathrm{d}x\mathrm{d}y\,. \tag{2.31}$$

If we restate our bound in terms of Fourier space variables we get

$$\int\left[\frac{|p|^2}{2} + \sqrt{\frac{\alpha}{2}}\Lambda^3\frac{\widehat{w}(p)}{\sqrt{\widehat{w}(0)}}\right]\langle\widehat{\Psi}(p), \widehat{\Psi}(p)\rangle\mathrm{d}p - \Lambda^4\widehat{w}(0)\int\langle\widehat{\Psi}(p), \widehat{\Psi}(p)\rangle\mathrm{d}p\,. \tag{2.32}$$

Choosing the function $\widehat{w}(p)$ to be $(2\pi)^3\Lambda^{-18/7}$ times the characteristic function of the ball of radius $\Lambda^{6/7}$, we have that $w(x) \leq 1$ and it remains to optimize (2.32) over all normalized states $\widehat{\Psi}(p)$. This is easily achieved by noting that the function

$$\frac{1}{2}|p|^2 + \sqrt{\frac{\alpha}{2}}\Lambda^3\frac{\widehat{w}(p)}{\sqrt{\widehat{w}(0)}} \tag{2.33}$$

is everywhere larger than $\Lambda^{12/7}$. Strictly speaking, the function $w(x)$ should be positive in order for the argument that led to (2.31) to be valid. This can be achieved with a different choice of $w(x)$, like the one in (2.24), that is more complicated but does not change the argument in an essential way.

3. Non-relativistic many-body energies

A problem that has to be addressed is the energy of N particles (bosons or fermions) interacting with the radiation field. If $E_0 = E_0(1)$ is the energy of one particle (which we estimated in the preceding section) then, ideally the energy, $E_0(N)$, of N particles (which trivially satisfies $E_N \leq NE$, since the N particles can be placed infinitely far apart) ought to be, *exactly*,

$$E_0(N) = NE_0 \tag{3.1}$$

in a correct QED. In other words, in the absence of nuclei and Coulomb potentials, there should be *no binding* caused by the field energy H_f. This is what we seem to observe experimentally, but this important topic does not seem to have been discussed in the QED literature.

Normally, one should expect binding, for the following mathematical reason: The first particle generates a field, $A(x)$, and energy E_0. The second particle can either try to generate a field $A(x+y)$, located very far away at y or the second particle can try to take advantage of the field $A(x)$, already generated by the first particle, and achieve an insertion energy lower than E_0.

Indeed, this second phenomenon happens for bosons, as expected. For fermions, however, the Paul principle plays a crucial role (even in the absence of Coulomb attractions). We show that $E_0(N) \geq CNE_0$ for fermions, but we are unable to show that the universal constant $C = 1$. Even if $C < 1$, the situation could still be saved by mass renormalization, which drives the bare mass to zero as Λ increases, thereby pushing particles apart.

3.1. Bosons

Theorem 1.3: This theorem concerns the ground state energy of N charged bosons. the Hamiltonian is given by 1.11 acting on the Hilbert space of symmetric functions tensored with the photon Fock space $\mathscr{F}$. It states, basically, that

$$C_1\sqrt{N}\sqrt{\alpha}\Lambda^{3/2} \leq E_0^{\text{boson}}(N) \leq C_2 N^{5/7}\alpha^{2/7}\Lambda^{12/7}.$$

The proof follows essentially that of the one particle case. The interesting fact is that it implies *binding* of *charged bosons* (in the absence of the Coulomb repulsion). The binding energy is defined by

$$\Delta E(N) = E_0(N) - NE_0(1)$$

and satisfies the bounds

$$\begin{aligned} \Delta E(N) &\geq C_1\sqrt{N}\sqrt{\alpha}\,\Lambda^{3/2} - C_2 N\alpha^{2/7}\Lambda^{12/7} \\ \Delta E(N) &\leq C_2 N^{2/7}\alpha^{2/7}\Lambda^{12/7} - C_1 N\sqrt{\alpha}\,\Lambda^{3/2} \end{aligned} \tag{3.2}$$

which can be made negative for appropriately chosen N and Λ. There will be binding for all large enough N, irrespective of the cutoff Λ. It also has to be remarked that the Coulomb repulsion will, in all likelihood, not alter this result since it has an effect on energy scales of the order of Λ and not $\Lambda^{12/7}$ or $\Lambda^{3/2}$.

3.2. Fermions

The real issue for physics is what happens with fermions. We cannot show that there is no binding but we can show that the energy is extensive as in Theorem 1.4. The Hamiltonian is the same as (1.11) but it acts on antisymmetric functions tensored with $\mathscr{F}$. (Spin can be ignored for present purposes.)

Rough sketch of the proof of Theorem 1.11. The difficulty in proving this theorem stems from the fact that the field energy is not extensive in any obvious way.

Define $\underline{X} = (x_1, \cdots, x_N)$ and define the function

$$n_j(\underline{X}, R) = \#\{x_i \neq x_j \ : \ |x_i - x_j| < R\} \, .$$

This function counts the number of electrons that are within a distance R of the jth electron. Note that this function is not smooth, so that all the following computations have to be modified. (See [6].) We save half of the kinetic energy and write

$$H = \frac{1}{4} \sum_{j=1}^{N} (p_j + \sqrt{\alpha} A(x_j))^2 + H' \, .$$

We apply the commutator estimate (2.11) to the pair

$$i[p_j + \sqrt{\alpha} A(x_j), \frac{1}{\sqrt{N_j(\underline{X}, R) + 1}} \Pi(x_j)]$$

and obtain the bound (with the caveat mentioned above), for all $\alpha > 0$,

$$H' \geq a\sqrt{\alpha}\Lambda^3 \sum_{j=1}^{N} \frac{1}{\sqrt{N_j(\underline{X}, R) + 1}} - a^2 \sum_{j=1}^{N} \frac{1}{N_j(\underline{X}, R) + 1} \mathrm{F}(x_j)^2 + H_f \tag{3.3}$$

The next two steps are somewhat nontrivial and we refer the reader to [6]. First one notes that the modes $F(x_i)$ and $F(x_j)$ are essentially orthogonal (i.e., they commute) if $|x_i - x_j| > \Lambda^{-1}$. Ignoring the technical details of how this is implemented, the key observation is that the last two terms in (3.3) can be estimated from below by $-N\Lambda$ provided $a = \Lambda^{-3/2}$.

The next ingredient is a new Lieb-Thirring type estimate involving the function $N_j(\underline{X}, R)$. It is here and only here that the Pauli exclusion principle is invoked.

Theorem 3.1 *On the space* $\wedge_{j=1}^{N} L^2\mathbb{R}^3; \mathbb{C}^q)$ *of antisymmetric functions*

$$\sum_{j=1}^{N} (p_j + \sqrt{\alpha} A(x_j))^2 \geq \frac{C}{q^{2/3}} \frac{1}{R^2} \sum_{j=1}^{N} N_j(\underline{X}, R)^{2/3} \tag{3.4}$$

with $C \geq 0.00127$. *An analogous inequality holds for the relativistic case as well:*

$$\sum_{j=1}^{N} |p_j + \sqrt{\alpha} A(x_j)| \geq \frac{C}{q^{1/3}} \frac{1}{R} \sum_{j=1}^{N} N_j(\underline{X}, R)^{1/3} \tag{3.5}$$

By using the kinetic energy previously saved together with (3.3) and the previous discussion, we get

$$H \geq \sum_{j=1}^{N} \left\{ N_j(\underline{X}, R)^{2/3} + \sqrt{\alpha} \Lambda^{3/2} \frac{1}{\sqrt{N_j(\underline{X}, R) + 1}} \right\} - N\Lambda .$$

By minimizing over N_j the desired estimate is obtained. The upper bound is fairly elementary and is omitted. ∎

4. Relativistic energy bounds

Theorem 1.5: Sketch of Proof.

An upper bound for E_0 is easy to obtain, but it is indirect. Note that

$$|p + \sqrt{\alpha} A(x)| \leq \varepsilon [p + \sqrt{\alpha} A(x)]^2 + (4\varepsilon)^{-1} \tag{4.1}$$

for any $\varepsilon > 0$. Take $\Psi = f(x) \otimes |0\rangle$ with $|0\rangle$ being the Fock space vacuum. Using (4.1)

$$\begin{aligned} (\Psi, H\Psi) &\leq \varepsilon \int_{\mathbb{R}^3} \{\alpha \langle 0|A(x)^2|0\rangle |f(x)|^2 + |\nabla f(x)|^2\} dx + \varepsilon^{-1} \\ &= \frac{\varepsilon \alpha \Lambda^2}{4\pi} + \int |\nabla f|^2 + \frac{1}{4\varepsilon} , \end{aligned} \tag{4.2}$$

since $\langle 0|A(x)^2|0\rangle = (2\pi)^{-3} \int_{|k|<\Lambda} |k|^{-1} dk = \Lambda^2/4\pi^2$. We can now let $f(x) \to V^{-\frac{1}{2}}$ and take $\varepsilon = (\pi/\alpha)^{1/2} \Lambda^{-1}$, whence

$$E_0 \leq (\alpha/4\pi)^{1/2} \Lambda . \tag{4.3}$$

Now we turn to the lower bound for H.

Step 1: Since $x \to \sqrt{x}$ is operator monotone,

$$T > T_1 = |p_1 + \sqrt{\alpha} A_1(x)| , \tag{4.4}$$

where the subscript 1 denotes the 1 component (i.e., the x-component) of a vector. By replacing T by T_1, we are now in a position to remove A_1 by a gauge transformation - but it has to be an *operator-valued gauge transformation*. The use of such a gauge transformation is a novelty, as far as we are aware, in QED.

To effect the gauge transformation, set

$$\varphi(x) = \frac{1}{\sqrt{2V}} \sum_{h,\lambda} \frac{\varepsilon^{\lambda}(k)_1}{\sqrt{(k)}} \left[a_\lambda(k) + a^*_\lambda(-k)\right] \frac{e^{ik_1x_1}-1}{ik_1} e^{ik_\perp x_\perp} \tag{4.5}$$

with $x_\perp = (x_2, x_3)$. Then $[A_j(x), \varphi(x)] = 0, \quad j = 1,2,3$ and $-i\frac{\partial}{\partial x_1} \exp\,[i\varphi(x)] = -A_1(x)$. The unitary $U = \exp\,[i\varphi(x)]$ is a gauge transformation, but it is *operator-valued.* We have

$$\begin{aligned} U^{-1}|p_1 + A_1(x)|U &= |p_1| \\ U^{-1}a_\lambda(k)U &= a_\lambda(k) + f_\lambda(k,x) \\ U^{-1}a^*_\lambda(k)U &= a^*_\lambda(k) + \bar{f}_\lambda(k,x) \\ \widetilde{H}_f = U^{-1}H_fU &= \sum_{k,\lambda} |k|[a^*_\lambda(k) + \bar{f}_\lambda(k,x)][a_\lambda(k) + f_\lambda(k,x)] \end{aligned} \tag{4.6}$$

with

$$f_\lambda(k,x) = \sqrt{\frac{\alpha}{2V}} \sum_{k,\lambda} \frac{\varepsilon_\lambda(k)_1}{|k|} \frac{e^{-ik_1x_1}-1}{k_1} e^{-ik_\perp x_\perp}\,. \tag{4.7}$$

Since $p_\perp$ does not appear in our new Hamiltonian,

$$\widetilde{H} = U^{-1}HU = |p_1| + \widetilde{H}_f\,, \tag{4.8}$$

the variable $x_\perp$ appears only as a parameter, and thus we can set $x_\perp$ = constant = (0,0), by translation invariance, and replace $\mathbb{R}^3$ by $\mathbb{R}^1 = \mathbb{R}$.

From now on $x_1 = x$ and, $p_1 = p = -i\,d/dx$.

Step 2: The dependence on x now appears in $\widetilde{H}_f$ instead of in the kinetic energy, $|p|$. For each x we can try to put $\widetilde{H}_f$ into its ground state, which is that of a displaced harmonic oscillator. But, since this state depends on x, to do so will require a great deal of kinetic energy, $|p_1|$.

Let Ψ be a normalized wave-function, i.e., a function on $L^2(\mathbb{R}) \otimes \mathscr{F}$. We write it as ψ_x where $\psi_x \in \mathscr{F}$. Thus, with $\langle\cdot,\cdot\rangle$ denoting the inner product on $\mathscr{F}$, we have $\int_{\mathbb{R}} \langle\psi_x, \psi_x\rangle dx = 1$.

Decompose $\mathbb{R}$ as the disjoint union of intervals of length ℓ/Λ, where ℓ is a parameter to be determined later. Denote these intervals by $I_j,\ j = 1,2,\ldots$. A simple Poincaré type inequality gives, for $g\,:\, L^2(\mathbb{R}) \to \mathbb{C}$,

$$(g, |p|g) \geq C_1 \frac{\Lambda}{\ell} \sum_j \int_{I_j} \{|g(x)|^2 - |\bar{g}_j|^2\} dx\,,$$

where $\bar{g}_j = \frac{\Lambda}{\ell} \int_{I_j} g(x)dx$ is the average of g in I_j. Then

$$(\Psi, |p|\Psi) \geq C_1 \frac{\Lambda}{\ell} \sum_j \int_{I_j} \{\langle\psi_x, \psi_x\rangle - \langle\bar{\psi}_j, \bar{\psi}_j\rangle\} dx\,. \tag{4.9}$$

Step 3: Next, we analyze $\widetilde{H}_f$. We think of this as an operator on $\mathscr{F}$, parameterized by $x \in \mathbb{R}$. We would like $\widetilde{H}_f$ to have a gap so we define

$$H_x = \frac{\Lambda}{2} \sum_{\varepsilon\Lambda \leq |k| \leq \Lambda} \sum_{\lambda} [a_\lambda^+(k) + \bar{f}_\lambda(k,x)] \cdot [\,h.c.\,] \tag{4.10}$$

Clearly, $\widetilde{H}_f \geq H_x$ and

$$(\Psi, \widetilde{H}\,\Psi) \geq \frac{\Lambda}{\ell} \sum_j \int_{I_j} \langle \psi_x, \psi_x \rangle - \langle \bar{\psi}_j, \bar{\psi}_j \rangle + \langle \psi_x, H_x \psi_x \rangle dx\,. \tag{4.11}$$

For each interval I_j we can minimize (4.11) subject to $\int_{I_j} \langle \psi_x, \psi_x \rangle dx$ fixed. This leads to

$$(h_j\,\psi)_x = e\,\psi_x \tag{4.12}$$

with

$$(h_j\,\psi)_x = \frac{\Lambda}{\ell}\,\psi_x - \frac{\Lambda}{\ell}\,\bar{\psi}_j + H_x\,\psi_x \tag{4.13}$$

Obviously, this eigenvalue problem (4.12, 4.13) is the same for all intervals I_j, so we shall drop the subscript j and try to find the minimum e.

A lower bound to h_j (and hence to e) can be found by replacing H_x by

$$\widehat{H}_x = \frac{\Lambda}{2}(1 - \Pi_x)\,,$$

where $\Pi_x = |g_x\rangle\langle g_x|$ is the projector onto the ground state, $|g_x\rangle$, of H_x.

If we substitute $\widehat{H}_x$ into (4.13) the corresponding eigenvalue equation (4.12) becomes soluble. Multiply (4.12) on the left by $\langle g_x|$, whence

$$\left(\frac{\Lambda}{\ell} - e\right) \langle g_x, \psi_x \rangle = \frac{\Lambda}{\ell} \langle g_x, \bar{\psi} \rangle \tag{4.14}$$

Then, substitute (4.14) into (4.13) and integrate $\int_I dx$ to find

$$\frac{1}{2}\Lambda^3 \ell^{-2} \left(\int \Pi_x dx \right) \bar{\psi} = \left(\frac{\Lambda}{\ell} - e\right)\left(\frac{\Lambda}{2} - e\right) \bar{\psi}\,. \tag{4.15}$$

We know that $e < \Lambda/2$ because we could take $\psi_x =$ constant as a trial function, and then use $\widetilde{H}_x \leq \Lambda/2$. Also, $e < \Lambda/\ell$, because we could take $\Psi = \delta_{x_0} |g_{x_0}\rangle$.

Step 4: Eq. (4.15) will give us a lower bound to e if we can find an upper bound to $Y = (\Lambda/\ell) \int_I \Pi_x dx$. To do this note that

$$\begin{aligned} Y^2 &\leq \text{Trace}\, Y^2 = \left(\frac{\Lambda}{\ell}\right)^2 \int_I \int_I |\langle g_x, g_y \rangle|^2 dx dy \\ &= \left(\frac{\Lambda}{\ell}\right)^2 \int_I \int_I \exp\{-\frac{\alpha}{2V} \sum_{\varepsilon\Lambda \leq |k| \leq \Lambda} \sum_{\lambda} |f_\lambda(k,x) - f_\lambda(k,y)|^2 dx dy\} \end{aligned} \tag{4.16}$$

Noting that $\sum_{\lambda=1}^{2} e_\lambda(k)_1^2 = 1 - k_1^2/k^2$, the quantity { } in (4.16) becomes (as $V \to \infty$)

$$\{\ \ \} = -\frac{2\alpha}{(2\pi)^3}\int_{\Lambda/2<|k|<\Lambda} \frac{1}{|k|^3 k_1^2}(k_\perp^2)[\sin\frac{k_1}{2}(x-y)] \tag{4.17}$$

After some algebra we find that

$$\left(\frac{1}{\ell}-\frac{e}{\Lambda}\right)\left(\frac{1}{2}-\frac{e}{\Lambda}\right) \le \frac{1}{2\ell}(\text{Trace } Y^2)^{1/2} \le \frac{1}{2\ell}\sqrt{K_\ell(\alpha)} \tag{4.18}$$

where

$$\begin{aligned} K_\ell(\alpha) &= \int_0^1\int_0^1 \exp\left[-\alpha\frac{\ell}{\pi^2}|x-y|\int_0^{|x-y|\ell/4}\left(\frac{\sin t}{t}\right)^2 dt\right] dxdy\,. \\ &\le \int_{-1/2}^{1/2} \exp[-\alpha x^2\ell^2/8\pi]\, dx\,. \end{aligned} \tag{4.19}$$

We find that

$$\begin{aligned} K_\ell(\alpha) &\sim 1-\alpha\ell^2/96\pi, \quad \ell^2\alpha \text{ small} \\ &\sim \sqrt{2}\pi(\alpha\ell^2)^{1/2}, \quad \ell^2\alpha \text{ large} \end{aligned} \tag{4.20}$$

If α is small we take $\ell \sim \alpha^{-1/2}$. If α is large we take $\ell = 2$. This establishes our theorem for $N = 1$. ■

Theorem 1.6: Sketch of Proof.

For $N > 1$ we can decompose $\mathbb{R}^3$ into cubic boxes $B_j, j = 1,2,3,\ldots$ of size $\ell\Lambda$ and "borrow" $\frac{1}{2}|p + A(x)|^2$ kinetic energy from each particle. That is, $H_N = H_N^{1/2} + \frac{1}{2}T_N$ with $T_N = \sum_{i=1}^{N} T(x_i)$. The Pauli principle will then yield an energy for $\frac{1}{2}T_N$ that is bounded below by (const.) $(n_{j-1})^{4/3}$, where n_j is the particle number in box B_j.

References

1. Bugliaro, L., Fröhlich, J., and Graf, G. M.: Stability of quantum electrodynamics with nonrelativistic matter, *Phys. Rev. Lett.* **77** (1996), 3494–3497.
2. Bach, V., Fröhlich, J., and Sigal, I. M.: Renormalization group analysis of spectral problems in quantum field theory, *Adv. Math.* **137** (1998), 205–298; Quantum electrodynamics of confined nonrelativistic particles, *Adv. Math.* **137** (1998), 299–395.
3. Fefferman, C.: Stability of Coulomb systems in a magnetic field, *Proc. Natl. Acad. Sci. USA* **92** (1995), 5006–5007.
4. Fefferman, C., Fröhlich, J., and Graf, G. M.: Stability of nonrelativistic quantum mechanical matter coupled to the (ultraviolet cutoff) radiation field, *Proc. Natl. Acad. Sci. USA* **93** (1996), 15009–15011; Stability of ultraviolet cutoff quantum electrodynamics with non-relativistic matter, *Commun. Math. Phys.* **190** (1997), 309–330.
5. Hogreve, H. H.: Math. Reviews **99e:81051a, b** Amer. Math. Soc. (1999).

6. Lieb, E. H. and Loss, M.: Remarks about the ultraviolet problem in QED, (in preparation).
7. Lieb, E. H., Loss, M., and Solovej, J. P.: Stability of Matter in Magnetic Fields, *Phys. Rev. Lett.* **75** (1995), 985–989.
8. Lieb, E. H., Siedentop, H., and Solovej, J. P.: Stability and Instability of Relativistic Electrons in Magnetic Fields, *J. Stat. Phys.* **89** (1997), 37-59.
9. Lieb, E. H. and Yamazaki, K.: Ground State Energy and Effective Mass of the Polaron, *Phys. Rev.* **111** (1958), 728–733.
10. Okamoto, T. and Yajima, K.: Complex scaling technique in nonrelativistic massive QED, *Ann. Inst. H. Poincaré Phys. Théor.* **42** (1985), 311–327.

Automorphic distributions for $SL(2,\mathbb{R})$

Wilfried Schmid
Department of Mathematics, Harvard University, Cambridge, MA 02138, USA

Abstract. Classical (holomorphic) modular forms on the upper half plane have boundary values along the real axis, as do (non-holomorphic) Maass forms. These boundary values are distributions which transform according to a factor of automorphy under the action of the discrete group which leaves the modular forms invariant. The main result of this paper is a characterization of the degree of regularity of such automorphic distributions. The boundary values of Maass forms, for example, are first derivatives of continuous functions, which are Hölder continuous of index depending on the eigenvalue corresponding to the Maass form in question. The existence of a continuous anti-derivative makes it possible to prove the functional equation for L-functions attached to Maass forms by integration along the *real* axis, rather than the *imaginary* axis, as in the usual argument. This new approach to the functional equation appears likely to generalize to higher rank groups that are beyond the reach of the classical arguments.

Keywords: automorphic representations, Maass forms, functional equation.

Mathematics Subject Classifications (2000): 11F66, 11F37, 11F70

Dedicated to the memory of my friend Moshé Flato

1. Introduction

The automorphic distributions to which the title refers arise as boundary values of classical modular forms and of Maass modular forms. In this paper I shall address the question of how regular these distributions are – in other words, how close they come to being continuous functions. My answer to this question leads to new proofs of certain known results. The new arguments look simpler to me, in some cases, than the classical arguments; more importantly, it appears that they can be generalized to higher dimensional cases which are beyond the reach of the classical arguments.

To explain the main idea, I consider initially an arbitrary real reductive matrix group $G_{\mathbb{R}}$, a finitely generated, discrete subgroup $\Gamma \subset G_{\mathbb{R}}$, and an irreducible unitary representation (π, V) of $G_{\mathbb{R}}$. One calls $v \in V$ a C^∞ vector if $g \mapsto \pi(g)v$ is a C^∞ map from $G_{\mathbb{R}}$ to V. The space of all C^∞ vectors $V^\infty \subset V$ is dense, and carries a natural Fréchet topology; the representation π restricts to a continuous representation on this Fréchet space. There is also a notion of distribution vector: A continuous linear functional on V'^{∞}, the space of C^∞ vectors for the irreducible unitary representation (π', V') dual to (π, V). Dualizing the dense embedding $V'^{\infty} \subset V'$,

Research supported in part by an NSF grant.

G. Dito and D. Sternheimer (eds.), Conférence Moshé Flato 1999, Vol. 1, 345 – 387.

one obtains an embedding of V into its space of distribution vectors $V^{-\infty}$. Thus $V^{\infty} \subset V \subset V^{-\infty}$, and dually, $V'^{\infty} \subset V' \subset V'^{-\infty}$.

I now suppose that (π, V) occurs as a discrete summand in $L^2(\Gamma\backslash G_{\mathbb{R}})$. Then (π', V') also occurs as a direct summand, since $L^2(\Gamma\backslash G_{\mathbb{R}})$ is self-dual. The inclusion

$$i : V' \hookrightarrow L^2(\Gamma\backslash G_{\mathbb{R}}) \tag{1.1}$$

sends the subspace $V'^{\infty} \subset V'$ to C^{∞} functions, which can be evaluated at any $g \in G_{\mathbb{R}}$. The evaluation map

$$\tau : v' \mapsto i(v')(e) \qquad (v' \in V'^{\infty}) \tag{1.2}$$

is continuous with respect to the topology of V'^{∞}, and thus defines a distribution vector for (π, V) – a Γ-invariant distribution vector since $i(v')$, for $v' \in V'^{\infty}$, is a Γ-invariant function:

$$\tau \in (V^{-\infty})^{\Gamma}. \tag{1.3}$$

The distribution vector τ completely determines the inclusion (1.1). To see this, note that $i(v')(g) = < \tau, \pi'(g)v' >$ for every v' in the dense subspace $V'^{\infty} \subset V'$ and every $g \in G_{\mathbb{R}}$.

The preceeding discussion applies in particular to the group $G_{\mathbb{R}} = SL(2,\mathbb{R})$ and any finitely generated, discrete subgroup Γ. Then Γ acts properly discontinuously on the upper half plane $\mathcal{H}$. One knows that there is a natural bijection between non-zero cuspidal modular forms $F(z)$ of weight k on $\Gamma\backslash\mathcal{H}$ and embeddings of the discrete series representation of lowest weight k,

$$V_k \hookrightarrow L^2(\Gamma\backslash SL(2,\mathbb{R}));\tag{1.4}$$

in effect, the modular form $F(z)$ is obtained by projecting the image of (1.4) onto a lowest $SO(2)$-weight vector and multiplying the resulting function by $(\operatorname{Im} z)^{-\frac{k}{2}}$. In this situation, the limit

$$\tau(x) = \lim_{y \to 0+} F(x+iy) \tag{1.5}$$

exists in the sense of distributions, and this distribution τ is a concrete realization of the distribution vector corresponding to the embedding (1.4) via the construction (1.2–3). My main technical result, specialized to the case at hand, asserts:

(1.6)

a) when the weight $k = 2\ell + 1$ is an odd integer, the distribution τ can be expressed as the $(\ell+1)$st derivative $\tau = \phi^{(\ell+1)}$ of a function ϕ which is Hölder continuous of index $\frac{1}{2}$;

b) when the weight $k = 2\ell$ is an even integer, the distribution τ can be expressed as the $(\ell+1)$st derivative $\tau = \phi^{(\ell+1)}$ of a function ϕ which is Hölder continuous of index α, for every $\alpha < 1$.

Recall that a function $\phi(x)$ is Hölder continuous of index α if there exists a constant $C > 0$ such that $|\phi(x) - \phi(y)| \leq C|x-y|^\alpha$ for all x, y; the weaker notion of local Hölder continuity of index α requires only that the constant C can be chosen in a locally uniform manner.

Next I consider a square-integrable Maass modular form on $\Gamma\backslash\mathscr{H}$ – in other words, a non-constant Γ-invariant eigenfunction $F(x,y)$ of the hyperbolic Laplace operator with finite L^2 norm:

$$\begin{aligned} y^2\left(\tfrac{\partial^2}{\partial x^2} + \tfrac{\partial^2}{\partial y^2}\right)F(x,y) &= \tfrac{1}{4}(\lambda^2 - 1)\,F(x,y), \\ \int_{\Gamma\backslash\mathscr{H}} |F(x,y)|^2\,\frac{dx\,dy}{y^2} &< \infty. \end{aligned} \tag{1.7}$$

Here λ is either purely imaginary or real, $-1 < \lambda < 1$, in both cases determined only up to sign. Square-integrable Maass modular forms correspond bijectively to embeddings

$$V_{\lambda,+1} \hookrightarrow L^2(\Gamma\backslash SL(2,\mathbb{R})) \tag{1.8}$$

of an even principal series representation (when $\lambda \in i\mathbb{R}$) or a representation of the complementary series (when $\lambda \in \mathbb{R}$, $0 < |\lambda| < 1$); the subscript $+1$ refers to the even parity condition, which simply means that these representations involve only even $SO(2)$-weights. Near the real axis, the Maass modular form $F(x,y)$ has an asymptotic expansion

$$F(x,y) \sim y^{\frac{1-\lambda}{2}} \textstyle\sum_{n\geq 0} \tilde{\tau}_{\lambda,n}(x)\,y^n + y^{\frac{1+\lambda}{2}} \sum_{n\geq 0} \tilde{\tau}_{-\lambda,n}(x)\,y^n\,, \tag{1.9}$$

with distribution coefficients $\tilde{\tau}_{\pm\lambda,n}$. The tilde in the notation $\tilde{\tau}_{\pm\lambda,n}$ has no particular significance for now; it simply makes the notation consistent with that in the later sections. The leading coefficients $\tilde{\tau}_{\pm\lambda,0}$ play the role of τ in the construction (1.2-3) for the embedding (1.8) – both represent the same distribution vector, but in the two different realizations $V_{\pm\lambda,+1}$ of the same abstract representation. In this situation, the main technical result asserts the following regularity property of the distributions $\tilde{\tau}_{\pm\lambda,0}$:

(1.10)

a) when the embedding (1.8) is cuspidal, $\tilde{\tau}_{\pm\lambda,0}$ can be expressed as the first derivative of a function ϕ which is Hölder continuous of index $\frac{1}{2}(1 \pm \mathrm{Re}\lambda)$;

b) when the embedding fails to be cuspidal, $\tilde{\tau}_{\lambda,0}$ with $\lambda > 0$ can be expressed as the first derivative of a function ϕ which is Hölder continuous of index λ, and $\tilde{\tau}_{-\lambda,0}$ as the second derivative of a function which is Hölder continuous of index α, for every $\alpha < 1$.

One calls the embedding (1.8) cuspidal when the $K_{\mathbb{R}}$-finite functions in its image are bounded. This condition holds automatically when λ is imaginary or Γ cocompact, or when Γ contains a principal congruence subgroup of $SL(2,\mathbb{Z})$.

This still leaves the case of the trivial representation, about which there is not much to say, and representations of the odd principal series. The latter always occur cuspidally in $L^2(\Gamma\backslash SL(2,\mathbb{R}))$. There are again two realizations of the distribution τ, just as in the case of the even principal series, and they can be expressed as first derivatives of functions that are Hölder continuous of index $\frac{1}{2}$. My main technical result, theorem 2.18 below, is a statement that uniformly covers all cases.

The boundary values $\tilde{\tau}_{\lambda,0}$ of Maass modular forms turn out to have intriguing properties. According to (1.10), they are first derivatives $\tilde{\tau}_{\lambda,0} = \phi_\lambda'$ of continuous functions ϕ_λ, except in the single case of a non-cuspidal Maass form and the negative choice among the two parameters $\pm\lambda$. The opposite extreme is that of a cuspidal form, with $\lambda > 0$ and $\Gamma \subset SL(2,\mathbb{Z})$ a subgroup of finite index. In this situation, the anti-derivative ϕ_λ is differentiable, with zero derivative, at all rational points; near rational points $\frac{p}{q}$, the difference $\phi_\lambda(x) - \phi_\lambda(\frac{p}{q})$ oscillates, with amplitude proportional to $|qx-p|^{\lambda+1}$ and frequency proportional to $|qx-p|^{-2}$. In particular, when ϕ_λ is real valued, as may be assumed, it assumes an infinite number of local maxima and local minima in every non-empty open interval. A conjecture of Selberg predicts that Maass forms with $\lambda > 0$ do not exist when Γ contains one of the principal congruence subgroups Γ_m, but such forms – even cuspidal Maass forms – do exist for certain other subgroups $\Gamma \subset SL(2,\mathbb{Z})$ of finite index. The wild behavior of the ϕ_λ can be used to establish Selberg's conjecture for small values of m. These matters will be explored in section 3.

Maass [10] attaches L-functions to Maass forms, and proves a functional equation for them. His proof of the functional equation follows the classical pattern, namely taking the Mellin transform along the imaginary axis and exploiting the transformation property under $z \mapsto -\frac{1}{z}$. To get the explicit form of the equation, Maass works with the analogue, involving Bessel functions, of the q-expansion of classical modular forms. The continuity of the anti-derivative ϕ_λ of the boundary distribution $\tilde{\tau}_{\lambda,0}$ can be used to give a different derivation of the functional equation: Instead of integrating the Maass form along the *imaginary* axis, one can take the Mellin transform of ϕ_λ along the *real* axis, and use the integrated transformation law for $\tilde{\tau}_{\lambda,0}$ under the transformation $x \mapsto -\frac{1}{x}$ of the real line. I shall present the details in section 4. In my approach the expansion of the Maass modular form on the upper half plane, in terms of Bessel functions, plays no role. Instead, the argument uses the interplay between Fourier analysis on the real line and the action of the Weyl group. What makes this interesting is the fact that the real axis is the specialization to $SL(2,\mathbb{R})$ of a maximal unipotent subgroup of a reductive group of arbitrary rank, whereas the imaginary axis is the analogue of a maximal split torus. One expects functional equations of automorphic representations to depend on an integration process over a space larger than a maximal split torus – a space of the same dimension, in fact, as a maximal unipotent subgroup. There are some other hints that the strategy of my argument does work in higher rank cases; I shall pursue this idea elsewhere.

The central question of this paper – the regularity of automorphic distributions – was also considered by Bernstein-Reznikov [3, 4], from a somewhat different point of view: They describe the degree of regularity in terms of a Sobolev notion based on the Hilbert space structure of the representation space V. I shall discuss the precise connection between their results and mine following the statement of the main theorem in the next section.

The results of this paper are several years old. I reported on them publicly in the summer of 1997. At that time my results were somewhat less general – they applied only to subgroups $\Gamma \subset SL(2,\mathbb{Z})$ of finite index – and the proofs I knew then were more intricate.

It is a pleasure to acknowledge my indebtedness to several colleagues. Curt McMullen suggested the appropriate setting for my arguments, namely finitely generated discrete subgroups of $SL(2,\mathbb{R})$. Gerald Folland answered my questions about the history of Serge Bernstein's theorem on the growth of Fourier coefficients of functions satisfying a Hölder condition. Peter Sarnak explained to me the existence of cuspidal Maass forms corresponding to small eigenvalues, for subgroups $\Gamma \subset SL(2,\mathbb{Z})$ not covered by Selberg's conjecture. Steve Miller alerted me to the special role played by Γ_2, the principal congruence subgroup of level 2. Dick Gross, finally, introduced me to the literature on functional equations for L-functions attached to automorphic representations of higher rank groups.

2. The main technical result

I begin by establishing notation that will be used throughout the paper. I always denote the ambient group by $G_{\mathbb{R}}$ and its maximal compact subgroup by $K_{\mathbb{R}}$:

$$G_{\mathbb{R}} = SL(2,\mathbb{R}), \qquad K_{\mathbb{R}} = SO(2). \tag{2.1}$$

As in the introduction, I fix a subgroup Γ,

$$\Gamma \subset G_{\mathbb{R}} \qquad \text{(finitely generated discrete subgroup)}, \tag{2.2}$$

a cocompact discrete subgroup, for example, or an arithmetically defined subgroup. Formally $\Gamma = \{e\}$ is allowed, but the results will be essentially obvious in this extreme case. Other subgroups that will be important:

$$\begin{aligned} N_{\mathbb{R}} &= \{n_x \mid x \in \mathbb{R}\}, & n_x &= \begin{pmatrix} 1 & x \\ 0 & 1 \end{pmatrix} \\ N_{\mathbb{R}}^- &= \{n_y^- \mid y \in \mathbb{R}\}, & n_y^- &= \begin{pmatrix} 1 & 0 \\ y & 1 \end{pmatrix} \\ A_{\mathbb{R}} &= \{a_t \mid t \in \mathbb{R}\}, & a_t &= \begin{pmatrix} e^t & 0 \\ 0 & e^{-t} \end{pmatrix}. \end{aligned} \tag{2.3}$$

The group $A_{\mathbb{R}}$ normalizes both $N_{\mathbb{R}}$ and $N_{\mathbb{R}}^-$, and

$$(2.4) \qquad G_{\mathbb{R}} = K_{\mathbb{R}} \cdot A_{\mathbb{R}} \cdot N_{\mathbb{R}}^-$$

is an Iwasawa decomposition of $G_{\mathbb{R}}$. I denote the Lie algebras of these groups by the corresponding German letters with the subscript $\mathbb{R}$, such as $\mathfrak{g}_{\mathbb{R}}$ and $\mathfrak{k}_{\mathbb{R}}$, and the complexified Lie algebras by the same letters without the subscript $\mathbb{R}$.

Principal series representations of $G_{\mathbb{R}}$, unitary or not, are parametrized by a complex number and a parity. Specifically, for $\lambda \in \mathbb{C}$ and $\varepsilon \in \{\pm 1\}$, I define

$$(2.5) \qquad \begin{aligned} V^{\infty}_{\lambda,\varepsilon} &= \text{space of all } f \in C^{\infty}(G_{\mathbb{R}}) \text{ such that} \\ & f((-1_{2\times 2})\, g\, a_t\, n^-) = \varepsilon\, e^{-(\lambda-1)t} f(g), \\ & \qquad \text{for all } g \in G_{\mathbb{R}},\ t \in \mathbb{R},\ n^- \in N_{\mathbb{R}}^-. \end{aligned}$$

The group $G_{\mathbb{R}}$ acts by left translation since $-1_{2\times 2}$ lies in the center of $G_{\mathbb{R}}$:

$$(2.6) \qquad \left(\pi_{\lambda,\varepsilon}(h) f\right)(g) = f(h^{-1}g), \quad \text{for } f \in V^{\infty}_{\lambda,\varepsilon},\ g,h \in G_{\mathbb{R}}.$$

Because of (2.4), restriction to $K_{\mathbb{R}}$ induces an isomorphism of $K_{\mathbb{R}}$-representations

$$(2.7) \qquad V^{\infty}_{\lambda,\varepsilon} \simeq \{ f \in C^{\infty}(K_{\mathbb{R}}) \mid f((-1_{2\times 2})k) \equiv \varepsilon f(k) \}.$$

Via the usual identification $\mathbb{R} \cup \{\infty\} \simeq G_{\mathbb{R}} / \left((\pm 1_{2\times 2}) A_{\mathbb{R}} N_{\mathbb{R}}^-\right)$, the upper triangular group N acts freely on $\mathbb{R} \simeq N$-orbit through the identity coset. Thus restriction from $G_{\mathbb{R}}$ to $N \simeq \mathbb{R}$ identifies $V^{\infty}_{\lambda,\varepsilon}$ with a subspace of $C^{\infty}(\mathbb{R})$, namely

$$(2.8) \qquad V^{\infty}_{\lambda,\varepsilon} \simeq \{ \phi \in C^{\infty}(\mathbb{R}) \mid (\operatorname{sgn} x)^{\frac{1-\varepsilon}{2}} |x|^{\lambda-1} \phi(\tfrac{1}{x}) \in C^{\infty}(\mathbb{R}) \};$$

to derive the particular form of the condition at infinity, one factors $n_x \in N$ as an Iwasawa product $n_x = k\, a_t\, n^-$ and appeals to the definition (2.5). In terms of the isomorphism (2.8), the action (2.6) can be written as follows:

$$(2.9) \qquad \begin{aligned} \left(\pi_{\lambda,\varepsilon}(g)\phi\right)(x) &= (\operatorname{sgn}(cx+d))^{\frac{1-\varepsilon}{2}} |cx+d|^{\lambda-1} \phi(\tfrac{ax+b}{cx+d}) \\ & \text{when } g^{-1} = \begin{pmatrix} a & b \\ c & d \end{pmatrix} \in G_{\mathbb{R}}. \end{aligned}$$

This, too, can be deduced from (2.5) when one factors $g^{-1} n_x$ as an Iwasawa product.

The space $V^{\infty}_{\lambda,\varepsilon}$ coincides with the space of C^{∞} vectors in any Banach realization of the underlying Harish-Chandra module – in effect, this is the specialization to $G_{\mathbb{R}}$ of a general result of Casselman-Wallach [6]. Dually the space of distribution vectors $V^{-\infty}_{\lambda,\varepsilon}$ corresponds, via restriction to $K_{\mathbb{R}}$, to the space of distributions on $K_{\mathbb{R}}$ that satisfy the parity condition specified by ε. I use the word "distribution" to refer to objects *dual to compactly supported smooth measures*, so

(2.10) distributions behave like functions under diffeomorphisms;

in particular, functions are distributions. With this convention, (2.5–9) remain valid when one replaces "C^∞ function" by "distribution" and the superscript ∞ by $-\infty$. Integration over $K_{\mathbb{R}}$ sets up a $G_{\mathbb{R}}$-invariant, nondegenerate pairing between $V^\infty_{\lambda,\varepsilon}$ and $V^\infty_{-\lambda,\varepsilon}$, hence

$$\text{(2.11)} \qquad V^{-\infty}_{\lambda,\varepsilon} \text{ is the continuous dual of } V^\infty_{-\lambda,\varepsilon}.$$

When $V^\infty_{\lambda,\varepsilon}$ is irreducible – equivalently, when $\lambda \notin 2\mathbb{Z} + \frac{1}{2}(\varepsilon+1)$ – the two $G_{\mathbb{R}}$-modules $V^\infty_{\lambda,\varepsilon}$ and $V^\infty_{-\lambda,\varepsilon}$ are isomorphic via the canonical intertwining operator; in that case, $V^{-\infty}_{\lambda,\varepsilon}$ is also isomorphic to the continuous dual of $V^\infty_{\lambda,\varepsilon}$ as a topological $G_{\mathbb{R}}$-module. However, the duality between $V^\infty_{\lambda,\varepsilon}$ and $V^{-\infty}_{\lambda,\varepsilon}$ involves the intertwining operator and is *not* given by the integration pairing over $K_{\mathbb{R}}$.

The statement of the main theorem involves a notion of fractional differentiability. To motivate the definition, I recall a classical theorem of Serge Bernstein [2]; for a modern account, see for example Folland [9]. Let ϕ be a continuous function on $\mathbb{R}/\mathbb{Z}$, with Fourier series

$$\text{(2.12)} \qquad \phi(x) = \Sigma_{-\infty<k<\infty}\, a_k e^{2\pi i k x}.$$

When ϕ is Hölder continuous of index α, $0 < \alpha \leq 1$,

$$\text{(2.13)} \qquad \begin{array}{llll} \text{a)} & s < 2\alpha & \Longrightarrow & \Sigma_{k\neq 0}\, |k|^s\, |a_k|^2 < \infty \\ \text{b)} & s < \alpha - \frac{1}{2} & \Longrightarrow & \Sigma_{k\neq 0}\, |k|^s\, |a_k| < \infty. \end{array}$$

Thus, when α is strictly greater than $\frac{1}{2}$, the Fourier series of ϕ converges absolutely. This latter statement is usually called Bernstein's theorem, but the proof readily extends to give the more general statement above. In view of (2.13a), Hölder continuity can be regarded as the analogue, in the context of continuous functions, of the L^2 notion of Sobolev regularity. The Hölder notion provides a finer measure of regularity than the Sobolev notion: Hölder continuity of any index $\alpha > \frac{1}{2}$, for example, implies finiteness of the Sobolev-$\frac{1}{2}$ norm; but conversely, it requires finiteness of the Sobolev-1 norm – i.e., the existence of an L^2 derivative – to ensure Hölder continuity of index $\frac{1}{2}$.

2.14 Definition. *A distribution τ, defined on an open interval $I \subset \mathbb{R}$ or on the circle $\mathbb{R}/\mathbb{Z}$, is of class C^α, $\alpha \in \mathbb{R}$, provided*
i) for $\alpha = k$ a non-negative integer, τ is a C^k function in the usual sense;
ii) for $\alpha = -k$ a negative integer, $\tau = \phi^{(k)}$ is the k-th derivative of a continuous function ϕ;
iii) for $\alpha = k + \beta$ with k non-negative integral and $0 < \beta < 1$, τ has k continuous derivatives, and the k-th derivative $\tau^{(k)}$ is locally Hölder continuous of index β;
iv) for $\alpha = -k + \beta$ with k positive integral and $0 < \beta < 1$, $\tau = \phi^{(k)}$ is the k-th derivative of a function ϕ that is locally Hölder continuous of index β.
By convention, $C^{<\alpha}(I)$ denotes the intersection $\bigcap_{\beta<\alpha} C^\beta(I)$.

In several respects, this notion of fractional differentiability behaves like the usual notion of differentiability; in particular,

(2.15)

a) $C^{\alpha}(I) \subset C^{<\alpha}(I) \subset C^{\beta}(I)$ if $\alpha > \beta$;

b) $\tau \in C^{\alpha}(I) \implies \tau' \in C^{\alpha-1}(I)$;

c) the implications (2.13a,b) remain valid for any $\alpha \in \mathbb{R}$ and any $\phi \in C^{\alpha}(\mathbb{R}/\mathbb{Z})$;

d) $\phi \in C^{\infty}(I)$, $\tau \in C^{\alpha}(I) \implies \phi\tau \in C^{\alpha}(I)$;

e) $F : I \xrightarrow{\sim} J$ diffeomorphically, $\tau \in C^{\alpha}(J) \implies F^{*}\tau \in C^{\alpha}(I)$.

Statements a)-c) hold for essentially formal reasons, as do d) and e) when $\alpha \geq 0$. For the general version of d) and e), one can argue by downward induction, using Leibniz' rule.

Now let (π, V) be an irreducible unitary representation of $G_{\mathbb{R}}$ which occurs as a direct summand in $L^2(\Gamma\backslash G_{\mathbb{R}})$. I define the Γ-invariant distribution vector $\tau \in (V^{-\infty})^{\Gamma}$ as in (1.1–3). Like any irreducible unitary representation, one can realize π as a subrepresentation of a not-necessarily-unitary principal series representation $\pi_{\lambda,\varepsilon}$, and correspondingly the dual representation π' as a quotient of the dual principal series representation $\pi_{-\lambda,\varepsilon}$. On the level of distribution vectors, $V^{-\infty} \hookrightarrow V_{\lambda,\varepsilon}^{-\infty}$, for some appropriately chosen $\lambda \in \mathbb{C}$, $\varepsilon \in \{\pm 1\}$. Thus

$$(V^{-\infty})^{\Gamma} \ni \tau \mapsto \tau_{\lambda} \in (V_{\lambda,\varepsilon}^{-\infty})^{\Gamma}. \tag{2.16}$$

Via the distribution analogue of (2.7), τ now determines a distribution on the circle:

$$\tau_{\lambda} \in C^{-\infty}(\mathbb{R}/\mathbb{Z}). \tag{2.17}$$

2.18 Theorem. *a) If the embedding (1.1) is cuspidal and λ not an odd integer, the distribution τ_{λ} lies in $C^{\frac{1}{2}(\operatorname{Re}\lambda-1)}(\mathbb{R}/\mathbb{Z})$.*
b) If $\lambda \in 2\mathbb{Z}+1$, in which case the embedding is cuspidal, τ_{λ} lies in $C^{<\frac{1}{2}(\lambda-1)}(\mathbb{R}/\mathbb{Z})$.
c) If the embedding is not cuspidal, τ_{λ} lies in $C^{\lambda-1}(\mathbb{R}/\mathbb{Z})$ or in $C^{<(-1)}(\mathbb{R}/\mathbb{Z})$, depending on whether $0<\lambda<1$ or $-1<\lambda<0$.

The distribution vector $\tau \in (V^{-\infty})^{\Gamma}$ corresponds also to a distribution $\tilde{\tau}_{\lambda} \in C^{-\infty}(\mathbb{R})$, via the distribution analogue of the isomorphism (2.8). Going from the "compact realization" (2.7) of $V_{\lambda,\varepsilon}^{\infty}$ to the "unbounded realization" (2.8) involves the coordinate change $x = \tan\theta$ and multiplication by a C^{∞} function. The same process – on the level of distributions – relates $\tilde{\tau}_{\lambda}$ to τ_{λ}. Thus, in view of (2.15), the theorem implies also the analogous statement about $\tilde{\tau}$:

2.19 Corollary. *The distribution $\tilde{\tau}_\lambda$ corresponding to a cuspidal embedding lies in $C^{\frac{1}{2}(\mathrm{Re}\lambda-1)}(\mathbb{R})$ when $\lambda \notin 2\mathbb{Z}+1$, and in $C^{<\frac{1}{2}(\lambda-1)}(\mathbb{R})$ when $\lambda \in 2\mathbb{Z}+1$. If the embedding is non-cuspidal and $0<\lambda<1$, then $\tilde{\tau}_\lambda \in C^{\lambda-1}(\mathbb{R})$ and $\tilde{\tau}_{-\lambda} \in C^{<(-1)}(\mathbb{R})$.*

Bernstein-Reznikov [3, 4] show that the distribution vector $\tau \in V^\infty$ corresponding to a cuspidal embedding $V' \hookrightarrow L^2(\Gamma\backslash G_\mathbb{R})$ lies in the Sobolev-α space for any $\alpha < -\frac{1}{2}$, relative to the Sobolev notion which they define in terms of the unitary structure of the Hilbert space V. When (π, V) belongs to the unitary principal series, realized as in (2.7), Bernstein-Reznikov's Sobolev notion agrees precisely with the usual Sobolev notion on the circle group $K_\mathbb{R}$. In this situation, part a) of theorem 2.18, coupled with Serge Bernstein's result (2.13a), directly implies Bernstein-Reznikov's regularity statement. For other unitary representations, theorem 2.18 and (2.13a) also imply the Bernstein-Reznikov result, but the translation process requires pinning down the relationship of the unitary structure of V and the L^2 structure of $K_\mathbb{R}$, via the embedding $V^{-\infty} \hookrightarrow V^{-\infty}_{\lambda,\varepsilon}$ that relates τ to τ_λ.

To what extent does theorem 2.18 have a converse? If $\pi_{\lambda,\varepsilon}$ is a unitary principal series representation and f_v the function on $\mathbb{R}/\mathbb{Z}$ corresponding to a vector $v \in V_{\lambda,\varepsilon}$ in the (unitary) representation space via (2.7), then $f_v \in L^2(\mathbb{R}/\mathbb{Z})$. The indefinite integral of an L^2 function is Hölder continuous of index $\frac{1}{2}$, as follows readily from the Schwarz inequality. This makes f_v a distribution of class $C^{\frac{1}{2}(\mathrm{Re}\lambda-1)}$, since $\lambda \in i\mathbb{R}$ by assumption. No principal series representation occurs discretely in $L^2(G_\mathbb{R})$, so the converse of theorem 2.18 fails for $\Gamma = \{e\}$ – but not by much, since $\pi_{\lambda,\varepsilon}$ does occur continuously. The failure reflects the fact that the trivial subgroup is "too small". To make this precise, I consider an arbitrary irreducible unitary representation (π, V) and a non-zero, Γ-invariant distribution vector $\tau \in (V^{-\infty})^\Gamma$. As before, I embed V^∞ into the space of C^∞ vectors of a (not necessarily unitary) principal series representation $V_{\lambda,\varepsilon}$, and let $\tau_\lambda \in C^{-\infty}(\mathbb{R}/\mathbb{Z})$ denote the image of τ via $V^{-\infty} \hookrightarrow V^{-\infty}_{\lambda,\varepsilon}$ and the distribution analogue of the isomorphism (2.7).

2.20 Theorem. *Let Γ be a finitely generated, discrete subgroup of finite co-volume.*
a) If $\tau_\lambda \in C^{<\frac{1}{2}(\mathrm{Re}\lambda-1)}(\mathbb{R}/\mathbb{Z})$, the distribution vector τ arises from a cuspidal embedding $V' \hookrightarrow L^2(\Gamma\backslash G_\mathbb{R})$ as in (1.1–3).
b) For $0<\lambda<1$, τ corresponds to a possibly non-cuspidal embedding $V' \hookrightarrow L^2(\Gamma\backslash G_\mathbb{R})$ provided $\tau_\lambda \in C^{<(\lambda-1)}(\mathbb{R}/\mathbb{Z})$ or $\tau_{-\lambda} \in C^{<(-1)}(\mathbb{R}/\mathbb{Z})$.

Note that $\tau_\lambda \in C^{<\alpha}(\mathbb{R}/\mathbb{Z})$ is a weaker condition than $\tau_\lambda \in C^\alpha(\mathbb{R}/\mathbb{Z})$; in fact, the proof will show that there is even more room to spare between the conclusion of theorem 2.18 and the hypotheses of theorem 2.20.

I shall take up a number of consequences of theorem 2.18, before turning to its proof in section 5, along with that of its converse 2.20.

3. Maass forms

Any non-zero, square-integrable Maass form with eigenvalue $\frac{1}{4}(1-\lambda^2)$ corresponds to an embedding of the even principal series or complementary series representation $V_{\lambda,+1}$ into $L^2(\Gamma\backslash G_{\mathbb{R}})$, and thus to a Γ-invariant distribution vector $\tau_\lambda \in V_{\lambda,+1}^{-\infty}$. I let $\tilde{\tau}_\lambda \in C^{-\infty}(\mathbb{R})$ denote the distribution obtained from τ_λ via the distribution analogue of the isomorphism (2.8). The proof of theorem 2.18 will show that $\tilde{\tau}_\lambda$ coincides with the distribution $\tilde{\tau}_{\lambda,0}$ in (1.9). I can assume

$$(3.1)\qquad \begin{aligned} &\text{either} \quad \text{a) the embedding } V_{\lambda,+1} \hookrightarrow L^2(\Gamma\backslash G_{\mathbb{R}}) \text{ is cuspidal and } |\mathrm{Re}\lambda| < 1 \\ &\qquad \text{or} \quad \text{b) } \lambda \in \mathbb{R} \text{ and } 0 < \lambda < 1, \end{aligned}$$

without essential loss of generality: Because $V_{\lambda,+1} \cong V_{-\lambda,+1}$, there are two distribution vectors τ_λ and $\tau_{-\lambda}$ – except when $\lambda = 0$ – but they are related by the standard intertwining operator; in the non-cuspidal case, λ is necessarily real, of absolute value strictly between 0 and 1, so b) amounts to a definite choice between τ_λ and $\tau_{-\lambda}$. In any case, for representations of the unitary principal series or complementary series, $\mathrm{Re}\lambda$ is strictly bounded by 1. The reason for the assumption (3.1) is that it implies the existence of a continuous first anti-derivative for $\tilde{\tau}_\lambda$,

$$(3.2)\qquad \tilde{\tau}_\lambda = \phi_\tau', \qquad \phi_\tau \in C^0(\mathbb{R}),$$

whether or not the embedding is cuspidal – see corollary 2.19. The function ϕ_τ is determined only up to an additive constant; however, in the most important special cases, there is a canonical choice, as we shall see.

The transformation law (2.9) for the action of $G_{\mathbb{R}}$ on the even principal series representation $V_{\lambda,+1}$ extends naturally to

$$(3.3)\qquad G_{\mathbb{R}}^{\pm} = SL^{\pm}(2,\mathbb{R}) = \{g \in GL(2,\mathbb{R}) \mid \det g = \pm 1\},$$

which normalizes $G_{\mathbb{R}}$. In fact, $PGL(2,\mathbb{R}) \cong SL^{\pm}(2,\mathbb{R})/\{\pm 1_{2\times 2}\}$ acts on $V_{\lambda,+1}$, but many of the interesting discrete groups Γ lie in $SL^{\pm}(2,\mathbb{R})$ rather than $PGL(2,\mathbb{R})$. The transformation law applies in particular to the distribution $\tilde{\tau}_\lambda$, except at the singularity of the factor of automorphy. To simplify the notation, I write the action of $G_{\mathbb{R}}^{\pm}$ on $V_{\lambda,+1}$ by juxtaposition. With this convention,

$$(3.4)\qquad \begin{aligned} &(g\tilde{\tau})_\lambda(x) = |cx+d|^{\lambda-1}\tilde{\tau}_\lambda\left(\tfrac{ax+b}{cx+d}\right) \qquad (x \neq -\tfrac{d}{c}) \\ &\qquad\qquad \text{when } g^{-1} = \begin{pmatrix} a & b \\ c & d \end{pmatrix} \in G_{\mathbb{R}}^{\pm}. \end{aligned}$$

The distribution vector $g\tau \in V_{\lambda,+1}^{-\infty}$ corresponds to an embedding of $V_{\lambda,+1}$ into $L^2(\Gamma^g\backslash G_{\mathbb{R}})$, with $\Gamma^g = g\Gamma g^{-1}$. It is therefore of the same type as τ, so

$$(3.5)\qquad g\tilde{\tau}_\lambda = \phi_{g\tau}', \qquad \phi_{g\tau} \in C^0(\mathbb{R}).$$

When the transformation law (3.4) is integrated, it takes the form

$$
(3.6)\qquad \begin{aligned} \phi_{g\tau}(x) \ &= \ \det g\,|cx+d|^{\lambda+1}\phi_\tau(\tfrac{ax+b}{cx+d}) \ + \\ &+ \ (\lambda+1)\, c\,\operatorname{sgn}(cx+d)\,\det g \int^{\frac{ax+b}{cx+d}} |a-ct|^{-\lambda-2}\,\phi_\tau(t)\,dt \ + \ \text{const.}; \end{aligned}
$$

here $g \in G^{\pm}_{\mathbb{R}}$ is the same matrix as in (3.4). What matters here is the fact that the coordinate change $t = \frac{ax+b}{cx+d}$ has derivative $\det g\,(cx+d)^{-2} = \det g\,(a-ct)^2$.

The equation (3.6) can be made more explicit for certain groups Γ, in particular for subgroups of finite index in $SL(2,\mathbb{Z})$. Any such subgroup contains another subgroup of finite index which is normal in $SL^{\pm}(2,\mathbb{Z})$. Thus, for the remainder of this section, I shall assume

$$
(3.7)\qquad \Gamma \subset SL^{\pm}(2,\mathbb{Z}) \ \text{ is a normal subgroup of finite index.}
$$

The quotient group $SL^{\pm}(2,\mathbb{Z})/\Gamma$ acts on the spaces

$$
(3.8)\qquad \begin{aligned} &\{\tau \in (V_{\lambda,+1}^{-\infty})^{\Gamma} \mid \tau_\lambda \in C^{\frac{1}{2}(\operatorname{Re}\lambda-1)}(\mathbb{R}/\mathbb{Z})\} && (\lambda \text{ arbitrary}) \\ &\{\tau \in (V_{\lambda,+1}^{-\infty})^{\Gamma} \mid \tau_\lambda \in C^{\lambda}(\mathbb{R}/\mathbb{Z})\} && (0<\lambda<1); \end{aligned}
$$

cf. (3.1). According to theorem 2.20, every non-zero τ in either of these two spaces corresponds to an embedding $V_{\lambda,+1} \hookrightarrow L^2(\Gamma\backslash G^{\pm}_{\mathbb{R}})$.

The cyclic group $\{n_k \mid k \in \mathbb{Z}\}$, in the notation of (2.3), maps to a – necessarily finite – subgroup of $SL^{\pm}(2,\mathbb{Z})/\Gamma$, so there exists a least positive integer m such that $n_m \in SL^{\pm}(2,\mathbb{Z})$. This makes the distribution $\tilde{\tau}_\lambda$ periodic of period m. It therefore has a Fourier series

$$
(3.9)\qquad \tilde{\tau}_\lambda(x) \ = \ \sum\nolimits_{-\infty<k<\infty} a_k\, e^{2\pi i \frac{k}{m} x},
$$

which can be integrated term-by-term:

$$
(3.10)\qquad \phi_\tau(x) \ = \ a_0 x \ + \ \sum\nolimits_{k\neq 0} \tfrac{m}{2\pi i k}\, a_k\, e^{2\pi i \frac{k}{m} x}.
$$

The coefficient $a_0 = a_{\tau,0}$ has special significance:

$$
(3.11)\qquad \begin{aligned} &\text{the embedding } V_{\lambda,+1} \hookrightarrow L^2(\Gamma\backslash G^{\pm}_{\mathbb{R}}) \text{ defined by } \tau \text{ is cuspidal} \\ &\qquad\qquad \text{if and only if } a_{\gamma\tau,0} = 0, \ \text{ for all } \gamma \in SL(2,\mathbb{Z}). \end{aligned}
$$

Indeed, because of (3.7), all the cusps of Γ occur at rational points or infinity – in other words, at points of the $SL(2,\mathbb{Z})$-orbit of infinity. We shall see that $a_{\tau,0} = 0$ is the cuspidality condition at ∞; see lemma 5.52 and the proof of lemma 5.23, in particular, (5.29). Thus $a_{\gamma,0} = 0$ for all $\gamma \in SL(2,\mathbb{Z})$ precisely when the cuspidality condition is satisfied at every cusp.

I shall take a more detailed look at the cuspidal case. The antiderivatives $\phi_{\gamma\tau}$ of the various $SL^{\pm}(2,\mathbb{Z})$-translates $\gamma\tau$ are then periodic with vanishing constant term,

$$\phi_{\gamma\tau}(x) = \sum_{k\neq 0} \tfrac{m}{2\pi i k}\, a_{\gamma\tau,k}\, e^{2\pi i \frac{k}{m} x} \qquad (\gamma\in SL^{\pm}(2,\mathbb{Z}))\,. \tag{3.12}$$

By the very definition of $V_{\lambda,+1}$, $\{\pm 1_{2\times 2}\}$=center of $G^{\pm}_{\mathbb{R}}$ acts trivially. Since the square of the matrix

$$w =_{\text{def}} \begin{pmatrix} 0 & 1 \\ -1 & 0 \end{pmatrix} \tag{3.13}$$

is the negative of the identity, $w^2\tau$ equals τ. Thus τ and $w\tau$ play symmetric roles. I claim

$$\phi_\tau(x) - \phi_\tau(0) = \phi_{w\tau}(-\tfrac{1}{x})\,|x|^{\lambda+1} - (\lambda+1)\int_{\frac{1}{|x|}}^{+\infty} t^{-\lambda-2}\,\phi_{w\tau}(-\operatorname{sgn}(x)t)\,dt\,. \tag{3.14}$$

To verify this identity, note that the integral converges and that both sides coincide when $x=0$. One can then appeal to (3.6), or directly to (3.4) by differentiating both sides. At first glance, the two terms on the right have the same order of magnitude. However,

$$(\lambda+1)\int_{\frac{1}{|x|}}^{+\infty} t^{-\lambda-2}\,\phi_{w\tau}(-\operatorname{sgn}(x)t)\,dt = O(|x|^{\operatorname{Re}\lambda+2}) \qquad \text{as } x\to 0, \tag{3.15}$$

because $\phi_{w\tau}$ is periodic with zero constant term. The leading term on the right oscillates, with amplitude and frequency proportional to $|x|^{\operatorname{Re}\lambda+1}$ and $|x|^{-2}$, respectively. This implies

$$\limsup_{a,b\to 0, a\neq b}\left(|\phi_\tau(a)-\phi_\tau(b)|\,|a-b|^{-\frac{\operatorname{Re}\lambda+1}{2}}\right) > 0\,. \tag{3.16}$$

In particular, ϕ_τ cannot be Hölder continuous of index strictly greater than $\frac{\operatorname{Re}\lambda+1}{2}$, showing that (2.18a) is sharp.

For $j>0$, let $\phi_\tau^{(-j)}$ denote the j-th anti-derivative of ϕ_τ, normalized by the requirement that all the $\phi_\tau^{(-j)}$ shall be periodic with zero constant term. Repeated integration by parts in (3.14) results in the identity

(3.17)
$$\begin{aligned}\phi_\tau(x) - \phi_\tau(0) &= \sum_{j=0}^{n}(-\operatorname{sgn}(x))^j\left(\prod\nolimits_{1\le i\le j}(\lambda+i)\right)\phi_{w\tau}^{(-j)}(-\tfrac{1}{x})\,|x|^{\lambda+j+1}\\ &\quad - (-\operatorname{sgn}(x))^n\left(\prod\nolimits_{0\le i\le n}(\lambda+i+1)\right)\int_{\frac{1}{|x|}}^{+\infty} t^{-\lambda-n-2}\,\phi_{w\tau}^{(-n)}(-\operatorname{sgn}(x)t)\,dt\,.\end{aligned}$$

Here, too, the last term has strictly lower order of magnitude than the others. In effect, (3.17) can be regarded as an asymptotic expansion of $\phi_\tau(x)-\phi_\tau(0)$ as $x\to 0$, with coefficients that oscillate.

The function ϕ_τ has the same qualitative behavior near all rational points – not surprisingly, since all the rational points correspond to cusps. I consider a particular rational number $\frac{p}{q}$, with p and q relatively prime and $q > 0$. Then there exists a matrix

$$\gamma = \begin{pmatrix} r & -s \\ -q & p \end{pmatrix} \in SL(2,\mathbb{Z}), \tag{3.18}$$

which sends the point $\frac{p}{q}$ to ∞. With this notation,

(3.19)

$$\begin{aligned} \phi_\tau(x) - \phi_\tau(\tfrac{p}{q}) \;&=\; \phi_{\gamma\tau}(\gamma x)\, q^{\lambda+1}\, |x - \tfrac{p}{q}|^{\lambda+1} \\ &- q(\lambda+1)\int_{\operatorname{sgn}(p-qx)\gamma x}^{+\infty} (qt + r\operatorname{sgn}(p-qx))^{-\lambda-2}\, \phi_{\gamma\tau}(\operatorname{sgn}(p-qx)\,t)\, dt\,. \end{aligned}$$

The verification is essentially the same as that of (3.14); one needs to observe that the expression $(qt + r\operatorname{sgn}(p-qx))$ takes the value $|qx-p|$ at the left endpoint of the interval of integration, and is therefore strictly positive over the entire interval – recall that $q > 0$. The second term on the right has strictly lower order of magnitude as $x \to \frac{p}{q}$, just as in the case of (3.14). Since

$$\gamma x \;=\; -\tfrac{1}{q^2}\left(1 - rq(x - \tfrac{p}{q})\right)\left(x - \tfrac{p}{q}\right)^{-1} \;\sim\; -\tfrac{1}{q^2}\left(x - \tfrac{p}{q}\right)^{-1} \quad \text{as } x \to \tfrac{p}{q}, \tag{3.20}$$

the dominating first term oscillates near $\frac{p}{q}$, with increasing frequency and decreasing amplitude as x approaches $\frac{p}{q}$. More precisely,

(3.21)

as x approaches $\frac{p}{q}$, $\phi_\tau(x) - \phi_\tau(\frac{p}{q})$ oscillates, with frequency proportional to $|qx-p|^{-2}$ and amplitude proportional to $|qx-p|^{\operatorname{Re}\lambda+1}$.

Though the oscillations have larger amplitude in the small as the denominator q increases, they also die out more quickly for larger q, as x moves away from $\frac{p}{q}$. The asymptotic expansion (3.17) has an analogue in the present setting, but we shall not pursue this further.

The identity (3.14) can be used to give elementary proofs of certain known facts. I consider an $SL^{\pm}(2,\mathbb{Z})$-invariant subspace

$$U \subset \{\tau \in (V_{\lambda,+1}^{-\infty})^{\Gamma} \mid \tau_\lambda \in C^{\frac{1}{2}(\operatorname{Re}\lambda-1)}(\mathbb{R}/\mathbb{Z})\}, \tag{3.22}$$

the entire ambient space, for example, or an $SL^{\pm}(2,\mathbb{Z})$-irreducible subspace. The ambient space is known to be finite dimensional – the discrete part of $L^2(\Gamma\backslash G_{\mathbb{R}})$ breaks up with finite multiplicities – but this will become clear momentarily. Thus, to keep the argument logically self-contained, I expressly assume $\dim U < \infty$. Since $SL^{\pm}(2,\mathbb{Z})$ acts on U via a finite quotient group, I can introduce an invariant inner product. Let $\tau_1, \tau_2, \dots, \tau_n$ be an orthonormal basis of U, and let $\tilde{\tau}_j \in C^{-\infty}(\mathbb{R})$ denote the image of τ_j under the distribution analogue of the isomorphism (2.8). I write $\Phi(x)$, $\Phi_w(x)$ for the column vectors with entries $\phi_{\tau_j}(x)$ and $\phi_{w\tau_j}(x)$, respectively:

$$\Phi(x) = \begin{pmatrix} \phi_{\tau_1}(x) \\ \phi_{\tau_2}(x) \\ \cdot \\ \cdot \\ \cdot \\ \phi_{\tau_n}(x) \end{pmatrix}, \qquad \Phi_w(x) = \begin{pmatrix} \phi_{w\tau_1}(x) \\ \phi_{w\tau_2}(x) \\ \cdot \\ \cdot \\ \cdot \\ \phi_{w\tau_n}(x) \end{pmatrix}. \tag{3.23}$$

Translation commutes with the integration process which produces the ϕ_{τ_j} from the $\tilde{\tau}_j$. Since $(n_1\tilde{\tau}_j)(x) = \tilde{\tau}_j(x-1)$, $\Phi(x-1) = A(n_1)\Phi(x)$, where $A(n_1)$ is the unitary matrix which represents the action of $n_1 \in SL^{\pm}(2,\mathbb{Z})$ on U, relative to the basis $\{\tau_1,\ldots,\tau_n\}$. The reflection $x \mapsto -x$ reverses the sign when one takes the anti-derivative of $\tilde{\tau}_j$, so $\Phi(-x) = -A(r)\Phi(x)$, where $A(r)$ represents reflection about the origin. Similarly $\tilde{\Phi}(x) = A(w)\Phi(x)$, with $A(w)$ representing the action of w. For future reference,

$$\begin{gathered} \Phi(x-1) = A(n_1)\Phi(x), \quad \Phi(-x) = -A(r)\Phi(x), \quad \Phi_w(x) = A(w)\Phi(x), \\ \text{with } A(n_1),\, A(r),\, A(w) \in U(m) \text{ representing} \\ n_1 = \begin{pmatrix} 1 & 1 \\ 0 & 1 \end{pmatrix}, \quad r = \begin{pmatrix} -1 & 0 \\ 0 & 1 \end{pmatrix}, \quad w = \begin{pmatrix} 0 & 1 \\ -1 & 0 \end{pmatrix}. \end{gathered} \tag{3.24}$$

The pointwise norm $\|\Phi(x)\|$ of the column vector $\Phi(x)$ is therefore invariant under reflection and translation by one, and is equal to $\|\Phi_w(x)\|$. In particular, it assumes its maximum at some point $x_{\max}$, $0 \le x_{\max} \le \frac{1}{2}$. Thus, using (3.14) component-by-component, I find

(3.25)

$$\begin{aligned} \sup\|\Phi(x)\| &= \|\Phi(x_{\max})\| \\ &\le \|\Phi(0)\| + \sup\|\Phi(x)\| \left(x_{\max}^{\mathrm{Re}\,\lambda+1} + |1+\lambda| \int_{\frac{1}{x\max}}^{+\infty} t^{-\mathrm{Re}\,\lambda-2}\, dt \right) \\ &\le \|\Phi(0)\| + x_{\max}^{\mathrm{Re}\,\lambda+1} \left(1 + \tfrac{|1+\lambda|}{|1+\mathrm{Re}\,\lambda|} \right) \sup\|\Phi(x)\|. \end{aligned}$$

When λ is positive and $\Phi(0) = 0$, this cannot be – even for $\lambda = 0$, since the oscillatory behavior makes the inequality strict unless $x_{\max} = 0$. This implies the crude bound

(3.26)

$$\dim\{\, \tau \in (V_{\lambda,+1}^{-\infty})^{\Gamma} \mid \tau_\lambda \in C^{\frac{1}{2}(\mathrm{Re}\,\lambda-1)}(\mathbb{R}/\mathbb{Z}) \,\} \;\le\; \#(SL^{\pm}(2,\mathbb{Z})/\Gamma) \quad \text{if } \lambda \in \mathbb{R}.$$

Indeed, if the bound on the dimension were to fail, one could construct an $SL^{\pm}(2,\mathbb{Z})$-invariant subspace U consisting of distribution vectors τ such that $\phi_\tau(0) = 0$. One can also assume $\lambda \ge 0$ without loss of generality, in which case

(3.25) gives the desired contradiction. For imaginary λ, one can argue similarly, using (3.19) instead of (3.14), but the resulting bounds are even less precise.

A well known conjecture of Selberg asserts that representations of the complementary series do not occur as summands in $L^2(\Gamma\backslash G_{\mathbb{R}})$ when Γ is one of the principal congruence subgroups of $SL(2,\mathbb{R})$. Non-cuspidal summands, if they were to occur, would correspond to residues of Eisenstein series. In the case of principal congruence subgroups, Eisenstein series are known not to have poles in the relevant region. Thus Selberg's conjecture is equivalent to the vanishing of the space

$$(3.27) \qquad \{\tau \in (V_{\lambda,+1}^{-\infty})^\Gamma \mid \tau_\lambda \in C^{\frac{1}{2}(\mathrm{Re}\lambda-1)}(\mathbb{R}/\mathbb{Z})\} \qquad 0<\lambda<1;$$

cf. theorem 2.20. I shall give a quick argument for the vanishing of this space when $\Gamma = \Gamma_2$, the principal congruence subgroup of level 2, even including $\lambda = 0$. Not that this is new – I merely wish to illustrate further how (3.14) can be used.

The reflection r, which does not lie in $SL(2,\mathbb{Z})$, commutes with w. Also $rn_1r^{-1}n_1^{-1} = n_{-2} \in \Gamma_2$, so r commutes with n_1 modulo Γ_2. Since w and n_1 generate $SL(2,\mathbb{Z})$, I conclude that the image of r in $SL^{\pm}(2,\mathbb{Z})/\Gamma_2$ commutes with

$$(3.28) \qquad SL(2,\mathbb{Z})/\Gamma_2 \;=\; SL(2,\mathbb{Z}/2\mathbb{Z}) \;\cong\; S_3\,.$$

Thus, if the space (3.27) is non-zero, there exists an $SL^{\pm}(2,\mathbb{Z})$-irreducible subspace U such that r acts on all of U either trivially or as multiplication by -1. Since $r\tau = \tau$ implies $\phi_\tau(x) = -\phi_\tau(-x)$ and hence $\phi(0) = 0$, the case of trivial r-action is dealt with already by (3.25). Let me assume, then, that $r\tau = -\tau$ for all $\tau \in U$, which makes all the functions ϕ_τ even. The group (3.28) has three distinct irreducible representations. For the trivial representation, I use a variant of (3.14),

$$(3.29) \qquad \begin{aligned} \phi_\tau(x) - \phi_\tau(1) \;=\; & \phi_{w\tau}(-\tfrac{1}{x})\,|x|^{\lambda+1} \;-\; \phi_{w\tau}(-1) \\ & +\; (\lambda+1)\int_1^{\frac{1}{|x|}} t^{-\lambda-2}\,\phi_{w\tau}(-\operatorname{sgn}(x)\,t)\,dt\,, \end{aligned}$$

which is obtained by evaluating (3.14) at $x=1$ and subtracting the result from (3.14). Since $\phi_\tau(x) = \phi_{w\tau}(x) = \phi_\tau(-x)$, the terms $\phi_\tau(1)$ and $\phi_{w\tau}(-1)$ cancel. For $|x|<1$, this implies

$$(3.30) \quad |\phi_\tau(x)| \;<\; \sup|\phi_\tau(x)|\left(|x|^{\lambda+1} \;+\; (\lambda+1)\int_1^{\frac{1}{|x|}} t^{-\lambda-2}dt\right) \;=\; \sup|\phi_\tau(x)|\,;$$

the inequality is strict since ϕ_τ cannot be constant for non-zero τ. Contradiction: ϕ_τ is periodic of period one. Next the non-trivial one dimensional representation of S_3. Now $\phi_\tau = -\phi_{n_1\tau} = -\phi_{w\tau}$, hence $\phi_\tau(x) = -\phi_\tau(x-1) = -\phi_{w\tau}(x) = \phi_\tau(-x)$.

This allows me to re-write (3.29) as follows:

(3.31)

$$\begin{aligned}\phi_\tau(x) - \phi_\tau(1) &= -\phi_\tau(\tfrac{1}{x})\,|x|^{\lambda+1} + \phi_\tau(1) \\ &\quad - (\lambda+1)\int_1^{\frac{1}{|x|}} t^{-\lambda-2}\,(\phi_\tau(t)-\phi_\tau(1))\,dt - \phi_\tau(1)(1-|x|^{\lambda+1}) \\ &= -(\phi_\tau(\tfrac{1}{x})-\phi_\tau(1))\,|x|^{\lambda+1} - (\lambda+1)\int_1^{\frac{1}{|x|}} t^{-\lambda-2}\,(\phi_\tau(t)-\phi_\tau(1))\,dt\,.\end{aligned}$$

Now, arguing as in (3.30), I find $|\phi_\tau(x)-\phi_\tau(1)| < \sup|\phi_\tau(x)-\phi_\tau(1)|$ for $|x|<1$. Impossible if $\tau \neq 0$ since ϕ_τ is periodic of period two. Finally the two dimensional irreducible representation. There exists a non-zero vector $\tau \in U$ that is fixed by w, and one can argue as in (3.30), since periodicity of period 2 suffices to derive a contradiction; the oscillatory behavior rules out a maximum at $x = \pm 1$.

It should be remarked that the vanishing of ϕ_τ directly implies the vanishing of $\tilde{\tau}_\lambda$ only away from the point at infinity. In the arguments above this is not a problem because τ is invariant under Γ, whose orbit through ∞ contains finite points. Alternatively, one can argue that a non-zero distribution supported at a single point cannot be of class C^α for $\alpha \geq -1$.

The group Γ_2 plays a special role – this is an observation that has been attributed to Mark Stern. As was argued above, if U is a finite-dimensional $SL^{\pm}(2,\mathbb{Z})/\Gamma_2$-module, then the eigenspaces of r are $SL(2,\mathbb{Z})$-invariant. This property characterizes Γ_2 among all normal subgroups $\Gamma \subset SL(2,\mathbb{Z})$. Indeed, if Γ is such a subgroup, and if r acts as multiplication by a scalar on an $SL^{\pm}(2,\mathbb{Z})/\Gamma$-module U, then the action of n_1 on U agrees with that of $n_1^{-1} = n_{-1} = rn_1r^{-1}$, and hence n_2 acts trivially. Similarly wn_2w^{-1} acts trivially. The two elements n_2 and wn_2w^{-1} of $SL(2,\mathbb{Z})$ generate Γ_2, so Γ_2 acts trivially on U. For emphasis,

3.32 Observation. *Let $\Gamma \subset SL(2,\mathbb{Z})$ be a normal subgroup, and U an $SL^{\pm}(2,\mathbb{Z})/\Gamma$-module. If r acts on U as multiplication by a scalar, then Γ_2 acts trivially.*

Coupled with the statement about Γ_2 that was proved earlier, this implies: If U is a non-zero, $SL(2,\mathbb{Z})$-invariant subspace as in (3.22), with $\Gamma \subset SL(2,\mathbb{Z})$ normal of finite index and $-1 < \lambda < 1$, then U contains both r-invariant and anti-invariant distributions. Equivalently, any non-zero, $SL(2,\mathbb{Z})$-invariant space of cuspidal Maass forms corresponding to an eigenvalue $\leq \frac{1}{4}$ must contain both even and odd forms. Since Γ_2 does not admit non-cuspidal, square-integrable Maass forms, the statement is correct even for spaces of square-integrable, but not necessarily cuspidal, Maass forms.

I finish this section with the formula which generalizes (3.14) to the non-cuspidal case. The parameter λ is then necessarily real and non-zero. I may and shall assume $\lambda > 0$ – cf. (3.1) – to ensure that $\tilde{\tau}_\lambda$ has a continuous anti-derivative

ϕ_τ. The Fourier series (3.9) can now have a non-zero constant term $a_{\tau,0}$. I write

$$\begin{aligned}\phi_\tau(x) &= a_{\tau,0}x + \psi_\tau(x), \\ \text{where } \psi_\tau(x) &= \textstyle\sum_{k\neq 0} \frac{m}{2\pi i k} a_k e^{2\pi i \frac{k}{m} x},\end{aligned} \tag{3.33}$$

so that $\tilde{\tau}_\lambda = a_{\tau,0} + \frac{d}{dx}\psi_\tau$, with ψ_τ periodic and having zero constant term. In this situation,

$$\begin{aligned}\psi_\tau(x) - \psi_\tau(0) &= \lambda^{-1} a_{w\tau,0}\, \mathrm{sgn}(x)\,|x|^\lambda - a_{\tau,0}x + \psi_{w\tau}(-\tfrac{1}{x})\,|x|^{\lambda+1} \\ &\quad - (\lambda+1)\int_{\frac{1}{|x|}}^{+\infty} t^{-\lambda-2}\, \psi_{w\tau}(-\,\mathrm{sgn}(x)t)\, dt.\end{aligned} \tag{3.34}$$

The proof is the same as that of (3.14): The two sides agree at $x = 0$, and the first derivatives are related by the transformation law for $\tilde{\tau}_\lambda$. When the term $|x|^\lambda$ occurs with non-zero coefficient, as it may, ϕ_τ cannot be Hölder continuous of index greater than λ. Thus theorem 2.18 is sharp in this case as well.

4. L-functions and functional equation

The L-functions attached to modular forms satisfy a functional equation, analogous to that satisfied by the Riemann zeta function. The functional equation is usually proved by taking the Mellin transform of the modular form along the imaginary axis and then applying the transformation law under the inversion $z \mapsto -\frac{1}{z}$. Maass [10] showed that the same strategy works in the context of Maass forms. The "q-expansion" of modular forms has an analogue in the Maass case, with coefficients that are Bessel functions, rather than exponentials. The Mellin transform of the relevant Bessel function is essentially a product of two Gamma functions, and that accounts for the particular form of the functional equation.

In this section I shall show that the functional equation can also be derived by integration along the *real axis*. For groups of higher rank, a maximally split torus and a maximal unipotent subgroup are the analogues of, respectively, the imaginary axis and the real axis. Formally at least, a maximal unipotent subgroup has the "right dimension" for the derivation of a functional equation attached to automorphic representations, whereas the maximally split torus is "too small". The argument I am about to present suggests that functional equations for L-functions in the higher rank case should be provable by integration along a maximal unipotent subgroup – they should reflect the interplay between harmonic analysis on $N_\mathbb{Z}/(N_\mathbb{Z} \cap G_\mathbb{Z})$ and the action of the Weyl group.

For classical – holomorphic – modular forms it is not so surprising, perhaps, that the contour of integration can be moved from one axis to the other. I shall therefore concentrate on the Maass case. The notation and hypotheses are the same as in the previous section. In particular, $\Gamma \subset SL(2,\mathbb{Z})$ denotes a normal subgroup of finite index. I consider a Γ-invariant distribution vector $\tau \in (V_{\lambda,+1}^{-\infty})^\Gamma$ which corresponds to a cuspidal embedding $V_{\lambda,+1}^{-\infty} \hookrightarrow L^2(\Gamma, G_{\mathbb{R}})$. I express its image $\tilde{\tau}_\lambda \in C^{-\infty}(\mathbb{R})$ as a Fourier series

$$(4.1) \qquad \tilde{\tau}_\lambda(x) = \textstyle\sum_{k\neq 0} a_k e^{2\pi i \frac{k}{m} x},$$

as in (3.9); the cuspidality assumption eliminates the constant term. One can write τ as the sum of an even and an odd part. Thus, without essential loss of generality, I may assume

$$(4.2) \qquad \tilde{\tau}_\lambda(-x) = \eta\, \tilde{\tau}_\lambda(x), \quad \text{with } \eta \in \{\pm 1\}.$$

This has the advantage of making the positive Fourier coefficients determine the negative ones as well. Consequently

$$(4.3) \qquad L(\tau, s) = \textstyle\sum_{k>0} a_k k^{\frac{\lambda}{2}-s},$$

the Maass L-function of the embedding $V_{\lambda,+1}^{-\infty} \hookrightarrow L^2(\Gamma, G_{\mathbb{R}})$ defined by τ, completely determines $\tilde{\tau}_\lambda$, and hence τ itself – recall that the behavior of $\tilde{\tau}_\lambda$ at finite points determines the behavior of the Γ-invariant distribution τ also at infinity.

The canonical intertwining operator $V_{\lambda,+1} \cong V_{-\lambda,+1}$ relates $\tilde{\tau}_\lambda$ to another Γ-invariant distribution vector $\tilde{\tau}_{-\lambda}$, which represents the same embedding. One can show that the intertwining operator sends $e^{2\pi i \frac{k}{m} x}$, $k \neq 0$, to $|k|^\lambda e^{2\pi i \frac{k}{m} x}$, up to some universal constant which reflects the normalization of the operator. The L-function is therefore independent of the choice of λ over $-\lambda$. The same conclusion can be drawn from the fact that (4.3) really agrees with Maass' definition, up to a non-zero constant. To see this, one needs to go into the details of the proof of theorem 2.18 – specifically the series expansion (5.29) and lemma 5.65. Maass' explicit version of (5.29), in terms of Bessel functions, then makes the connection between the coefficients a_k in (4.1) and the coefficients used by Maass to define his L-function.

Theorem 2.18 and (2.15c) give an estimate for the abscissa of absolute convergence of the L-function: If $\delta > 0$,

(4.4) the series (4.3) converges uniformly and absolutely for $\mathrm{Re}\, s \geq 1 + \delta$.

Maass' functional equation asserts that $L(\tau, s)$ extends meromorphically to the entire complex plane, and

$$(4.5) \qquad \begin{aligned} &\left(\tfrac{m}{\pi}\right)^s \Gamma\!\left(\tfrac{s}{2} + \tfrac{\lambda}{4} + \tfrac{1-\eta}{4}\right) \Gamma\!\left(\tfrac{s}{2} - \tfrac{\lambda}{4} + \tfrac{1-\eta}{4}\right) L(\tau, s) \\ &\qquad = \eta \left(\tfrac{m}{\pi}\right)^{1-s} \Gamma\!\left(\tfrac{1-s}{2} + \tfrac{\lambda}{4} + \tfrac{1-\eta}{4}\right) \Gamma\!\left(\tfrac{1-s}{2} - \tfrac{\lambda}{4} + \tfrac{1-\eta}{4}\right) L(w\tau, 1-s) \end{aligned}$$

[10]. This version of the functional equation clearly displays the symmetry between λ and $-\lambda$. It is not the simplest way of writing the formula, however.

Recall the definition (3.13) of the matrix w and the definition (3.10) of the anti-derivative ϕ_τ of $\tilde{\tau}_\lambda$. At first glance, the integral defining the Mellin transform

$$(4.6) \qquad s \longmapsto M(\phi_\tau, s) = \int_0^{+\infty} \phi_\tau(x)\, x^{s-1}\, dx$$

converges in the two disjoint regions $\operatorname{Re} s > 0$ and $\operatorname{Re} s < 0$ for x near 0 and ∞, respectively. However, the fact that the continuous function ϕ_τ is periodic, with zero constant term, makes $M(\phi_\tau, s)$ well-defined and holomorphic at least in the region $0 < \operatorname{Re} s < 1$.

4.7 Lemma. *The Mellin transform $M(\phi_\tau, s)$ extends meromorphically to the region $-2-\operatorname{Re}\lambda < \operatorname{Re} s < 1$, where it has no singularities except for a first order pole at $s=0$, with residue $\phi_\tau(0)$. On this region, $sM(\phi_\tau, s) = -\eta(s+\lambda+1)M(\phi_{w\tau}, -s-\lambda-1)$.*

When one applies integration by parts purely formally, the assertion of the lemma gets transformed into the equation $M(\tau,s) = \eta M(w\tau, 1-s-\lambda)$ which "follows", again purely formally, from the transformation law (2.9). Before turning to the proof, I shall show that the lemma implies Maass' formula (4.5).

Let $\phi_\tau^{(-1)}$ denote the first anti-derivative of ϕ_τ, normalized so as to have zero constant term. By definition, $M(\phi_\tau^{(-1)}, s)$ is holomorphic on the strip $0 < \operatorname{Re} s < 1$, but the identity

$$
\begin{aligned}
\int_0^{+\infty} \phi_\tau^{(-1)}(x) x^{s-1}\, dx \;&=\; s^{-1}\phi_\tau^{(-1)}(0) \;+ \\
&+ \int_0^1 (\phi_\tau^{(-1)}(x) - \phi_\tau^{(-1)}(0)) x^{s-1}\, dx + \int_1^{+\infty} \phi_\tau^{(-1)}(x) x^{s-1}\, dx
\end{aligned}
\tag{4.8}
$$

shows how to extend $M(\phi_\tau^{(-1)}, s)$ meromorphically to $-1 < \operatorname{Re} s < 1$, with a first order pole at $s=0$, where the residue equals $\phi_\tau^{(-1)}(0)$. I claim:

$$
M(\phi_\tau, s) \;=\; -(s-1)\, M(\phi_\tau^{(-1)}, s-1) \qquad (0 < \operatorname{Re} s < 1)\,. \tag{4.9}
$$

When $\phi_\tau^{(-1)}(0) = 0$, this follows from integration by parts; the vanishing of $\phi_\tau^{(-1)}(0)$ makes the boundary term drop out. The assertion is known for the function $e^{2\pi i x}$ – see (4.11) below; the general case follows by taking a suitable linear combination of $\phi_\tau^{(-1)}$ and $e^{2\pi i x}$.

The Fourier series for $\phi_\tau^{(-1)}$ converges uniformly and absolutely, as follows from theorem 2.18 and (2.15c). Thus, in the notation of (4.1–2), for $0 < \operatorname{Re} s < 1$,

$$
\begin{aligned}
M(\phi_\tau^{(-1)}, s) \;&=\; -\tfrac{m^2}{4\pi^2} \sum\nolimits_{k\neq 0} \frac{a_k}{k^2} \int_0^{+\infty} e^{2\pi i \frac{k}{m} x} x^{s-1}\, dx \\
&=\; -\tfrac{m^2}{4\pi^2} \sum\nolimits_{k>0} \frac{a_k}{k^2} \int_0^{+\infty} \left(e^{2\pi i \frac{k}{m} x} + \eta\, e^{-2\pi i \frac{k}{m} x} \right) x^{s-1}\, dx\,.
\end{aligned}
\tag{4.10}
$$

A standard Gamma identity asserts

$$
\int_0^{+\infty} e^{iax} x^{s-1}\, dx \;=\; a^{-s}\Gamma(s)\, e^{i\frac{\pi}{2}s} \qquad (0 < a,\ 0 < \operatorname{Re} s < 1)\,. \tag{4.11}
$$

When combined with (4.3) and (4.10), this leads to the identity

$$(4.12)\qquad \begin{aligned} M(\phi_\tau^{(-1)},s) &= -m^{2+s}(2\pi)^{-2-s}\Gamma(s)\left(e^{i\frac{\pi}{2}s}+\eta\, e^{-i\frac{\pi}{2}s}\right)\textstyle\sum_{k>0}a_k k^{-2-s} \\ &= -m^{2+s}(2\pi)^{-2-s}\Gamma(s)\left(e^{i\frac{\pi}{2}s}+\eta\, e^{-i\frac{\pi}{2}s}\right)L(\tau,s+2+\tfrac{\lambda}{2}). \end{aligned}$$

At this point, analytic continuation, (4.9), and the identity $s\Gamma(s)=\Gamma(s+1)$ imply

$$(4.13)\qquad M(\phi_\tau,s) = \tfrac{1}{i}m^{1+s}(2\pi)^{-1-s}\Gamma(s)\left(e^{i\frac{\pi}{2}s}-\eta\, e^{-i\frac{\pi}{2}s}\right)L(\tau,s+1+\tfrac{\lambda}{2}),$$

for $0<\operatorname{Re}s<1$. Lemma 4.7 asserts that $sM(\phi_\tau,s)$ is holomorphic on $-2-\operatorname{Re}\lambda<\operatorname{Re}s<1$ and invariant under the simultaneous substitutions $s\mapsto -s-\lambda-1$, $\tau\mapsto w\tau$, up to a factor η. But $L(\tau,s)$ is known to be holomorphic for $\operatorname{Re}s>1$, so the symmetry of $sM(\phi_\tau,s)$ continues $L(\tau,s)$ meromorphically to the entire plane, and implies

(4.14)

$$\begin{aligned} &\left(\tfrac{m}{2\pi}\right)^{1+s}\Gamma(s+1)\left(e^{i\frac{\pi}{2}s}-\eta\, e^{-i\frac{\pi}{2}s}\right)L(\tau,s+1+\tfrac{\lambda}{2}) \\ &\quad= \eta\left(\tfrac{m}{2\pi}\right)^{-s-\lambda}\Gamma(-s-\lambda)\left(e^{-i\frac{\pi}{2}(s+\lambda+1)}-\eta\, e^{i\frac{\pi}{2}(s+\lambda+1)}\right)L(w\tau,-s-\tfrac{\lambda}{2}). \end{aligned}$$

Next, I substitute $s-1-\frac{\lambda}{2}$ for s, multiply both sides by $im^{\frac{\lambda}{2}}(2\pi)^{-\frac{\lambda}{2}}$, and re-write the exponential terms:

$$(4.15)\qquad \begin{aligned} &i\left(\tfrac{m}{2\pi}\right)^{s}\Gamma(s-\tfrac{\lambda}{2})\left(e^{i\frac{\pi}{2}(s-\frac{\lambda}{2})}+\eta\, e^{-i\frac{\pi}{2}(s-\frac{\lambda}{2})}\right)L(\tau,s) \\ &\quad= \left(\tfrac{m}{2\pi}\right)^{1-s}\Gamma(1-s-\tfrac{\lambda}{2})\left(e^{i\frac{\pi}{2}(s+\frac{\lambda}{2})}-\eta\, e^{-i\frac{\pi}{2}(s+\frac{\lambda}{2})}\right)L(w\tau,1-s). \end{aligned}$$

I must show that this implies the functional equation (4.5).

Continuing the derivation, I observe that $e^{i\frac{\pi}{2}z}+\eta\, e^{-i\frac{\pi}{2}z}=2e^{i\frac{\pi}{4}(1-\eta)}\sin(\frac{\pi}{2}(z+\frac{1+\eta}{2}))$ and similarly for $-\eta$ in place of η, and $e^{i\frac{\pi}{2}\eta}=i\eta$. I also use the argument doubling formula

$$\Gamma(\tfrac{z}{2})\Gamma(\tfrac{z}{2}+\tfrac{1}{2})=2^{1-z}\pi^{\frac{1}{2}}\Gamma(z)$$

and cancel various like terms, to put (4.15) into the equivalent form

(4.16)

$$\begin{aligned} &\left(\tfrac{m}{\pi}\right)^{s}\Gamma(\tfrac{s}{2}-\tfrac{\lambda}{4})\Gamma(\tfrac{s}{2}-\tfrac{\lambda}{4}+\tfrac{1}{2})\sin\left(\pi(\tfrac{s}{2}-\tfrac{\lambda}{4}+\tfrac{1+\eta}{4})\right)L(\tau,s) \\ &= \eta\left(\tfrac{m}{\pi}\right)^{1-s}\Gamma(\tfrac{1}{2}-\tfrac{s}{2}-\tfrac{\lambda}{4})\Gamma(1-\tfrac{s}{2}-\tfrac{\lambda}{4})\sin\left(\pi(\tfrac{s}{2}+\tfrac{\lambda}{4}+\tfrac{1-\eta}{4})\right)L(w\tau,1-s). \end{aligned}$$

For the last step, I recall that $\Gamma(z)\Gamma(1-z)\sin(\pi z)=\pi$, hence

$$(4.17)\qquad \begin{aligned} &\left(\tfrac{m}{\pi}\right)^{s}\frac{\Gamma(\frac{s}{2}-\frac{\lambda}{4})\Gamma(\frac{s}{2}-\frac{\lambda}{4}+\frac{1}{2})}{\Gamma(\frac{s}{2}-\frac{\lambda}{4}+\frac{1+\eta}{4})\Gamma(1-\frac{s}{2}+\frac{\lambda}{4}-\frac{1+\eta}{4})}L(\tau,s) \\ &= \eta\left(\tfrac{m}{\pi}\right)^{1-s}\frac{\Gamma(\frac{1}{2}-\frac{s}{2}-\frac{\lambda}{4})\Gamma(1-\frac{s}{2}-\frac{\lambda}{4})}{\Gamma(\frac{s}{2}+\frac{\lambda}{4}+\frac{1-\eta}{4})\Gamma(1-\frac{s}{2}-\frac{\lambda}{4}-\frac{1-\eta}{4})}L(w\tau,1-s). \end{aligned}$$

For both choices of η, i.e., $\eta = 1$ and $\eta = -1$, the appropriate terms cancel, and the equation becomes equivalent to (4.5). Thus lemma 4.7 does imply Maass' functional equation.

Proof of lemma 4.7. The identity (4.2) is equivalent to $(r\tau)(x) = \eta\,\tau(x)$; cf. (3.24). Since r and w commute, $w\tau$ also satisfies the symmetry relation (4.2). The passage from $\tilde{\tau}_\lambda$ to ϕ_τ reverses parity with respect to $x \mapsto -x$, so

(4.18) $$\phi_\tau(-x) \;=\; -\eta\,\phi_\tau(x)\,, \qquad \phi_{w\tau}(-x) \;=\; -\eta\,\phi_{w\tau}(x)\,.$$

I now re-interpret the integral (4.6) as a sum of two integrals, over the intervals $(0,1]$ and $[1,+\infty)$. As was argued already,

(4.19) $$s \longmapsto \int_1^{+\infty} \phi_\tau(x)\,x^{s-1}\,dx \quad \text{is holomorphic on } \{\operatorname{Re} s < 1\}\,.$$

On the other subinterval, I use (3.14): For $0 < \operatorname{Re} s < 1$,

(4.20)
$$\begin{aligned}
\int_0^1 \phi_\tau(x)\,x^{s-1}\,dx \;-\; s^{-1}\,\phi_\tau(0) \;&=\; \int_0^1 (\phi_\tau(x)-\phi_\tau(0))\,x^{s-1}\,dx \\
&= -\eta \int_0^1 \left(\phi_{w\tau}(\tfrac{1}{x})\,x^{\lambda+1} - (\lambda+1)\int_{\frac{1}{x}}^{+\infty} t^{-\lambda-2}\,\phi_{w\tau}(t)\,dt\right) x^{s-1}\,dx \\
&= -\eta \int_1^{+\infty} \left(\phi_{w\tau}(x)\,x^{-\lambda-1} - (\lambda+1)\int_x^{+\infty} t^{-\lambda-2}\,\phi_{w\tau}(t)\,dt\right) x^{-s-1}\,dx \\
&= -\eta \int_1^{+\infty} \phi_{w\tau}(x)\,x^{-\lambda-s-2}\,dx \\
&\qquad\qquad + \;\eta\,(\lambda+1)\int_{t=1}^{+\infty}\int_{x=1}^{t} t^{-\lambda-2}\,x^{-s-1}\,\phi_{w\tau}(t)\,dx\,dt \\
&= -\eta\,(1+\tfrac{\lambda+1}{s})\int_1^{+\infty} \phi_{w\tau}(x)\,x^{-\lambda-s-2}\,dx \;+\; \eta\,\tfrac{\lambda+1}{s}\int_1^{+\infty} x^{-\lambda-2}\,\phi_{w\tau}(x)\,dx\,.
\end{aligned}$$

At the middle step, I have applied the change of variables $x \mapsto \frac{1}{x}$; the penultimate step involves a switch in order of integration, and the final step simply rearranges the terms. I now multiply through by s and use (3.14) at $x=1$ to express the last integral on the right in terms of $\phi_\tau(0)$, $\phi_\tau(1)$, and $\phi_{w\tau}(1)$,

(4.21)
$$\begin{aligned}
&s\int_0^1 \phi_\tau(x)\,x^{s-1}\,dx \;-\; \phi_\tau(1) \;= \\
&\qquad -\,\eta\left((\lambda+s+1)\int_1^{+\infty} \phi_{w\tau}(x)\,x^{-\lambda-s-2}\,dx \;-\; \phi_{w\tau}(1)\right).
\end{aligned}$$

The right hand side represents a holomorphic function on the region $\operatorname{Re} s > -\operatorname{Re}\lambda - 2$, in complete analogy to (4.19). This, coupled with (4.19), implies the first assertion of the lemma, except possibly the identification of the residue at

$x=0$. The fact that the residue is $\phi_\tau(0)$ already follows from the very first equality in (4.20), since $\phi_\tau(x)-\phi_\tau(0)$ has order of magnitude $|x|^{\mathrm{Re}\lambda+1}$ as $x\to 0$. I now multiply the equation (4.21) by η, interchange the roles of ϕ_τ and $\phi_{w\tau}$, and replace s by $-s-\lambda-1$. Adding the resulting equation to (4.21), I find

$$(4.22)\qquad s\int_0^{+\infty}\phi_\tau(x)x^{s-1}\,dx \;=\; -\,\eta\,(\lambda+s+1)\int_0^{+\infty}\phi_{w\tau}(x)x^{-\lambda-s-2}\,dx,$$

with both sides holomorphic at least on the strip $-\mathrm{Re}\lambda-2<\mathrm{Re}\,s<1$. This completes the proof of the lemma. □

5. Proof of theorems 2.18 and 2.20

I use the notation and conventions of §2. In particular, (π,V) is an irreducible unitary representation, i an embedding of the dual $G_{\mathbb{R}}$-module V' into $L^2(\Gamma\backslash G_{\mathbb{R}})$, and τ the Γ-invariant distribution vector defined by (1.1–3). For any $v'\in V'^{\infty}$,

$$(5.1)\qquad G_{\mathbb{R}}\ni g\;\mapsto\;\langle\tau\,,\,\pi'(g)v'\rangle$$

is the function $i(v')\in L^2(\Gamma\backslash G_{\mathbb{R}})\cap C^\infty(G_{\mathbb{R}})$. Indeed, for $g=e$, this amounts to the definition of τ, and the general case then follows, with $\pi'(g)v'$ in place of v'.

When v' is a $K_{\mathbb{R}}$-isotypic vector, (5.1) can be regarded as function on the upper half plane

$$(5.2)\qquad \mathcal{H}\;=\;\{z=x+iy\in\mathbb{C}\mid y>0\}\;\simeq\;G_{\mathbb{R}}/K_{\mathbb{R}}\,.$$

For this purpose, I parametrize the groups $A_{\mathbb{R}}$, $N_{\mathbb{R}}$ as in (2.3) and set

$$(5.3)\qquad k_\theta\;=\;\begin{pmatrix}\cos\theta & \sin\theta\\ -\sin\theta & \cos\theta\end{pmatrix},$$

so that $K_{\mathbb{R}}=\{k_\theta\mid\theta\in\mathbb{R}/2\pi\mathbb{Z}\}$. Every $g\in G_{\mathbb{R}}$ can then be expressed as product

$$(5.4)\qquad g\;=\;n_x\,a_t\,k_\theta\,,$$

with $x,t\in\mathbb{R}$ and $\theta\in\mathbb{R}/2\pi\mathbb{Z}$ uniquely determined. Now let $v'\in V'$ be a non-zero vector such that

$$(5.5)\qquad \pi'(k_\theta)v'\;=\;e^{ir\theta}\,v'\,,$$

for some $r\in\mathbb{Z}$. In this situation, the function $F_{\tau,v'}:\mathcal{H}\to\mathbb{C}$,

$$(5.6)\qquad F_{\tau,v'}(x,y)\;=\;\langle\tau\,,\,\pi'(n_xa_t)v'\rangle\,,\qquad\text{with } e^{2t}=y\,,$$

determines the function (5.1) and is determined by it.

The function $F_{\tau,v'}$ satisfies a second-order differential equation, which reflects the fact that the Casimir operator acts as a scalar in the irreducible unitary representation π'. In terms of the basis

$$e = \begin{pmatrix} 0 & 1 \\ 0 & 0 \end{pmatrix}, \quad f = \begin{pmatrix} 0 & 0 \\ 1 & 0 \end{pmatrix}, \quad h = \begin{pmatrix} 1 & 0 \\ 0 & -1 \end{pmatrix} \tag{5.7}$$

of the Lie algebra $\mathfrak{g}_{\mathbb{R}}$, the Casimir operator – suitably normalized – can be expressed as

$$\begin{aligned} \Omega &= 2ef + 2fe + h^2 = 4ef + h^2 - 2h \\ &= -4e(e-f) + 4e^2 + h^2 - 2h. \end{aligned} \tag{5.8}$$

I choose $\lambda \in \mathbb{C}$ so that Ω acts on V' by the constant $\lambda^2 - 1$; in particular, then,

$$\pi'(\Omega)v' = (\lambda^2 - 1)v'. \tag{5.9}$$

Note that λ is determined only up to sign.

For the statement of the next lemma, I temporarily drop the assumption that τ arise from an embedding of (π', V') into $L^2(\Gamma\backslash G_{\mathbb{R}})$. To signal this change in hypotheses, I let $\sigma \in V^{-\infty}$ denote an arbitrary distribution vector, and write $F_{\sigma,v'}$ for the function defined by (5.6), but with σ in place of τ.

5.10 Lemma. $\left(y^2(\frac{\partial^2}{\partial x^2} + \frac{\partial^2}{\partial y^2}) - iry\frac{\partial}{\partial x} - \frac{1}{4}(\lambda^2 - 1)\right) F_{\sigma,v'}(x,y) = 0.$

Proof. I combine the definition (5.6) of $F_{\sigma,v'}$ with (5.8–9):

$$\begin{aligned} (\lambda^2 - 1)\, F_{\sigma,v'}(x,y) &= -4\langle \tau, \pi'(n_x a_t)\, \pi'(e(e-f))\, v' \rangle \\ &\quad + 4\langle \tau, \pi'(n_x a_t)\, \pi'(e^2)\, v' \rangle \\ &\quad + \langle \tau, \pi'(n_x a_t)\, \pi'(h^2 - 2h)\, v' \rangle. \end{aligned} \tag{5.11}$$

Here, as before, y equals e^{2t}. Since $e - f$ generates the one parameter subgroup $\{k_\theta \mid \theta \in \mathbb{R}\}$, (5.5) is equivalent to

$$\pi'(e-f)v' = \tfrac{d}{d\theta}\left(\pi'(k_\theta)v'\right)\big|_{\theta=0} = irv'. \tag{5.12}$$

For similar reasons,

$$\begin{aligned} \pi'(n_x a_t)\, \pi'(h)\, v' &= \tfrac{\partial}{\partial t}\left(\pi'(n_x a_t)\, v'\right) \\ &= 2y\tfrac{\partial}{\partial y}\left(\pi'(n_x a_t)\, v'|_{e^{2t}=y}\right) \end{aligned} \tag{5.13}$$

and

$$\begin{aligned} \pi'(n_x a_t)\, \pi'(e)\, v' &= \pi'(n_x)\, \pi'((\mathrm{Ad}\, a_t)e)\, \pi'(a_t)\, \pi'(e)\, v' \\ &= y\pi'(n_x)\, \pi'(e)\, \pi'(a_t)\, \pi'(e)\, v' \\ &= y\tfrac{\partial}{\partial x}\left(\pi'(n_x a_t)\, v'\right). \end{aligned} \tag{5.14}$$

In (5.13–14), which do not depend on (5.12), v' can be replaced by $\pi(h)v'$ and $\pi(e)v'$, respectively, so that

$$(5.15)\qquad \begin{aligned} \pi'(n_x a_t)\,\pi'(h^2)\,v' &= 4\,(y\tfrac{\partial}{\partial y})^2 F_{\sigma,v'}(x,y)\,, \\ \pi'(n_x a_t)\,\pi'(e^2)\,v' &= (y\tfrac{\partial}{\partial x})^2 F_{\sigma,v'}(x,y)\,. \end{aligned}$$

At this point, (5.11–15) imply the lemma. □

One can formally re-write the differential equation (5.10) as first order system with respect to the variable y. Since

$$(5.16)\qquad \left(y\tfrac{\partial}{\partial y} - \tfrac{1-\lambda}{2}\right)\left(y\tfrac{\partial}{\partial y} - \tfrac{1+\lambda}{2}\right) = \left(y^2\tfrac{\partial^2}{\partial y^2} + \tfrac{1-\lambda^2}{4}\right),$$

the equation can be put into the equivalent form

$$(5.17)\qquad \left(\tfrac{\partial}{\partial y} - \tfrac{1-\lambda}{2y}\right)\left(\tfrac{1+\lambda}{2} - y\tfrac{\partial}{\partial y}\right)F_{\sigma,v'} = \left(y\tfrac{\partial^2}{\partial^2 x} - ir\tfrac{\partial}{\partial x}\right)F_{\sigma,v'}\,.$$

At this point, integration with respect to y gives

5.18 Corollary. $y^{\frac{\lambda-1}{2}}\left(\tfrac{1+\lambda}{2} - y\tfrac{\partial}{\partial y}\right)F_{\sigma,v'} = \int y^{\frac{\lambda-1}{2}}\left(y\tfrac{\partial^2}{\partial^2 x} - ir\tfrac{\partial}{\partial x}\right)F_{\sigma,v'}\,dy.$

Note that the derivation of corollary 5.18 involves λ^2 rather than λ itself. Thus the corollary remains valid with $-\lambda$ in place of λ.

When π belongs to the discrete series or limits of the discrete series, the function $F_{\sigma,v'}$ – with v' appropriately chosen – satisfies a first order differential equation. To state it, I suppose that v' is a $K_{\mathbb{R}}$-weight vector corresponding to the extreme weight. In other words, if r is the weight as in (5.5), the other $K_{\mathbb{R}}$-weights are $r+2\ell$, $\ell \in \mathbb{N}$ (lowest weight case), or $r-2\ell$, $\ell \in \mathbb{N}$ (highest weight case). The extreme weight r determines the action of the Casimir operator: The two choices $\lambda = \pm(1-|r|)$ of the constant λ satisfy the equation (5.9).

5.19 Lemma. *Under the hypotheses just stated, with $\lambda = 1-|r|$, and with $\eta = +1$ or $\eta = -1$ depending on whether r is the highest or lowest weight for π',*

$$y^{\frac{\lambda-1}{2}}F_{\sigma,v'} = -\int y^{\frac{\lambda-1}{2}}\,i\eta\,\tfrac{\partial}{\partial x}F_{\sigma,v'}\,dy.$$

Proof. A short computation shows

$$(5.20)\qquad \operatorname{Ad}k_\theta(h+i\eta(e+f)) = 2\eta\,(h+i\eta(e+f))\,,$$

which implies that $\pi'(h+i\eta(e+f))v'$ has $K_{\mathbb{R}}$-weight $r+2\eta$. According to the hypotheses, $r+2\eta$ is not a weight, so

$$(5.21)\qquad \begin{aligned} 0 &= \pi'(h+i\eta(e+f))v' = \pi'(h-i\eta(e-f)+2i\eta e)v' \\ &= \pi'(h+2i\eta e)v' + \eta r v' = \pi'(h+2i\eta e)v' - (1-\lambda)v'; \end{aligned}$$

at the last two steps, I have used (5.12), the equality $\lambda = 1 - |r|$, and the fact that the lowest weight is positive, the highest weight negative. I now argue as in the proof of lemma 5.10, to find

$$\left(\tfrac{\partial}{\partial y} + i\eta \tfrac{\partial}{\partial x}\right) F_{\sigma,v'} = \tfrac{1-\lambda}{2y} F_{\sigma,v'}, \tag{5.22}$$

the Cauchy-Riemann equations in disguise. The lemma follows by integration. □

I return to the original definition of the function (5.6), with $\tau \in V^{-\infty}$ corresponding to an embedding of (π', V') into $L^2(\Gamma\backslash G_{\mathbb{R}})$. The parameter λ is related to π' by (5.9), and thus is determined only up to sign.

5.23 Lemma. *The functions $F_{\tau,v'}(x,y)$, $y\frac{\partial}{\partial x}F_{\tau,v'}$, and $y\frac{\partial}{\partial y}F_{\tau,v'}$ are bounded, provided the embedding $V' \hookrightarrow L^2(\Gamma\backslash G_{\mathbb{R}})$ is cuspidal. In the non-cuspidal situation – in which case $\lambda \in \mathbb{R}$, $0 < |\lambda| < 1$ – these three functions are bounded by a constant multiple of* $\max\{y^{\frac{1-|\lambda|}{2}}, y^{\frac{|\lambda|-1}{2}}\}$.

Proof. In more concrete, but equivalent form this is certainly known. However, I shall sketch the argument to make the point that one need not use properties of Bessel functions and other very specific information.

For the first part of the proof, it is better to interpret the function (5.6) more invariantly, as a section of a $G_{\mathbb{R}}$-equivariant line bundle on $G_{\mathbb{R}}/K_{\mathbb{R}} \cong \mathcal{H}$. To simplify the notation, I drop the subscripts τ, v', and define

$$F : G_{\mathbb{R}} \longrightarrow \mathbb{C}, \qquad F(g) = \langle \tau, \pi'(g)v' \rangle. \tag{5.24}$$

The $K_{\mathbb{R}}$-isotopy condition (5.5) implies

$$F(gk_\theta) = e^{ir\theta} F(g), \tag{5.25}$$

which allows me to regard F as section of the $G_{\mathbb{R}}$-equivariant line bundle

$$\mathbf{L}_{-r} \to G_{\mathbb{R}}/K_{\mathbb{R}} \cong \mathcal{H}, \qquad \mathbf{L}_{-r} = G_{\mathbb{R}} \times_{K_{\mathbb{R}}} \mathbb{C}_{-r}, \tag{5.26}$$

(fiber product), where $\mathbb{C}_{-r}$ denotes the one dimensional $K_{\mathbb{R}}$-module dual to $\mathbb{C}v'$. By construction, the section is Γ-invariant and satisfies the $G_{\mathbb{R}}$-invariant, second order, elliptic differential equation

$$\Omega F = (\lambda^2 - 1)F; \tag{5.27}$$

cf. (5.8). Lastly, F is square-integrable over the quotient of $G_{\mathbb{R}}/K_{\mathbb{R}} \cong \mathcal{H}$ by the action of Γ,

$$\int_{\Gamma\backslash\mathcal{H}} |F|^2\, dm < \infty, \tag{5.28}$$

relative to the hyperbolic volume form dm.

The hypotheses on Γ – discrete in $G_{\mathbb{R}}$ and finitely generated – imply that there at most finitely many elliptic conjugacy classes. These correspond to points in $\Gamma\backslash\mathcal{H}$ near which the quotient looks like the quotient of a disc centered at the origin in $\mathbb{C}$, modulo the action of a finite rotation group. There also may be finitely many cusps, i.e., Γ-orbits of parabolic fixed points on the boundary $\mathbb{R}\cup\{\infty\}$ of $\mathcal{H}$. For a discussion of the nature of $\Gamma\backslash\mathcal{H}$ see, for example, [1].

Now let $U_1, U_2, \ldots, U_m$ be punctured neighborhoods of the cusps, and $\tilde{U}$ the inverse image of their union in $\mathcal{H}$. Then there exists $\delta > 0$ with the following property: For any $z \in \mathcal{H} - \tilde{U}$, the disc (in the hyperbolic metric) $\Delta(z,\delta)$ centered at z, of radius δ maps to its image in $\Gamma\backslash\mathcal{H}$ with fibers of bounded cardinality – namely, the maximal branch index at the elliptic fixed points. Standard elliptic estimates then bound the value of F at z, and also of $y\frac{\partial F}{\partial x}$ and $y\frac{\partial F}{\partial y}$ at z, in terms of δ, the global L^2-norm of F, and the maximal branch index – the assertion about the derivatives uses the fact that the vector fields $y\frac{\partial}{\partial x}$, $y\frac{\partial}{\partial y}$ have constant length.

What has been said so far proves the assertion of the lemma away from the cusps. Since the assertion is invariant under translation by real numbers and under the inversion $z \mapsto -\frac{1}{z}$, it suffices to establish the bound for $y \gg 1$ in case there is a cusp at infinity. Let me suppose then that Γ contains an infinite cyclic group of translations fixing the point at infinity. I may as well suppose that $z \mapsto z+1$ generates this group. I now regard F again as function of the variables x, y, as in (5.6), and express $F(x,y)$ as a Fourier series in the variable x,

$$F(x,y) = \sum\nolimits_{-\infty<k<\infty} a_k(y)\, e^{2k\pi i x}. \tag{5.29}$$

For y_0 sufficiently large, the half-strip $y \geq y_0$, $0 \leq x < 1$ maps to $\Gamma\backslash\mathcal{H}$ in one-to-one fashion, so the square-integrability condition (5.28) implies

$$\sum\nolimits_{-\infty<k<\infty} \int_{y_0}^{\infty} y^{-2}\, |a_k(y)|^2\, dy < \infty. \tag{5.30}$$

The differential equation (5.10) is equivalent to an ordinary differential equations for the Fourier coefficients,

$$a_k''(y) = \left(4\pi^2k^2 - 2\pi r k y^{-1} + \tfrac{1}{4}(\lambda^2-1)y^{-2}\right) a_k(y). \tag{5.31}$$

Changing variables from y to $|k|\,y$, one finds that

(5.32)

$$a_k(y) = \begin{cases} c_k\, b_+(ky) & \text{if } k>0 \\ c_k\, b_-(-ky) & \text{if } k<0 \end{cases}$$

where $b_+(y)$ and $b_-(y)$ are solutions of the equation

$$b_\pm''(y) = \left(4\pi^2 \mp 2\pi r y^{-1} + \tfrac{1}{4}(\lambda^2-1)y^{-2}\right) b_\pm(y)$$

and the c_k are complex constants. The equation satisfied by $b_\pm(y)$ can be transformed into the confluent hypergeometric equation, whose solutions are known.

However, one can argue qualitatively, as follows. The theory of asymptotic expansions – see [8], for example – produces two solutions, one asymptotic to $y^{\mp\frac{r}{2}}e^{2\pi y}$ as $y \to +\infty$, the other asymptotic to $y^{\pm\frac{r}{2}}e^{-2\pi y}$. Only exponential decay is compatible with the finiteness condition (5.30), so

$$(5.33) \qquad b_{\pm}(y) \sim y^{\pm\frac{r}{2}}e^{-2\pi y} \text{ and } b'_{\pm}(y) \sim -2\pi y^{\pm\frac{r}{2}}e^{-2\pi y}, \quad \text{as } y \to +\infty .$$

Comparing (5.32–33) and noticing that the series (5.29) converges absolutely for any particular y, one finds that

$$(5.34) \qquad F(x,y) \sim a_0(y), \quad y\tfrac{\partial}{\partial y}F(x,y) \sim y a'_0(y), \quad y\tfrac{\partial}{\partial x}F(x,y) \sim 0$$

uniformly in x, as $y \to +\infty$.

The embedding $V' \hookrightarrow L^2(\Gamma\backslash G_{\mathbb{R}})$ corresponding to τ is said to be cuspidal at ∞ if $a_0(y) \equiv 0$, and cuspidal without qualification if the analogous condition holds at every cusp. At first glance, the notion of cuspidality appears to depend on the choice of the $K_{\mathbb{R}}$-isotypic vector (5.5) used to define F. One can show directly that the notion is well-defined – the functions F corresponding to different choices of v' are related by the so-called "Maass operators", certain very simple differential operators – but this will also become obvious as a consequence of the proof of theorem 2.18. In any case, (5.34) implies the lemma when the embedding is cuspidal.

The parameter λ in (5.9) can be purely imaginary, or real with $0 < |\lambda| < 1$, or a non-zero integer; in the first situation π' belongs to the unitary principal series or the limits of the discrete series, in the second situation π' belongs to the complementary series, and in the third π' is a discrete series representation or possibly the trivial one dimensional representation. We may disregard the case of the trivial representation, since then F is a constant function. Solving (5.31) for $a_0(y)$ gives

$$(5.35) \qquad a_0(y) = \begin{cases} c_+ y^{\frac{1}{2}(1+\lambda)} + c_- y^{\frac{1}{2}(1-\lambda)} & \text{if } \lambda \neq 0 \\ c_+ y^{\frac{1}{2}} \log y + c_- y^{\frac{1}{2}} & \text{if } \lambda = 0 . \end{cases}$$

For $\lambda \in i\mathbb{R}$, only $a_0(y) \equiv 0$ is compatible with (5.30), and the embedding is cuspidal at ∞. If $\lambda \in \mathbb{R}$ and $0 < |\lambda| < 1$, square-integrability forces $c_+ = 0$ or $c_- = 0$ depending on the sign of λ; this is the possibly non-cuspidal case referred to in the lemma, and the assertion of the lemma then follows from (5.34–35). In the case of the discrete series, the first order differential equation implicit in lemma 5.19 can be satisfied only when $c_+ = 0$. However, in the setting of lemma 5.19, λ is a strictly negative integer, so c_- must also vanish if F is to be square-integrable. Thus, in this remaining case, the embedding must again be cuspidal at ∞. □

I shall need an analogue of lemma 5.23 that applies to certain anti-derivatives of the function $F_{\tau,v'}$. The hypotheses are the same as those for lemma 5.23. As usual,

$[a]$ shall denote the largest integer less than or equal to $a \in \mathbb{R}$. Note that $[\frac{|\mathrm{Re}\,\lambda|+1}{2}] = 0$ when the embedding τ fails to be cuspidal, since then $\lambda \in \mathbb{R},\ 0 < |\lambda| < 1$.

5.36 Lemma. *There exist C^∞ functions $G_k(x,y)$, $k = 1,2,\dots,[\frac{|\mathrm{Re}\,\lambda|+1}{2}]$ such that $\frac{\partial^k}{\partial x^k} G_k(x,y) = F_{\tau,v'}(x,y)$ and, for $0 < y \le \frac{1}{2}$,*

$$|G_k(x,y)|,\ |y\tfrac{\partial}{\partial x} G_k(x,y)|,\ |y\tfrac{\partial}{\partial y} G_k(x,y)| \ \le\ \begin{cases} C y^k & \text{if } k < \frac{|\mathrm{Re}\,\lambda|+1}{2} \\ C y^k\,|\log y| & \text{if } k = \frac{|\mathrm{Re}\,\lambda|+1}{2} \end{cases},$$

with a positive constant $C = C(x,k,\tau,v')$ that can be chosen locally uniformly in the variable x. If the embedding τ is non-cuspidal, there exists a C^∞ function $G_1(x,y)$, such that $\frac{\partial}{\partial x} G_1(x,y) = F_{\tau,v'}(x,y)$ and, for $0 < y \le \frac{1}{2}$,

$$|G_1(x,y)|,\ |y\tfrac{\partial}{\partial x} G_1(x,y)|,\ |y\tfrac{\partial}{\partial y} G_1(x,y)| \ \le\ C y^{\frac{1+|\lambda|}{2}}\,|\log y|,$$

again with $C = C(x,\tau,v') > 0$ depending locally uniformly on x.

Proof. Since π' is irreducible unitary, the parameter λ must be real or purely imaginary. The statement holds vacuously if $\mathrm{Re}\,\lambda = 0$, and becomes obvious if π is the trivial representation. Also, the assertion of the lemma remains unchanged when λ is replaced by $-\lambda$. I can therefore assume

$$\lambda \in \mathbb{R},\ \lambda > 0, \quad \pi \neq \text{trivial representation}. \tag{5.37}$$

Let $\mathfrak{n} = \mathbb{C}e$ denote the one dimensional Lie algebra spanned by e. Then $\mathfrak{n}$ acts on the space of C^∞ vectors V'^∞. It is known, and can also be checked by direct computation, that the (Lie algebra) $\mathfrak{n}$-homology of V'^∞ – for any infinite dimensional, irreducible representation π' – is finite dimensional and concentrated in degree zero. In other words, the action map $\mathfrak{n} \otimes V'^\infty \to V'^\infty$ has zero kernel and finite dimensional cokernel. Dually,

$$\pi(e) : V^{-\infty} \to V^{-\infty} \quad \text{is surjective.} \tag{5.38}$$

This makes it possible to inductively choose $\sigma_k \in V^{-\infty}$, $k = 0,1,2,\dots$, so that

$$\sigma_0 = \tau, \quad \text{and} \quad \pi(e)\sigma_{k+1} = -\sigma_k \quad \text{for all } k. \tag{5.39}$$

I define $G_k = F_{\sigma_k,v'}$; here, as in (5.10), $F_{\sigma_k,v'}$ denotes the function (5.6), but with σ_k taking the place of τ. Then $G_0 = F_{\tau,v'}$ and

$$\begin{aligned} \tfrac{\partial}{\partial x} G_k(x,y) &= \tfrac{\partial}{\partial u} \langle \sigma_k, \pi'(n_{x+u} a_t) v' \rangle \Big|_{u=0} \qquad (y = e^{2t}) \\ &= \langle \sigma_k, \pi'(e)\pi'(n_x a_t) v' \rangle \\ &= \langle \pi(-e)\sigma_k, \pi'(n_x a_t) v' \rangle \ =\ G_{k-1}(x,y). \end{aligned} \tag{5.40}$$

I first consider the cuspidal case. In view of (5.37), the assertion of the lemma is then vacuous unless $\lambda \geq 1$. This forces π to belong to the discrete series or the limits of the discrete series.

I prove the bounds on $|G_k(x,y)|$, $|y\frac{\partial}{\partial x}G_k(x,y)|$, and $|y\frac{\partial}{\partial y}G_k(x,y)|$ by induction on k. Lemma 5.23 establishes the bounds for $k=0$. For the induction step, I fix the integer k, $0<k\leq\frac{\lambda+1}{2}$, and suppose that the bounds are known at the previous step. The bound for $|y\frac{\partial}{\partial x}G_k(x,y)|$ follows immediately from (5.40) and the inductive hypothesis. Until now the choice of the (non-zero) $K_{\mathbb{R}}$-isotypic vector v' has been arbitrary. Taking finite linear combinations, one can draw the following conclusion from the induction hypothesis: If $0\leq\ell<k$,

$$|\langle\sigma_\ell,\pi'(n_xa_t)w'\rangle| \leq C(x,\sigma_\ell,w')\,e^{2\ell t} \quad \text{for } t\ll-1 \text{ and } w'\in V_0'; \tag{5.41}$$

here V_0' denotes the space of $K_{\mathbb{R}}$-finite vectors in V' – in other words, the linear span of the $K_{\mathbb{R}}$-isotypic vectors. Similarly, the bound for $|G_k(x,y)|$ would follow from

$$|\langle\sigma_\ell,\pi'(n_xa_t)w'\rangle| \leq \begin{cases} C(x,\sigma_\ell,w')\,e^{2kt} & \text{if } k<\frac{\lambda+1}{2} \\ C(x,\sigma_\ell,w')|t|\,e^{2kt} & \text{if } k=\frac{\lambda+1}{2}\,, \end{cases} \tag{5.42}$$

in both cases for $t\ll-1$ and arbitrary $w'\in V_0'$. Since

$$\begin{aligned} y\tfrac{\partial}{\partial y}G_k(x,y) &= \tfrac{1}{2}\tfrac{\partial}{\partial s}\langle\sigma_k,\pi'(n_x,a_{t+s})v'\rangle\Big|_{s=0} \\ &= \tfrac{1}{2}\langle\sigma_k,\pi'(n_xa_t)\pi'(h)v'\rangle, \end{aligned} \tag{5.43}$$

and since V_0' is $\pi'(\mathfrak{g})$-invariant, (5.42) also implies the bound for $|y\frac{\partial}{\partial y}G_k(x,y)|$. I claim:

(5.44) the assertion (5.42) is correct when $w'\in\pi'(\mathfrak{n})V_0'$.

Indeed, this follows from the bound for $|y\frac{\partial}{\partial x}G_k(x,y)|$, which has already been established, and the identity

$$\begin{aligned} y\tfrac{\partial}{\partial x}G_k(x,y) &= e^{2t}\tfrac{\partial}{\partial u}\langle\sigma_k,\pi'(n_{x+u}a_t)v'\rangle\Big|_{u=0} \\ &= e^{2t}\langle\sigma_k,\pi'(n_x)\pi'(e)\pi'(a_t)v'\rangle \\ &= \langle\sigma_k,\pi'(n_xa_t)\pi'(e)v'\rangle. \end{aligned} \tag{5.45}$$

It now suffices to establish (5.42) for a collection of vectors w' whose images span $V_0'/\pi'(\mathfrak{n})=H_0(\mathfrak{n},V_0')$. Since $\mathfrak{a}=\mathbb{C}h$ normalizes $\mathfrak{n}$, it acts on $V_0'/\pi'(\mathfrak{n})$. In the current situation, it was remarked, π' belongs to the discrete series or limits of the discrete series. For representations of this type,

(5.46)
a) $\dim\left(V_0'/\pi'(\mathfrak{n})\right) = 1$;
b) $\pi'(h)$ acts on $V_0'/\pi'(\mathfrak{n})$ as multiplication by $\lambda-1$;
c) the image of any non-zero $K_{\mathbb{R}}$-isotypic vector spans $V_0'/\pi'(\mathfrak{n})$.

These statements are known, and also follow from the discussion of related matters in the proof of lemma 5.65 below. I conclude that it suffices to verify (5.42) for the single vector $w' = \frac{1}{2}(\pi'(h) + 1 - \lambda)v'$, with v' as in (5.5). Now

$$\langle \sigma_k, \pi'(n_x a_t)(\pi'(h) + 1 - \lambda)v' \rangle = -2\left(\tfrac{1-\lambda}{2} - y\tfrac{\partial}{\partial y}\right) G_k(x,y), \tag{5.47}$$

so the bounds (5.42), for all $w' \in V_0'$, follow from the single bound

$$\left|\left(\tfrac{1-\lambda}{2} - y\tfrac{\partial}{\partial y}\right) G_k(x,y)\right| \leq \begin{cases} C(x,k,\tau,v')\, y^k & \text{if } k < \frac{\lambda+1}{2} \\ C(x,k,\tau,v')\, y^k\, |\log y| & \text{if } k = \frac{\lambda+1}{2}, \end{cases} \tag{5.48}$$

for $0 < y \leq \frac{1}{2}$. Recall that $G_k = F_{\sigma_k, v'}$. Corollary (5.18) remains valid with $-\lambda$ in place of λ, as was remarked below the statement of the corollary. Hence, for $0 < y \leq \frac{1}{2}$,

$$\begin{aligned} &\left| y^{-\frac{\lambda+1}{2}} \left(\tfrac{1-\lambda}{2} - y\tfrac{\partial}{\partial y}\right) G_k(x,y)\right| - \left| 2^{\frac{\lambda+1}{2}} \left(\tfrac{1-\lambda}{2} - \tfrac{1}{2}\tfrac{\partial}{\partial y}\right) G_k(x,\tfrac{1}{2})\right| \\ &\quad \leq \left| y^{-\frac{\lambda+1}{2}} \left(\tfrac{1-\lambda}{2} - y\tfrac{\partial}{\partial y}\right) G_k(x,y) - 2^{\frac{\lambda+1}{2}} \left(\tfrac{1-\lambda}{2} - \tfrac{1}{2}\tfrac{\partial}{\partial y}\right) G_k(x,\tfrac{1}{2})\right| \\ &\quad = \left| \int_y^{\frac{1}{2}} s^{-\frac{\lambda+1}{2}} \left(s\tfrac{\partial^2}{\partial x^2} - ir\tfrac{\partial}{\partial x}\right) G_k(x,s)\, ds \right| \\ &\quad = \left| \int_y^{\frac{1}{2}} s^{-\frac{\lambda+1}{2}} \left(s\tfrac{\partial}{\partial x} - ir\right) G_{k-1}(x,s)\, ds \right| \\ &\quad \leq D \int_y^{\frac{1}{2}} s^{k-1-\frac{\lambda+1}{2}}\, ds \leq \begin{cases} D\frac{2}{\lambda+1-2k}\, y^{k-\frac{\lambda+1}{2}} & \text{if } k < \frac{\lambda+1}{2} \\ D\,|\log y| & \text{if } k = \frac{\lambda+1}{2}, \end{cases} \end{aligned} \tag{5.49}$$

with $D = (1+|r|)C(x,k-1,\tau,v')$. The inequality (5.48) now follows when I bring the second term on the left all the way to the other side of the inequality and multiply through by $y^{\frac{\lambda+1}{2}}$. This implies the lemma in the cuspidal case.

Turning to the non-cuspidal case, I continue with the hypotheses (5.37) and the definition (5.39). In the current situation, $0 < \lambda < 1$. Lemma 5.24 provides bounds

$$|G_0(x,y)|,\ |y\tfrac{\partial}{\partial x}G_0(x,y)|,\ |y\tfrac{\partial}{\partial y}G_0(x,y)| \leq C y^{\frac{\lambda-1}{2}} \qquad (0 < y \leq 1), \tag{5.50}$$

with $C = C(\tau, v')$ – recall that $G_0(x,y) = F_{\tau,v'}(x,y)$. This immediately implies the bound for $|y\frac{\partial}{\partial x}G_1(x,y)|$ since $\frac{\partial}{\partial x}G_1(x,y) = G_0(x,y)$. I now follow the pattern of inequalities (5.49), simultaneously with the parameter λ and its negative; now the two cases are similar since the exponent of y in the first term is strictly negative with either choice of sign. Skipping some of the intermediary steps and using

(5.50), I find

(5.51)

$$\begin{aligned}&\left|y^{-\frac{1\pm\lambda}{2}}\left(\tfrac{1\mp\lambda}{2}-y\tfrac{\partial}{\partial y}\right)G_1(x,y)\right|-\left|2^{\frac{1\pm\lambda}{2}}\left(\tfrac{1\mp\lambda}{2}-\tfrac{1}{2}\tfrac{\partial}{\partial y}\right)G_1(x,\tfrac{1}{2})\right|\\ &\quad\leq\left|\int_y^{\frac{1}{2}}s^{-\frac{1\pm\lambda}{2}}\left(s\tfrac{\partial}{\partial x}-ir\right)G_0(x,s)\,ds\right|\\ &\quad\leq D\int_y^{\frac{1}{2}}s^{\frac{\lambda-1}{2}-\frac{1\pm\lambda}{2}}\,ds\ \leq\ \begin{cases}D\,|\log y| & \text{for the upper choice of sign}\\ D\lambda^{-1}y^{\lambda} & \text{for the lower choice of sign}\,,\end{cases}\end{aligned}$$

for $0<y\leq\frac{1}{2}$, and with $D=(1+|r|)C(\tau,v)$. I multiply through by $y^{\frac{1\pm\lambda}{2}}$, rearrange terms, and take appropriate linear combinations of the two resulting inequalities to get the desired conclusion. □

5.52 Lemma. *1) If* $\mathrm{Re}\,\lambda\leq 1$ *and if* τ *corresponds to a cuspidal embedding, the limit*

$$\lim_{y\to 0^+}\left(y^{\frac{\lambda-1}{2}}\left(\tfrac{1+\lambda}{2}-y\tfrac{\partial}{\partial y}\right)F_{\tau,v'}(x,y)\right)$$

exists in $C^{-\infty}(\mathbb{R})$*, equipped with the strong distribution topology, and the limit lies in* $C^{\frac{1}{2}(\mathrm{Re}\lambda-1)}(\mathbb{R})$ *if* $\frac{1}{2}(\mathrm{Re}\,\lambda-1)\notin\mathbb{Z}$*, or in* $C^{<\frac{1}{2}(\mathrm{Re}\lambda-1)}(\mathbb{R})$ *if* $\frac{1}{2}(\mathrm{Re}\,\lambda-1)\in\mathbb{Z}$.
2) If $\lambda\in\mathbb{R}$, $0<\lambda<1$*, with* τ *corresponding to a non-cuspidal embedding, the limits*

$$\lim_{y\to 0^+}\left(y^{\frac{\lambda-1}{2}}\left(\tfrac{1+\lambda}{2}-y\tfrac{\partial}{\partial y}\right)F_{\tau,v'}(x,y)\right),\quad \lim_{y\to 0^+}\left(y^{\frac{-\lambda-1}{2}}\left(\tfrac{1-\lambda}{2}-y\tfrac{\partial}{\partial y}\right)F_{\tau,v'}(x,y)\right)$$

exist and lie in $C^{\lambda-1}(\mathbb{R})$ *and* $C^{<(-1)}(\mathbb{R})$*, respectively.*
3) If π *belongs to the limits of the discrete series – in which case* $\lambda=0$ *– also*

$$\lim_{y\to 0^+}\left(y^{\frac{1}{2}}F_{\tau,v'}(x,y)\right)$$

exists in the strong distribution topology, and the limit lies in $C^{-\frac{1}{2}}(\mathbb{R})$.

The reason for the exceptional statement for the limits of the discrete series will become apparent in the course of the proof of lemma 5.65 below. Formally, the assertion of the lemma remains correct even without hypothesis that $\mathrm{Re}\,\lambda\leq 1$ – in the other cases, the limits can be shown to be zero.

Proof. Note that $1-\mathrm{Re}\,\lambda\geq 0$. I set $k=[\frac{1-\mathrm{Re}\lambda}{2}]$ and let $G_k(x,y)$ denote the function whose existence is asserted in lemma 5.36 when $k>0$; when $k=[\frac{1-\mathrm{Re}\lambda}{2}]=0$, I set $G_0(x,y)=F_{\tau,v'}(x,y)$. In both cases I regard $y\mapsto G_k$ as a function of the variable y, with values in the space of distributions on the real line. In view of (5.18), for

$u > 0$,

(5.53)

$$\lim_{y\to 0^+}\Big(y^{\frac{\lambda-1}{2}}\big(\tfrac{1+\lambda}{2}-y\tfrac{\partial}{\partial y}\big)G_k(x,y)\Big) =$$
$$= u^{\frac{\lambda-1}{2}}\big(\tfrac{1+\lambda}{2}-u\tfrac{\partial}{\partial y}\big)G_k(x,u) - \lim_{y\to 0^+}\int_y^u s^{\frac{\lambda-1}{2}}\Big(s\tfrac{\partial^2}{\partial x^2}-ir\tfrac{\partial}{\partial x}\Big)G_k(x,s)\,ds$$
$$= u^{\frac{\lambda-1}{2}}\big(\tfrac{1+\lambda}{2}-u\tfrac{\partial}{\partial y}\big)G_k(x,u) - \tfrac{\partial}{\partial x}\Big(\int_0^u s^{\frac{\lambda-1}{2}}\Big(s\tfrac{\partial}{\partial x}-ir\Big)G_k(x,s)\,ds\Big),$$

provided the integral on the right hand side converges, locally uniformly in the variable x. I suppose initially that the embedding is cuspidal. According to (5.36) and (5.23), for $0 < y \le \frac{1}{2}$, $|G_k(x,y)|$ and $|y\frac{\partial}{\partial x}G_k(x,y)|$ are bounded by Cy^k or by $Cy^k|\log y|$, depending on whether $\frac{1-\mathrm{Re}\,\lambda}{2}$ is non-integral or integral; the constant $C = C(x,k,\tau,v')$ can be chosen locally uniformly in x. Recall that $1-\mathrm{Re}\,\lambda \ge 0$ by assumption, and $k = [\frac{1-\mathrm{Re}\,\lambda}{2}]$. Now, for $0 < y_1 < y_2 \le \frac{1}{2}$,

(5.54)

$$\Big|\int_{y_1}^{y_2} s^{\frac{\lambda-1}{2}}\Big(s\tfrac{\partial}{\partial x}-ir\Big)G_k(x,s)\,ds\Big|$$
$$\le \begin{cases} (1+|r|)C\displaystyle\int_{y_1}^{y_2} s^{k-\frac{1-\mathrm{Re}\,\lambda}{2}}\,ds \;\le\; Dy_2^{[\frac{1-\mathrm{Re}\,\lambda}{2}]-\frac{1-\mathrm{Re}\,\lambda}{2}+1} & \text{if } \frac{1-\mathrm{Re}\,\lambda}{2}\notin\mathbb{Z} \\ (1+|r|)C\displaystyle\int_{y_1}^{y_2} |\log s|\,s^{k-\frac{1-\mathrm{Re}\,\lambda}{2}}\,ds \;\le\; Dy_2|\log y_2| & \text{if } \frac{1-\mathrm{Re}\,\lambda}{2}\in\mathbb{Z}. \end{cases}$$

Here D is a positive constant with the same type of dependence on parameters as C, in particular, locally uniform dependence on x. This does imply the locally uniform convergence of the last integral in (5.53). It also implies: For $0 < u \le \frac{1}{2}$,

$$\text{(5.55)}\qquad \begin{aligned} &\text{a)}\quad \phi_u(x) =_{\text{def}} -\int_0^u s^{\frac{\lambda-1}{2}}\Big(s\tfrac{\partial}{\partial x}-ir\Big)G_k(x,s)\,ds \in C^0(\mathbb{R}),\\ &\text{b)}\quad |\phi_u(x)| \le Du^{[\frac{1-\mathrm{Re}\,\lambda}{2}]-\frac{1-\mathrm{Re}\,\lambda}{2}+1} \quad\text{or}\quad \le Du|\log u|, \end{aligned}$$

depending on whether $\frac{1-\mathrm{Re}\,\lambda}{2}$ is non-integral or integral. It follows that the limit on the left hand side of (5.53) is the first derivative of a continuous function,

(5.56)

$$\lim_{y\to 0^+}\Big(y^{\frac{\lambda-1}{2}}\big(\tfrac{1+\lambda}{2}-y\tfrac{\partial}{\partial y}\big)G_k(x,y)\Big) = \tfrac{\partial}{\partial x}\psi_u(x),$$
$$\text{where}\quad \psi_u(x) =_{\text{def}} u^{\frac{\lambda-1}{2}}\int_0^x\big(\tfrac{1+\lambda}{2}-u\tfrac{\partial}{\partial y}\big)G_k(s,u)\,ds + \phi_u(x).$$

Since partial differentiation is continuous with respect to the strong distribution topology, I can take the k-th partial derivative with respect to x and conclude

$$\text{(5.57)}\qquad \lim_{y\to 0^+}\Big(y^{\frac{\lambda-1}{2}}\big(\tfrac{1+\lambda}{2}-y\tfrac{\partial}{\partial y}\big)F_{\tau,v'}(x,y)\Big) = \tfrac{\partial^{k+1}}{\partial x^{k+1}}\psi_u(x).$$

As defined, $\psi_u(x)$ depends on u. However, the partial derivative of $\psi_u(x)$ with respect to x is a distribution in the variable x alone, so $a(u_1,u_2) = \psi_{u_1}(x) - \psi_{u_2}(x)$ depends on u_1 and u_2, but not on x. I conclude that the difference $\psi_u(x_1) - \psi_u(x_2)$ depends on x_1 and x_2 alone, not on u. Thus, for $x, x+\delta \in \mathbb{R}$, with $0 < \delta \leq \frac{1}{2}$,

(5.58)
$$\begin{aligned} |\,\psi_u(x+\delta) - \psi_u(x)\,| &= |\,\psi_\delta(x+\delta) - \psi_\delta(x)\,| \\ &\leq |\,\phi_\delta(x+\delta)\,| + |\,\phi_\delta(x)\,| + \delta^{-\frac{1-\mathrm{Re}\lambda}{2}} \left| \int_x^{x+\delta} (\tfrac{1+\lambda}{2} - \delta \tfrac{\partial}{\partial y})\, G_k(s,\delta)\, ds \right| . \end{aligned}$$

To simplify the notation, I set $\alpha = [\frac{1-\mathrm{Re}\lambda}{2}] - \frac{1-\mathrm{Re}\lambda}{2} + 1$ when $\frac{1-\mathrm{Re}\lambda}{2} \notin \mathbb{Z}$, and let α denote an arbitrary number strictly between 0 and 1 when $\frac{1-\mathrm{Re}\lambda}{2} \in \mathbb{Z}$; in either case, $0 < \alpha < 1$. All three terms on the right of (5.58) are bounded by multiples of δ^α – the first two because of (5.55b) and the third because of (5.36) in case $k > 0$, or because of (5.23) in case $k = 0$. Conclusion: For fixed $u > 0$, the function $\psi_u(x)$ is locally Hölder continuous of index α. In view of (5.57), this establishes the first assertion of the lemma.

When the embedding corresponding to τ is non-cuspidal and $0 < \lambda < 1$, most of the preceeding argument still applies, with $k = 0$ and accordingly, with $G_k = F_{\tau,\nu'}$. However, I have to make the following modifications: In the present situation, according to (5.23),

$$(5.59)\quad |F_{\tau,\nu'}(x,y)|,\ |y\tfrac{\partial}{\partial x}F_{\tau,\nu'}(x,y)\,|,\ |y\tfrac{\partial}{\partial y}F_{\tau,\nu'}(x,y)\,| \leq C y^{\frac{\lambda-1}{2}} \quad \text{if } 0<y\leq 1,$$

for some $C = C(\tau,\nu')$. This results in weaker bounds, namely y_2^λ instead of $y_2^{\frac{1+\mathrm{Re}\lambda}{2}}$ in (5.54) (recall $k = [\frac{1-\mathrm{Re}\lambda}{2}] = 0$), with u^λ instead of $u^{\frac{1+\mathrm{Re}\lambda}{2}}$ in (5.55b), and with multiples of δ^λ bounding the three terms in (5.58). This makes $\psi_u(x)$ Hölder continuous of index λ, and implies the first non-cuspidal case in the statement of the lemma. For the remaining non-cuspidal case, I use the equations (5.53–59) with $-\lambda$ in place of λ, with $k = 1$, and with $G_1(x,y)$ chosen as in the second part of lemma 5.36. The analogue of (5.53) in the present situation is

(5.60)
$$\begin{aligned} &\lim_{y\to 0^+} \left(y^{-\frac{\lambda+1}{2}} (\tfrac{1-\lambda}{2} - y\tfrac{\partial}{\partial y}) G_1(x,y) \right) = \\ &= u^{-\frac{\lambda+1}{2}} (\tfrac{1-\lambda}{2} - u\tfrac{\partial}{\partial y}) G_1(x,u) - \tfrac{\partial}{\partial x} \left(\int_0^u s^{-\frac{\lambda+1}{2}} \left(s\tfrac{\partial}{\partial x} - ir \right) G_1(x,s)\, ds \right), \end{aligned}$$

The weaker bounds on $|G_1(x,y)|$ and $|y\frac{\partial}{\partial x}G_1(x,y)|$ result in the inequality

(5.61)
$$\begin{aligned} &\left| \int_{y_1}^{y_2} s^{-\frac{\lambda+1}{2}} \left(s\tfrac{\partial}{\partial x} - ir \right) G_k(x,s)\, ds \right| \\ &\qquad \leq (1+|r|)C \int_{y_1}^{y_2} |\log s|\, ds \leq D y_2 |\log y_2|, \end{aligned}$$

which now replaces (5.54). In the definitions of $\phi_u(x)$ and $\psi_u(x)$, the quantity λ must be replaced by $-\lambda$. Otherwise the earlier argument applies unchanged, and shows that $\psi_u(x)$ is locally Hölder continuous of index α, for any α strictly between 0 and 1, as asserted by the lemma.

Only the limits of the discrete series remain to be dealt with. I first suppose that v' is an extreme weight vector, so that $r=-1$ or $r=1$, depending on whether π' has a highest or lowest weight; in either case, (5.9) forces $\lambda=0$. I argue again as in (5.53–59), but now using lemma 5.19 instead of corollary 5.18, and with $k=0$, $G_0(x,y)=F_{\tau,v'}(x,y)$, $\lambda=0$. Now (5.53) gets replaced by

(5.62)
$$\lim_{y\to 0^+}\left(y^{-\frac{1}{2}}F_{\tau,v'}(x,y)\right) = u^{-\frac{1}{2}}F_{\tau,v'}(x,u)-\tfrac{\partial}{\partial x}\left(i\eta\int_0^u s^{-\frac{1}{2}}F_{\tau,v'}(x,s)\,ds\right),$$

and (5.54–59) get modified accordingly. The limit is now the first derivative of $\psi_u(x)$, and $\psi_u(x)$ is locally Hölder continuous of index $\frac{1}{2}$. This establishes the third part of the lemma, in the special case of an extreme weight vector v'. The general case then follows formally; the reduction of the general to the special case involves ideas used in the proof of lemma 5.36. I shall skip the details, since the special case suffices for the proof of theorem 2.18. □

To make the connection with the statement of theorem 2.18, I suppose that (π,V) has been identified with a subrepresentation of a principal series representation. Since the Casimir operator acts as a multiple of the identity in any principal series representation, it must act by the same constant on V and on the ambient principal series representation. From (5.8) and the definition (2.5), one sees that $\pi_{\lambda,\varepsilon}(\Omega)=(\lambda^2-1)1_{V_{\lambda,\varepsilon}}$. Comparing this with (5.9) and noting that the Casimir operator is formally self-adjoint, I conclude that the constant λ in the statement of theorem 2.18 coincides with $\pm\lambda$ in this section. Since $\pm\lambda$ play symmetric roles, I may as well suppose that the constants λ in theorem 2.18 and in lemma 5.52 agree. Thus

$$V^{-\infty}\hookrightarrow V^{-\infty}_{\lambda,\varepsilon},\qquad V^{\infty}_{-\lambda,\varepsilon}\twoheadrightarrow V'^{\infty}, \tag{5.63}$$

with $\varepsilon=+1$ or $\varepsilon=-1$, depending on whether V has only even or only odd $K_{\mathbb{R}}$-weights. When π is a representation of the principal or complementary series, the maps (5.63) are isomorphisms, and $\pm\lambda$ occur on an equal footing. On the other hand, (5.63) forces

(5.64)
$$\lambda\in\mathbb{Z}_{\leq 0},\ \ \lambda\equiv\tfrac{\varepsilon+1}{2}\mod(2),\quad\text{if }\pi\text{ belongs to the discrete series or limits of the discrete series,}$$

and $(\lambda,\varepsilon)=(1,+1)$ when π is the trivial representation. In the latter two cases, the symmetry between λ and $-\lambda$ is broken. However, the hypothesis $\mathrm{Re}\lambda\leq 1$ of lemma 5.52 is satisfied whenever embeddings and quotient maps (5.63) exist. Recall that τ_λ denotes the image in $(V^{-\infty}_{\lambda,\varepsilon})^\Gamma$ of the distribution vector (1.2–3).

5.65 Lemma. *When* $(\lambda,\varepsilon)\neq(0,-1)$, $\lim_{y\to 0^+}\left(y^{\frac{\lambda-1}{2}}\left(\frac{1+\lambda}{2}-y\frac{\partial}{\partial y}\right)F_{\tau,v'}(x,y)\right)$ *is a non-zero multiple of the image of* τ_λ *in* $C^{-\infty}(\mathbb{R})$, *via the distribution analogue of the isomorphism (2.8). In the remaining case* $(\lambda,\varepsilon)=(0,-1)$, *the limit* $\lim_{y\to 0^+}y^{-\frac{1}{2}}F_{\tau,v'}(x,y)$ *is a non-zero multiple of the image of* τ_λ *in* $C^{-\infty}(\mathbb{R})$, *again via the distribution analogue of the isomorphism (2.8).*

Note that the existence of the limits referred to in the preceeding statement has already been shown in the proof of lemma 5.52, where τ was assumed to arise from an embedding $V'\hookrightarrow L^2(\Gamma\backslash G_\mathbb{R})$. The proof of the current lemma also establishes the existence of the limits, with respect to the strong distribution topology, without any particular hypotheses on the distribution vector $\tau\in V^{-\infty}$; this will be used in the proof of theorem 2.20.

Proof. It is possible to argue by direct computation. However, I shall give an argument that uses representation-theoretic facts valid in any semisimple group $G_\mathbb{R}$; these facts can, of course, be verified directly in the special case $G_\mathbb{R}=SL(2,\mathbb{R})$. I let V_0 and V_0' denote the subspaces of $K_\mathbb{R}$-finite vectors in V and V', respectively – in other words, the linear span of all $K_\mathbb{R}$-isotypic vectors in the two spaces. The complexified Lie algebra $\mathfrak{g}$ and its universal enveloping algebra $U(\mathfrak{g})$ act on V_0 and V_0'. For

$$v'\in V_0',\qquad v\in V_0, \tag{5.66}$$

the function $t\mapsto\langle v,\pi'(a_t)v'\rangle$ has a convergent series expansion near $t=-\infty$, in powers of e^{2t},

$$\begin{aligned}\langle v,\pi'(a_t)v'\rangle &= e^{(1-\lambda)t}\left(a_0(v,v')+a_1(v,v')e^{2t}+a_2(v,v')e^{4t}+\dots\right)\\ &\quad+e^{(1+\lambda)t}\left(b_0(v,v')+b_1(v,v')e^{2t}+b_2(v,v')e^{4t}+\dots\right)\end{aligned} \tag{5.67a}$$

when π is a principal series or complementary series representation and $(\lambda,\varepsilon)\neq(0,+1)$,

$$\begin{aligned}\langle v,\pi'(a_t)v'\rangle &= t\,e^{t}\left(a_0(v,v')+a_1(v,v')e^{2t}+a_2(v,v')e^{4t}+\dots\right)\\ &\quad+e^{t}\left(b_0(v,v')+b_1(v,v')e^{2t}+b_2(v,v')e^{4t}+\dots\right)\end{aligned} \tag{5.67b}$$

when π is the principal series representation with $(\lambda,\varepsilon)=(0,+1)$, and

$$\langle v,\pi'(a_t)v'\rangle = e^{(1-\lambda)t}\left(a_0(v,v')+a_1(v,v')e^{2t}+a_2(v,v')e^{4t}+\dots\right) \tag{5.67c}$$

when π is a representation of the discrete series or limits of the discrete series and $\lambda \leq 0$, as in (5.64) [7]. The function $t \mapsto \langle v, \pi'(a_t)v' \rangle$, with v and v' $K_{\mathbb{R}}$-isotypic, satisfies an ordinary differential equation in the variable $y = e^{2t}$, with a regular singular point at $y = 0$. The indicial equation has distinct roots precisely when $\lambda \neq 0$, and this explains the appearance of the linear term in (5.67b); the fact that no linear term appears in (5.67c) with $\lambda = 0$ – in the other case when the two roots coincide – reflects the reducibility of the principal series representation $V_{0,-1}$. The coefficients are uniquely determined, except in the situation (5.67a) with $\lambda \in \mathbb{Z}$. This exceptional case never occurs for an irreducible unitary representation, and so the coefficients *are* uniquely determined in the cases of interest.

The series (5.67) can be differentiated term-by-term, since the change of variables $y = e^{2t}$ transforms them into slight variants of power series. Also, certain properties of the coefficients hold for formal reasons. Specifically, we shall use:

$$\text{(5.68)}\qquad \begin{aligned} a_0(\pi(h)v, v') &= (\lambda - 1)\, a_0(v, v'), & a_0(\pi(f)v, v') &= 0, \\ a_0(v, \pi'(h)v') &= (-\lambda + 1)\, a_0(v, v'), & a_0(v, \pi'(e)v') &= 0. \end{aligned}$$

Indeed, the two identities on the left reflect the fact that h is the infinitesimal generator of the one-parameter group $\{a_t\}$. As for the other two, $\operatorname{Ad} a_t(f) = e^{-2t} f$, so $a_0(\pi(f)v, v')$ is the coefficient of $e^{(-\lambda-1)t}$, or of te^{-t} if $(\lambda, \varepsilon) = (0, +1)$, in the series expansion of $t \mapsto \langle \pi(f)v, \pi'(a_t)v' \rangle$, but this term does not occur at all; $a_0(v, \pi'(e)v')$ vanishes for the same reason. These assertions hold uniformly for the series (5.67a-c), with $\lambda = 0$ in the case of (5.67b).

The series may not converge when v is only a C^∞ vector, and not a $K_{\mathbb{R}}$-finite vector, as has been assumed so far. However,

(5.69)

a) for $v' \in V_0'$ fixed, the linear functions $v \mapsto a_j(v, v')$ and $v \mapsto b_j(v, v')$ are continuous in the topology of V^∞;

b) for $v \in V^\infty$ and $v' \in V_0'$, the series (5.67) remain valid in the weaker sense of asymptotic expansions for $t \to -\infty$,

with the k-th remainder bounded by a multiple of the same power of e^t as that of the next term, and a coefficient that is one of the semi-norms defining the topology of V^∞ [11]. Because of a), the asymptotic expansions can still be differentiated term-by-term.

I set $\mathfrak{n} = \mathbb{C}e$, $\mathfrak{n}^- = \mathbb{C}f$; this notation is consistent with the conventions of §2. Since a_0 is a bilinear function of its two arguments, (5.68) implies that it induces a bilinear pairing between $V_0/\pi(\mathfrak{n}^-)V_0$ and $V_0'/\pi'(\mathfrak{n})V_0'$. Since $\{a_t\}$ normalizes $\mathfrak{n}$ and $\mathfrak{n}^-$, h operates on both quotients. For $\mu \in \mathbb{C}$, I define

$$\text{(5.70)}\qquad \begin{aligned} (V_0'/\pi'(\mathfrak{n})V_0')_\mu &=_{\text{def}} (\mu + 1) - \text{eigenspace of } \pi'(h), \\ (V_0/\pi(\mathfrak{n}^-)V_0)_\mu &=_{\text{def}} (\mu - 1) - \text{eigenspace of } \pi'(h); \end{aligned}$$

the reason for the shifts by ± 1 will become apparent soon. Now, invoking (5.68) once more, I find

(5.71)

$$a_0(\ ,\ )\ \text{induces a bilinear pairing}\ \left(V_0/\pi(\mathfrak{n}^-)\right)_\lambda \times \left(V_0'/\pi'(\mathfrak{n})V_0'\right)_{-\lambda} \longrightarrow \mathbb{C}.$$

For the next statement, I set $\varepsilon = +1$ or $\varepsilon = -1$, depending on whether π has even or odd $K_{\mathbb{R}}$-weights. Then, according to Casselman's reciprocity theorem,

(5.72)
$$\begin{aligned} \left(V_0/\pi(\mathfrak{n}^-)V_0\right)_\lambda &\cong \operatorname{Hom}_{G_{\mathbb{R}}}(V^\infty, V_{\lambda,\varepsilon}^\infty), \\ \left(V_0'/\pi'(\mathfrak{n})V_0'\right)_{-\lambda} &\cong \operatorname{Hom}_{G_{\mathbb{R}}}(V'^\infty, V_{\lambda,\varepsilon}'^{\ \infty}), \end{aligned}$$
and the pairing (5.71) is non-degenerate.

In the case of the group $G_{\mathbb{R}} = SL(2,\mathbb{R})$, embeddings into (possibly non-unitary) principal series representations are unique up to scaling. Thus, in the situation (5.63),

(5.73)
$$\dim\left(V_0/\pi(\mathfrak{n}^-)V_0\right)_\lambda = \dim\left(V_0'/\pi'(\mathfrak{n})V_0'\right)_{-\lambda} = 1.$$

The isomorphism (5.72) can be described explicitely, as follows. I fix a vector $v' \in V_0'$ whose image generates $\left(V_0'/\pi'(\mathfrak{n})V_0'\right)_{-\lambda}$. Then $v \mapsto a_0(v,v')$ spans the dual space of $\left(V_0/\pi(\mathfrak{n}^-)V_0\right)_\lambda$, and

(5.74a)
$$\begin{aligned} V^\infty \hookrightarrow V_{\lambda,\varepsilon}^\infty &\cong \{ f \in C^\infty(\mathbb{R}) \mid (\operatorname{sgn} x)^{\frac{1+\varepsilon}{2}} |x|^{\lambda-1} f(\tfrac{1}{x}) \in C^\infty(\mathbb{R}) \} \\ V^\infty \ni v \longmapsto f_v &\in C^\infty(\mathbb{R}),\ f_v(x) = a_0(\pi(n_{-x})v, v'); \end{aligned}$$

this involves the isomorphism (2.8), the notation (2.3), and (5.69a), which ensures that $a_0(\pi(n_{-x})v,v')$ is well-defined for $v \in V^\infty$. The embedding is usually described in terms of the "compact realization" (2.7) of $V_{\lambda,\varepsilon}^\infty$, but the two statements are equivalent because of the formal properties (5.68) of the bilinear function a_0. Embeddings into principal series representations extend continuously to the spaces of distribution vectors. In particular,

(5.74b)
$$\text{the inclusion (5.74a) extends continuously to } V^{-\infty} \hookrightarrow V_{\lambda,\varepsilon}^{-\infty}.$$

The general facts I have used here are proved in [5, 11, 6]. The inclusion (5.74b) is unique up to scaling, and thus coincides with (5.63), again up to scaling. The function corresponding to a non-zero $K_{\mathbb{R}}$-isotypic vector $v \in V_{\lambda,\varepsilon}^\infty$ via the "compact realization" (2.7), suitably renormalized, is a character of $K_{\mathbb{R}}$, and therefore takes the value 1 at the identity. In terms of the "unbounded realization" (2.8), this means that $f_v(0) = 1$. Thus, by (5.68) and (5.73–74), any $K_{\mathbb{R}}$-isotypic vector in V maps to a generator of $\left(V_0/\pi(\mathfrak{n}^-)V_0\right)_\lambda$. Since π and π' play symmetric roles in (5.68),

(5.75)

the image of any non-zero $K_{\mathbb{R}}$-isotypic vector in V' spans $\left(V_0'/\pi'(\mathfrak{n})V_0'\right)_{-\lambda}$.

In particular, the vector (5.5) which was used in the definition of $F_{\tau,v'}$ can and shall play the role of v' in (5.74). Multiplying v' by a suitable multiple, I can arrange that (5.74b) coincides exactly with (5.63), not just up to a scale factor.

Distributions do not have values at points. For that reason, one cannot describe (5.74b) quite as explicitly as the embedding on the level of C^∞-vectors. However, one can get a similar expression after smoothing distribution vectors by integration against a test function. For $\psi \in C_c^\infty(\mathbb{R})$, I define

$$\text{(5.76a)} \qquad \pi_{N_{\mathbb{R}}}(\psi) = \int_{\mathbb{R}} \psi(x)\,\pi(n_x)\,dx \in \operatorname{End}(V^\infty).$$

The integral is well defined because the $\operatorname{End}(V^\infty)$-valued function $x \mapsto \psi(x)\pi(n_x)$ is continuous in the strong operator topology and has compact support. I claim:

(5.76b) $\quad \pi_{N_{\mathbb{R}}}(\psi)$ extends continuously to $V^{-\infty}$, and is a smoothing operator,

in the sense that the extension is a linear map $V^{-\infty} \to V^\infty$, continuous with respect to the strong distribution topology on $V^{-\infty}$ and the C^∞ topology on V^∞. To see this, I pick $v \in V^\infty$ and let $f_v \in C^\infty(\mathbb{R})$ denote the function which corresponds to v via the embedding $V^\infty \hookrightarrow V^\infty_{\lambda,\varepsilon}$ and the isomorphism (2.8). Recall that $N_{\mathbb{R}} \cong \mathbb{R}$ acts on f_v via translation – cf. (2.9). Thus, with $w = \pi_{N_{\mathbb{R}}}(\psi)v$,

$$\text{(5.77)} \qquad f_w(x) = \int_{\mathbb{R}} f_v(x-u)\,\psi(u)\,du.$$

Since $\psi \in C_c^\infty(\mathbb{R})$, the smoothing property is apparent except for the behavior of f_w at ∞. To deal with the missing point, I let $g_v, g_w \in C^\infty(K_{\mathbb{R}})$ denote the functions which correspond to v and $w = \pi_{N_{\mathbb{R}}}(\psi)v$ via the "compact realization" (2.7) of $V^\infty_{\lambda,\varepsilon}$. They are related to f_v and f_w by the formulas

(5.78)

$$f_v(\tan\theta) = (\cos\theta)^{1-\lambda} g_v(k_\theta)\,, \quad f_w(\tan\theta) = (\cos\theta)^{1-\lambda} g_w(k_\theta) \quad \left(-\tfrac{\pi}{2} < \theta < \tfrac{\pi}{2}\right),$$

which is a consequence of the definition (2.5) and the identity

$$\text{(5.79)} \quad k_\theta = n_x a_t n^- \quad \text{when} \quad -\tfrac{\pi}{2} < \theta < \tfrac{\pi}{2},\ x = \tan\theta\,,\ e^{-t} = \cos\theta\,,\ n^- \in N^-_{\mathbb{R}}.$$

With this change of variables, the equation (5.77) takes the form

$$g_w(k_\theta) = \int_{\mathbb{R}} \left(\frac{\cos\theta}{\cos\alpha(\theta,u)} \right)^{\lambda-1} g_v(k_{\alpha(\theta,u)})\, \psi(u)\, du, \tag{5.80}$$

$$\text{where} \quad \alpha(\theta,u) = \tan^{-1}(\tan\theta - u).$$

Near $\theta = \frac{\pi}{2}$, the function $\cot\theta$ is a coordinate, and

$$\cot\alpha(\theta,u) = (1-u\cot\theta)^{-1}\cot\theta, \quad \alpha(\tfrac{\pi}{2},u) = \tfrac{\pi}{2}. \tag{5.81}$$

It follows that $\alpha(\theta,u)$ is a real analytic function, and $\frac{\cos\theta}{\cos\alpha(\theta,u)}$ takes the value 1 at $\theta = \frac{\pi}{2}$, independently of the value of u. These facts, coupled with the hypothesis that ψ have compact support, imply the smoothing property of $\pi_{N_\mathbb{R}}$ also at the point at infinity.

I continue with the notation f_v, f_w, for $v \in V^\infty$ and $w = \pi_{N_\mathbb{R}}(\psi)v$. From the formulas (5.74a), (5.77), one sees that

$$a_0(\pi_{N_\mathbb{R}}(\psi)v, v') = f_w(0) = \int_{\mathbb{R}} f_v(u)\, \psi(-u)\, du. \tag{5.82}$$

In this identity, I let v run through a sequence $\{v_n\}$ in V^∞ which converges to the distribution τ in the topology of $V^{-\infty}$. The corresponding sequence of functions f_{v_n} then converges in $C^{-\infty}(\mathbb{R})$ to f_τ – i.e., the image of τ via (5.74b); note that f_τ is the image of τ_λ under the distribution analogue of (2.8). In the statement of corollary 2.19, this distribution is denoted by $\tilde{\tau}$. As $v_n \to \tau$, the sequence of C^∞ vectors $\pi_{N_\mathbb{R}}(\psi)v_n$ converges in V^∞ to $\pi_{N_\mathbb{R}}(\psi)\tau$; here I am using the smoothing property of $\pi_{N_\mathbb{R}}$. Taking limits, I find

$$a_0(\pi_{N_\mathbb{R}}(\psi)\tau, v') = \int_{\mathbb{R}} f_\tau(u)\, \psi(-u)\, du. \tag{5.83}$$

The integral on the right is to be taken in the symbolic sense: I formally write the natural pairing between distributions and smooth measures as an integral.

I now substitute $\pi_{N_\mathbb{R}}(\psi)\tau \in V^\infty$ for the C^∞ vector v in the asymptotic expansion (5.70b). I shall show:

$$\lim_{t\to-\infty} \left(e^{(\lambda-1)t}(1+\lambda-\tfrac{\partial}{\partial t})\, \langle \pi_{N_\mathbb{R}}(\psi)\tau, \pi'(a_t)v' \rangle \right) =$$

$$= \begin{cases} 2\lambda\, a_0(\pi_{N_\mathbb{R}}(\psi)\tau, v') & \text{if } \lambda \neq 0, \\ -a_0(\pi_{N_\mathbb{R}}(\psi)\tau, v') & \text{if } (\lambda,\varepsilon) = (0,+1), \\ 0 & \text{if } (\lambda,\varepsilon) = (0,-1). \end{cases} \tag{5.84}$$

This is evidently correct on the level of leading terms; the only potential obstacle would be a non-zero coefficient $b_j(\pi_{N_\mathbb{R}}(\psi)\tau, v')$ with $0 < j < -\mathrm{Re}\,\lambda$. However,

$\operatorname{Re}\lambda > -1$ unless π is a discrete series representation, in which case all the b_j vanish identically. In view of (5.6),

$$\begin{aligned}\langle \pi_{N_{\mathbb{R}}}(\psi)\tau, \pi'(a_t)v' \rangle &= \int_{\mathbb{R}} \psi(x) \langle \pi(n_x)\tau, \pi'(a_t)v' \rangle \, dx \\ &= \int_{\mathbb{R}} \psi(x) \langle \tau, \pi'(n_{-x}a_t)v' \rangle \, dx \\ &= \int_{\mathbb{R}} \psi(-x) \langle \tau, \pi'(n_x a_t)v' \rangle \, dx = \int_{\mathbb{R}} \psi(-x) F_{\tau,v'}(x,y) \, dx,\end{aligned} \tag{5.85}$$

with $y = e^{2t}$, as usual. At this point, (5.83–84) imply

$$\begin{aligned}\lim_{y\to 0^+} \left(y^{\frac{\lambda-1}{2}} \left(\tfrac{1+\lambda}{2} - y \tfrac{\partial}{\partial y} \right) \int_{\mathbb{R}} \psi(-x) F_{\tau,v'}(x,y) \, dx \right) &= \\ = c(\lambda,\varepsilon) \int_R \psi(-x) f_\tau(x) \, dx;\end{aligned} \tag{5.86}$$

the constant $c(\lambda,\varepsilon)$ equals λ, $-\frac{1}{2}$, or 0, depending on whether $\lambda \neq 0$, $(\lambda,\varepsilon) = (0,+1)$, or $(\lambda,\varepsilon) = (0,-1)$.

The limit mentioned in the statement of the lemma exists, in the strong distribution topology, because $\pi_{N_{\mathbb{R}}}(\psi)$ is continuous with respect to that topology. Since ψ is an arbitrary test function, (5.86) implies the first part of the lemma – recall the definition of f_τ as the image of τ_λ under the distribution analogue of (2.8). The second assertion of the lemma follows from

$$\lim_{y\to 0^+} \left(y^{-\frac{1}{2}} \psi(-x) F_{\tau,v'}(x,y) \, dx \right) = \int_R \psi(-x) f_\tau(x) \, dx, \tag{5.87}$$

which is entirely analogous to (5.86). □

Taken together, lemmas 5.52 and 5.65 show that τ_λ has the degree of regularity asserted by theorem 2.18, except possibly near the point at infinity. But the $G_{\mathbb{R}}$ acts transitively on $\mathbb{R} \cup \{\infty\}$, so no point plays an exceptional role.

Proof of theorem 2.20. I must show that when τ satisfies the hypotheses of the theorem, the functions

$$F_{\tau,v'}(x,y) = \langle \tau, \pi'(n_x a_t)v' \rangle, \qquad \text{with } e^{2t} = y, \tag{5.88}$$

for some non-zero, $K_{\mathbb{R}}$-isotypic vector $v' \in V'$, is square-integrable over $\mathscr{H} \cong \Gamma\backslash G_{\mathbb{R}}$. Note that $F_{\tau,v'}$ is a well-defined, C^∞ function, because v' is a C^∞ vector. Much of the discussion at the beginning of this section applies to this function – anything that does not depend on knowing the square-integrability of $F_{\tau,v'}$ in the first place. The fact that τ is a distribution vector makes $F_{\tau,v'}$ have at most polynomial growth,

$$|F_{\tau,v'}(x,y)| \leq C \max\{y^a, y^{-a}\}, \qquad \text{for some } C > 0, a > 0. \tag{5.89}$$

To see this, one can express τ as $\pi(\Omega_K)^m(v)$, an integral power of the Casimir operator of $K_{\mathbb{R}}$ applied to a vector v in the unitary representation space V. This makes $F_{\tau,v'}$ a – possibly high – derivative of a bounded function satisfying an elliptic differential equation. From here on, one can argue as in the proof of Lemma 5.23.

The square-integrability of $F_{\tau,v'}$ is entirely a matter of the behavior at the cusps: Because of the conditions on Γ, there are finitely many cusps, and away from the cusps, $\Gamma\backslash G_{\mathbb{R}}$ is compact. The group $G_{\mathbb{R}}$ acts transitively on $\mathbb{R}\cup\{\infty\}$, so nothing is lost by looking only at a potential cusp at the point at infinity. Further, I may assume that $z \mapsto z+1$ generates the action of $\Gamma \cap N_{\mathbb{R}}$. I can then express $F_{\tau,v'}$ as a Fourier series

$$F_{\tau,v'}(x,y) \;=\; \textstyle\sum_{-\infty<k<\infty} a_k(y)\, e^{2\pi i k x}\,, \tag{5.90}$$

as in (5.29). The Fourier coefficients $a_k(y)$ satisfy the differential equation (5.31), for the same reason as in the earlier case. The boundedness condition (5.89) implies that the $a_k(y)$, for $k \neq 0$, decay exponentially – again, as explained in the proof of lemma 5.23. Thus square-integrabilty near the cusp at infinity hinges on $a_0(y)$: In the notation of (5.35), the square-integrability around the cusp at infinity is equivalent to

(5.91)

a) $c_+ \;=\; c_- \;=\; 0$ if τ corresponds to a cuspidal embedding of a representation π' of the unitary principal series or complementary series;

b) $c_+ \;=\; 0$ if τ corresponds to a possibly non-cuspidal embedding of a complementary series representation and $\lambda > 0$;

c) $c_- \;=\; 0$ if τ corresponds to a possibly non-cuspidal embedding of a complementary series representation and $\lambda < 0$;

d) $c_- \;=\; 0$ if τ corresponds to an embedding of a representation of the discrete series or limits of the discrete series;

in the latter case λ is a non-negative integer and c_+ vanishes automatically, as a consequence of the first order equation (5.22).

As before, I write $\tilde{\tau}_\lambda$ for the image of τ via $V^{-\infty} \hookrightarrow V^{-\infty}_{\lambda,\varepsilon}$ and the distribution analogue of (2.8). This is a distribution, periodic of period 1,

$$\tilde{\tau}_\lambda(x) \;=\; \textstyle\sum_{-\infty<k<\infty} a_k\, e^{2\pi i k x}\,. \tag{5.92}$$

According to the hypotheses, τ_λ – and therefore also $\tilde{\tau}_\lambda$ – is the n-th derivative of a continuous function, with

$$n \;=\; \begin{cases} 1 & \text{if } \pi \text{ belongs to the principal or complementary series} \\ [\frac{3}{2}-\frac{\lambda}{2}] & \text{if } \pi \text{ belongs to the discrete series or its limits;} \end{cases} \tag{5.93}$$

in the first case $-1 < \mathrm{Re}\,\lambda < 1$, and in the latter λ is a non-positive integer. Thus I can write

$$\tilde{\tau}_\lambda(x) = a_0 + \psi_\lambda^{(n)}(x), \tag{5.94}$$

where ψ_λ is continuous, periodic, with zero constant term.

The expression (5.94) determines $\tilde{\tau}_\lambda$ only away from the point at infinity. Let me show how to give meaning to $\psi_\lambda^{(n)}$ not only as a distribution on the real line, but as vector in $V_{\lambda,\varepsilon}^{-\infty}$. That means extending the distribution $x \mapsto (\mathrm{sgn}\,x)^{\frac{1-\varepsilon}{2}} |x|^{\lambda-1} \psi_\lambda^{(n)}(\frac{1}{x})$ across $x = 0$ – see (2.8). To simplify the discussion, I look at positive x separately; the case of negative x can be handled the same way. For $j > 0$, I let $\psi_\lambda^{(-j)}$ denote the j-th anti-derivative of the continuous function ψ_λ; it becomes well-defined when one requires it to be periodic with zero constant term. Using the formal identity

(5.95)

$$x^\alpha h'(\tfrac{1}{x}) = \tfrac{d}{dx}\left(-x^{\alpha+2}h(\tfrac{1}{x}) - (\alpha+2)x^{\alpha+3}h^{(-1)}(\tfrac{1}{x})\right) + (\alpha+2)(\alpha+3)x^{\alpha+2}h^{(-1)}(\tfrac{1}{x})$$

repeatedly, I find

$$x^{\lambda-1}\psi_\lambda^{(n)}(\tfrac{1}{x}) = \textstyle\sum_{0\le i,j,\ell\le n} a_{i,j,\ell}\, \frac{d^\ell}{dx^\ell}\left(x^{\lambda-1+2n+i}\,\psi_\lambda^{(-j)}(\tfrac{1}{x})\right), \tag{5.96}$$

with suitable coefficients $a_{i,j,\ell}$. Since $\mathrm{Re}\,\lambda - 1 + 2n + i > 0$ in the range of summation, (5.96) exhibits $(\mathrm{sgn}\,x)^{\frac{1-\varepsilon}{2}} |x|^{\lambda-1} \psi_\lambda^{(n)}(\frac{1}{x})$ as a distribution even across $x = 0$, a distribution which is also the n-th derivative of a continuous function. The contribution of the term a_0 in (5.94), in terms of the coordinate system at infinity, is

(5.97)

$$(\mathrm{sgn}\,x)^{\frac{1-\varepsilon}{2}} |x|^{\lambda-1} a_0 = \begin{cases} a_0 (\mathrm{sgn}\,x)^{\frac{1-\varepsilon}{2}+n} \frac{d^n}{dx^n}\left(\frac{1}{\lambda(\lambda-1)\ldots(\lambda+1-n)}|x|^{\lambda-1+n}\right) & \text{if } \lambda \neq 0 \\ a_0 (\mathrm{sgn}\,x)^{\frac{1+\varepsilon}{2}} \frac{d}{dx}\log|x| & \text{if } \lambda = 0; \end{cases}$$

in the cases covered by theorem 2.20, this term does not have the required degree of regularity except in the situation b), with $0 < \lambda < 1$. The expression (5.94) cannot detect a delta function or derivative of a delta function at the point at infinity. However, a delta function has degree of regularity exactly $C^{<(-1)}$, and is not allowed by the hypotheses of theorem 2.20, except in the situation b), with $-1 < \lambda < 0$, and conceivably also when λ is a strictly negative integer.

To complete the argument, I must show that the hypotheses of theorem 2.20 translate into the conditions (5.91). Recall the definition (5.35). Because of the existence of the limit in 5.65 in the strong distribution topology, the Fourier coefficients in (5.90) and (5.92) are matched in the limiting process. Specifically

$$a_0 = 0 \iff \begin{cases} c_- = 0 & \text{if } (\lambda,\varepsilon) \neq (0,+1) \\ c_+ = 0 & \text{if } (\lambda,\varepsilon) = (0,+1); \end{cases} \tag{5.98}$$

the factor $\log y$ in (5.35) when $(\lambda,\varepsilon) = (0,+1)$ accounts for the reversal of c_+ and c_- in the second case. The coefficient $c_\pm$ not identified by (5.98) disappears in the limiting process when one looks only at finite values of x. On general grounds, it must correspond to a distribution supported at the point at infinity, and invariant under the action of $N_{\mathbb{R}}$. The only such distribution is the delta function at infinity, hence

(5.99)

$$\tau_\lambda \text{ involves the delta function at } \infty \iff \begin{cases} c_+ = 0 & \text{if } (\lambda,\varepsilon) \neq (0,+1) \\ c_- = 0 & \text{if } (\lambda,\varepsilon) = (0,+1)\,. \end{cases}$$

One can see (5.98–99) also directly by computing $F_{\tau,\nu'}$ in terms of τ and ν' from the defining formula (5.6). In the case of a representation of the discrete series or limits of the discrete series, (5.99) holds vacuously since c_+ is known to vanish; on the other hand, the delta function cannot arise as the boundary value of a holomorphic or anti-holomorphic function. The translation of the hypotheses of theorem 2.20 in the previous paragraph now connects directly with (5.91) and (5.98–99). □

References

1. Beardon, A. F.: *The Geometry of Discrete Groups*, Graduate Texts in Mathematics 91, Springer, New York-Berlin, 1983.
2. Bernstein, S.: Sur la convergence absolue des séries trigonometriques, *C. R. Acad. Sci. Paris* **158** (1914), 1661–1664.
3. Bernstein, J. N. and Reznikov, A.: Sobolev norms of automorphic functionals and Fourier coefficients of cusp forms, *C. R. Acad. Sci. Paris Sér. I Math.* **327** (1998), 111–116.
4. Bernstein, J. N. and Reznikov, A.: Analytic continuation of representation and estimates of automorphic forms, *Ann. of Math.* **150** (1999), 329–352.
5. Casselman, W.: Jacquet modules for real reductive groups, in: *Proceedings of the International Congress of Mathematicians, Helsinki, 1978*, Acad. Sci. Fennica, Helsinki, 1980, pp. 557–563.
6. Casselman, W.: Canonical extensions of Harish-Chandra modules to representations of G, *Canad. J. Math.* **41** (1989), 385–438.
7. Casselman, W. and Miličić, D.: Asymptotic behavior of matrix coefficients, *Duke Math. J.* **49** (1982), 869–930.
8. Coddington, E. A. and Levinson, N.: *Theory of Ordinary Differential Equations*, McGraw-Hill, New York, 1955.
9. Folland, G. B.: *Real Analysis*, Wiley-Interscience, New York, 1984.
10. Maass, H.: Über eine neue Art von nichtanalytischen automorphen Funktionen und die Bestimmung Dirichletscher Reihen durch Funktionalgleichungen, *Math. Ann.* **121** (1949), 141–183.
11. Wallach, N. R.: Asymptotic expansions of generalized matrix entries of representations of real reductive groups, in: R. Herb, R. Lipsman, and J. Rosenbeg, (eds.), *Lie Group Representations I*, Lecture Notes in Math. 1024, Springer, Berlin-New York, 1983, pp. 287–369.

Nonlinear relativistic evolution equations: Survey of a new approach

Jacques C. H. Simon
Laboratoire Gevrey de mathématique physique
Université de Bourgogne, B.P. 47870, 21078 Dijon cedex, France
e-mail: `jacques.simon@u-bourgogne.fr`

Erik Taflin
AXA, 23 avenue Matignon, 75008 Paris, France
e-mail: `erik.taflin@ceremade.dauphine.fr`, `erik.taflin@u-bourgogne.fr`

Abstract. We review the theory of global existence and scattering for both nonlinear Klein-Gordon and Maxwell-Dirac equations in the framework of nonlinear representations of the Poincaré group.

Keywords: relativistic nonlinear wave equation, linearization map, Cauchy problem, scattering, nonlinear representation of Lie groups and algebras

Mathematics Subject Classifications (2000): 35L70, 35P25, 22E70, 81R20

1. Introduction

The Conférence Moshé Flato is an opportunity for us to report on the results obtained in the field of relativistic nonlinear wave equations during a collaboration composed of Moshé Flato, its initiator, and the authors. It started in 1977 (see [1]) from some heuristic observations showing that relativistic covariance implies constraints that could give information on the solutions of wave equations. The present account concentrates on the results we have obtained since then. The reader can find references to related work in the papers listed at the end of the paper.

For several years, we came back periodically to classical electrodynamics, while we were refining the techniques for nonlinear Klein-Gordon equation (NLKG). For NLKG, we have solved the Cauchy scattering and asymptotic completeness problems and constructed the nonlinear representations of the Poincaré group for arbitrary C^∞-coupling for any space dimension $n \geq 2$ and for small initial data [9]. The nonlinear representation is linearized by wave operators.

The equations of classical electrodynamics are the coupled Maxwell-Dirac equations. By introducing long-range corrections and using Hamilton-Jacobi techniques, we first proved in [2] the existence of global solutions in time. We then solved the Cauchy problem, the problem of scattering and asymptotic completeness for small initial data. We constructed the nonlinear representation of the Poincaré group and also the corresponding asymptotic representations (see[3]). To our surprise, we discovered that the asymptotic representations are not linear representations!

G. Dito and D. Sternheimer (eds.), Conférence Moshé Flato 1999, Vol. 1, 389 – 402.

In this article, we are going to expose the main ideas which will be developed in a way that is easy to understand. In particular no inequalities will be displayed and calculations will be explained at a heuristic level.

We shall use the following notations for the Minkowski space: $\eta^{\mu\nu}, 0 \leq \mu, \nu \leq 3$, is the metric tensor $\eta^{00} = 1$, $\eta^{ii} = -1$, $\eta^{ij} = 0$, for $1 \leq i, j \leq 3$, $i \neq j$;

$$\square = \eta^{\mu\nu}\partial_\mu\partial_\nu, \qquad \Delta = \sum_{i=1}^{3} \partial_i\partial_i.$$

The Dirac 4×4 complex matrices γ^μ, $0 \leq \mu \leq 3$, satisfy

$$\gamma^\mu\gamma^\nu + \gamma^\nu\gamma^\mu = 2\eta^{\mu\nu}, \quad 0 \leq \mu, \nu \leq 3.$$

We shall denote by P_μ, $M_{\mu\nu}$, $0 \leq \mu, \nu \leq 3$, $\mu < \nu$, a standard basis of the Lie algebra of the Poincaré group of the Minkowski space.

2. Relativistic covariance and nonlinear representations

Let us now describe the concept of relativistic covariance. We denote by $\mathscr{P}$ the Poincaré group acting on the 4-dimensional Minkowski space by $gx = \Lambda x - a$, where $g = (\Lambda, a)$, $\Lambda \in SL(2, \mathbb{C})$, $a \in \mathbb{R}^4$, and Λ is acting on $\mathbb{R}^4$ as an element of $SO(1,3)$.

Let $A(\varphi) = 0$ be a nonlinear partial differential equation, where the unknown φ is an element of some space $\mathscr{E}$ of functions from the Minkowski space to a real or complex vector space V of finite dimension, realized as a first-order equation in t. In addition, we suppose that we are given a linear representation L of the Poincaré group on V. Let $\mathscr{E}_{sol}$ be the set of solutions, which usually is not a linear space. A linear representation R of $\mathscr{P}$ on $\mathscr{E}$ is defined by

$$(R_g\varphi)(m) = L_g\varphi(g^{-1}m), \quad m \in \mathbb{R}^4.$$

The field equation is said to be manifestly covariant if $R_g\mathscr{E}_{sol} \subset \mathscr{E}_{sol}$.

Assume now that the field equation can be solved uniquely from the initial Cauchy data. This means that there exists a vector space E of functions from $\mathbb{R}^3$ to V such that the mapping $\pi : \mathscr{E}_{sol} \to E$ defined by

$$(\pi\varphi)(x) = \varphi(0, x), \quad x \in \mathbb{R}^3,$$

is a bijection. We denote by $\boldsymbol{i} : E \to \mathscr{E}_{sol}$ its inverse.

At this level we are aware that we have made the "almost unrealistic" assumption that the Cauchy problem is already solved for large Cauchy data — this is to simplify the discussion.

The mapping $U_g : E \to E$ defined by

$$U_g\psi = \pi R_g \boldsymbol{i}\psi, \quad g \in \mathscr{P},$$

has the obvious properties:

$$U_e = Id_E, \tag{2.1}$$

where e is the unit element in $\mathscr{P}$;

$$U_{gg'} = U_g U_{g'}, \quad g, g' \in \mathscr{P}. \tag{2.2}$$

In addition, if $A(0) = 0$, we have

$$U_g(0) = 0, \quad g \in P. \tag{2.3}$$

We have a group action of $\mathscr{P}$ on E having 0 as a fixed point. If the Cauchy problem is solved only for small initial data, one realizes that we have a local action U of $\mathscr{P}$ on a neighborhood of 0 in E, having 0 as a fixed point. This is what we shall call in the following a nonlinear representation U of $\mathscr{P}$. We say that U is C^0 (resp. C^∞, analytic) if the mapping

$$(g, \varphi) \mapsto U_g \varphi,$$

is C^0 (resp. C^∞, analytic).

The nonlinear representation U contains all the information about the solutions of the field equation since they can be written as

$$\varphi(t,x) = (U_{\exp(tP_0)}\, \varphi_0)(x).$$

3. Nonlinear Lie algebra representations

3.1. Definitions

Assuming that U is a C^∞-nonlinear representation of $\mathscr{P}$ on E, the infinitesimal generators $T_X = \frac{d}{dt} U_{\exp(tX)} |_{t=0}$, $X \in \mathfrak{p}$, $\mathfrak{p}$ being the Lie algebra of $\mathscr{P}$, are to be considered as vector field which integrate to the group action. In fact

$$\frac{d}{dt}(U_{\exp(tP_0)}\varphi) = T_{P_0}(U_{\exp(tP_0)}\varphi) \tag{3.1}$$

is the field equation.

Let us introduce the bracket

$$[T_X, T_Y]\varphi = (D(T_X)(\varphi))(T_Y\varphi) - (D(T_Y)(\varphi))(T_X\varphi). \tag{3.2}$$

We have

$$T_X + T_Y = T_{X+Y}, \tag{3.3}$$
$$T_{[X,Y]} = [T_X, T_Y], \tag{3.4}$$
$$T_X(0) = 0. \tag{3.5}$$

When these three properties are satisfied on says that T is a nonlinear representation of $\mathfrak{p}$ in E.

Let us now look at two examples of field equations and their related nonlinear Lie algebra representation.

3.2. NLKG

An example of the nonlinear Klein Gordon (NLKG) equation for the scalar field $\varphi : \mathbb{R}\times\mathbb{R}^3 \to \mathbb{C}$ is the equation

$$(\Box + m^2)\varphi(t,x) = P\big(\varphi(t,x), \frac{\partial}{\partial t}\varphi(t,x), \nabla\varphi(t,x)\big), \quad m^2 > 0. \tag{3.6}$$

For the sake of simplicity, we shall suppose that P is a C^∞-function such that the equation is manifestly covariant.

We define $a_\varepsilon(t) = \dot\varphi(t) - \varepsilon i\omega(-i\nabla)\varphi(t)$, $\varepsilon = \pm 1$, and $\omega(-i\nabla) = (m^2 - \Delta)^{\frac{1}{2}}$. Equation (3.6) can now be rewritten as

$$\frac{d}{dt}\, a(t) = i\omega(-i\nabla)(a_+(t), -a_-(t)) + \big(F(a(t)), F(a(t))\big), \tag{3.7}$$

where $F(a(t)) = P(\varphi, \dot\varphi, \nabla\varphi) = (2i\omega(-i\nabla))^{-1}\, \nabla(a_+(t) - a_-(t))$.
Let $E_\infty = \mathscr{S}(\mathbb{R}^3, \mathbb{C}) \oplus \mathscr{S}(\mathbb{R}^3, \mathbb{C})$ and define

$$T^1_{P_0}(f_+, f_-) = i\omega(-i\nabla)(f_+, -f_-), \tag{3.8}$$

$$T^1_{Pi}(f_+, f_-) = \partial_i(f_+, f_-), \tag{3.9}$$

$$T^1_{M_{ij}}(f_+, f_-)(x) = (x_i\partial_j - x_j\partial_i)(f_+(x), f_-(x)), \tag{3.10}$$

$$T^1_{M_{0j}}(f_+, f_-)(x) = (i\omega(-i\nabla)x_j f_+(x), -i\omega(-i\nabla)x_j f_-(x)), \tag{3.11}$$

where $1 \le i \le 3$, $1 \le j \le 3$, and $(f_+, f_-) \in E_\infty$. This defines a linear Lie algebra representation of the Lie algebra $\mathfrak{p}$ on E_∞, which we extend to the universal enveloping algebra $\mathscr{U}(\mathfrak{p})$.

Let $E = L^2(\mathbb{R}^3, \mathbb{C}) \oplus L^2(\mathbb{R}^3, \mathbb{C})$ and let $\{X_1, \ldots, X_{10}\}$ be a basis of $\mathfrak{p}$. The set Π of elements of $\mathscr{U}(\mathfrak{p})$ of the form $y = X_1^{i_1} \cdots X_{10}^{i_{10}}$ is a basis of $\mathscr{U}(\mathfrak{p})$. Set $|y| = i_1 + \cdots + i_{10}$.

Let E_i, $i \ge 0$, be the completed space of E_∞ for the norm

$$\| f \|_{E_i} = \Big(\sum_{y\in\Pi, |y|\le i} \| T^i_y f \|^2_E \Big)^{\frac{1}{2}}. \tag{3.12}$$

T^1 is the differential of a continuous linear representation U^1 of $\mathscr{P}$ on E, E_∞ is the space of differentiable vectors of U^1, and E_i, $i \ge 0$, is invariant under U^1.

We now define $T_X = T^1_X + \hat T_X$, $X \in \mathfrak{p}$, by

$$\hat T_{P_0}(f) = (F(f), F(f)), \tag{3.13}$$

$$\hat T_{P_i}(f) = 0, \quad \hat T_{M_{ij}}(f) = 0, \tag{3.14}$$

$$\hat T_{M_{0i}}(f) = (x_j F(f), x_j F(f)), \tag{3.15}$$

where $1 \leq i \leq 3$, $1 \leq j \leq 3$, $f \in E_\infty$.

We showed that T_X, $X \in \mathfrak{p}$, is semi-Lipschitz-Ian for $N \geq 0$ sufficiently large and then, by using contraction methods, we showed that there exists a local group action U of $\mathscr{P}$ on E_N such that

$$\frac{d}{dt}(U_{\exp(tX)}(a))_{t=0} = T_X(U_{exp(tX)}(a)), \quad a \in E_\infty.$$

3.3. Maxwell-Dirac equations

We next consider the example of the Maxwell-Dirac (MD) equations. Here $A^\mu : \mathbb{R}^4 \to \mathbb{R}^4$ is the electromagnetic vector potential and $\psi : \mathbb{R}^4 \to \mathbb{C}^4$ is the electron field. The MD equations read

$$\Box A^\mu = \bar{\psi}\gamma^\mu \psi, \quad \text{where } \bar{\psi} = \psi^* \gamma^0, \tag{3.16}$$

$$(i\gamma^\mu \partial_\mu + m)\psi = A_\mu \gamma^\mu \psi, \tag{3.17}$$

$$\partial_\mu A^\mu = 0. \tag{3.18}$$

By writing $(A_\mu(t))(x) = A_\mu(t,x)$ and $(\psi(t))(x) = \psi(t,x)$, we can express Equations (3.16) and (3.17) as a first-order system

$$\frac{d}{dt}(A_\mu(t),\dot{A}_\mu(t)) =(\dot{A}_\mu(t),\Delta A_\mu(t)) + (0, \bar{\psi}(t)\gamma^\mu \psi(t)), \tag{3.19}$$

$$\frac{d}{dt}\psi(t) =\mathscr{D}\psi(t) - iA_\mu(t)\gamma^0\gamma^\mu \psi(t), \tag{3.20}$$

where $\mathscr{D} = -\sum_{j=1}^3 \gamma^0\gamma^j\partial_j + i\gamma^0 m$ is the free Dirac Hamiltonian. The gauge condition (3.18) takes now the form

$$\dot{A}^0(t) + \sum_{i=1}^3 \partial_i A^i(t) = 0 \tag{3.21}$$

and

$$\Delta A^0(t)+ \mid \psi(t) \mid^2 + \sum_{i=1}^3 \partial_i \dot{A}^i(t) = 0. \tag{3.22}$$

We shall now describe the space of initial data. Let $\frac{1}{2} < \rho < 1$, and let M^ρ be the space of functions $(f,\dot{f}) : \mathbb{R}^3 \to \mathbb{R}^4 \times \mathbb{R}^4$ for which the norm

$$\| (f,\dot{f}) \|_{M^\rho}= \left(\||\, \nabla \,|^\rho f \|_{L^2}^2 + \||\, \nabla \,|^{\rho-1} \dot{f} \|_{L^2} \right)^{1/2} \tag{3.23}$$

is finite, and let $D = L^2(\mathbb{R}^3, \mathbb{C}^4)$. We define $E^\rho = M^\rho \oplus D$, and $M^{0\rho}$ is the set of $(f,\dot{f}) \in M^\rho$ such that

$$\dot{f}_0 = \sum_{i=1}^3 \partial_i f_i \quad \text{and} \quad f_0 = -\sum_{i=1}^3 |\, \nabla \,|^{-2} \partial_i \dot{f}_i. \tag{3.24}$$

Define $E^{0\rho} = M^{0\rho} \oplus D$.

It is a well-known fact that the free Maxwell-Dirac system: $\Box b_\mu = 0$, $(i\gamma^\mu \partial_\mu + m)\beta = 0$, has a unique solution for initial data $(c, \dot{c}, \alpha) \in E^\rho$ (with the Lorentz condition satisfied if $(c, \dot{c}, \alpha) \in E^{0\rho}$).

We then define by covariance a representation U^1 of $\mathscr{P}$ on E^ρ, as in the introduction, by $U_g^1(c, \dot{c}, \alpha) = (c_g, \dot{c}_g, \alpha_g)$ with

$$\begin{aligned} c_g(x) &= V_\Lambda b(\Lambda^{-1}(0,x) + a), \\ \dot{c}_g(x) &= V_\Lambda \partial_t (b(\Lambda^{-1}(t,x) + a))_{t=0}, \\ \alpha_g(x) &= \Gamma_\Lambda \alpha(\Lambda^{-1}(0,x) + a), \end{aligned}$$

where $g = (\Lambda, a) \in \mathscr{P}$, V is the vector representation of $\mathscr{P}$, and Γ is the $D(\frac{1}{2}, 0) \oplus D(0, \frac{1}{2})$ representation of $\mathscr{P}$.

We denote by E_∞^ρ the space of differentiable vectors for the representation U^1. Let $T^1 = dU^1$ be the differential of U^1, i.e,

$$T_X^1 \varphi = \frac{d}{dt}(U^1_{\exp(tX)}\varphi)_{t=0}, \quad X \in \mathfrak{p}, \varphi \in E_\infty^\rho.$$

One can easily check that if $(f, \dot{f}, \alpha) \in E_\infty^\rho$, then

$$T^1_{P_0}(f, \dot{f}, \alpha) = (\dot{f}, \Delta f, \mathscr{D}\alpha), \tag{3.25}$$

$$T^1_{P_i}(f, \dot{f}, \alpha) = \partial_i (f, \dot{f}, \alpha), \quad 1 \le i \le 3, \tag{3.26}$$

$$\begin{aligned} T^1_{M_{ij}}(f, \dot{f}, \alpha)(x) =& -(x_i \partial_j - x_j \partial_i)(f, \dot{f}, \alpha)(x) \\ &+ (n_{ij} f(x), n_{ij}\dot{f}(x), \sigma_{ij}\alpha(x)), \quad 1 \le i < j \le 3, \end{aligned} \tag{3.27}$$

$$\begin{aligned} T^1_{M_{0i}}(f, \dot{f}, \alpha)(x) =& (x_i \dot{f}(x), \sum_{j=1}^{3} \partial_j (x_i \partial_j f)(x), x_i \mathscr{D}\alpha(x)) \\ &+ (n_{0i} f(x), n_{0i}\dot{f}(x), \sigma_{0i}\alpha(x)), \quad 1 \le i \le 3, \end{aligned} \tag{3.28}$$

where $\sigma_{\mu\nu} = \frac{1}{2}\gamma_\mu \gamma_\nu$, $n_{\mu\nu} \in \mathfrak{so}(1,3)$.

We can extend T^1 to the universal enveloping Lie algebra $\mathscr{U}(\mathfrak{p})$ of $\mathfrak{p}$. Let $\{X_1, \dots, X_{10}\}$ be a basis of $\mathfrak{p}$, and let Π be the set of element of $\mathscr{U}(\mathfrak{p})$ of the form $Y = X_1^{i_1} \cdots X_{10}^{i_{10}}$. Set $|Y| = i_1 + \cdots + i_{10}$. Let E_i^ρ, $i \ge 0$, be the completed space of E_∞^ρ for the norm

$$\| u \|_{E_i^\rho} = \Big(\sum_{Y \in \Pi,\ |Y| \le i} \| T_Y^1 u \|^2_{E^\rho} \Big)^{\frac{1}{2}}$$

We have $E_\infty^\rho = \bigcap_{i \ge 0} E_i^\rho$. We define $E_i^{0\rho} = E^{0\rho} \cap E_i^\rho$, $0 \le i \le +\infty$. One finds that $D \cap E_\infty^\rho = \mathscr{S}(\mathbb{R}^3, \mathbb{C}^4)$, the Schwartz space of C^∞-functions decreasing rapidly together with all their derivatives at infinity, and that $M_\infty^\rho = E_\infty^\rho \cap M^\rho$ contains long range-potentials. All this is linear (free) electrodynamics.

To define the nonlinear representation, for $u = (f, \dot{f}, \alpha) \in E^{\rho}_{\infty}$, let

$$\begin{aligned} T^2_{P_0}(u) =& (0, \bar{\alpha}\gamma\alpha, -if_\mu \gamma^0 \gamma^\mu \alpha), & (3.29)\\ T^2_{P_i}(u) =& 0, \quad 1 \leq i \leq 3, & (3.30)\\ T^2_{M_{ij}}(u) =& 0, \quad 1 \leq i < j \leq 3, & (3.31)\\ (T^2_{M_{0i}}(u))(x) =& x_i (T^2_{P_0} u)(x), \quad 1 \leq i \leq 3. & (3.32) \end{aligned}$$

One then verifies that $T_X = T^1_X + T^2_X$, $X \in \mathfrak{p}$ is a nonlinear Lie algebra representation of $\mathfrak{p}$ on E^{ρ}_{∞}.

The gauge condition $\partial_\mu A^\mu = 0$ (or equivalently (3.21) and (3.22)) can then be written, for $u = (f, \dot{f}, \alpha) \in E^{\rho}_{\infty}$, as

$$\dot{f}_0 = \sum_{i=1}^{3} \partial_i f_i, \qquad \Delta f_0 = \sum_{i=1}^{3} \partial_i \dot{f}_i - |\alpha|^2. \tag{3.33}$$

Let V^{ρ}_{∞} be the set of elements of E^{ρ}_{∞} satisfying (3.33). It is invariant under T_X, $X \in \mathfrak{p}$. The map $F : (f, \dot{f}, \alpha) \mapsto (g, \dot{g}, \alpha)$ sending E^{ρ}_{∞} to E^{ρ}_{∞} and defined by

$$g_0 = f_0 - \Delta^{-1} |\alpha|^2, \quad g_i = f_i,\ 1 \leq i \leq 3, \quad \text{and} \quad \dot{g} = \dot{f},$$

is a diffeomorphism from $E^{0\rho}_{\infty}$ onto V^{ρ}_{∞}. Therefore the conjugate of T by F is a nonlinear Lie algebra representation of $\mathfrak{p}$ on $E^{0\rho}_{\infty}$.

4. The classical scattering theory

One obtains in the case of NLKG a local group action from a nonlinear Lie algebra representation by standard techniques. More difficult problems are:

1) Find an invariant domain $\mathscr{U}_\infty$ for this local action.
2) Extend this action to a group action.
3) Find the asymptotic properties of $U_{\exp(tP_0)}\varphi$, $\varphi \in \mathscr{U}_\infty$, when $t \to \pm\infty$.

Usually one tries to solve these problems simultaneously by linearizing the group action. The classical wisdom in classical field theory without long-range phenomena is to say that when $t \to \pm\infty$, the time evolution $t \to U_{\exp(tP_0)}(\varphi)$ of the solution behaves like the evolution of a free field.

This is translated to the case of the NLKG equation, by saying that there exist nonlinear operators $\Omega_\pm$ on domains $\mathscr{U}^{\pm}_{\infty}, with \Omega_\pm : \mathscr{U}^{\pm}_{\infty} \to \mathscr{U}_\infty$, $\mathscr{U}^{\pm}_{\infty} \subset E_\infty$, $\mathscr{U}_\infty \subset E_\infty$, $\mathscr{U}^{\pm}_{\infty}$ being invariant under U^1, and such that

$$U_g \Omega_\pm = \Omega_\pm U^1_g, \quad g \in \mathscr{P}, \tag{4.1}$$

and

$$\lim_{t \to \pm\infty} (U_{\exp(tP_0)} \Omega_\pm(\varphi) - U^1_{\exp(tP_0)}(\varphi)) = 0, \tag{4.2}$$

the limit being taken in E.

In the following, in order to keep the notation simple, we shall only consider the case where $t \to +\infty$; we shall write

$$V_t = U_{\exp(tP_0)}, \quad V_t^1 = U^1_{\exp(tP_0)}, \qquad \Omega = \Omega_+.$$

Properties a) and b) above are usually obtained by solving the Yang-Feldman-Källén (YFK) equation for Ω

$$\Omega = I - \int_0^\infty V^1_{-s} \hat{T}_{P_0}(\Omega V_s^1) ds, \tag{4.3}$$

where $\hat{T}_{P_0} = T_{P_0} - T^1_{P_0}$. Now assume that Ω is a solution of (4.3). By writing

$$\Omega V_t^1 = V_t^1 - \int_t^\infty V^1_{-s+t} \circ \hat{T}_{P_0} \circ \Omega \circ V_s^1 ds.$$

then, if φ a C^∞-vector for V^1, we have

$$\begin{aligned}\frac{d}{dt}\Omega \circ V_t^1(\varphi) &= T^1_{P_0} \circ V_t^1(\varphi) + \hat{T}_{P_0} \circ \Omega \circ V_t^1(\varphi) \\ &\quad - T^1_{P_0} \circ \Big(\int_0^\infty V^1_{-s} \circ \hat{T}_{P_0} \circ \Omega \circ V_s^1 ds\Big) \circ V_t^1(\varphi) \\ &= T^1_{P_0} \circ \Omega \circ V_t^1(\varphi) + \hat{T}_{P_0} \circ \Omega \circ V_t^1(\varphi).\end{aligned}$$

Therefore

$$\frac{d}{dt}(\Omega V_t^1(\varphi)) = T_{P_0}(\Omega V_t^1(\varphi)),$$

and, since $\frac{d}{dt}(V_t \Omega(\varphi)) = T_{P_0}(V_t \Omega(\varphi))$, it follows that

$$V_t(\Omega(\varphi)) = \Omega(V_t^1(\varphi)).$$

Moreover

$$V_t(\Omega(\varphi)) - V_t^1(\varphi) = \Omega(V_t^1(\varphi)) - V_t^1 \varphi = -V_t^1 \int_t^\infty V^1_{-s} \hat{T}_{P_0}(\Omega(V_s^1(\varphi))) ds \tag{4.4}$$

and we see that the existence of Ω in formula (4.3) is equivalent (if U_t^1 is unitary) to the convergence of the integral

$$\int_t^\infty \hat{T}_{P_0}(V_s(\varphi)) ds \tag{4.5}$$

in Equation (4.4).

In n space dimensions, it is well known (cf. [5]) that $\|V_t^1\varphi\|_{L^\infty} \leq C(\varphi)(1+|t|)^{-n/2}$ and V_t^1 is unitary in E. Hence we have $\|T_{P_0}^2(V_s^1\varphi)\|_E \leq C_2(\varphi)t^{-n/2}$ (and, consequently, $\|T_{P_0}^k(V_s^1\varphi)\|_E \leq C_k(\varphi)t^{-(k-1)n/2}$). This gives a better convergence of the integral in (4.5) in higher dimension.

- For $n \geq 3$, one can take advantage of the absolute convergence of the integral in (4.5) when constructing a solution by iteration. The cubic case was solved by C. Morawetz and W. Strauss [6] and the quadratic case by: J. Simon [7], S. Klainerman [4], J. Simon and E. Taflin [8].
- For $n = 2$ and in the presence of quadratic terms, the integral in (4.5) is not absolutely convergent in general and more information is needed to prove the existence of the integral!

In order to understand the problem in a space of dimension 2, let us go one step further at the heuristic level by assuming that Ω can be written as a formal power series on the space of initial data at the origin.

$$\Omega = I + \sum_{n>1} \Omega^n,$$

where Ω^n is a monomial of degree n from E_N to E_N for some $N \geq 0$. One sees right away that the YFK equation can be solved recursively

$$\Omega^2 = \int_0^\infty V_{-s}^1 \circ T_{P_0}^2 \circ V_s^1 ds \tag{4.6}$$

$$\Omega^3 = 2\int_0^\infty V_{-s}^1 T_{P_0}^2((\Omega^2 \circ V_s^1) \otimes V_s^1)ds + \int_0^\infty V_{-s}^1 \circ T_{P_0}^3 \circ V_s^1 ds, \tag{4.7}$$

where $T_{P_0}^2$ is here identified to a bilinear symmetric function.

The first integral in the right-hand side can be proved to be absolutely divergent (for general T^2, for spaces of dimension 2). One way to turn around this problem is to compute Ω^2 by another method, as in [5], for the case of space of dimension 3.

Let $L_\mu = T_{P\mu}^1$, $0 \leq \mu \leq 2$ and define

$$\Omega^2 = (L^0 T_{P0}^2 + T_{P0}^2(I \otimes L^0 + L^0 \otimes I)) \circ Q^{-1},$$

where

$$Q = \left(m^2 - (I \otimes L^\mu + L^\mu \otimes I)(I \otimes L_\mu + L_\mu \otimes I)\right). \tag{4.8}$$

One verifies that this formula makes sense on the space E_∞ of C^∞-vectors of U^1, and that

$$U_g^2 + U_g^1\Omega^2 = \Omega^2 U_g^1, \quad g \in \mathscr{P}.$$

Consequently, Equations (4.1) and (4.2) are satisfied and the YFK equation is satisfied at order 2. But now $\| \Omega^2 U_t^1 \|_E \sim t^{-1}$ since $U_t^1 Q = Q U_t^1$, and since we are in a space of dimension two.

Formula (4.8) does not come out by hocus-pocus, in fact it is the squared mass-operator and it can be derived by cohomological considerations which we shall not develop here (see [7, 9] for details). In other cases (see [8]), for example when the linear part involves several distinct masses, it can be more convenient to calculate directly the leading terms of the integral appearing in (4.4).

To obtain that $\Omega : E_N \to E_N$ for some E_N, so that its invertibility is guaranteed, one needs a nonlinear equivalent of the linear representation of the universal enveloping algebra $\mathscr{U}(\mathfrak{p}) \ni Y \to T_Y^1$. This is obtained by defining

$$T_{YX} = DT_Y.T_X, \quad Y \in \mathscr{U}(\mathfrak{p}), X \in \mathfrak{p}, \tag{4.9}$$

$T_I = Id_{E_N}$. The evolution equation for $T_Y(t)(\phi(t))$ in E_0, for $| Y | \leq N$ where $Y(t) = \exp(t\, ad_{P_0} Y)$ and $ad_{P_0} Z = [P_0, Z]$, $Z \in \mathscr{U}(\mathfrak{p})$, then permits to study the evolution of $\phi(t)$ in E_N (see [9]).

5. Classical electrodynamics: A less conventional scattering

Let us go back to the coupled MD equations. As we shall see the classical wisdom of asymptotic free field breaks down, since here the integral in (4.4) is really divergent, so the limit 4.2 does not exist. This is due to the fact that U^1 just does not give the right asymptotic dynamics in the presence of long-range phenomena. By choosing a modified dynamics, not only the integral becomes convergent, but also provides global solutions [2]. We have defined in §3.3 the linear Lie algebra representation T^1 of $\mathfrak{p}$ associated to the non-coupled equations and the nonlinear representation T of $\mathfrak{p}$ associated to the MD equations. We now define a new nonlinear Lie algebra representation $T^{(+)}$ of $\mathfrak{p}$, called the asymptotic representation, by $T^{(+)} = T^1 + T^{(+)2}$ in the following way:

$$T^{(+)2}_{M\mu\nu}(0) = 0, \quad 0 \leq \mu < \nu \leq 3, \tag{5.1}$$

and for $u = (f, \dot{f}, \alpha) \in E_\infty$,

$$(T^{(+)2}_{P_\mu}(u))^\wedge(k) = i\big(0, 0, \sum_{\varepsilon=\pm} (\Phi_{\varepsilon, P_\mu}(f, \dot{f})(k) P_\varepsilon(k) \hat{\alpha}(k)\big). \tag{5.2}$$

Let us explain the notation of (5.2). Here $f \to \hat{f}$ is the Fourier transform

$$\hat{f}(k) = (2\pi)^{-\frac{3}{2}} \int_{\mathbb{R}^3} e^{-ikx} f(x) dx.$$

$P_\varepsilon(k)$ is the orthogonal projector onto the subspace of $D = L^2(\mathbb{R}^3, \mathbb{C}^4)$ constituted by elements of energy sign equal to ε:

$$\begin{aligned}(P_\varepsilon(-i\partial)\alpha)^\wedge(k) &= P_\varepsilon(k)\hat{\alpha}(k)\\ &= \frac{1}{2}\Big(I+\varepsilon(-\sum_{\partial=1}^{3}\gamma^0\gamma^j k_j + m\gamma^0)\omega(k)^{-1}\Big)\hat{\alpha}(k),\end{aligned} \tag{5.3}$$

where $\omega(k) = (m^2 + |k|^2)^{\frac{1}{2}}$. The phase Φ_{ε,P_μ} is given by the following relation:

$$\Phi_{\varepsilon,X}(f,\dot{f})(k) = \int_0^\infty a^\nu l_\varepsilon^\mu \chi_0(m\tau) B_{\nu\mu}^{(+)1}(\tau l_\varepsilon)d\tau, \tag{5.4}$$

where $X = a^\nu P_\nu$ for $0 \le \nu \le 3$, $l_\varepsilon^0 = \omega(k)$, and $l_\varepsilon^j = -\varepsilon k_j$ for $1 \le j \le 3$; $\chi_0 \in C^\infty([0,1])$ is a cut-off function satisfying $0 \le \chi_0 \le 1$, $\chi_0(s) = 1$ if $s \ge 2$; and finally

$$B_\mu^{(+)1}(t) = \cos((-\Delta)^{\frac{1}{2}}t)f_\mu + (-\Delta)^{\frac{1}{2}}\sin(t\Delta)^{\frac{1}{2}}t)\dot{f}_\mu, \tag{5.5}$$

$$B_{0\mu}^{(+)}(t) = \frac{d}{dt}B_\mu^{(+)1}(t), \tag{5.6}$$

$$B_{j\mu}^{(+)}(t) = \partial_j B_\mu^{(+)1}(t), \quad 1 \le j \le 3,\ 0 \le \mu \le 3. \tag{5.7}$$

Formulas (5.4)–(5.7) define a C^∞-function $k \mapsto \Phi_{\varepsilon,X}(f,\dot{f})(k)$.

One verifies that $T^{(+)}$ given by this long definition is a nonlinear representation of $\mathfrak{p}$ on E_∞^ρ. This representation is sufficiently regular that we have the following result.

Theorem 5.1 *Let $N \ge 4$, then there exists a unique continuous nonlinear representation $(g,u) \mapsto U_g^{(+)}u$ from $\mathscr{P}_0 \times E_N^\rho$ to $E_{N'}^\rho$, which is C^∞ from $\mathscr{P} \times E_\infty^\rho$ to E_∞^ρ, and such that*

$$T_X^{(+)}(u) = \frac{d}{dt}\Big(U_{\exp(tX)u}^{(+)}\Big)_{t=0}, \quad u \in E_\infty.$$

Moreover, $E_k^{0\rho}$, $k \ge N$, is invariant under $U^{(+)}$.

In fact the representation $U^{(+)}$ can be computed explicitly and it is of the form:

$$U_g^{(+)}(u) = (U_g^{1M}(u), U_g^{(+)D}(u)), \quad u = (f,\dot{f},\alpha),$$

where $U^{1M}(u)$ is the component on M^ρ of U^1 and

$$(U_g^{(+)D}(u))^\wedge(k) = \sum_{\varepsilon=\pm} e^{i\varphi_g(u,-\varepsilon k)}P_\varepsilon(k)(U_g^{1D}\alpha)^\wedge(k),$$

where U^{1D} is the restriction of U^1 to D. The map $(g,u,k) \mapsto \varphi_g(u,-\varepsilon k)$ is C^∞ from $\mathscr{P} \times E_\infty^\rho \times \mathbb{R}^3$ to $\mathbb{R}$.

φ_g can be written explicitly down and its expression is given in [3].

One can prove that changing the cut-off χ_0 result in changing $U^{(+)}$ to an equivalent representation (conjugated by an invertible nonlinear mapping). In addition $U^{(+)}$ is not equivalent to U^1 on $E_\infty^{0\rho}$ by a C^2-equivalence, hence $U^{(+)}$ is truly nonlinear.

Now we can state the *modified scattering* theorem as follows.

Theorem 5.2 *There exists a neighborhood $\mathcal{U}_\infty$ (resp. $\mathcal{O}_\infty^{\pm}$) of 0 in V_∞^ρ (resp. $E_\infty^{0\rho}$), a diffeomorphism $\Omega_+ : \mathcal{O}_\infty^+ \to \mathcal{U}_\infty$, and a C^∞-nonlinear representation of $\mathcal{P}$, $U : \mathcal{P} \times \mathcal{U}_\infty \to \mathcal{U}_\infty$, such that:*

1) $\frac{d}{dt}U_{\exp(tX)}(u) = T_X(U_{\exp(tX)}(u))$, $X \in \mathfrak{p}$, $t \in \mathbb{R}$, $u \in \mathcal{U}_\infty$;

2) $\Omega_+ U_g^{(+)} = U_g \Omega_+$;

3)
$$\lim_{t\to\infty}\Big(\| U^M_{\exp(tP_0)}\Omega_+(u) - U^{1M}_{\exp(tP_0)}(f,\dot f) \|_{M^\rho} + \| U^D_{\exp(tP_0)}(\Omega_+(u)) - \Sigma_{\varepsilon=\pm} e^{is_\varepsilon(u,t,-i\partial)} P_\varepsilon(-i\partial) U^{1D}_{\exp(tP^0)}\alpha \|_D \Big) = 0,$$

where $u = (f, \dot f, \alpha) \in \mathcal{O}_\infty^{(+)}$ and $s_\varepsilon(u,t,-i\partial)$ is the inverse Fourier transform of the operator defined by multiplication by the function $\hat s_\varepsilon(u,t,k)$.

Furthermore, the function $\hat s_\varepsilon$ admits an explicit expression that we shall present. First, consider $A^{(+)} = A^{(+)1} + A^{(+)2}$ defined by

$$A^{(+)1}(t)(x) = \chi_0((t^2 - |x|^2)^{\frac12})(B^{(+)1}(t))(x)$$
$$A^{(+)2}(t)(x) = \chi((t^2 - |x|^2)^{\frac12})(B^{(+)2}(t))(x)$$

if $t \geq |x|$, and

$$(A^{(+)1}(t))(x) = \chi_0(0)(B^{(+)1}(t))(x)$$
$$(A^{(+)2}(t))(x) = 0$$

if $t < |x|$, where χ_0 and $B^{(+)1}$ were defined in (5.5) and $\chi \in C^\infty(\mathbb{R})$ satisfies $0 \leq \chi \leq 1$, $\chi(\tau) = 0$ if $\tau \leq 1$, and $\chi(\tau) = 1$ if $\tau \geq 2$. Second, we define

$$B^{+2}(t) = -\int_t^\infty |\nabla|^{-1} \sin(|\nabla|(t-s)) J_\mu^{(+)}(s,\cdot)ds,$$

where

$$J_\mu^{(+)}(t,x) = (m/t)^3(\omega(t,q(t,x)/m)^5 \times \\ \times \sum_{\pm}((P_\varepsilon(-i\partial)\alpha)^\wedge(\varepsilon q(t,x)/m))^+\gamma^0\gamma^\mu(P_\varepsilon(-i\partial)\alpha)^\wedge(\varepsilon q(t,x)/m),$$

for $t > 0, |x| < t$, with $q(t,x) = -mx/(t^2 - |x|^2)^{\frac{1}{2}}$, and where $J^+_\mu(t,x) = 0$ elsewhere.

Now let $\Theta(H,y) = \int_{L(y)} H_\mu(z)dz^\mu$, $y \in \mathbb{R}^4$, where

$$L(y) = \{z \in \mathbb{R}^4 \mid z = sy,\ 0 \leq s \leq 1\}.$$

At last the explicit form for $s^{(+)}_\varepsilon$ is

$$s^{(+)}_\varepsilon(u,t,k) = -\Theta\big(A^{(+)}(u),(t,-\varepsilon tk/\omega(k))\big).$$

As said before, the representation U is not equivalent to a linear representation. In fact, if one tries to compute "by force" $\Omega = I + \sum_{n\geq 2}\Omega^n$ such that $U_g\Omega = \Omega U^1_g$ one sees that the calculations lead to divergences of integrals of Fourier transforms at $k = 0$. This is what one calls infrared divergences. This shows that it is a misconception to think that an electron can be stripped of its electromagnetic field, even asymptotically in time. Indeed, in the equivalence $U_g\Omega_+ = \Omega_+ U^{(+)}_g$ one considers that the electron is dressed with its long-range electromagnetic potential asymptotically in time. Since $U^{(+)}$ is not equivalent to U^1, the electron cannot in general be undressed.

From the point of view of the observables, the asymptotic representation $U^{(+)}$ is indistinguishable from the linear one. To see that, let $u = (f,\dot{f},\alpha) \in \mathscr{U}_\infty$, and consider the field $(A(t),\dot{A}(t),\psi(t)) = U_{\exp(tP^0)}\Omega_+(u)$ and the free field $(A_L(t),\dot{A}_L(t),\psi_L(t)) = U^1_{\exp(tP^0)}(u)$. The electromagnetic field is denoted by $F_{\mu\nu} = \partial_\mu A_\nu - \partial_\nu A_\mu$. For the energy-momentum tensor density

$$t^{\mu\nu} = -F^{\mu\nu}F^\nu_\alpha + \frac{1}{4}g^{\mu\nu}F_{\alpha\beta}F^{\alpha\beta} + \frac{1}{2}(\bar{\psi}\gamma^\mu(i\partial^\nu - A^\nu))\psi + (\overline{(i\partial^\nu - A^\nu)\psi}\gamma^\mu\psi)$$

and the current density $j^\mu = \bar{\psi}\gamma^\mu\psi$, one can compute the corresponding densities $t^{\mu\nu}_L$ and j^μ_L for the free field $(A_L\dot{A}_L\psi_L)$. Then it can be proved that

$$\lim_{t\to\infty}\int_{\mathbb{R}^3} |\, t^{\mu\nu}(t,x) - t^{\mu\nu}_L(t,x)\,|\, dx = 0$$

and

$$\lim_{t\to\infty}\int_{\mathbb{R}^3} |\, j^\mu(t,x) - j^\mu_L(t,x)\,|\, dx = 0.$$

This means that for these observables, the total energy-momentum and the total current, the field $(A,\dot{A},\psi)$ behave asymptotically in time like the free field $(A_L,\dot{A}_L,\psi_L)$.

References

1. Flato, M., Pinczon, G., and Simon, J.: Non-linear representations of Lie groups, *Ann. Sci. École Norm. Sup.(4)* **10** (1977), 405–418.
2. Flato, M., Simon J., and Taflin E.: On global solutions of the Maxwell-Dirac equations, *Comm. Math. Phys.* **112** (1987), 21–49.
3. Flato, M., Simon, J. C H., and Taflin, E.: Asymptotic completeness, global existence and the infrared problem for the Maxwell-Dirac equations, *Mem. Amer. Math. Soc.* **127** no. 606, (1997), 1–311.
4. Klainerman, S.: Global existence of small amplitude solutions to non-linear Klein-Gordon equations in four space-time dimensions, *Comm. Pure Appl. Math.* **38** (1985), 631–641.
5. Morawetz, C. and Strauss, W.: Decay and scattering of solutions of a nonlinear relativistic wave equation, *Comm. Pure Appl. Math.* **25** (1972), 1–31.
6. Morawetz, C. and Strauss, W.: On a nonlinear scattering operator, *Comm. Pure Appl. Math.* **26** (1973), 47–54.
7. Simon J. C. H.: A wave operator for nonlinear Klein-Gordon equation, *Lett. Math. Phys.* **7** (1983), 387–398.
8. Simon, J. C. H. and Taflin, E.: Wave operators and analytic solutions for systems of nonlinear Klein-Gordon equations and nonlinear Schrödinger equations, *Comm. Math. Phys.* **99** (1985), 541–562.
9. Simon, J. C. H. and Taflin, E.: The Cauchy problem for nonlinear Klein-Gordon equations, *Comm. Math. Phys.* **152** (1993), 433–478.

Quantum field theories on an algebraic curve

Leon A. Takhtajan
Department of Mathematics, SUNY at Stony Brook
Stony Brook, NY 11794-3651, USA
e-mail: `leontak@math.sunysb.edu`

Abstract. We formulate quantum field theories on an algebraic curve and outline a "paradigm" interpreting Ward identities as reciprocity laws.

Keywords: algebraic curves, Clifford and Heisenberg algebras, quantum field theories, Ward identities, reciprocity laws

Mathematics Subject Classifications (2000): 14H05, 14H81, 81R10

Dedicated to the memory of Moshé Flato

1. Introduction

1.1. Overview

Recent development in mathematical physics is characterized by applications of Quantum Field Theory (QFT) to various areas of mathematics. They include the following.[1]

1) Application of Topological QFT to Geometry and Topology: $d = 3$ Jones-Witten theory [1], $d = 4$ Donaldson and Seiberg-Witten theories [2, 3].

2) Application of CFT and vertex operator algebras to the "monstrous moonshine" [4].

3) Application of CFT to Complex and Algebraic Geometry: $d = 2$ quantum gravity and intersection theory on moduli spaces of stable curves [5, 6, 7], and Weil-Petersson geometry [8, 9].

In this talk[2], we are interested in applying QFT to Arithmetics, i.e., in developing QFT methods for algebraic number fields and fields of algebraic functions. On several occasions, I discussed these topics with Moshé Flato, who had deep thoughts about possible relations between QFT and Arithmetics. Thus, Moshé came up with a beautiful idea to use factoring of polynomials in several variables

[1] This is not a complete list, of course; it rather reflects author's interests in topics somewhat related to the subject of the paper.

[2] The paper is an expanded version of the talk, given at the Conférence Moshé Flato, September 5-8 1999, Dijon, France. Detailed presentation will be published elsewhere.

G. Dito and D. Sternheimer (eds.), Conférence Moshé Flato 1999, Vol. 1, 403 – 416.

for the special quantization of the Nambu bracket (called Zariski quantization), developed in our paper [10].

1.2. Historic background

In the 1920s, E. Artin and H. Hasse introduced "Calculus" to Arithmetics, and in 1930-1940s, C. Chevalley and A. Weil introduced infinite-dimensional methods. Recent development can be succinctly described as follows.

- In 1968, J. Tate [11] defined residues of differentials on algebraic curves in terms of the traces of certain linear operators on infinite-dimensional spaces of adeles, and gave a new proof of the residue theorem.
- In 1987, E. Arbarello, C. de Concini and V. Kac [12] interpreted Tate's approach in terms of central extensions of infinite-dimensional Lie algebras and proved A. Weil reciprocity law using the infinite-wedge representation.
- In 1987-1988, D. Kazhdan [13] and E. Witten [14, 15] proposed an adelic formulation of free fermions on an algebraic curve.
- In 1990, H. Garland and G. Zuckerman [16], constructed local Fock spaces for multiplicative bosons on a Riemann surface.
- In 1988-1991, A. Beilinson, B. Feigin, and B. Mazur [17] developed a representation theory approach for CFT on algebraic curves.
- In 1989-1991, A. Raina [18, 19], developed an algebro-geometric approach for free fermions on Riemann surfaces and gave a new proof of Fay's trisecant identity.
- In 1994, A. Beilinson and V. Drinfeld, in the course of their work on geometric Langlands correspondence, introduced "chiral algebras" [20, 21] on algebraic curves as mathematical objects that imitate operator product expansions of quantum fields.

Results of papers [17, 20, 21] have been recently summarized by D. Gaitsgory [22].

Primarily based on [13, 14, 15], we interpret this development as emergence of a "new paradigm", that introduces reciprocity laws and Class Field Theory for algebraic function fields in one variable (see [23] for a review) as Ward Identities — quantum Noether conservation laws in Conformal Field Theory (see [14, 24] for a review).

Conservation laws in Physics include Noether symmetries, gauge symmetries, and discrete symmetries. Here are the simplest examples of conservation laws in Arithmetics.

(a) Artin-Whaples product formula

$$\prod_{p\in S}|r|_p = 1, \ r\in\mathbb{Q},$$

where S is the set of all places on $\mathbb{Q}$, consisting of all primes $p = 2,3,5,\ldots$, and an "infinite prime" $p=\infty$, and $|\ |_p$ is the p-adic absolute value (it is the usual absolute value on $\mathbb{R}$ for $p=\infty$).

(b) Gauss quadratic reciprocity law

$$(\frac{p}{q})(\frac{q}{p}) = (-1)^{\frac{p-1}{2}\frac{q-1}{2}},$$

where p, q are distinct odd primes, and (p/q) is the Legendre symbol: $(p/q)=1$ or $(p/q)=-1$ depending on whether p is quadratic residue or non residue modulo p.

(c) Cauchy residue theorem

$$\sum_{P\in X}\mathrm{Res}_P f dg = 0,$$

where X is compact Riemann surface, and $f,g\in\mathbb{C}(X)$ are meromorphic functions on X.

(d) A. Weil reciprocity law

$$\prod_{Q\in(g)} f(Q)^{n_Q} = \prod_{P\in(f)} g(P)^{n_P}, \ f,g\in\mathbb{C}(X),$$

where

$$(f) = \sum_{P\in X} n_P\cdot P, \ (g) = \sum_{Q\in X} n_Q\cdot Q$$

are divisors of meromorphic functions f and g with the property $(f)\cap(g)=\emptyset$.

2. Mathematical set-up

2.1. Definitions

Let X be a complete, irreducible, nonsingular, algebraic curve of genus g over an algebraically closed field k of characteristic zero, and let $K=k(X)$ be the field of rational functions on X — a finitely-generated algebraic extension of k of transcendence degree 1 (see, e.g., [25, 26]). When $k=\mathbb{C}$ — the field of complex numbers, X is a compact Riemann surface, and $K=\mathbb{C}(X)$ is the field of meromorphic functions on X. Points $P\in X$ correspond to regular discrete valuations v_P of K over k, i.e., surjective homomorphisms $v: K^*\to\mathbb{Z}$ of the multiplicative group of the field K into the additive group of rational integers $\mathbb{Z}$, that are trivial on the subgroup k^* in K^*, and satisfy the property

$$v(x+y)\geq\min\{v(x),v(y)\}, \ \text{for all } x,y\in K^*.$$

Let $\mathscr{K}$ be the canonical line bundle (invertible sheaf) over X, and let $\mathscr{L}$ be a spin structure for X — a line bundle of degree $g-1$ over X, such that $\mathscr{L}^2 = \mathscr{K}$. For any line bundle $\mathscr{D}$ of degree 0 over X set $\mathscr{L}^{\pm} = \mathscr{L} \otimes \mathscr{D}^{\pm 1}$. Denote by $\mathscr{K}_X$, $\mathscr{L}_X$, and $\mathscr{L}_X^{\pm}$ the corresponding infinite-dimensional k - vector spaces of rational sections over X of the line bundles $\mathscr{K}$, $\mathscr{L}$, and $\mathscr{L}^{\pm}$. These are the spaces of rational differentials on X, rational $1/2$ differentials on X and rational "chiral" $1/2$ differentials on X, correspondingly. Set $\mathscr{L}(D)_X = \mathscr{L}_X^+ \oplus \mathscr{L}_X^-$.

2.2. Local data

The following local objects at every $P \in X$ are canonically associated with the trivial line bundle $\mathscr{O}_X$ over X, line bundle $\mathscr{L}$ and with the rank 2 vector bundle $\mathscr{L}(D) = \mathscr{L}^+ \oplus \mathscr{L}^-$.

(i) Local field at $P \in X$ — a completion K_P of the field K with respect to the valuation v_P.

(ii) Local ring $\mathscr{O}_P = \{u \in K_P \mid v_P(u) \geq 0\}$.

(iii) Unique maximal ideal $\mathfrak{p} = \{u \in \mathscr{O}_P \mid v_P(u) > 0\}$ in K_P.

(iv) Residue field $k = \mathscr{O}_P/\mathfrak{p}$.

(v) Group of units $U_P = \{u \in K_P \mid v_P(u) = 0\}$ in $\mathscr{O}_P$, $P \in X$, with the property $U_P \simeq k^* \times U_P^1$, where $U_P^1 = 1 + \mathfrak{p}$.

(vi) k - vector spaces L_P and $L(D)_P$ — completions at $P \in X$ of $\mathscr{L}_X$ and $\mathscr{L}(D)_X$, correspondingly, with respect to the valuation v_P.

(vii) Subspaces $\mathscr{O}(L)_P$ and $\mathscr{O}(L(D))_P$ of regular elements in L_P and $L(D)_P$, correspondingly.

k - vector spaces K_P, L_P, and $L(D)_P$ are complete in the $\mathfrak{p}$-adic topology, the residue field k is discrete in the quotient topology in K_P, $P \in X$, and these vector spaces are Tate vector spaces — topological vector spaces with a base of neighborhoods of zero given by mutually commensurable subspaces.

k - vector spaces L_P, K_P and $L(D)_P$ and the group K_P^* come equipped with the following canonical structures.

- L_P has non-degenerate, continuous, symmetric, k - bilinear form $(\ ,\ )_P : L_P \otimes L_P \to k$,
$$(\phi_1, \phi_2)_P = \operatorname{Res}_P(\phi_1 \phi_2).$$
The pair L_P, $(\ ,\ )_P$ defines a Clifford algebra Cliff_P.

- $L(D)_P$ has non-degenerate, continuous, symmetric, k - bilinear form $(\ ,\ )_P : L(D)_P \otimes L(D)_P \to k$,
$$(\psi_1 + \bar{\psi}_1, \psi_2 + \bar{\psi}_2)_P = \operatorname{Res}_P(\psi_1 \bar{\psi}_2 + \bar{\psi}_1 \psi_2),$$

where $\psi \in L_P^+$, $\bar{\psi} \in L_P^-$. The pair $L(D)_P$, $(\ ,\ )_P$ defines a Clifford algebra $\mathrm{Cliff}(D)_P$.

- K_P, as abelian Lie algebra $\mathfrak{g}_P = \mathfrak{gl}_1(K_P)$ — "geometric current algebra", has continuous, skew-symmetric, k - bilinear form — canonical Lie algebra two-cocycle given by the residue symbol,

$$c_P(u,v) = -\mathrm{Res}_P(u\,dv).$$

  The pair $\mathfrak{g}_P$, c_P defines a central extension $\widehat{\mathfrak{g}}_P$ — "geometric affine algebra".

- K_P^* as abelian Lie group $G_P = \mathrm{GL}_1(K_P)$ — "geometric loop group", has continuous, skew-symmetric, k - bilinear form — canonical group two-cocycle, given by Tate's tame symbol,

$$(f,g)_P = (-1)^{v_P(f)v_P(g)} \frac{f^{v_P(g)}}{g^{v_P(f)}} \mod \mathfrak{p} \in k^*.$$

  The pair G_P, $(\ ,\ )_P$ defines a central extension $\widehat{G}_P$ — "geometric affine group".

2.3. Global data

Global versions of the local objects introduced in the previous section are the following.

- The ring of adeles for K,

$$\mathbb{A}_X = \coprod_{P \in X} K_P$$

  — restricted Cartesian product of spaces K_P over $P \in X$ with respect to subspaces $\mathscr{O}_P$. An element $a = \{a_P\}_{P \in X} \in \mathbb{A}_X$, if $a_P \in \mathscr{O}_P$ for all except finitely many $P \in X$.

- Diagonal embedding $K \ni f \mapsto \{f|_P\}_{P \in X} \in \mathbb{A}_X$ of the global field K into the ring of adeles $\mathbb{A}_X$.

- The group of ideles for K — the group of units in the ring of adeles $\mathbb{A}_X$,

$$\mathbb{J}_X = \coprod_{P \in X} K_P^*$$

  — restricted Cartesian product of multiplicative groups K_P^* with respect to subgroups U_P.

- Diagonal embedding $K^* \hookrightarrow \mathbb{J}_X$ of the multiplicative group of the global field into the group of ideles $\mathbb{J}_X$.

- Restricted Cartesian product of spaces L_P,

$$\mathbb{L}_X = \coprod_{P \in X} L_P,$$

with respect to subspaces $\mathscr{O}(L)_P$.

- Diagonal embedding $\mathscr{L}_X \hookrightarrow \mathbb{L}_X$ of the space of rational sections of $\mathscr{L}$ over X into the space of adeles $\mathbb{L}_X$.

- Restricted Cartesian product of the spaces $L(D)_P$,

$$\mathbb{L}(D)_X = \coprod_{P \in X} L(D)_P,$$

with respect to subspaces $\mathscr{O}(L(D))_P$.

- Diagonal embedding $\mathscr{L}(D)_X \hookrightarrow \mathbb{L}(D)_X$ of the space of rational sections of $\mathscr{L}(D)$ over X into the space of adeles $\mathbb{L}(D)_X$.

2.4. Artin-Whaples and Weil examples

1. For $x = \{x_P\}_{P \in X} \in \mathbb{J}_X$ define

$$||x|| = \prod_{P \in X} |x_P|_P,$$

where $|x_P|_P = c^{v_P(x_P)}$ for some fixed $0 < c < 1$ is the $\mathfrak{p}$-adic metric for the valuation v_P. The Artin-Whaples product formula [28]

$$||fx|| = ||x||,$$

where $x \in \mathbb{J}_X$ and $f \in K^* \hookrightarrow \mathbb{J}_X$, is an example of "conservation law" stating that "the sum over X of residues of the meromorphic differential df/f is zero".

2. For $a = \{a_P\}_{P \in X}, b = \{b_P\}_{P \in X} \in \mathbb{J}_X$ define the global symbol

$$(a,b)_X = \prod_{P \in X} (a_P, b_P)_P \in k^*.$$

The A. Weil reciprocity law (see, e.g. [26])

$$(f,g)_X = 1,$$

for all $f, g \in K^*$, is an example of "conservation law"

$$\frac{(fa, gb)_X}{(f,b)_X (a,g)_X} = (a,b)_X,$$

for all $a, b \in \mathbb{A}_X$.

3. QFT set-up

3.1. Definitions

In correspondence with different vector spaces of global rational sections $\mathscr{L}_X$, $\mathscr{L}(D)_X$, $K = k(X)$, and with the group K^*, we consider the following quantum field theories on algebraic curve X.

a) Majorana-Weyl (one-component) free fermions $\longleftrightarrow \mathscr{L}_X$.

b) Charged (two-component) free fermions $\longleftrightarrow \mathscr{L}(D)_X$.

c) Additive $\mathfrak{gl}_1$ - bosons $\longleftrightarrow K$.

d) Multiplicative GL_1 - bosons $\longleftrightarrow K^*$.

Specifically, for every $P \in X$, let $\mathscr{G}_P$ be either the Clifford algebra Cliff_P or $\mathrm{Cliff}(D)_P$, or the Lie algebra $\widehat{\mathfrak{g}}_P$, or the Lie group $\widehat{G}_P$. Also, let $\mathscr{G}_P^+$ be, correspondingly, either the Clifford subalgebra $\Lambda^*\mathscr{O}(L)_P$ or $\Lambda^*\mathscr{O}(L(D))_P$, or the Lie subalgebra $\mathscr{O}_P \oplus \{0\}$, or the Lie subgroup $U_P^1 \times \{1\}$.

By a QFT with symmetry algebra or group $\mathscr{G}$ on an algebraic curve X, we understand the following local and global data.

3.2. Local QFT

It consists of an irreducible highest weight $\mathscr{G}_P$-module $\mathscr{F}_P$ for $P \in X$ — fermionic or bosonic Fock space containing distinguished vector $\mathbf{1}_P$, which is either annihilated or is invariant with respect to the action of the corresponding algebra or group $\mathscr{G}_P^+$.

The correspondence between quantum observables and states in CFT asserts that $\mathscr{F}_P$ is the space of quantum observables at $P \in X$.

3.2.1.
For Majorana-Weyl fermions, set

$$\mathscr{F}_P = \Lambda^* L_P^- \simeq \mathrm{Cliff}_P / \mathrm{Cliff}_P \cdot \mathscr{O}(L)_P,$$

with the left action of Cliff_P, where

$$L_P = \mathscr{O}(L)_P \oplus L_P^-,$$

is a decomposition of L_P into maximal isotropic subspaces with respect to the symmetric bilinear form $(\, , \,)_P$, introduced in Section 2.1. Though subspace L_P^- — the complement of $\mathscr{O}(L)_P$ in L_P, depends on the choice of a uniformizer at P, the corresponding Cliff_P-module $\mathscr{F}_P$ is defined canonically.

In the case when the spin structure $\mathscr{L}$ has no regular global sections, i.e., $h^0(X,\mathscr{L}) = 0$, the subspace L_P^- can be canonically defined as follows. By the Riemann-Roch theorem, $h^0(X,\mathscr{L}(nP)) = n$ for all $n \in \mathbb{N}$, so that there exist global rational sections $s_P^{(n)}$ of $\mathscr{L}$ with the only pole at $P \in X$ of order n. The subspace L_P^- is the localization of the subspace $H^0(X, \mathscr{L}(*P))$ at P and has basis $\{v_P^{(n)} = s_P^{(n)}|_P\}_{n\in\mathbb{N}}$. The pairing $(\ ,\ )_P : \mathscr{O}(L)_P \otimes L_P^- \to k$ is non-degenerate and continuous with respect to the $\mathfrak{p}$-adic topology on L_P and discrete topology on k, so that the k - vector space $\mathscr{O}(L)_P$ is dual to L_P^-: $\mathscr{O}(L)_P = (L_P^-)^\vee$. The k - vector space $\mathscr{F}_P^\vee = \Lambda^* \mathscr{O}(L)_P$ is a dual Cliff$_P$-module to the Fock module $\mathscr{F}_P$.

3.2.2.
For additive $\mathfrak{gl}_1$-bosons, set

$$\mathscr{F}_P = \mathrm{Sym}^* K_P^- \simeq W_P / W_P \cdot \mathscr{O}_P,$$

with the left action of $\widehat{\mathfrak{g}}_P$, where

$$K_P = \mathscr{O}_P \oplus K_P^-,$$

is a decomposition of K_P into isotropic subspaces with respect to the skew-symmetric bilinear form c_P, introduced in Section 2.1. Here W_P is the Weyl algebra — a quotient of $U(\widehat{\mathfrak{g}}_P)$ by the ideal generated by the element $(0,1) - \mathbf{1}$. Though the subspace K_P^- — the complement of $\mathscr{O}_P$ in K_P, depends on the choice of a uniformizer at P, the corresponding $\mathfrak{gl}_1$-module $\mathscr{F}_P$ is defined canonically.

The subspace K_P^- can be defined as follows. By the Riemann-Roch theorem, $H^0(X, \mathscr{K}(nP)) = g + n - 1$, so that there exist differentials of second kind $\theta_P^{(n)}$, with the only pole at P of order $n \geq 2$. Since $\mathrm{Res}_P \theta_P^{(n)} = 0$, there exist $v_P^{(n)} \in K_P$ such that $dv_P^{(n)} = \theta_P^{(n)}|_P$, and we define K_P^- as a k - vector space with basis $\{v_P^{(n)}\}_{n\in\mathbb{N}}$. The pairing $c_P : \mathfrak{p} \otimes K_P^- \to k$ is non-degenerate (note that k is the kernel of the bilinear form c_P), and continuous with respect to the $\mathfrak{p}$-adic topology on L_P and discrete topology on k, so that the k - vector space $\mathfrak{p}$ is dual to K_P^-: $\mathfrak{p} = (K_P^-)^\vee$. The k - vector space $\mathscr{F}_P^\vee = \mathrm{Sym}^* \mathfrak{p}$ is a dual $\widehat{\mathfrak{g}}_P$-module to the Fock module $\mathscr{F}_P$.

Local Fock spaces for charged fermions and GL_1-bosons are constructed in a similar way, the bosonic Fock space construction being equivalent to the Heisenberg system representation of Garland-Zuckerman [16]. There is also an algebraic version of Fermi-Bose correspondence between Fock spaces for charged fermions and for $\mathfrak{gl}_1$-bosons (the latter should be extended to include "zero modes").

3.2.3. *Applications*
Here we indicate applications of local QFT of $\mathfrak{gl}_1$-bosons to local CFT (Class Field Theory). We consider only the case when $\mathrm{char}\, k = 0$. Then one has the following (see, e.g., [27]).

(i) For every $P \in X$

$$K_P \simeq k((t)),$$

formal Laurent series in t, the uniformizer at $P \in X$.

(ii) All finite algebraic extensions of $k((t))$ are abelian with cyclic Galois group.

(iii) The algebraic closure K_P^{alg} of K_P is isomorphic to the union of the fields $K_P^{(n)} = k((t^{1/n}))$,

$$K_P^{alg} = \bigcup_{n\in\mathbb{N}} K_P^{(n)}.$$

(iv) The completion $\mathbb{K}_P$ of the field K_P^{alg} with respect to the valuation $\mathbf{v}_P$ extending the valuation v_P of K (the valuation $\mathbf{v}_P$ takes values in $\mathbb{Q}^*$) is isomorphic to the field $k((t^{1/\infty}))$, consisting of formal power series

$$f = \sum_{r_n\in\mathbb{Q}} c_n t^{r_n}, \text{ where } c_1 \neq 0 \text{ and } r_n \nearrow +\infty \text{ as } n \to \infty,$$

so that $\mathbf{v}_P(f) = r_1$.

(v) The Galois group of the extension K_P^{alg}/K_P — the absolute Galois group G, is isomorphic to $\hat{\mathbb{Z}}$ — a completion of the abelian group $\mathbb{Z}$ with respect to the Krull topology. The Galois group G has generator F — a geometric Frobenius element, defined by $F(\sqrt[n]{t}) = \omega_n\sqrt[n]{t}$, where $\omega_n \in k$ are n-th roots of unity in k, satisfying the property $\omega_n = \omega_{mn}^m$ for all $m, n \in \mathbb{N}$ (see [27]). The correspondence $\hat{\mathbb{Z}} \ni \nu \mapsto F^\nu \in G$ establishes an isomorphism $\hat{\mathbb{Z}} \simeq G$.

Let $\mathfrak{G}_P$ be the "absolute" affine Lie algebra — a central extension of the abelian Lie algebra $\mathbb{K}_P$ with respect to the two-cocycle c_P. It can be described (depending on the choice of the uniformizer) as the "absolute" Heisenberg algebra with generators $\{\alpha_r\}_{r\in\mathbb{Q}}$ and C, and relations

$$[\alpha_r, \alpha_s] = r\delta_{r,-s}C, \; [\alpha_r, C] = 0,$$

for all $r, s \in \mathbb{Q}$. The corresponding irreducible $\mathfrak{G}_P$-module is the "absolute" bosonic Fock space $\mathbb{F}_P = k[x_r]_{r\in\mathbb{Q}, r>0}$ of polynomials in infinitely many variables parameterized by a positive $r \in \mathbb{Q}$. The absolute Galois group G acts on the Lie algebra $\mathfrak{G}_P$ and on the Fock space $\mathbb{F}_P$, and direct sums of its invariant subspaces determine the Fock spaces $\mathscr{F}_P^{(n)}$ of additive $\mathfrak{gl}_1$-bosons for $K_P^{(n)}$.

This simple observation establishes a connection between local QFT and local CFT for the field $K_P = k((t))$ when $\operatorname{char} k = 0$. Of course, local CFT for this case is rather trivial. Similar relations exist for a really interesting case when $\operatorname{char} k = p$, i.e., $k = \overline{\mathbb{F}_p}$ — an algebraic closure of the finite field $\mathbb{F}_p$ of p elements.

3.3. Global QFT

Global QFT on algebraic curve X consists of the following data.

I Local QFT — a pair of algebra (group) $\mathscr{G}_P$ and local Fock space $\mathscr{F}_P$ with distinguished vector $\mathbf{1}_P$ for every $P \in X$.

II Global algebra (group) $\mathbb{G}_X$ — a restricted direct sum (direct product or graded tensor product) of local Lie algebras $\mathscr{G}_P$ (Lie groups or Clifford algebras) over all $P \in X$, with the diagonal embedding of the subalgebra (subgroup) $\mathscr{G}_X$, generated by rational sections over X.

III Irreducible highest weight $\mathbb{G}_X$-module — global Fock space $\mathbb{F}_X$ of quantum observables with distinguished vector $\mathbf{1}_X$ — a restricted graded-symmetric tensor product of local Fock spaces $\mathscr{F}_P$ over all $P \in X$ such that $\mathbf{1}_X = \otimes_{P\in X}\mathbf{1}_P$.

IV The k-linear functional $<\ \ >: \mathbb{F}_X \to k$ — the expectation value of quantum observables, that satisfies the following properties.

IV-1 $<\mathbf{1}_X>=1$

IV-2 $<\hat{f}\cdot v>=0$ — for additive QFT,
$<\hat{f}\cdot v>=<v>$ — for multiplicative QFT,
for all $f \in \mathscr{G}_X$ and $v \in \mathbb{F}_X$, where $\hat{f} \in \mathrm{End}_k\,\mathbb{F}_X$.

These are Ward identities, that express quantum symmetries of the theory.

3.3.1. *Applications*
Additive Ward identity for the Lie algebra case:

$$0=<\widehat{[f,g]}\cdot v>=c_X(f,g)<v>+<[\hat{f},\hat{g}]\cdot v>,$$

for all $f,g \in \mathscr{G}_X$ and $v \in \mathbb{F}_X$, implies that

$$c_X(f,g)=\sum_{P\in X} c_P(f,g)=0$$

— an additive "reciprocity law". The additive Ward identity for the Clifford algebra case:

$$0=<\widehat{fg+gf}\cdot v>=(f,g)_X<v>+<(\hat{f}\hat{g}+\hat{g}\hat{f})\cdot v>,$$

implies the additive conservation law

$$(f,g)_X=\sum_{P\in X}(f,g)_P=0,$$

for all $f,g \in \mathscr{G}_X$.

Similarly, the multiplicative Ward identity:

$$<v>=<\widehat{fg}\cdot v>=c(f,g)_X<\hat{f}(\hat{g}\cdot v)>,$$

for all $f,g \in \mathscr{G}_X$ and $v \in \mathbb{F}_X$, implies that

$$(f,g)_X=\prod_{P\in X}(f,g)_P=1$$

— a multiplicative "reciprocity law".

3.4. Examples

If the spin structure $\mathscr{L}$ for the algebraic curve X has no global regular sections, i.e., $h^0(X,\mathscr{L})=0$, or the degree 0 line bundle $\mathscr{D}$ over X is such that $h^0(X,\mathscr{L}\otimes\mathscr{D})=0$, then one can canonically construct a QFT of Majorana-Weyl and charged fermions on X.

Start with the case of Majorana-Weyl fermions and consider, for every $P\in X$, the k - vector spaces L_P^- and their dual spaces $(L_P^-)^\vee$, introduced in Section 3.2. For every $P\in X$ denote by $\{u_P^{(n)}\}_{n\in\mathbb{N}}$ the basis in $\mathscr{O}_P=(L_P^-)^\vee$, dual to the basis $\{v_P^{(n)}\}_{n\in\mathbb{N}}$ in L_P^- with respect to pairing $(\ ,\ )_P$. Let

$$\mathbb{F}_X^\vee=\bigotimes_{P\in X}\mathscr{F}_P^\vee$$

be the dual space to the global Fock space $\mathbb{F}_X$ (here $\otimes$ is a $\mathbb{Z}/2\mathbb{Z}$-graded symmetric tensor product, which is unrestricted over X), and consider a vector $\Omega\in\mathbb{F}_X^\vee$ in the following form of the "filled Dirac sea":

$$\Omega=\mathbf{1}_X^\vee+\sum_{n=1}^{\infty}\sum_{P_1\in X}\cdots\sum_{P_n\in X}\sum_{i_1=1}^{\infty}\cdots\sum_{i_n=1}^{\infty}c_{P_1\cdots P_n}^{i_1\cdots i_n}u_{P_1}^{(i_1)}\otimes\cdots\otimes u_{P_n}^{(i_n)}.$$

Consider the linear functional on $\mathbb{F}_X$, associated with $\Omega\in\mathbb{F}_X^\vee$,

$$<v>=(\Omega,v),\ v\in\mathbb{F}_X.$$

Denoting by $\hat{f}^\vee$ the dual Cliff-action, we see that the Ward identity IV-2 for $<\ >$ is equivalent to the following condition

$$\hat{f}^\vee\Omega=0 \text{ for all } f\in\mathscr{L}_X.$$

By Riemann-Roch theorem, it follows from $h^0(X,\mathscr{L})=0$, that every $f\in\mathscr{L}_X$ admits a "simple fraction decomposition", i.e., it is a linear combination of sections $s_P^{(n)}$ with $P\in(f)_\infty$ — the polar divisor of f. Therefore, it is sufficient to verify IV-2 only for sections $s_P^{(n)}$, and simple "calculus of residues" gives the following result.

Theorem 1 *Let a spin structure $\mathscr{L}$ for an algebraic curve X be such that $h^0(X,\mathscr{L})=0$. Then there exists a unique expectation value functional $<\ >:\mathbb{F}_X\to k$ satisfying the additive Ward identities, and it has the form $<v>=(\Omega,v)$, $v\in\mathbb{F}_X$, where*

$$\Omega=\exp\{\sum_{m,n=1}^{\infty}\sum_{P,Q\in X}c_{PQ}^{mn}u_P^{(m)}\otimes u_Q^{(n)}\}\in\mathbb{F}_X^\vee,$$

and

$$c_{PQ}^{mn}=Res_P(v_P^{(m)}v_Q^{(n)}).$$

The first statement of the theorem was established in [14]. The second statement gives the generating function for "free fermion correlation functions" (cf. [18, 19]) in a closed form. In particular, setting $v_P = v_P^{(1)}$, $P \in X$, we get the following expression

$$< \phi(P_1)\cdots\phi(P_n) >= (\Omega, v_{P_1}\otimes\cdots\otimes v_{P_n}) = \begin{cases} 0 & n \text{ is odd;} \\ \mathrm{Pf}(< \phi(P_i)\phi(P_j) >) & n \text{ is even,} \end{cases}$$

for the n-point correlation function of fermion field operators $\phi_i = \phi(P_i)$ on X.

An analogous construction works for the case of charged fermions, when $h^0(X, \mathscr{L}\otimes\mathscr{D}) = 0$.

For additive $\mathfrak{gl}_1$-bosons, the expectation value functional is not unique and depends on the choice of differentials of the second kind on X. Still, similar arguments work for this case. Namely, for every $P \in X$ choose the basis $\{u_P^{(n)}\}_{n\in\mathbb{N}}$ in $\mathfrak{p}$, dual to the basis $\{v_P^{(n)}\}_{n\in\mathbb{N}}$ in K_P^-, and define the dual Fock space $\mathbb{F}_X^\vee$ as (unrestricted) symmetric tensor product of the dual Fock spaces $\mathscr{F}_P^\vee$ over all $P \in X$. Consider a vector $\Omega \in \mathbb{F}_X^\vee$ in the form of the "filled Dirac sea". Contrary to the previous case, the simple fraction expansion is no longer valid for rational functions on X. However, one can consider a stronger condition than IV-2, by requiring the same equation to be valid for "multi-valued" rational functions on X (cf. [29]). By definition, $\{f_P\}_{P\in X} \in \mathbb{A}_X$ is a multi-valued rational function on X, if there exists a rational differential θ on X (which is necessarily of the second kind) such that $df_P = \theta|_P$ for all $P \in X$. The multi-valued functions — "abelian integrals", act on the dual Fock space $\mathbb{F}_X^\vee$ and admit simple fraction expansion, so that we can use the previous arguments and determine a vector Ω. Thus, denoting by J the free boson field (current) on X, we get

$$< J(P)J(Q) >= \mathrm{Res}_P(\theta_P d^{-1}\theta_Q).$$

Another way to construct additive $\mathfrak{gl}_1$-bosons is to apply the bosonization procedure to charged free fermions with $h^0(X, \mathscr{L}\otimes\mathscr{D}) = 0$. Thus the obtained QFT's are parameterized by $\mathrm{Pic}^{g-1}(X) - \Theta$, where Θ is the canonical theta-divisor (symmetric with respect to the involution $\mathscr{D} \mapsto \mathscr{K}\otimes\mathscr{D}^{-1}$).

The "exponentiated" version of these constructions allows to define a QFT of multiplicative bosons and to prove A. Weil reciprocity law.

3.5. Further development

Local QFT's discussed here can be also defined when $\mathrm{char}\, k = p > 0$. Thus, additive $\mathfrak{gl}_1$ - bosons correspond to the Artin-Schreier extensions, whereas multiplicative GL_1 - bosons correspond to the Kummer extensions of local fields, with global QFT's incorporating Artin reciprocity law for the norm residue symbol.

There is some evidence that these methods may also work for number fields, providing a QFT foundation for the Gauss quadratic reciprocity law and its generalizations. We plan to return to these issues elsewhere.

Acknowledgements. At different stages of this work I appreciated stimulating discussions with M. Flato, I. Frenkel and M. Semenov-Tian-Schansky. This research was supported by the NSF grant DMS-9802574, and was partially conducted for the Clay Mathematics Institute, when the author was a CMI Prize Fellow.

References

1. Witten, E.: Quantum field theory and the Jones polynomial, *Comm. Math. Phys.* **121** (1989), 351–399.
2. Donaldson, S.: The Seiberg-Witten equations and 4-manifolds topology, *Bull. Amer. Math. Soc. (N.S.)* **33** (1996), 45–70.
3. Witten, E.: Monopoles and four-manifolds, *Math. Res. Lett.* **1** (1994), 769–796.
4. Frenkel, I., Lepowsky, J., and Meurman, A.: *Vertex Operator Algebras and the Monster*, Academic Press Inc., Boston, MA, 1988.
5. Witten, E.: Two-dimensional quantum gravity and intersection theory on moduli space, *Surveys in Diff. Geom.* **1** (1991), 243–310.
6. Kontsevich, M.: Intersection theory on the moduli space of curves and the matrix Airy function, *Comm. Math. Phys.* **147** (1992), 1–23.
7. Manin Yu. I. and Zograf, P.: Invertible cohomological field theories and Weil-Petersson volumes, preprint math.AG/9902051.
8. Zograf P. and Takhtajan, L.: Action of the Liouville equation is a generating function for the accessory parameters and potential of the Weil-Petersson metric on the Teichmüller space, *Functional Anal. Appl.* **19** (1985), 219–220.
9. Takhtajan, L.: Topics in quantum geometry of Riemann surfaces: two-dimensional quantum gravity, In: L. Castellani and E. Wess, (eds), *Quantum groups and their applications in physics (Varenna, 1994)*, Proc. Internat. School Phys. Enrico Fermi, 127, IOS, Amsterdam, 1996, pp. 541–579.
10. Dito, G., Flato, M., Sternheimer, D., and Takhtajan, L.: Deformation quantization and Nambu mechanics, *Comm. Math. Phys.* **183** (1997), 1–22.
11. Tate, J.: Residues of differentials on curves, *Ann. Sci. École Norm. Sup.* **1** (1968), 149–159.
12. Arbarello, E., De Concini C., and Kac, V.: The infinite wedge representation and the reciprocity law for algebraic curves, in: *Theta functions—Bowdoin 1987, Part 1 (Brunswick, ME, 1987)*, Proc. Sympos. Pure Math. **49**, Amer. Math. Soc., Providence, RI, 1989, pp. 171–190.
13. Kazhdan, D.: Lecture on the AMS Summer Research Institute "Theta Functions", Bowdoin college, Maine 1987 (unpublished).
14. Witten,E.: Quantum field theory, Grassmannians, and algebraic curves, *Comm. Math. Phys.* **113** (1988), 529–600.
15. Witten, E.: Free fermions on an algebraic curve, in: *The mathematical heritage of Hermann Weyl (Durham, NC, 1987)*, Proc. Sympos. Pure Math. **48**, Amer. Math. Soc., Providence, RI, 1988, pp. 329–344.
16. Garland, H. and Zuckerman, G.: Representations of Heisenberg systems and vertex operators, in: *Complex geometry and Lie theory (Sundance, UT, 1989)*, Proc. Sympos. Pure Math., **53**, Amer. Math. Soc., Providence, RI, 1991, pp. 135–150.
17. Beilinson, A., Feigin, B., and Mazur, B.: Introduction to algebraic field theory on curves, unpublished manuscript.
18. Raina, A.: Fay's trisecant identity and conformal field theory, *Comm. Math. Phys.* **122** (1989), 625–641.
19. Raina, A.: An algebraic geometry study of the $b-c$ system with arbitrary twist fields and arbitrary statistics, *Comm. Math. Phys.* **140** (1991), 373–397.
20. Beilinson, A. and Drinfeld, V.: Chiral algebras I., unpublished manuscript.

21. Beilinson, A. and Drinfeld, V: Quantization of Hitchin's integrable system and Hecke eigensheaves, unpublished manuscript.
22. Gaitsgory, D.: Notes on 2d conformal field theory and string theory. In: P. Deligne et al (eds), *Quantum Fields and Strings. A Course for Mathematicians*, Amer. Math. Soc., Providence, RI, 1999, pp. 1017–1089.
23. Artin, E. and J. Tate, J.: *Class Field Theory*, W. A. Benjamin Inc., New York-Amsterdam 1968.
24. Belavin, A., Polyakov A., and Zamolodchikov, A.: Infinite conformal symmetry in two-dimensional quantum field theory, *Nuclear Phys. B* **241** (1984), 333–380.
25. Iwasawa, K.: *Algebraic Functions*, Translations of Mathematical Monographs, 118, Amer. Math. Soc., Providence, RI, 1993.
26. Serre, J.-P.: *Groupes Algébriques et Corps de Classes*, Hermann, Paris, 1959.
27. Serre, J.-P.: *Local Fields*, GTM **67**, Springer-Verlag, New York-Berlin, 1979.
28. Artin, E.: *Algebraic Numbers and Algebraic Functions*, Gordon and Breach Science Publishers, New York-London-Paris 1967.
29. Sen, S. and Raina, A.: Grassmannians, multiplicative Ward identities and theta-function identities, *Phys. Lett. B* **203** (1988), 256–262.

Program of the Conférence Moshé Flato 1999

Université de Bourgogne, Dijon, September 5 – 8 1999

Inaugural meeting

Under the auspices of Édith Cresson, former Prime Minister

Chaired by Hubert Curien, Academy of Sciences, former Minister of Research

Sunday September 5, Cellier de l'Espace Grévin, Dijon

15:30 Opening by Bernard Laurin, President of the University
Speech by Hubert Curien, Academy of Sciences

Messages from
Édith Cresson, read by Hubert Curien
Yvette Chassagne, read by Chantalle Reimeringer
Clarice Schwinger and Kay & Leon Cooper, read by Noriko Sakurai
Ed Witten, read by Ruth Boulet-Kerins
GU ChaoHao (Shanghai and Academia Sinica), read by RAO BoPeng

Speeches by:
Huzihiro Araki (Tokyo and Kyoto)
Jerôme Lichnerowicz (André's son)
Bernard Loiseau (Chef of *La Côte d'Or*, Saulieu)
Bengt Nagel (KTH, former president of the Nobel Committee for Physics)
Konrad Osterwalder (Rector of ETH, Zürich)

Music:
Chopin: *Nocturne in C Minor and Andante, Pianato, Grande Polonaise*
interpreted by Eugen Indjic, piano

Université de Bourgogne's medal awarded to I. Flato-Izakson, A. Connes, M. Kontsevich, Y. Aharonov, and B. Nagel.

17:00 *Short break*

Speeches by Irit Flato-Izakson and Joseph Sack, with slide show
Moshé d'après lui-même, in English, by Daniel Sternheimer

Montage video:
Extracts from the *Colloque Louis de Broglie,* November 6–7, 1987
Polac's TV program *Faut-il brûler le CNRS ?* November 1985
The scientific council of UAP, September 1987
Seminar in Dijon, 1995

Music:
Michio Miyagi: *Haru no Umi* and Kozaburo Hirai: *Sakura Sakura*
interpreted by Tomoko Ojima, piano

Songs by Civia Millender and Margalit Izakson, accompaniment at the guitar by Évariste: *Golden Jerusalem* (Naomi Shemer, 1967); *Ma mie* (Évariste);
Song for peace (lyrics by Yair Rosenblum, sung by Y. Rabin just before his assassination)

Short break

19:00 *Apéritif, followed by a buffet by Lenôtre (and Loiseau)*
23:00 *End*

Applications of sciences to finance and insurance

Under the auspices of Denis Kessler
President of the Fédération Française des Sociétés d'Assurances

Chaired by Dominique Perrin, President of the Université de Marne-la-Vallée

Monday September 6, in the University, Bâtiment Sciences Mirande

9:15 – 9:30	Opening of the special session and of the scientific conference by Arthur Jaffe and Daniel Sternheimer
9:30 – 10:10	Introduction: Yvette Chassagne, Honorary Chairperson of UAP (presented by Roland Lejart, former General Director of UAP) *Le Conseil scientifique de l'UAP et ses réalisations* Erik Taflin (AXA) *A new method for optimal equity allocation and portfolio selection*
10:10 – 10:50	Jean-Claude Simon† PDG, A2iA and Emeritus professor, Université Paris VI *La reconnaissance automatique de l'écriture manuscrite*
10:50 – 11:30	Hélyette Geman Professor of Finance, Université Paris IX and ESSEC *From Bachelier and Lundberg to insurance derivatives*
12:00 – 14:00	Lunch at *Clos du Roy* in Chenôve

Quantization, deformations, and symmetries

Monday September 6

9:15 – 9:30	Opening of the special session and of the scientific conference by Arthur Jaffe and Daniel Sternheimer
9:30 – 14:00	Get together session with informal discussions and organization of seminars and poster sessions; lunch
14:30 – 15:25	Yakir Aharonov (U. South Carolina and Tel-Aviv) *A new approach to quantum mechanics* Chair: Paul Sally (Chicago)
15:40 – 16:35	Elliott Lieb (Princeton) *Stability of matter and the self-energy of an electron in non-perturbative quantum electrodynamics* Chair: Marcel Guenin (Geneva)
	– *Coffee break* –
17:15 – 18:10	Sergio Ferrara (CERN and UCLA) *Superconformal field theories, multiplet shortening, and the* $AdS_5/SCFT_4$ *correspondence* Chair: Robert Finkelstein (UCLA)
18:25 – 19:20	Christian Frønsdal (UCLA) *Singletons and neutrinos* Chair: Piotr Rączka (Warsaw)

Tuesday September 7

9:15 – 10:10	Leon Takhtajan (SUNYSB) *Quantum field theories on an algebraic curve* Chair: J.-P. Labesse (Director of CIRM)
	– *Coffee break* –
10:35 – 11:30	Maxim Kontsevich (IHES) *Geometry of the moduli space of non-commutative algebras* Chair: Hideki Omori (Science University of Tokyo)
11:45 – 12:40	Alain Connes (Collège de France) *Cyclic cohomology and Hopf algebras* Chair: Murray Gerstenhaber (U. Pennsylvania)
13:00 – 15:00	*Lunch and seminars by T. Kimura (Boston), M. Bordemann and S. Waldmann (Freiburg)*

15:00 – 15:55	Simone Gutt (ULB and Metz) *Variations on deformation quantization* Chair: Alan Weinstein (Berkeley)
16:10 – 17:05	Boris Fedosov (Moscow and Potsdam) *On G-trace and G-index in deformation quantization* Chair: Ryszard Nest (Copenhagen)
	– Coffee break –
17:45 – 18:40	Louis Boutet de Monvel (Paris 6) *Real and complex star products* Chair: Yvette Kosmann-Schwarzbach (École polytechnique)
18:50 – 20:00	*Seminars by A. Karabegov (Erevan), A. Asada (Shinshu), and R. Léandre (Nancy); poster sessions*

Wednesday September 8

9:15 – 10:10	Jacques C. H. Simon (U. Bourgogne) *Nonlinear relativistic evolution equations:* *Survey of a new approach* Chair: Ivan Todorov (Sofia)
	– Coffee break –
10:35 – 11:30	Bertram Kostant (MIT) *A cubic Dirac operator and a generalization* *of the Bott-Borel-Weil theorem* Chair: Nicolai Reshetikhin (Berkeley)
11:45 – 12:40	Wilfried Schmid (Harvard) *L-functions attached to Maass modular forms* Chair: Gérard Schiffmann (Strasbourg)
13:00 – 15:00	*Lunch and seminars by P. Rączka (Warsaw), J. Niederle (Prague), and M. Kontsevich (IHES)*
15:00 – 15:55	Jürg Fröhlich (ETHZ) *Classical limits of Bose systems* Chair: Jiri Niederle (Prague)
16:10 – 17:05	Ludwig Faddeev (St. Petersburg and Helsinki) *The modular double of a quantum group* Chair: Yuri Novozhilov (St. Petersburg)
17:30 – 19:00	Closure by Gilles Bertrand, vice president for research and former president of the University, and cocktail

List of participants

Abe, Kojun: *Shinshu University, Japan*
Aharonov, Yakir: *University of South Carolina, U.S.A. and Tel Aviv University, Israel*
Anderson, Robert: *University of Georgia, U.S.A.*
Andler, Martin: *Université de Versailles St Quentin, France*
Angelopoulos, Evghenios: *Université de Bourgogne, France and National & Technical University, Greece*
Araki, Huzihiro: *Science University of Tokyo, Japan*
Arnal, Didier: *Université de Metz, France*
Asada, Akira: *Shinshu University, Japan*
Asin, Santos: *The Abdus Salam International Centre, Italy*
Baguis, Pierre: *Université Libre de Bruxelles, Belgium*
Beau, Daniel: *Université de Bourgogne, France*
Bertrand, Gilles: *Université de Bourgogne, France*
Bidegain, Frédéric: *Université de Metz, France*
Bieliavsky, Pierre: *Université Libre de Bruxelles, Belgium*
Bonneau, Philippe: *Université de Bourgogne, France*
Bordemann, Martin: *University of Freiburg, Germany*
Boutet de Monvel, Anne: *Université Paris VII, France*
Boutet de Monvel, Louis: *Université Paris VI, France*
Buffenoir, Eric: *Université Montpellier II, France*
Cahen, Michel: *Université Libre de Bruxelles, Belgium*
Cambefort, Jean-Louis: *Université de Bourgogne, France*
Charpentier, Eric: *Université Bordeaux I, France*
Chloup, Véronique: *Université de Metz, France*
Choquet-Bruhat, Yvonne: *Université Paris VI and Académie des Sciences, France*
Conley, Charles: *University of North Texas, U.S.A.*
Connes, Alain: *Collège de France, France*
Cortet, Jean-Claude: *Université de Bourgogne, France*
Dito, Giuseppe: *Université de Bourgogne, France*
Droz-Vincent, Philippe: *Université Paris VI, France*
Dubois-Violette, Michel: *Université Paris XI, France*
Dufour, Jean-Paul: *Université Montpellier II, France*
El Gradechi, Amine: *Université d'Artois, France*
Faddeev, Ludwig: *St. Petersburg branch of the Steklov Mathematical Institute, Russia and University of Helsinki, Finland*
Fedosov, Boris: *University of Potsdam, Germany*
Ferrara, Sergio: *C.E.R.N., Switzerland and University of California at Los Angeles, U.S.A.*
Finkelstein, Robert: *University of California at Los Angeles, U.S.A.*

Fjeld, Morten: *E.T.H.Z., Switzerland*
Frabetti, Alessandra: *Université Louis Pasteur, Strasbourg, France*
Fröhlich, Jürg: *E.T.H.Z., Switzerland*
Frønsdal, Christian: *University of California at Los Angeles, U.S.A.*
Fukui, Kazuhiko: *Kyoto Sangyo University, Japan*
Galindo, Alberto: *Universidad Complutense, Spain*
Gammella, Angela: *Université de Metz, France*
Gazeau, Jean-Pierre: *Université Paris VII, France*
Geman, Hélyette: *Université Paris IX, France*
Gerstenhaber, Murray: *University of Pennsylvania, U.S.A.*
Ginocchio, Maurice: *Université Paris VII, France*
Giorgi, Giordano: *Rome University, Italy*
Gréa, Jean: *Université Lyon I, France*
Gross, Maurice: *Université Paris VII, France*
Guenin, Marcel: *Université de Genève, Switzerland*
Gutt, Simone: *Université de Metz, France*
and Université libre de Bruxelles, Belgium
Halbout, Gilles: *Université Louis Pasteur, Strasbourg, France*
Hamachi, Kentaro: *RIMS, Kyoto University, Japan*
Herbig, Hans-Christian: *University of Freiburg, Germany*
Horowitz, Joël: *Université libre de Bruxelles, Belgium*
Irac-Astaud, Michèle: *Université Paris VII, France*
Ito, Keiichi: *Setsunan University, Japon*
Jaffe, Arthur: *Harvard University, U.S.A.*
Jakobsen, Hans Plesner: *Copenhagen University, Denmark*
Jauslin, Hans: *Université de Bourgogne, France*
Jurzak, Jean-Paul: *Université de Bourgogne, France*
Karabegov, Alexander: *Joint Institute for Nuclear Research, Russia*
Kawamura, Katsunori: *RIMS, Kyoto University, Japan*
Kerbrat, Hélène: *Université Lyon I, France*
Kerbrat, Yvan: *Université Lyon I, France*
Kerner, Richard: *Université Paris VI, France*
Kibler, Maurice: *Université Lyon I, France*
Kimura, Takashi: *Boston University, U.S.A.*
Kontsevich, Maxim: *Institut des Hautes Études Scientifiques, France*
Kosmann-Schwarzbach, Yvette: *École Polytechnique, France*
Kostant, Bertram: *Massachussets Institute of Technology, U.S.A.*
Kreimer, Dirk: *University of Mainz, Germany*
Labesse, Jean-Pierre: *Centre International de Rencontres Mathématiques (CIRM)*
and Université Paris VII, France
Lafforêt, Bernard: *Infotel, France*
Lapidus, Michel: *University of California at Riverside, U.S.A.*
Laurin, Bernard: *Université de Bourgogne, France*

Léandre, Rémi: *Université Nancy I, France*
Leichtnam, Eric: *Université Paris VI, France*
Lejart, Roland: *former General Director of U.A.P., France*
Lemaire, Luc: *Université libre de Bruxelles, Belgium*
Lesimple, Marc: *University of Padova , Italy*
Lieb, Elliott: *Princeton University, U.S.A.*
Maeda, Yoshiaki: *Keio University, France*
Magnin, Louis: *Université de Bourgogne, France*
Maillard, Jean-Marie: *Université de Bourgogne, France*
Manchon, Dominique: *Université Nancy I, France*
Marle, Charles-Michel: *Université Paris VI, France*
Martin, Christiane: *Université de Bourgogne, France*
Martin, Isabelle: *Université de Bourgogne, France*
Mathieu, Olivier: *Université Lyon I, France*
Matveev, Vladimir: *Université de Bourgogne, France*
Michéa, Sébastien: *Yonsei University, South Korea*
Mickelsson, Jouko: *Royal Institute of Technology, Sweden*
Mielnik, Bogdan: *University of Warsaw, Poland*
Molin, Paul: *Université de Bourgogne, France*
Moreno, Carlos: *Université de Bourgogne, France*
and Universidad Complutense, Spain
Moser-Jauslin, Lucy: *Université de Bourgogne, France*
Nagel, Bengt: *Royal Institute of Technology, Sweden*
Nakanishi, Nobutada: *Gifu Keizai University, Japan*
Nest, Ryszard: *Copenhagen University, Denmark*
Neumaier, Nikolai: *University of Freiburg, Germany*
Niederle, Jiri: *Czech Academy of Sciences, Czech Republic*
Novozhilov, Yuri: *St Petersburg University, Russia*
Ojima, Izumi: *RIMS, Kyoto University, Japan*
Omori, Hideki: *Science University of Tokyo, Japan*
Osterwalder, Konrad: *E.T.H.Z., Switzerland*
Parmentier, Serge: *Université Lyon I, France*
Pauly, Marc: *Université Paris XI, France*
Perrin, Dominique: *Université de Marne-la-Vallée, France*
Piard, Alain: *Université de Bourgogne, France*
Picco, Marco: *Université Paris VI, France*
Pillin, Mathias: *King's College, U.K.*
Pinczon, Georges: *Université de Bourgogne, France*
Polyak, Michael: *Tel-Aviv University, Israel*
Rączka, Piotr: *University of Warsaw, Poland*
Rawnsley, John: *University of Warwick, U.K.*
Reimeringer, Chantalle: *AXA, France*
Rentschler, Rudolf: *Université Paris VI, France*

Reshetikhin, Nicolai: *University of California at Berkeley, U.S.A.*
Rideau, Guy: *Université Paris VII, France*
Rieffel, Marc: *University of California at Berkeley, U.S.A.*
Robinson, Bencion: *Center for Technological Education, Israel*
Roos, Paul: *Kluwer Academic Publishers, The Netherlands*
Sally, Paul: *University of Chicago, U.S.A.*
Schiffmann, Gérard: *Université Louis Pasteur, Strasbourg, France*
Schlein, Peter: *University of California at Los Angeles, U.S.A.*
Schlichenmaier, Martin: *University of Mannheim, Germany*
Schmid, Wilfried: *Harvard University, U.S.A.*
Semenov-Tian-Shansky, Michael: *Université de Bourgogne, France*
Simon, Jacques: *Université de Bourgogne, France*
Simon, Jean-Claude: *A2iA, France*
Smirnov, Fedeor: *Université Paris VI, France*
Snellman, Håkan: *Royal Institute of Technology, Sweden*
Sternheimer, Daniel: *Université de Bourgogne, France*
Strasburger, Aleksander: *University of Warsaw, Poland*
Taflin, Erik: *AXA, France*
Takhtajan, Leon: *SUNY at Stony Brook, U.S.A.*
Todorov, Ivan: *Academy of Sciences, Bulgaria*
Ushirobira, Rosane: *Université de Bourgogne, France*
Volkov, Alexandre: *St. Petersburg branch of the Steklov Mathematical Institute, Russia and Vrije Universiteit Brussel, Belgium*
Wade, Aissa: *The Abdus Salam International Centre, Italy*
Waldmann, Stefan: *University of Freiburg, Germany*
Walter, Martin: *University of Freiburg, Germany*
Weinstein, Alan: *University of California at Berkeley, U.S.A.*
Xu, Ping: *Pennsylvania State University, U.S.A.*
Yoshioka, Akira: *Science University of Tokyo, Japan*

Committees

Organizers

Jean-Claude Cortet, Université de Bourgogne, France

Yvette Kosmann-Schwarzbach, École polytechnique, France

Wilfried Schmid, Harvard University, U.S.A.

Jacques Simon, Université de Bourgogne, France

Daniel Sternheimer (Coordinator), CNRS and Université de Bourgogne, France

Scientific committee

Michel Cahen, Université libre de Bruxelles, Belgium

Alain Connes, Collège de France, France

Ludwig Faddeev, POMI, Russia

Murray Gerstenhaber, University of Pennsylvania, U.S.A.

Richard Kerner, Université Paris VI, France

Maxim Kontsevich, I.H.E.S., France

Yvette Kosmann-Schwarzbach, École polytechnique, France

Wilfried Schmid, Harvard University, U.S.A.

Jacques Simon, Université de Bourgogne, France

Daniel Sternheimer, CNRS and Université de Bourgogne, France

Erik Taflin, AXA, France

The local organization was undertaken by the Laboratoire Gevrey de Mathématique physique and especially by Marylise Debret, Giuseppe Dito, Daniel Sternheimer, Philippe Bonneau, and Jean-Claude Cortet.

Index

The page number refers to the beginning of a relevant contribution and boldface page numbers refer to volume two.

FSC
www.fsc.org
MIX
Papier | Fördert gute Waldnutzung
FSC® C083411

Kolibri 360 GmbH
Ferdinand-Jühlke-Straße 7
99095 Erfurt, Deutschland
produktsicherheit@kolibri360.de